Environmental Resources: Planning and Engineering

Environmental Resources: Planning and Engineering

Edited by
Stacy Keach

Willford Press
www.willfordpress.com

Published by Willford Press,
118-35 Queens Blvd., Suite 400,
Forest Hills, NY 11375, USA

ISBN: 978-1-68285-423-5

Cataloging-in-Publication Data

Environmental resources : planning and engineering / edited by Stacy Keach.
p. cm.
Includes bibliographical references and index.
ISBN 978-1-68285-423-5
1. Environmental sciences. 2. Environmental policy. 3. Natural resources.
3. Environmental engineering. I. Keach, Stacy.
GE105 .E58 2018
363.7--dc23

For information on all Willford Press publications
visit our website at www.willfordpress.com

Willford Press

Printed in China.

Contents

Preface VII

Chapter 1 **Coliform Bacteria as Indicators of Diarrheal Risk in Household Drinking Water** 1
Joshua S. Gruber, Ayse Ercumen, John M. Colford, Jr.

Chapter 2 **Hyperspectral Detection of a Subsurface CO_2 Leak in the Presence of Water Stressed Vegetation** 15
Gabriel J. Bellante, Scott L. Powell, Rick L. Lawrence, Kevin S. Repasky, Tracy Dougher

Chapter 3 **Streamflow Impacts of Biofuel Policy-Driven Landscape Change** 30
Sami Khanal, Robert P. Anex, Christopher J. Anderson, Daryl E. Herzmann

Chapter 4 **Improved Health among People Living with HIV/AIDS who Received Packages of Proven Preventive Health Interventions, Amhara, Ethiopia** 40
Ciara E. O'Reilly, Ethel V. Taylor, Tracy Ayers, Ribka Fantu, Sisay Alemayehu Abayneh, Barbara Marston, Yordanos B. Molla, Tegene Sewnet, Fitsum Abebe, Robert M. Hoekstra, Robert Quick

Chapter 5 **The Role of Water Channel Proteins in Facilitating Recovery of Leaf Hydraulic Conductance from Water Stress in *Populus trichocarpa*** 49
Joan Laur, Uwe G. Hacke

Chapter 6 **Modeling the Impact of Soil and Water Conservation on Surface and Ground Water based on the SCS and Visual Modflow** 59
Hong Wang, Jian-en Gao, Shao-long Zhang, Meng-jie Zhang, Xing-hua Li

Chapter 7 **Vertical Profiles of Soil Water Content as Influenced by Environmental Factors in a Small Catchment on the Hilly-Gully Loess Plateau** 71
Bing Wang, Fenxiang Wen, Jiangtao Wu, Xiaojun Wang, Yani Hu

Chapter 8 **Sensitivity and Tolerance of Riparian Arthropod Communities to Altered Water Resources along a Drying River** 83
Kevin E. McCluney, John L. Sabo

Chapter 9 **Iodine Deficiency and Excess Coexist in China and Induce Thyroid Dysfunction and Disease** 96
Yang Du, Yanhui Gao, Fangang Meng, Shoujun Liu, Zhipeng Fan, Junhua Wu, Dianjun Sun

Chapter 10 **Effects of Climate Variability and Accelerated Forest Thinning on Watershed-Scale Runoff in Southwestern USA Ponderosa Pine Forests** 107
Marcos D. Robles, Robert M. Marshall, Frances O'Donnell, Edward B. Smith, Jeanmarie A. Haney, David F. Gori

Chapter 11 **Photosynthetic Diffusional Constraints Affect Yield in Drought Stressed Rice Cultivars during Flowering** 122
Marco Lauteri, Matthew Haworth, Rachid Serraj, Maria Cristina Monteverdi, Mauro Centritto

Chapter 12 **Transcriptome-Wide Profiling and Expression Analysis of Diploid and Autotetraploid *Paulownia tomentosa* × *Paulownia fortunei* under Drought Stress** 134
Enkai Xu, Guoqiang Fan, Suyan Niu, Zhenli Zhao, Minjie Deng, Yanpeng Dong

Chapter 13 **Estimated Dietary Intake of Radionuclides and Health Risks for the Citizens of Fukushima City, Tokyo, and Osaka after the 2011 Nuclear Accident** 150
Michio Murakami, Taikan Oki

Chapter 14 **Cloning of *Gossypium hirsutum* Sucrose Non-Fermenting 1-Related Protein Kinase 2 Gene (*GhSnRK2*) and its Overexpression in Transgenic *Arabidopsis* Escalates Drought and Low Temperature Tolerance** 162
Babatunde Bello, Xueyan Zhang, Chuanliang Liu, Zhaoen Yang, Zuoren Yang, Qianhua Wang, Ge Zhao, Fuguang Li

Chapter 15 **Isoprene Emissions from Downy Oak under Water Limitation during an Entire Growing Season: What Cost for Growth?** 180
Anne-Cyrielle Genard-Zielinski, Elena Ormeño, Christophe Boissard, Catherine Fernandez

Chapter 16 **Interact to Survive: *Phyllobacterium brassicacearum* Improves Arabidopsis Tolerance to Severe Water Deficit and Growth Recovery** 192
Justine Bresson, François Vasseur, Myriam Dauzat, Marc Labadie, Fabrice Varoquaux, Bruno Touraine, Denis Vile

Chapter 17 **Systems Responses to Progressive Water Stress in Durum Wheat** 204
Dimah Z. Habash, Marcela Baudo, Matthew Hindle, Stephen J. Powers, Michael Defoin-Platel, Rowan Mitchell, Mansoor Saqi, Chris Rawlings, Kawther Latiri, Jose L. Araus, Ahmad Abdulkader, Roberto Tuberosa, David W. Lawlor, Miloudi M. Nachit

Chapter 18 **AtRD22 and AtUSPL1, Members of the Plant-Specific BURP Domain Family Involved in *Arabidopsis thaliana* Drought Tolerance** 225
Vokkaliga Thammegowda Harshavardhan, Le Van Son, Christiane Seiler, Astrid Junker, Kathleen Weigelt-Fischer, Christian Klukas, Thomas Altmann, Nese Sreenivasulu, Helmut Bäumlein, Markus Kuhlmann

Chapter 19 **Experimental Comparison of the Reproductive Outcomes and Early Development of the Offspring of Rats Given Five Common Types of Drinking Water** 239
Hui Zeng, Wei-qun Shu, Ji-an Chen, Lin Liu, Da-hua Wang, Wen-juan Fu, Ling-qiao Wang, Jiao-hua Luo, Liang Zhang, Yao Tan, Zhi-qun Qiu, Yu-jing Huang

Permissions

List of Contributors

Index

Preface

Environmental resources are scarce and are shared by all flora and fauna. Environmental resource management seeks to conserve available resources for their equitable use. Processes that are implemented for such management include the sustainable use of resources, integrated land management, coastal management, ecoforestry, etc. This book explores all the important aspects related to environmental resources in the present day scenario. Coherent flow of topics, student-friendly language and extensive use of examples make this book an invaluable source of knowledge.

After months of intensive research and writing, this book is the end result of all who devoted their time and efforts in the initiation and progress of this book. It will surely be a source of reference in enhancing the required knowledge of the new developments in the area. During the course of developing this book, certain measures such as accuracy, authenticity and research focused analytical studies were given preference in order to produce a comprehensive book in the area of study.

This book would not have been possible without the efforts of the authors and the publisher. I extend my sincere thanks to them. Secondly, I express my gratitude to my family and well-wishers. And most importantly, I thank my students for constantly expressing their willingness and curiosity in enhancing their knowledge in the field, which encourages me to take up further research projects for the advancement of the area.

Editor

1

Coliform Bacteria as Indicators of Diarrheal Risk in Household Drinking Water

Joshua S. Gruber*, Ayse Ercumen, John M. Colford, Jr.

Division of Epidemiology, University of California, Berkeley, California, United States of America

Abstract

Background: Current guidelines recommend the use of *Escherichia coli* (EC) or thermotolerant ("fecal") coliforms (FC) as indicators of fecal contamination in drinking water. Despite their broad use as measures of water quality, there remains limited evidence for an association between EC or FC and diarrheal illness: a previous review found no evidence for a link between diarrhea and these indicators in household drinking water.

Objectives: We conducted a systematic review and meta-analysis to update the results of the previous review with newly available evidence, to explore differences between EC and FC indicators, and to assess the quality of available evidence.

Methods: We searched major databases using broad terms for household water quality and diarrhea. We extracted study characteristics and relative risks (RR) from relevant studies. We pooled RRs using random effects models with inverse variance weighting, and used standard methods to evaluate heterogeneity and publication bias.

Results: We identified 20 relevant studies; 14 studies provided extractable results for meta-analysis. When combining all studies, we found no association between EC or FC and diarrhea (RR 1.26 [95% CI: 0.98, 1.63]). When analyzing EC and FC separately, we found evidence for an association between diarrhea and EC (RR: 1.54 [95% CI: 1.37, 1.74]) but not FC (RR: 1.07 [95% CI: 0.79, 1.45]). Across all studies, we identified several elements of study design and reporting (e.g., timing of outcome and exposure measurement, accounting for correlated outcomes) that could be improved upon in future studies that evaluate the association between drinking water contamination and health.

Conclusions: Our findings, based on a review of the published literature, suggest that these two coliform groups have different associations with diarrhea in household drinking water. Our results support the use of EC as a fecal indicator in household drinking water.

Editor: Sudha Chaturvedi, Wadsworth Center, United States of America

Funding: The authors have no funding or support to report.

Competing Interests: The authors have declared that no competing interests exist.

* Email: jsgruber@gmail.com

Introduction

Globally, drinking water has been established as a primary transmission pathway for diarrhea pathogens. [1,2] In industrialized countries, centrally treated drinking water distribution systems have largely eliminated outbreaks of waterborne diseases, such as typhoid fever and cholera. [3] In developing countries, there is a large body of evidence that improving the microbial quality of drinking water by household treatment and safe storage reduces diarrhea. [4–6] Yet, evidence directly linking diarrheal illness to measured fecal contamination in drinking water remains inconclusive. [3,7,8]

In general, it is not feasible to test water for all known waterborne pathogens to assess whether it is safe for drinking. [2,9–11] Instead, since the early 1900s there has been heavy reliance on fecal indicator organisms as measures of drinking water quality. [1,12] Current World Health Organization (WHO) guidelines recommend *Escherichia coli* (EC) and/or thermotolerant ("fecal") coliforms (FC) as indicators of the effectiveness of disinfection processes, and as index organisms for the potential presence of fecal contamination and waterborne pathogens; [2,10,12,13] previous WHO guidelines use categories of EC and FC concentrations to define levels of disease risk from drinking water. [14] While EC are considered the most suitable indicator organism due to their specificity to fecal sources of contamination, FC are also recommended as an acceptable surrogate; this recommendation exists despite the recognition that the FC group includes coliform species of environmental origin, and is therefore not likely specific to fecal contamination. [2,12,13] Despite heavy reliance on EC and FC to assess the microbiological safety of drinking water, it has yet to be shown that either of these specific indicators is associated with waterborne illness. [8]

A previous systematic review and meta-analysis evaluated the evidence for the link between household drinking water quality, measured by fecal indicator organisms, and health; [8] the authors found no evidence of an association between diarrhea and indicators of drinking water contamination (EC, FC and *fecal streptococci*). This review, however, was limited by the availability, size and quality of articles published at the time the search was conducted (2001). [8] Specifically, the previous review included only three relatively small studies that reported using EC, and therefore focused mostly on FC as a measure of water quality. Given the limited number of studies available to the authors, they were also unable to evaluate the performances of EC and FC separately. Since the previous review, several studies have been published using EC as a measure of water quality (n = 6) [15–21], motivating our updated review with this newly available evidence. In addition, this larger body of evidence allows us to evaluate the performance of EC and FC separately; we hypothesized that the two coliform groups would have different associations with diarrhea given their different specificities for fecal contamination. Given the widespread use of EC and FC as measures of water quality by researchers and policy makers, and the limited evidence regarding the association of these proxy measures with actual disease outcomes, we believe that our systematic review and meta-analysis will make a significant contribution to the body of knowledge on the performance of indicator organisms in household drinking water, and have implications for their recommended use for future guidelines and research. In addition, based on the findings of our systematic review, we are able to make several recommendations for the design and reporting of future studies that assess the relationship between drinking water contamination and health.

Methods

Search Strategy

We searched MEDLINE, EMBASE, and Web of Knowledge databases for relevant articles. We used broad search terms for water, water quality and indicator organisms, diarrhea and household (point-of-use) sampling (Table 1). Titles and abstracts from the search were examined, and the full texts of relevant articles were reviewed. The bibliographies of relevant articles identified from full-text reviews were scanned to identify additional relevant studies. Searches were limited to articles published in peer-reviewed journals, in English, Spanish, German or Turkish (languages spoken by the authors). No restrictions were placed on date of publication. The final search was conducted on February 27, 2013. A protocol was not registered for this systematic review.

Inclusion and Exclusion Criteria

We reviewed articles that reported data on the association between exposure to *Escherichia coli* (EC) or thermotolerant ("fecal") coliforms (FC) in household drinking water, and the occurrence of diarrhea under non-outbreak conditions; we excluded other, less commonly used indicator organisms (e.g., total coliforms, *fecal streptococci*). Our inclusion criterion was that the study collected exposure and outcome data at the household level or at the point of use; studies were excluded if they only reported measures of source water quality (e.g., shared wells, distribution system "nodes"), or used ecologic outcome data (e.g., health ministry reports). No further restrictions were placed on: definitions of diarrhea, recall period, study design, age groups, study location (developed/developing country), study setting (urban/rural), or drinking water source (as long as exposure to indicator bacteria was measured at the point of use).

Study Reporting and Data Extraction

For each relevant study two authors (JG and AE) independently extracted basic study characteristics, including year of publication, sample size, age groups, study location, setting and design, indicator organism and enumeration methods, and diarrhea definition and enumeration methods into standardized forms, and resolved any discrepancies. Where possible, a relative risk (RR) such as odds ratio, cumulative incidence ratio, incidence density ratio or prevalence ratio was extracted from relevant studies. When no effect measure was reported by the authors, raw data were extracted to calculate an appropriate RR and confidence interval using standard methods. [22] If authors reported both raw data and unadjusted RRs we confirmed the reported estimates using the raw data; we used the RRs reported by the authors, unless otherwise noted. If authors reported results for both indicators (EC and FC) the results were treated as two separate studies in the meta-analysis. All RRs were extracted such that a value greater than unity indicated increased risk of illness among the group exposed to contaminated water. We also noted whether authors controlled for additional variables (by study design, stratified analysis or regression methods) and whether they accounted for correlation (clustering) of outcomes in their analysis.

Exposure Thresholds

There are multiple ways to define comparison groups in water quality studies. For example, authors often use a threshold to define exposed and non-exposed groups based on indicator organism counts measured in 100 ml samples of water (i.e., exposed: ≥10 EC/100 ml; non-exposed: <10 EC/100 ml; with RRs calculated to compare these two groups); however, threshold levels can vary between studies. Current WHO guidelines recommend that drinking water that is safe for human consumption should have no detectable EC or FC in any 100 ml sample (i.e., a 1 EC/FC threshold); however, there is also evidence from the literature that much higher indicator organism concentrations are required to cause disease. For example, a study conducted in the Philippines on source water quality [23] previously reported a "threshold effect", where no association with diarrhea was found at the 1 EC, 10 EC or 100 EC cutoffs, but an association was found at the 1000 EC cutoff; it is plausible that a similar threshold effect could be present for point-of-use drinking water. Given the limited number of studies included in the previously published systematic review and meta-analysis, the authors made no attempt to differentiate indicator organism-diarrhea associations at different threshold levels. [8] In our review, we attempted to extract RRs for different thresholds at the 1 EC or FC, 10 EC or FC, 100 EC or FC and 1000 EC or FC levels, when the authors reported these data.

Categorical Exposures

As an alternative approach to exposure categorization, previous WHO guidelines defined disease risk from drinking water based on categories of indicator organism counts measured in 100 ml samples of water: 0 EC or FC, safe; 1–10 EC or FC, low risk; 11–100 EC or FC, intermediate risk; 101–1000 EC or FC, high risk;> 1000 EC or FC, very high risk. [14] Investigators commonly use these guidelines to define exposure categories in studies of drinking water quality; each category is treated as an independent exposure group, and RRs are calculated for each elevated category relative to the 0 EC or FC exposure group. In our review, we extracted RRs for these categorical exposures to explore the presence of a dose-response (increased disease risk associated with increasing exposure levels) and to evaluate previous WHO guidelines that defined health risk based on these exposure categories. Within

Table 1. Systematic Review Search Terms. [a]

Water Exposure	Household Sampling	Disease Outcome
water quality	household	Diarrhea
water microbiology	point of use	Diarrheoa
water pollution	Pou	gastrointestinal illness
water contamination	point-of-use	gastrointestinal disease
water supply	drinking water	gastrointestinal infection
drinking water	consumption	Dysentery
"indicator bacteria"	tap water	"HCGI"
"indicator organism"	well water	"highly credible gastrointestinal illness"
"thermotolerant coliforms"	domestic	"AGI"
"thermo-tolerant coliforms"		"AGII"
"fecal coliforms"		"acute gastrointestinal illness"
escherichia coli		
"e. coli"		
"fecal bacteria"		
"fecal contamination"		
"microbiological indicators"		

[a]Search terms combined using Boolean logic: within column terms combined using "or" statements; "and" statements were used to combine terms between columns.

studies that evaluated dose-responses, we extracted *p*-values for tests of trend. In studies that did not report *p*-values for tests of trend but reported results for different categorical exposure levels, we extracted the appropriate data, where possible, and conducted our own tests of trend using previously published methods that estimate a log-linear dose-response while accounting for correlations between RRs from a single study (*glst* command: STATA 12, STATA Corp, College Station Texas). [24]

Meta-analysis

For consistency with the previously published review, we conducted meta-analyses to calculate a summary measure pooled across all studies which provided an extractable RR, combining all EC and FC studies. [8] The previous review was not explicit about which results were used for meta-analyses if data could be extracted at multiple thresholds. In our review, we estimated a summary measure across all studies using the lowest extractable threshold from each study (e.g., if a study reported raw data that could be combined at the 1 EC, 10 EC and 100 EC thresholds, we included the RR calculated for the 1 EC threshold in the main meta-analysis). In addition, we conducted a sub-group analysis to explore the type of indicator organism used in each study as a source of heterogeneity in the main analysis; for this sub-group analysis we stratified studies by indicator used (EC and FC) and re-pooled summary measures separately for each indicator. In a secondary meta-analysis, we explored evidence for a threshold effect in household drinking water (see "Exposure Thresholds" above). For this analysis, we restricted meta-analyses to studies that reported extractable RRs at the 1 EC, 10 EC and 100 EC thresholds, respectively – this analysis was only feasible for studies that reported extractable data for all three threshold levels.

Summary measures were calculated by meta-analysis using random effects models with inverse variance weighting (STATA 12). [25] Heterogeneity across studies was evaluated using the Mantel-Haentzel χ^2 test; we considered a *p*-value on the χ^2 statistic <0.2 as evidence of heterogeneity. We used funnel plots to evaluate publication bias; plot asymmetry was interpreted as evidence of "small study bias", for which publication bias is a likely contributor. [26,27]

Results

Systematic Review

Database searches were completed on February 27, 2013, from which we identified 5,801 titles and abstracts; 20 relevant articles were identified for review (Figure 1). Of the 20 relevant articles, 14 presented data in an extractable format (see Methods) and were included in meta-analyses. One article [28] presented results for both EC and FC; these results were treated as separate studies in the meta-analyses (resulting in 15 total studies).

Study Characteristics: All Studies

The 20 relevant studies were published between 1977 and 2012 (Table 2). Two studies were conducted in a developed country (Canada), the rest in low-resource settings; no studies were conducted among populations with access to centralized water-treatment and distribution systems. The study settings included a mix of urban/peri-urban (n = 7) and rural (n = 13), and the study populations included a range of age groups (Table 2). The included articles used cohort (n = 14) and case-control (n = 6) study designs (Tables 3 and 4). We identified one randomized control trial [16] that reported data on the direct association between water quality and diarrhea; we considered this study a cohort design for the purposes of the review since water quality was not randomized. We note that, while cohort designs traditionally measure an exposure and then follow a population over time to measure outcomes, the cohort designs we identified in this review measured exposure (EC or FC): (i) simultaneously with outcomes through cross-sectional sampling; (ii) at intervals during ongoing diarrhea surveillance, or; (iii) the timing and analysis of water samples relative to diarrhea measurement was mixed or not clearly reported (Tables 3 and 4). The included case-control studies measured exposure after disease occurrence by design; however one study [19] utilized a nested case control design with

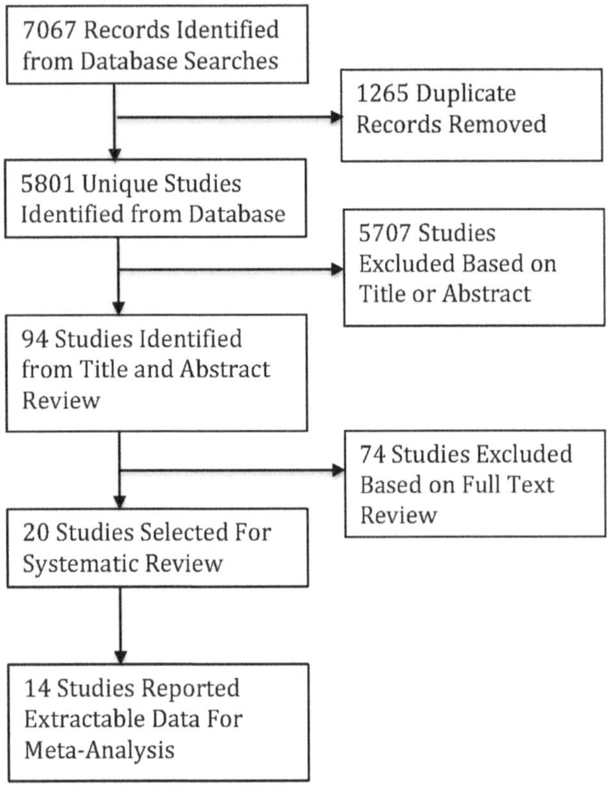

Figure 1. Systematic review flow chart.

risk set sampling in which disease and EC exposure were measured on the same day.

Included studies measured EC (n = 9), FC (n = 10), or both (n = 1) in drinking water samples collected at the point of use. Overall, studies varied in their reporting of enumeration and incubation methods, and we could not always confirm the indicator organism enumerated based on the reported information (Tables 3 and 4). In at least one study, authors reported using EC, but a description of their methods suggested they measured FC. [29] With this exception, we relied on the authors' reporting or deferred to the categorization used by the previous review. [8]

Most studies used a definition of diarrhea that is consistent with current recommendations: three or more loose or watery stools in a 24-hour period (Tables 3 and 4). [30,31] Four studies included the presence of blood or mucus in stool as part of their definition of diarrhea, [15,17,18,32] and two studies [15,17] reported dysentery as an outcome separate from "general" diarrhea; we did not include dysentery as a separate outcome in our review. Cohort studies in low-resource settings relied primarily on household surveillance and self-reported symptoms (e.g., maternal recall) for outcome assessment, using a range of recall periods (daily, up to one month; Tables 3 and 4); the two Canadian studies relied on monthly calendars maintained in the household and reported by telephone. Case-control studies mostly identified cases through hospitals and clinics, and selected controls from community samples; however, two studies identified cases and controls based on concurrent [19] or prior [33] household diarrhea surveillance conducted by the investigators.

Among the studies identified in our review, eight studies reported results that controlled for potential confounders; the other 12 studies found no evidence for confounding, or did not report attempting to control for confounding (Tables 5 and 6) – we did not differentiate between these studies in meta-analyses. In general, reporting on the independence of observations or methods to account for clustered outcomes was poor (Tables 5 and 6). Only five of the 20 studies reported using methods to account for correlated outcomes (e.g., generalized estimating equations; GEE). In general, the RRs and confidence intervals we calculated from available raw data were consistent with those reported by the authors, whether or not they reported accounting for correlated outcomes. However, in one of these studies, that reported using GEE, we calculated confidence intervals from extracted raw data that were substantially more conservative than those reported by the authors; [15] we included the more conservative confidence intervals for this study in our meta-analyses (Table 5).

Disease-Indicator Associations

Tables 5 and 6 summarize results from EC and FC studies, respectively. Among the ten EC studies, eight reported data that quantified the relationship between indicator exposure and diarrhea (Table 5). Of the eight studies with quantitative results, six reported threshold results at the 1 EC/100 ml cutoff [15,18,20,21,28,33] and all six reported RRs greater than unity; however, only one study was able to rule random error out as an explanation for their results (Table 5). [15] The remaining two studies, among the eight that quantified a relationship between EC and diarrhea, reported significant relationships between diarrhea and linear increases in exposure to mean log10 EC concentrations; the results were very consistent across these two studies (Table 5). [16,19] The two studies that did not have extractable, quantifiable results reported that no "association" or "correlation" was found between EC and diarrhea. [17,34] However, we could not confirm whether the authors' conclusions were based on the size of the effect estimate (i.e., similar to unity), precision (i.e., "non-significant" confidence intervals or *p*-values), or both.

Among the 11 FC studies (including FC results from Genthe et al. 1997 as separate from their EC results), nine studies reported data that quantified the relationship between indicator exposure and diarrhea (Table 6). Results from these nine FC studies were less consistent than results from EC studies. Using the lowest extractable threshold, five studies reported estimates that suggest increases in the risk of diarrhea with exposure to FC [32,35–38], and four reported effects equivalent to unity or reduced diarrhea risk with exposure to FC. [28,39–41] Henry and Rahim (1990) reported FC effects for two distinct study areas differentiated by "adequate" versus "poor" sanitation infrastructure; while a single summary measure for this study suggested an increased risk of diarrhea with exposure to FC, the results from the two separate study populations were very different (Table 6; FC counts were associated with an increase in illness in the area with improved sanitation and a reduction in illness in the area with unimproved sanitation). Of the two FC studies that did not report extractable data, results were also mixed: one reported no difference in FC isolation from water in case and control households, [29] the other reported that households with diarrhea had higher overall geometric mean FC levels compared to control households. [42]

Summary Measures

For consistency with the previous review, we conducted meta-analyses across all studies (combining both EC and FC studies), using the lowest extractable threshold for each study (Tables 5 and 6). Across all studies, our summary estimate suggests a non-significant increase in diarrhea risk from exposure to drinking water contaminated with EC or FC (RR 1.26 [95% CI: 0.98,

Table 2. Selected Characteristics of Included Studies.

First Author	Year	Location	Setting	Study Size[a]	Ages	Indicator
deAceituno	2012	Honduras	Rural	1020	All	EC
Levy	2012	Ecuador	Rural	115	NR	EC
Gundry	2009	S. Africa/Zimbabwe	Rural	254	1–2 y	EC
Brown	2008	Cambodia	Rural	1196	All	EC
Jensen	2004	Pakistan	Rural	209	<5 y	EC
Bhargava	2003	Bangladesh	Rural	99	1–10 y	FC
Strauss	2001	Canada	Rural	647	All	EC
Raina	1999	Canada	Rural	531	All	EC
Genthe	1997	S. Africa	Urban	316	"Pre-school"	Both
Jagals	1997	S. Africa	Urban	100	NR	FC
Vanderslice	1993	Philippines	Urban	254	<2 y	FC
Knight	1992	Malaysia	Rural	196	4–59 m	FC
Han	1991	Myanmar	Urban	208	6–29 m	FC
Henry	1990	Bangladesh	Rural	92	6–18 m	FC
Henry and Rahim	1990	Bangladesh	Urban	137	1–6 y	FC
Echeverria	1987	Thailand	Urban	NR	<5 y	FC[b]
Esrey	1986	Lesotho	Rural	545	1–60 m	FC
Lloyd-Evans	1984	Gambia	Urban	20	6–36 m	EC
Black	1982	Bangladesh	Rural	40	5–18 m	EC
Rajasekaran	1977	India	Rural	1091	<5 y	FC

EC: Escherichia Coli; FC: Fecal Coliform.
[a]May not reflect the analytic sample used by the authors;
[b]Authors report EC but methods suggest FC.

1.63]; Figure 2). However, we also found evidence of significant heterogeneity ($X^2 = 36.13$, degrees of freedom (d.f.) = 12, p< 0.001), and evidence of publication bias (Figure 3).

In order to explore a potential source of heterogeneity across all studies, we stratified results by indicator organism (EC vs. FC) and re-estimated summary measures in these sub-groups. Across all EC studies, the summary measure suggests a significant association between exposure to EC in household drinking water and diarrhea (RR: 1.54 [95% CI: 1.37, 1.74]); Figure 4). While we found no evidence of heterogeneity across EC studies ($X^2 = 1.65$, d.f. = 5, p = 0.90), and minimal evidence of publication bias (Figure 5), we note that the EC summary measure was disproportionately influenced by one study [15] ("% weight", Figure 4). We therefore conducted a sensitivity analysis by repeating the EC analysis without the study by Brown et al. Removing this study did not impact the magnitude of the summary measure, but did reduce the precision of the estimate (RR 1.48 [95% CI: 1.02, 2.15]); Figure S1); there was no evidence of heterogeneity after removing Brown et al. from the analysis ($X^2 = 1.59$, d.f. = 4, p = 0.81). Across all FC studies, our summary effect measure suggests no association between FC exposure in drinking water and diarrhea (RR: 1.07 [95% CI: 0.79, 1.45]); Figure 4); we also found evidence of significant heterogeneity ($X^2 = 2.30$, d.f. = 6, p = 0.06), and publication bias (Figure 6).

Threshold Effects

Two EC studies provided extractable data that allowed us to explore threshold effects at three different levels (1 EC, 10 EC, and 100 EC per 100 ml; Table 5) [15,18]. RRs were calculated by comparing outcomes in the exposed group (≥EC threshold) to outcomes in the control group (<EC threshold). In one study, [15] increasing the threshold from 1EC to 10EC suggested a slight increase in diarrhea risk; there was no further increase in risk when the threshold was increased to 100EC (Table 5). In the other study, increasing the threshold beyond 1EC did not increase the magnitude of the effect estimates (Table 5). [18] We pooled estimates across these two studies, and found no evidence that increasing thresholds beyond WHO drinking water guidelines (1 EC/100 mL) was associated with an increased risk of diarrhea (Figure 7). [2] A third EC study reported finding no threshold effect at 1000 EC/100 ml, but did not report a RR. [19]

Categorical Exposure and Dose-Response

Four EC studies reported investigating a dose-response relationship. [15,18,19,21] Only three of these studies evaluated the presence of a dose-response using similar exposure categories (<1 EC (referent), 1–10 EC, 11–100 EC, 100+, or 101–1000 EC and 1000+; Table 5). Of these three studies, one did not report extractable data, but reported finding no evidence of a dose-response effect. [19] Results from the other two studies [15,18] suggest a qualitative dose response (increased risk of diarrhea with increasing exposure categories; Table 5); formal tests of trend, however, had mixed results (Brown et al: p<0.001 provides evidence of a dose response; Jensen et al: p = 0.26 does not provide evidence of a dose response). A fourth study [21] used different EC exposure categorizations. The results from this study qualitatively suggested increasing risk across 0.1–1.5 EC and 1.6–700 EC exposure categories; however, the authors found no evidence of a linear trend (p = 0.45; Table 5).

Table 3. Summary of EC Study Exposure and Outcome Measurements.

First Author	Year	Study Design	Outcome [a]	Recall	Measured EC Relative to Outcome	Enumeration Method	Medium	Temp (°C)[f]
deAceituno	2012	RCT	Diarrhea	7-day	After (<1 week)	MPN	IDEXX Colilert	NR
Levy	2012	CC	Diarrhea	Daily	Same Day	MF	BD Difco ml agar[e] (EPA 1604)	30
Gundry	2009	Cohort	Diarrhea	Daily	Before (<2 weeks)	MPN	IDEXX Colilert	NR
Brown	2008	Cohort	Diarrhea	7-day	After (<1 week)	MF	"selective medium"[d]	NR
Jensen	2004	Cohort	Diarrhea	7-day	Before/After (<3 days)	MF	Hach m-ColiBlue 24 Broth PourRite	35
Strauss	2001	Cohort	AGI[b]	Monthly Diary	Twice over 28-d surveillance	MF	m-FC-BCIG agar	44.5
Raina	1999	Cohort	GI Illness[c]	Monthly Diary	5 times over 1-y surveillance	NR	"standard methods" (MOE 1992)	NR
Genthe	1997	CC	Diarrhea	Hospital	After	NR	"standard methods" (APHA 1989)	NR
Lloyd-Evans	1984	CC	Diarrhea	7-day	After (1–2 months)	MF	Millipore (DHSS 1969)	NR
Black	1982	Cohort	Diarrhea	NR	After (<1 month)	Plating	MacConkey's Agar	37

EC: *Escherichia Coli*; CC: Case Control; MF: Membrane Filtration; MPN: Most Probable Number; NR: Not Reported; C: Celsius; DHSS: Department of Health & Social Security: *The bacteriological examination of water* supplies; APHA: American Public Health Association: *Standard Methods for the Examination of Water and Wastewater*; MOE: *Ministry of Environment: Ontario's Drinking Water Objectives*.
[a]Unless noted, used a definition consistent with World Health Organization recommendation of 3 or more loose/watery stools in 24 hour period;
[b]Acute Gastrointestinal Illness: 1) vomiting or liquid diarrhea or 2) nausea or soft, loose diarrhea combined with abdominal cramps;
[c]Diarrhea with or with out vomiting;
[d]selective medium: "containing chromogenic and fluorogenic b-glucuronide and b-galactoside substrates";
[e]Multiple EC enumeration methods used – we report the method with the largest sample size and for which the authors report an extractable RR;
[f]Incubation temperature.

Table 4. Summary of FC Study Exposure and Outcome Measurements.

First Author	Year	Study Design	Outcome [a]	Recall	Measured FC Relative to Outcome	Enumeration Method	Medium	Temp (°C)[e]
Bhargava	2003	Cohort	GI Morbidity [b]	1-month	After (<1 month)	MF	*DIFCO* m-FC media	45
Genthe	1997	CC	Diarrhea	Hospital	After	NR	"standard methods" (APHA 1989)	NR
Jagals	1997	CC	Diarrhea [c]	Clinic	After	MF	"standard methods" (APHA 1992)	44.5
Vanderslice	1993	Cohort	Diarrhea [c]	7-day	2–5 times over 1 year	MF	m-FC agar (APHA 1985)	44.5
Knight	1992	CC	Diarrhea	Clinic	After (4–14 days)	MF	*Paqualab* (ELE International 1986)	NR
Han	1991	Cohort	Diarrhea	7-day	Averaged over surveillance	MF	m-FC agar (WHO 1983)	NR
Henry	1990	Cohort	Diarrhea	7-day	Averaged over surveillance	NR	"standard methods" (APHA 1981)	NR
Henry and Rahim	1990	Cohort	Diarrhea	14-day	Unclear (<6 months)	dip-slides	*ORION Hygicult* Agar Slides [d]	37
Echeverria	1987	CC	Diarrhea	Hospital	After	MF	m-FC agar (APHA 1975)	44
Esrey	1986	Cohort	Diarrhea [c]	24-hour	Unclear	MF	Faecal Coliform Broth (AJPH 1975)	NR
Rajasekaran	1977	Cohort	Diarrhea	NR	Unclear (<15 days)	MPN	IMViC (APHA 1965)	44

FC: Fecal Coliform; CC: Case Control; GI: Gastrointestinal; MF: Membrane Filtration; MPN: Most Probable Number; NR: Not Reported; C: Celsius; APHA: American Public Health Association: *Standard Methods for the Examination of Water and Wastewater*; WHO: World Health Organization: *Guidelines for Drinking Water Quality*; IMViC: Indole, methyl red, Voges–Proskauer and citrate [a] [a] Unless otherwise noted, study used a definition consistent with current World Health Organization recommendation of 3 or more loose/watery stools in a 24 hour period; [b] Morbidity included: diarrhea, dysentery, vomiting, stomachache, acidity, typhoid, cholera; [c] Did not report a definition; [d] Report enumerating *Enterobacteriaceae* – included as FC for consistency with previous review. [e] Incubation temperature

Table 5. EC Study Results.

First Author	Year	Adjusted Analysis	RR	Threshold Results [a] RR (95% CI)	Categorical/Dose Response: [b]RR (95% CI)	Alternate Results: RR (95% CI)
deAceituno	2012	Yes	OR [c]	-	-	Log10 increase: 1.26 (1.08,1.46)
Levy	2012	NR	OR [c]	1000 EC: "no effect"	no "dose-response" (1–10 EC, 11–100 EC, 101–1,000 EC, 1,000+ EC)	Log10 increase: 1.29 (1.02,1.65)
Gundry	2009	Yes	NR	-	-	"no association"
Brown [d]	2008	No	PR [c]	1 EC: 1.55 (1.36,1.76) 10 EC: 1.70 (1.52,1.90) 100 EC: 1.67 (1.51,1.84) 1000 EC: 1.37 (1.21,1.54)	1–10 EC: 0.98 (0.81–1.20) 11–100 EC: 1.35 (1.13–1.61) 101–1000 EC: 1.83 (1.58–2.11) 1001+EC: 1.81 (1.54–2.11) [e]	-
Jensen	2004	No	IDR	1 EC: 1.32 (0.77,2.27) 10 EC: 1.25 (0.83,1.88) 100 EC: 1.23 (0.85,1.80)	1–10 EC: 1.17 (0.55–2.47) 11–100 EC: 1.26 (0.70–2.28) 100+EC: 1.45 (0.81–2.59) [f]	-
Strauss	2001	Yes	OR [c]	1 EC: 1.52 (0.33,6.92)	0.1–1.5 EC: 0.85 (0.10,7.19) 1.6–700 EC: 2.69 (0.34,21.56) [g]	-
Raina	1999	Yes	OR [c]	1 EC: 2.11 (0.90,4.94)	-	-
Genthe	1997	Yes	OR	1 EC: 1.26 (0.59,2.72)	-	-
Lloyd-Evans	1984	No	OR	1 EC: 3.86 0.33,45.57)	-	-
Black	1982	NR	NR	-	-	"no correlation found"

EC: *Escherichia Coli*; RR: Relative Risk; OR: Odds Ratio; PR: Prevalence Ratio; IDR: Incidence Density Ratio; NR: Not Reported; CI: Confidence Interval;
[a]Compares risk in groups exposed to ≥EC value, to <EC value;
[b]Reference group is <1 EC for all categories;
[c]Authors report using methods to account for correlated (clustered) outcomes;
[d]RRs and 95% CIs calculated from raw data;
[e]*p*-value for linear trend <0.01;
[f]p-value for linear trend =0.26;
[g]p-value for linear trend =0.45.

Table 6. FC Study Results.

First Author	Year	Adjusted Analysis	RR	Threshold Results [a] RR (95% CI)	Alternate Results RR (95% CI)
Bhargava	2003	Yes	NA	-	ML Coefficient: 0.204 (SE: 0.08) *p-value* <0.05
Genthe	1997	Yes	OR	1 FC: 0.84 (0.46,1.52)	
Jagals	1997	NR	NA	-	Geometric mean FC higher in homes with diarrhea (compared to control homes)
Vanderslice	1993	Yes	NA	-	Probit coefficient: −0.002 *p-value* >0.10
Knight	1992	Yes	OR	1 FC: 1.45 (0.57,4.76)	-
Han	1991	No	PR	Medium: 0.72 (0.56,0.94) [b]	Medium-High: 0.73 (0.52,1.02) [c] High: 0.72 (0.52,1.00) [c]
Henry	1990	No	IDR	Low: 1.03 (0.75,1.42) [b]	-
Henry and Rahim	1990	No	CIR	10^4 FC: 1.26 (0.59,2.72) [d]	Improved Area: 2.58 (0.70,9.54) [d] Unimproved Area: 0.86 (0.34–2.18) [d]
Echeverria	1987	NR	NR	-	FC isolated from water as often in homes with diarrhea as in control homes
Esrey	1986	No	PR	10 FC: 1.53 (0.69,3.40) 100 FC: 1.46 (0.62,3.45)	10–100 FC: 1.41 (0.53,3.76) [e]>100 FC: 1.64 (0.64,4.18) [e]
Rajasekaran	1977	No	PR	10 FC: 2.93 (1.10,7.81)	-

FC: Fecal Coliform; RR: Relative Risk; OR: Odds Ratio; PR: Prevalence Ratio; IDR: Incidence Density Ratio; NR: Not Reported; CI: Confidence Interval; ML: maximum likelihood
[a]Compares risk in groups exposed to ≥FC value, to <FC value;
[b]Based on median of samples or arbitrary cutoffs;
[c]Compared to "Low" group – Authors do not report specific FC concentrations used to determined categories;
[d]Reference group is <1000 colony forming units FC per gram;
[e]Reference group <10 FC.

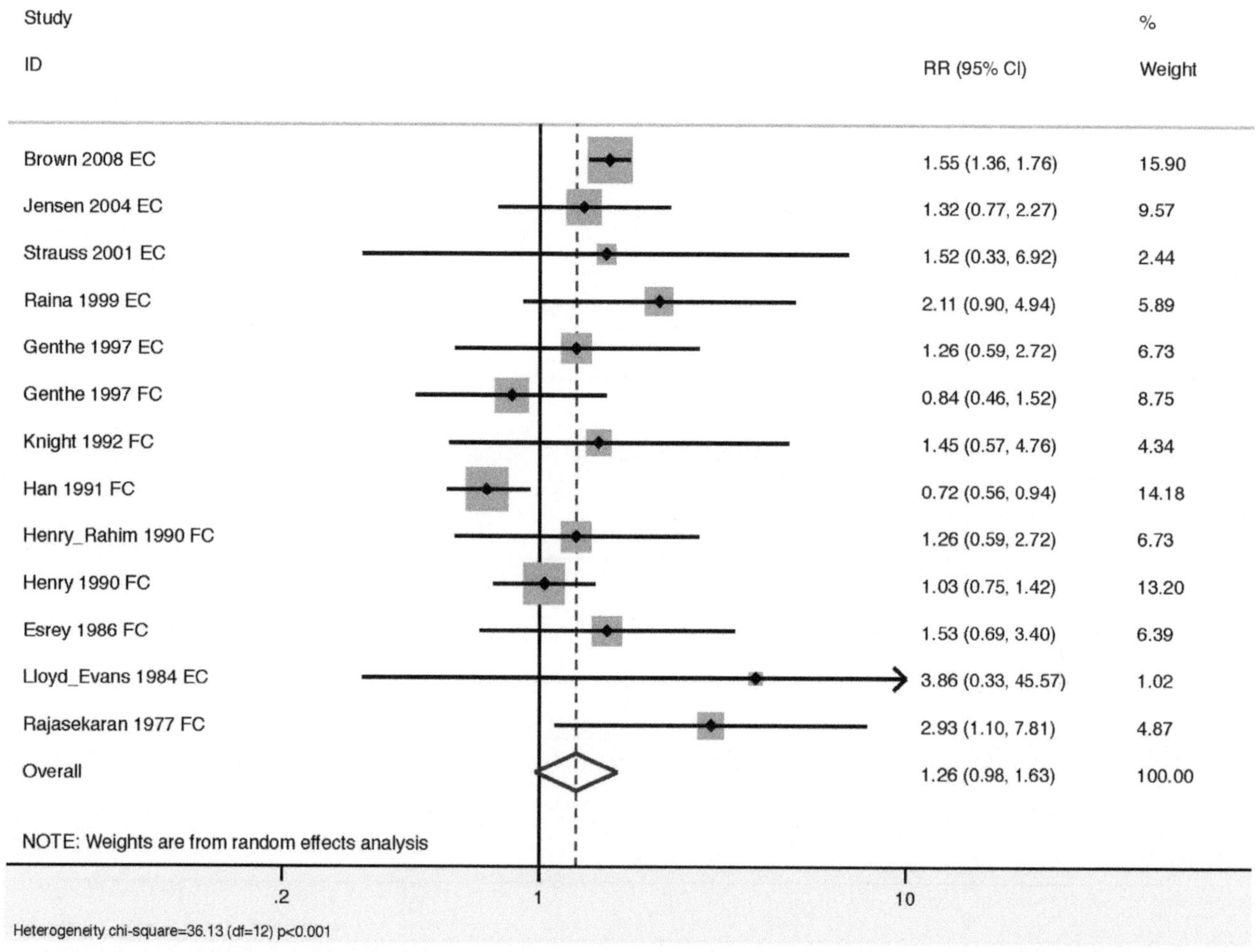

Figure 2. Forest plot for all included studies.

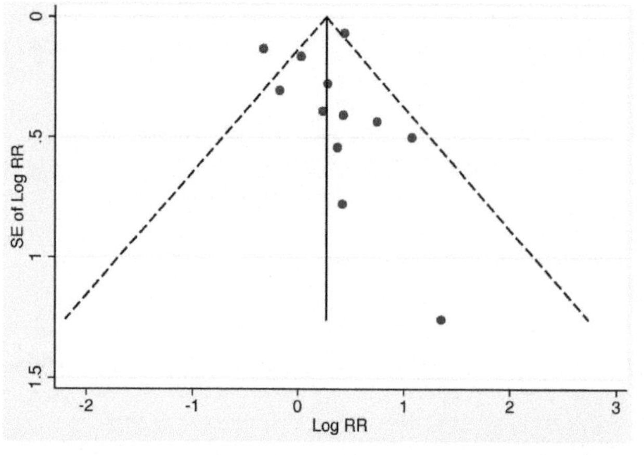

Figure 3. Funnel plot for all included studies (EC and FC studies). Studies are plotted relative to size (SE of Log RR), and reported log relative risk (Log RR); studies higher on the vertical axis are larger, and studies further to the right have larger RRs. The solid vertical line is the log of the summary RR for all studies, and the dotted lines are "pseudo 95% confidence intervals" for the summary measure. The absence of smaller studies with small or null RRs (lower left region of the plot) creates plot asymmetry, and provides evidence of possible publication bias.

Discussion

Summary of Findings

We conducted a systematic review and meta-analysis to evaluate the relationship between diarrheal illness and the presence of EC or FC indicators in household drinking water. Consistent with a prior review, [8] we did not find conclusive evidence for an association between contaminated drinking water and diarrhea when EC and FC results were combined (previous review: RR 1.12 [95% CI: 0.85, 1.48]; our review: RR 1.26, [95% CI: 0.98, 1.63]; Figure 2). We also found evidence of significant heterogeneity and publication bias across all studies (Figure 2, Figure 3). After stratifying studies by EC and FC to explore the choice of indicator organism as a source of heterogeneity, we found that EC studies reported consistent effect estimates that suggest an increased risk of diarrhea with exposure to contaminated drinking water with no evidence of heterogeneity (pooled RR 1.54 [95% CI: 1.37, 1.74]; Table 5, Figure 4). In contrast, FC studies reported mixed results with significant heterogeneity and provided minimal evidence of an association with diarrhea (pooled RR 1.07 [95% CI: 0.79, 1.45]; Table 6, Figure 4). We found no evidence that exposure thresholds greater than 1 EC/100 mL (e.g., 10 EC/ 100 mL, 100 EC/100 mL) were associated with increased diarrheal disease, but this evidence is based on a limited number of studies. Results regarding a dose-response relationship between EC in household drinking water and diarrhea were limited and

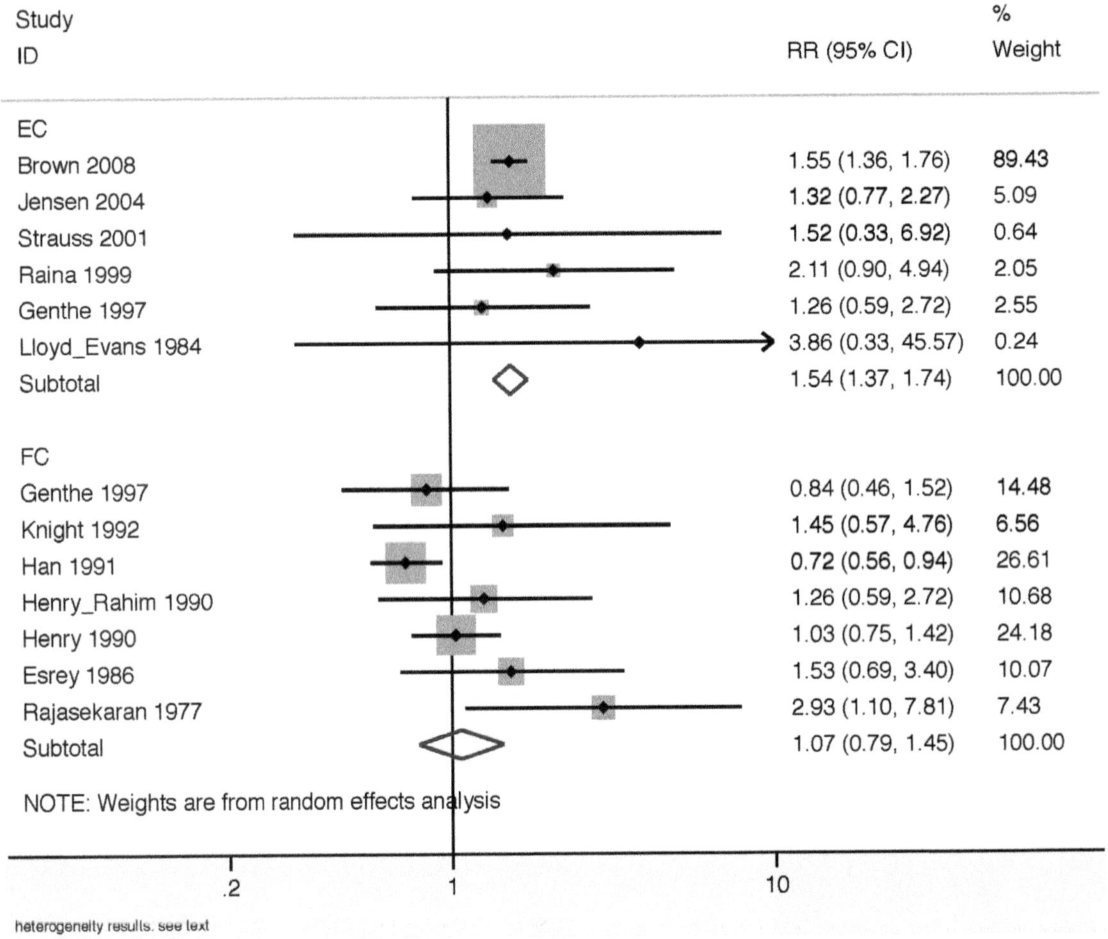

Figure 4. Forest plot for all studies stratified by indicator organism. Top panel, EC: *Escherichia coli*; Bottom panel, FC: thermotolerant ("fecal") coliforms.

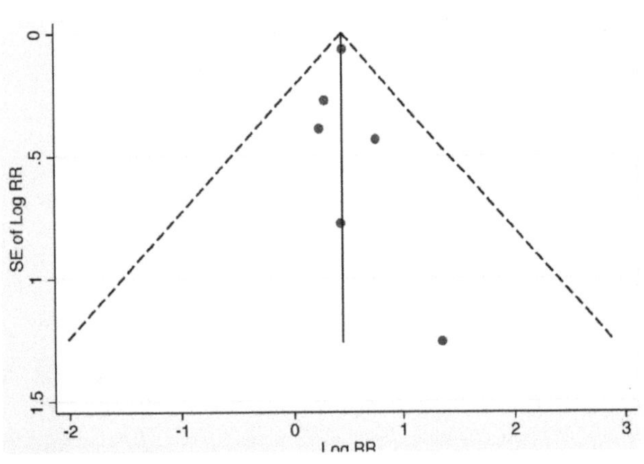

Figure 5. Funnel plot for EC studies. Studies are plotted relative to size (SE of Log RR), and reported log relative risk (Log RR); studies higher on the vertical axis are larger, and studies further to the right have larger RRs. The solid vertical line is the log of the summary RR for all studies, and the dotted lines are "pseudo 95% confidence intervals" for the summary measure. Studies on this plot are generally symmetrical, with the exception of one small study with a large RR, providing minimal evidence of publication bias.

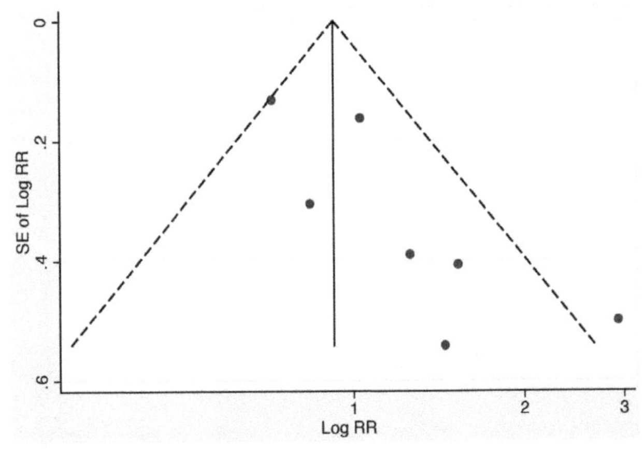

Figure 6. Funnel plot for all FC studies. Studies are plotted relative to size (SE of Log RR), and reported log relative risk (Log RR); studies higher on the vertical axis are larger, and studies further to the right have larger RRs. The solid vertical line is the log of the summary RR for all studies, and the dotted lines are "pseudo 95% confidence intervals" for the summary measure. The absence of smaller studies with small or null RRs (lower left region of the plot) creates plot asymmetry, and provides evidence of possible publication bias.

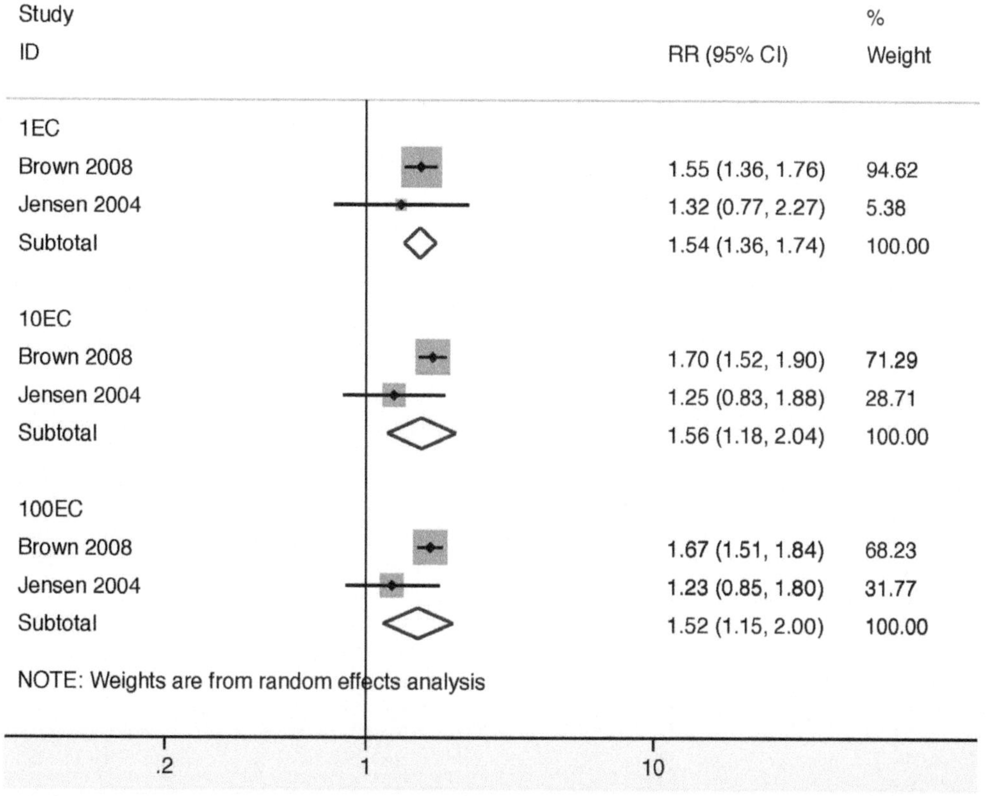

Figure 7. Forest plot for two studies with extractable data at multiple thresholds. Data were pooled across the two studies at the 1 EC/100 ml threshold (top panel), 10 EC/100 ml threshold (middle panel), and 100 EC/100 ml threshold (bottom panel). EC: *Escherichia coli.*

inconclusive. In addition to the results of the meta-analysis, we identified several elements of study design and reporting that could improve future studies evaluating the relationship between water quality and health.

Implications for Choice of Indicator Organism to Assess Microbiological Safety of Drinking Water

Current guidelines recommend two specific uses for indicator organisms: (i) fecal indicators (index organisms), used to indicate the presence of fecal contamination, potential pathogens, and possible health risk, and; (ii) process indicators used to evaluate the effectiveness of a water treatment process. [1,2,13]. While the performance of coliform bacteria as process indicators has been described previously, [2,43] and is outside the scope of this review, the results from our review have implications for the use of EC and FC as index organisms in household drinking water. Currently, EC is considered the "most suitable" fecal indicator; however, FC is generally recognized as a suitable surrogate. [1,2,12,13] Our findings support the continued use of EC as a fecal indicator (index organism) to assess water quality at the point of use, but do not support the use of FC as a surrogate as it does not demonstrate a clear association with diarrheal disease outcomes.

Our findings that EC and FC perform differently as indicators of diarrhea risk from contaminated drinking water are not surprising. While EC is generally of fecal origin, [10] and likely to be present simultaneously with diarrheagenic pathogens in recently contaminated drinking water, FC are known to include coliforms of environmental origin (e.g., *Klebsiella*), in addition to EC. [1,2,13] The consistency with which FC methods will be specific to human pathogens will therefore vary with the environments in which water samples are being collected. Indeed, our results show that the magnitude and direction of the effect estimates for FC were highly inconsistent (Figure 4, Table 6); even within a single study [37] the authors found highly divergent results using FC in two separate study areas (Table 6). In contrast, the magnitude and direction of effect estimates across the EC studies included in our review were consistent across diverse study regions, and when pooled provided evidence of a significant association with diarrhea; these results are consistent with the higher specificity of EC for fecal contamination, compared to FC (Table 5, Figure 4). The presence of a dose-response between EC and diarrhea could have provided further evidence to support our findings that drinking water contaminated with EC is associated with diarrhea. However, dose-response results from our review were inconclusive, and represent an avenue for future research.

While our summary measures suggest that the presence of EC in household drinking water is significantly associated with diarrhea, it is important to note that only one individual EC study in our review was able to rule out chance as a possible explanation of their findings [15]. It has been demonstrated that indicator levels can vary considerably even over short periods of time in both household and source water, [44,45] and that pathogens and indicator organisms have inconsistent correlations. [46] The studies in this review relied on intermittent grab samples to classify household water quality. Even daily grab samples likely provided only a crude assessment of a household's exposure to fecal contamination and diarrheagenic pathogens, which likely contributed to the lack of precision in the individual studies included in our review. While the imprecision of results from individual studies supports a previous assertion that fecal indicator organisms are "blunt" tools to characterize water quality, [19] our findings suggest that they have continued utility in health and water research. Indeed, while recent WHO drinking water guidelines have shifted focus from recommendations using specific indicator levels to define water safety to a more integrated approach centered on water safety plans, these recommendations are largely directed towards the management of water distribution systems. [2,13] In practice, regional organizations and researchers working in areas without access to drinking water distribution systems rely heavily on indicator organisms to assess the risk of disease from drinking water, and to measure the effectiveness of household water treatment and safe storage interventions; this is particularly true in resource poor settings with limited access to advanced laboratory facilities. Our results suggest that EC have continued value in both of these applications.

Threshold Effects for Diarrhea Risk

The findings from our review do not support the existence of a "threshold effect" in contaminated household drinking water. A previous study evaluating the effect of contaminated *source* water on diarrhea [23] found no evidence of an association between EC and diarrhea at the 1 EC threshold level; however, significant associations with diarrhea were observed at the 1000 EC threshold level. The authors concluded that their findings of a "threshold effect" supported earlier assertions that relaxing drinking water guidelines (beyond the 1 EC threshold) might be acceptable in developing countries. [47] Contrary to these findings, in our review of household drinking water we found consistently elevated risks of diarrhea at the 1 EC threshold, with a significant summary measure when pooled across all studies. These results support current WHO guidelines recommending that all drinking water intended for human consumption contain zero EC/100 ml, regardless of location. [2] Beyond the 1 EC threshold, our results did not suggest increased risk of diarrhea at elevated cutoffs, but this evidence was based on a limited number of studies.

Recommendations for Future Water Quality Studies

The studies included in our review covered several decades, and varied considerably with respect to methodological quality and completeness of reporting. In this section, we highlight several findings from our systematic review that could improve the design and reporting of future studies that seek to evaluate the relationship between water quality and health.

The studies in our systematic review used very consistent definitions of diarrhea, relying primarily on self- or caregiver reporting of diarrhea symptoms (Tables 3 and 4). However, we note that diarrhea recall periods and identification strategies varied greatly (e.g., daily recall, one-month recall, cases presenting to clinics; Tables 3 and 4). Reliance on self-reported diarrhea symptoms as a health measure could have introduced bias into individual studies through two primary mechanisms, which have implications for the use of self-reported diarrhea as an outcome in future studies. First, as with any self-reported, subjective outcome, there is a potential for differential reporting relative to exposure status [48]. If study-specific reporting biases were differential with respect to drinking water contamination, it is possible that the results of individual studies (and therefore our review) could be biased. However, there is no reason to suspect that individuals would be aware of the indicator status of their drinking water when reporting diarrhea symptoms, which would decrease the potential for this bias. Second, longer follow-up periods can bias diarrhea reporting due to poor outcome recall, which could have biased findings towards the null if non-differential with respect to exposure, and/or compounded biases associated with differential reporting in individual studies. [49] The inclusion of objective health outcome measures (e.g., pathogen-specific antibody responses) could improve future studies by reducing reliance on self-reported diarrhea symptoms and their associated biases. [50,51] In addition, while our review focused exclusively on diarrhea, expanding the range of water-related health outcomes studied and reviewed (e.g., indicators of growth and malnutrition) could also improve our understanding of the health impacts of drinking water contamination beyond gastrointestinal illnesses.

Very few studies in our review were able to establish temporality between exposures (indicator organism) and outcomes (diarrhea), and therefore mostly relied on the assumption that drinking water samples were representative of an exposure period that was relevant to measured disease outcomes. While this assumption might be reasonable, or the only option given the practicalities of field conditions, it does leave open the possibility of reverse causality (i.e., diarrhea caused by other environmental pathways led to increased fecal contamination of drinking water). The design of future water quality studies should attempt to measure exposure in advance of disease outcomes, and all studies should be explicit about the timing of exposure and disease measurements in their reporting. These recommendations would all but preclude the use of "traditional" case-control studies, where cases and controls (i.e., disease status) are identified first, followed by visits to households to collect water samples to ascertain exposure status. Instead, a cohort design that measures exposure to contaminated drinking water first, followed by surveillance visits after an appropriate incubation period to measure health outcomes, is a more suitable study design. Decisions regarding follow-up frequency, recall period, and measures of disease occurrence (i.e., incidence, prevalence) will vary depending on study context and logistics, but should be carefully considered and explicitly reported. [49,52]

In studies evaluating the relationship between drinking water quality and health, confounding bias is a concern: there are numerous factors that could cause both drinking water contamination and diarrhea in a given study setting. Overall, the studies in our review were not consistent in reporting adjusted results (Tables 5 and 6), and there was no consistent set of factors used across studies when authors did report attempting to control for confounding. Incomplete control of confounding in the included studies could have biased individual study results, and therefore the findings of our review. Future studies should be explicit about their attempts to measure and control for confounding factors. Additionally, several studies did not account for correlated outcomes in their analyses, which could have influenced the precision of summary estimates. However, for EC studies we note the consistency of the direction and magnitude of the associations reported across diverse study locations, and across studies that did and did not adjust for confounding and/or account for clustering. Regardless, reporting on these issues could be improved upon in future studies.

Our systematic review included published, peer-reviewed studies. As with any systematic review, publication bias, where smaller studies with small or null ("non-positive") results are systematically excluded from the published literature, could have impacted our findings. In our review, we found evidence of publication bias among FC studies (smaller studies with smaller effect sizes appear to be absent from the funnel plot; lower left region, Figure 6); however, inclusion of smaller FC studies with RRs at or less than unity would have strengthened our findings of no association between FC and diarrhea. We found minimal evidence to suggest publication bias among EC studies (Figure 5);

however, we cannot rule out publication bias as a possible explanation of our findings – the funnel plot suggests that most studies were balanced around the summary measure with the exception of one small study with a large positive finding. Taken together, there is evidence that smaller studies with small ("non-positive") effects sizes are systematically missing from the literature on the relationship between water quality and diarrheal illness.

In our review, we pooled RRs across relevant studies using meta-analysis to investigate the association of fecal indicator measures in drinking water with diarrhea. The limitations of meta-analysis have been discussed [53]; we caution against interpreting the summary measures from our review as "true" underlying causal effects, but rather consider them additional pieces of evidence regarding the relationship between fecal indicators in drinking water and diarrhea. The evidence from our pooled summary measures, taken together with biologic plausibility and the consistency of results across studies, provides evidence to support a relationship between diarrhea and the presence of EC in point-of-use drinking water; we do not, however, find evidence to support an equivalent relationship between diarrhea and FC.

Conclusions

We reviewed the available literature to identify studies that evaluated the relationship between microbial indicators of drinking water quality and diarrhea. We found that studies using *Escherichia coli* (EC) as an indicator of household drinking water quality reported consistent effect estimates, that when pooled suggested a significant association with increased diarrheal illness. Results from studies using thermotolerant ("fecal") coliforms (FC), on the other hand, were inconsistent, and suggested no association with diarrhea when pooled. In this review, we also note several areas where the design and reporting of the included studies could have been improved, and make recommendations for future studies. The results from our review suggest that EC has value as a fecal indicator organism, but that use of FC should be considered carefully in contexts where an association with diarrheal disease outcomes is important.

Acknowledgments

The authors would like to acknowledge Professor Kara Nelson for generously reviewing this manuscript, and providing key feedback.

Author Contributions

Conceived and designed the experiments: JG JC. Performed the experiments: JG AE. Analyzed the data: JG. Contributed reagents/materials/analysis tools: JG AE. Wrote the paper: JG AE JC.

References

1. Dufour AP (2003) Assessing microbial safety of drinking water: Improving approaches and methods: International Water Assn.
2. WHO (2011) Guidelines for drinking-water quality, fourth edition. Geneva: World Health Organization.
3. Cutler D, Miller G (2005) The role of public health improvements in health advances: The twentieth-century United States. Demography 42: 1–22.
4. Arnold BF, Colford JM Jr (2007) Treating water with chlorine at point-of-use to improve water quality and reduce child diarrhea in developing countries: a systematic review and meta-analysis. The American journal of tropical medicine and hygiene 76: 354.
5. Clasen T, Schmidt WP, Rabie T, Roberts I, Cairncross S (2007) Interventions to improve water quality for preventing diarrhoea: systematic review and meta-analysis. bmj 334: 782.
6. Fewtrell L, Kaufmann RB, Kay D, Enanoria W, Haller L, et al. (2005) Water, sanitation, and hygiene interventions to reduce diarrhoea in less developed countries: a systematic review and meta-analysis. The Lancet infectious diseases 5: 42–52.
7. National Research Council (2004) Indicators for Waterborne Pathogens. Washington, DC: The National Academies Press.
8. Gundry S, Wright J, Conroy R (2004) A systematic review of the health outcomes related to household water quality in developing countries. J Water Health 2: 1–13.
9. Gerba C (2000) Indicator Microogranisms. Environmental Microbiology. San Diego, CA: Academic Press.
10. Leclerc H, Mossel D, Edberg S, Struijk C (2001) Advances in the bacteriology of the coliform group: their suitability as markers of microbial water safety. Annual Reviews in Microbiology 55: 201–234.
11. Reynolds KA, K.D Mena, Gerba CP (2008) Risk of waterborne illness via drinking water in the United States. Reviews of Environmental Contamination and Toxicology 192: 117–158.
12. Tallon P, Magajna B, Lofranco C, Leung KT (2005) Microbial indicators of faecal contamination in water: a current perspective. Water, Air, and Soil Pollution 166: 139–166.
13. Ashbolt NJ, Grabow WOK, Snozzi M (2001) Indicators of microbial water quality. In: Fewtrell L, Bartram J, editors.Water Quality: Guidelines, Standards and Health. London, UK: IWA Publishing,.
14. WHO (1997) Guidelines for drinking-water quality, 2nd edition. Geneva: World Health Organization.
15. Brown JM, Proum S, Sobsey MD (2008) Escherichia coli in household drinking water and diarrheal disease risk: evidence from Cambodia. Water Science and Technology 58: 757–763.
16. de Aceituno AMF, Stauber CE, Walters AR, Sanchez REM, Sobsey MD (2012) A Randomized Controlled Trial of the Plastic-Housing BioSand Filter and Its Impact on Diarrheal Disease in Copan, Honduras. American Journal of Tropical Medicine and Hygiene 86: 913–921.
17. Gundry SW, Wright JA, Conroy RM, Du Preez M, Genthe B, et al. (2009) Child dysentery in the Limpopo Valley: A cohort study of water, sanitation and hygiene risk factors. Journal of Water and Health 7: 259–266.
18. Jensen PK, Jayasinghe G, van der Hoek W, Cairncross S, Dalsgaard A (2004) Is there an association between bacteriological drinking water quality and childhood diarrhoea in developing countries? Trop Med Int Health 9: 1210–1215.
19. Levy K, Nelson KL, Hubbard A, Eisenberg JN (2012) Rethinking indicators of microbial drinking water quality for health studies in tropical developing countries: case study in northern coastal Ecuador. Am J Trop Med Hyg 86: 499–507.
20. Raina PS, Pollari FL, Teare GF, Goss MJ, Barry DA, et al. (1999) The relationship between E. coli indicator bacteria in well-water and gastrointestinal illnesses in rural families. Can J Public Health 90: 172–175.
21. Strauss B, King W, Ley A, Hoey JR (2001) A prospective study of rural drinking water quality and acute gastrointestinal illness. BMC Public Health 1: 8.
22. Rothman K, Greenland S, (1998) Modern Epidemiology. Philadelphia: Lippincott-Raven Publishers.
23. Moe C, Sobsey M, Samsa G, Mesolo V (1991) Bacterial indicators of risk of diarrhoeal disease from drinking-water in the Philippines. Bulletin of the World Health Organization 69: 305.
24. Greenland S, Longnecker MP (1992) Methods for trend estimation from summarized dose-response data, with applications to meta-analysis. American Journal of Epidemiology 135: 1301–1309.
25. Egger M, Smith G Davey, Altman D (2001) Systematic Reviews in Healthcare. Meta-analysis in Context. London: BMJ Books.
26. Egger M, Smith GD, Schneider M, Minder C (1997) Bias in meta-analysis detected by a simple, graphical test. bmj 315: 629–634.
27. Sterne JAC, Egger M (2001) Funnel plots for detecting bias in meta-analysis: Guidelines on choice of axis. Journal of Clinical Epidemiology 54: 1046–1055.
28. Genthe B, Strauss N, Seager J, Vundule C, Maforah F, et al. (1997) The effect of type of water supply on water quality in a developing community in South Africa. Water Science and Technology 35: 35–40.
29. Echeverria P, Taylor DN, Seriwatana J (1987) Potential sources of enterotoxigenic Escherichia coli in homes of children with diarrhoea in Thailand. Bulletin of the World Health Organization 65: 207–215.
30. Baqui AH, Black RE, Yunus M, Hoque ARA, Chowdhury H, et al. (1991) Methodological issues in diarrhoeal diseases epidemiology: definition of diarrhoeal episodes. International journal of epidemiology 20: 1057.
31. WHO (2013) Diarrhoeal disease; fact sheet.

32. Rajasekaran P, Dutt PR, Pisharoti KA (1977) Impact of Water Supply on the Incidence of Diarrhoea and Shigellosis Among Children in Rural Communities in Madurai. Indian J Med Res 66: 189–199.
33. Lloyd-Evans N, Pickering HA, Goh SGJ, Rowland MGM (1984) Food and Water Hygiene and Diarrhoea in Young Gambian Children: a Limited Case Control Study. Transactions of the Royal Society of Tropical Medicine and Hygiene 78: 209–211.
34. Black RE, Brown KH, Becker S, Alim AR, Merson MH (1982) Contamination of weaning foods and transmission of enterotoxigenic Escherichia coli diarrhoea in children in rural Bangladesh. Trans R Soc Trop Med Hyg 76: 259–264.
35. Bhargava A, Bouis HE, Hallman K, Hoque BA (2003) Coliforms in the water and hemoglobin concentration are predictors of gastrointestinal morbidity of Bangladeshi children ages 1–10 years. American journal of human biology 15: 209–219.
36. Esrey SA, Habicht J, Casella G, Miliotis MD, Kidd AH, et al. (1986) Infection, Diarrha, and Growth Rates of Young Children Following the Installation of Village Water Supplies in Lesotho. Interanation Symptosium on Water-Related Health Issues November: 11–16.
37. Henry FJ, Rahim Z (1990) Transmission of diarrhoea in two crowded areas with different sanitary facilities in Dhaka, Bangladesh. J Trop Med Hyg 93: 121–126.
38. Knight SM, Toodayan W, Caique WC, Kyi W, Barnes A, et al. (1992) Risk factors for the transmission of diarrhoea in children: a case-control study in rural Malaysia. Int J Epidemiol 21: 812–818.
39. Han AM, Oo KN, Aye T, Hlaing T (1991) Bacteriologic studies of food and water consumed by children in Myanmar: 2 Lack of association between diarrhoea and contamination of food and water. J Diarrhoeal Dis Res 9: 91–93.
40. Henry F, Huttly S, Patwary Y, Aziz K (1990) Environmental sanitation, food and water contamination and diarrhoea in rural Bangladesh. Epidemiology and infection London, New York NY 104: 253–259.
41. Vanderslice J, Briscoe J (1993) All Coliforms Are Not Created Equal - a Comparison of the Effects of Water Source and in-House Water Contamination on Infantile Diarrheal Disease. Water Resources Research 29: 1983–1995.
42. Jagals P, Grabow WOK, Williams E (1997) The effects of supplied water quality on human health in an urban development with limited basic subsistence facilities. Water Sa 23: 373–378.
43. Sobsey MD (1989) Inactivation of health-related microorganisms in water by disinfection processes. Water Science & Technology 21: 179–195.
44. Levy K, Hubbard AE, Nelson KL, Eisenberg JN (2009) Drivers of water quality variability in northern coastal Ecuador. Environmental science & technology 43: 1788–1797.
45. Levy K, Nelson KL, Hubbard A, Eisenberg JN (2008) Following the water: a controlled study of drinking water storage in northern coastal Ecuador. Environmental health perspectives 116: 1533.
46. Wu J, Long S, Das D, Dorner S (2011) Are microbial indicators and pathogens correlated? A statistical analysis of 40 years of research. J Water Health 9: 265–278.
47. Feachem RG, McGarry MG, Mara DD (1977) Water, wastes and health in hot climates: John Wiley and Sons Ltd.
48. Wood L, Egger M, Gluud LL, Schulz KF, Jüni P, et al. (2008) Empirical evidence of bias in treatment effect estimates in controlled trials with different interventions and outcomes: meta-epidemiological study. bmj 336: 601–605.
49. Arnold BF, Galiani S, Ram PK, Hubbard AE, Briceño B, et al. (2013) Optimal recall period for caregiver-reported illness in risk factor and intervention studies: A multicountry study. American journal of epidemiology 177: 361–370.
50. Crump JA, Mendoza CE, Priest JW, Glass RI, Monroe SS, et al. (2007) Comparing serologic response against enteric pathogens with reported diarrhea to assess the impact of improved household drinking water quality. American Journal of Tropical Medicine and Hygiene 77: 136.
51. Schmidt W-P, Cairncross S (2009) Household Water Treatment in Poor Populations: Is There Enough Evidence for Scaling up Now? Environmental science & technology 43: 986–992.
52. Schmidt W-P, Arnold BF, Boisson S, Genser B, Luby SP, et al. (2011) Epidemiological methods in diarrhoea studies—an update. Int J Epidemiol 40: 1678–1692.
53. Berk RA, Freedman DA (2003) Statistical assumptions as empirical commitments: Law, Punishment, and Social Control: Essays in Honor of Sheldon Messinger, 2nd edn. Aldine de Gruyter.

2

Hyperspectral Detection of a Subsurface CO_2 Leak in the Presence of Water Stressed Vegetation

Gabriel J. Bellante[1]*, Scott L. Powell[1], Rick L. Lawrence[1], Kevin S. Repasky[2], Tracy Dougher[3]

1 Department of Land Resources and Environmental Sciences, Montana State University, Bozeman, Montana, United States of America, **2** Department of Electrical and Computer Engineering, Montana State University, Bozeman, Montana, United States of America, **3** Department of Plant Sciences and Plant Pathology, Montana State University, Bozeman, Montana, United States of America

Abstract

Remote sensing of vegetation stress has been posed as a possible large area monitoring tool for surface CO_2 leakage from geologic carbon sequestration (GCS) sites since vegetation is adversely affected by elevated CO_2 levels in soil. However, the extent to which remote sensing could be used for CO_2 leak detection depends on the spectral separability of the plant stress signal caused by various factors, including elevated soil CO_2 and water stress. This distinction is crucial to determining the seasonality and appropriateness of remote GCS site monitoring. A greenhouse experiment tested the degree to which plants stressed by elevated soil CO_2 could be distinguished from plants that were water stressed. A randomized block design assigned Alfalfa plants (Medicago sativa) to one of four possible treatment groups: 1) a CO_2 injection group; 2) a water stress group; 3) an interaction group that was subjected to both water stress and CO_2 injection; or 4) a group that received adequate water and no CO_2 injection. Single date classification trees were developed to identify individual spectral bands that were significant in distinguishing between CO_2 and water stress agents, in addition to a random forest classifier that was used to further understand and validate predictive accuracies. Overall peak classification accuracy was 90% (Kappa of 0.87) for the classification tree analysis and 83% (Kappa of 0.77) for the random forest classifier, demonstrating that vegetation stressed from an underground CO_2 leak could be accurately discerned from healthy vegetation and areas of co-occurring water stressed vegetation at certain times. Plants appear to hit a stress threshold, however, that would render detection of a CO_2 leak unlikely during severe drought conditions. Our findings suggest that early detection of a CO_2 leak with an aerial or ground-based hyperspectral imaging system is possible and could be an important GCS monitoring tool.

Editor: Rongrong Ji, Xiamen University, China

Funding: This research was funded by the U.S. Department of Energy and the National Energy Technology Laboratory through Award number: DE-FC26-05NT42587. Neither the United States Government nor any agency thereof, nor any of their employees, makes any warranty, express or implied, or assumes any legal liability or responsibility for the accuracy, completeness, or usefulness of any information, apparatus, product, or process disclosed, or represents that its use would not infringe privately owned rights. The views and opinions of authors expressed herein do not necessarily state or reflect those of the United States Government or any agency thereof. The funders had no role in study design, data collection and analysis, decision to publish, or preparation of the manuscript.

Competing Interests: The authors have declared that no competing interests exist.

* Email: gabriel.bellante@gmail.com

Introduction

Geologic Carbon Sequestration

Post-industrial atmospheric carbon dioxide (CO_2) concentration has risen from 280 ppm to over 380 ppm [1–6]. Atmospheric CO_2 absorbs and reemits radiation energy from the Earth's surface causing a warming of the surface environment [7]. Global warming could drastically reshape the Earth's climate and environment [8–12]. Geologic carbon sequestration (GCS) is a potential climate change mitigation strategy that captures point source CO_2 emissions from industrial sources and stores them in large sub-surface reservoirs. GCS, conceptually, has the potential to store gigatonnes (Gt) of CO_2, which would otherwise be released into the atmosphere, into specialized geologic formations for long-term storage [13–19].

There is a mandate to monitor GCS sites for CO_2 leakage to ensure the efficacy of this technology, given that it is on the brink of commercial-scale deployment [14,19,20–22]. Remote sensing is being investigated as a possible cost-effective, large-area monitoring method to detect surface CO_2 leaks at GCS sites [23–24]. A leak from a GCS site would not only compromise the viability of this technology as a climate mitigation strategy, but it could also threaten the safety of the surrounding environment and inhabitants at the surface. The 1986 eruption of CO_2 at Lake Nyos, Western Camaroon, killed over 1,700 people and demonstrated that a large pneumatic CO_2 explosion can have devastating consequences [25]. Natural sources of CO_2 leakage from the Long Valley Caldera in California have caused extensive forest mortality [26–28].

Elevated CO_2 levels in soil are known to cause anoxic conditions in plant roots [26,29], thereby interfering with plant respiration and inducing a stress response that could possibly be remotely sensed using aerial imagery [24]. Vegetated areas at GCS sites, presumably, could act as bellwethers to signal operational inefficiencies in hazardous CO_2 leak scenarios. Vegetation status and seasonality will determine when the remote detection of vegetation stress caused by elevated soil CO_2 would be feasible. Soil water availability is highly spatially variable, and water stressed vegetation could appear spectrally similar to vegetation stressed by elevated soil CO_2. Drought is another common environmental occurrence that can have lasting impacts on whole land regions and can be short-lived or persist for years. Regardless of duration and spatial extent, monitoring GCS sites with remote sensing data would require discrimination between water and CO_2 stress to accurately identify CO_2 leaks.

Remote Sensing as a Monitoring Technique for GCS Sites

A variety of monitoring techniques are being investigated for efficiency and accuracy at detecting a CO_2 leak from a subsurface storage reservoir [30]. Most methods tend to be expensive and resource intensive, while providing only limited spatial coverage. Remote sensing, alternatively, has the potential to provide near instantaneous monitoring over large swaths of land. Aerial imaging over GCS sites would be relatively inexpensive and might be performed with limited labor requirements. Regions of elevated soil CO_2, if accurately detected with remote sensing equipment, could be further investigated with instruments on the ground to properly diagnose the leak source. Early leak detection would allow site managers to take immediate remediation measures to prevent further CO_2 leakage, while minimizing safety risk and economic loss. GCS monitoring could tolerate false positives, within reason, whereas, a false negative resulting in a missed CO_2 leak could have serious consequences.

Hyperspectral Remote Sensing for Plant Stress Detection

Hyperspectral sensors collect nearly continuous spectral data across narrow channel widths throughout the visible and infrared portion of the electromagnetic (EM) spectrum (400–2500 nm). Spectral signatures or spectra can be used to detect subtle patterns in vegetation reflectance to better understand the physiological condition of plants. Plant pigment concentrations, leaf cellular structure, and leaf moisture content can be discerned with hyperspectral data to assess the overall health of vegetation and characterize plant stress [31–32].

Hyperspectral remote sensing has been used to detect and characterize numerous types of vegetation stressors within the visible and infrared portion of the EM spectrum. Hyperspectral data have been analyzed to model plant stress caused by elevated soil CO_2 [33], water stress [34–36], insect and pest invasion [37–40], heavy metal contamination [22], salinity stress [41], nutrient levels and crop status [42–43]; ozone exposure [44]; and natural gas leaks [45]. Certain spectral regions are known to be especially sensitive, because they relate to the specific chemical and physical responses of plants to stress.

Low reflectance in the visible portion of the EM spectrum (400–700 nm) is determined primarily by chlorophyll *a* and *b* pigment abundance and absorption, while high reflectance in the NIR (750–2500 nm) is governed by the structural, spongy mesophyll contained in plant leaves [31,46–49]. Chlorophyll absorption is highest in the visible blue (400–500 nm) and middle red (near 680 nm), therefore these regions are not sensitive to subtle changes in chlorophyll content, because the spectral signal is saturated [46,50]. Absorption by chlorophyll pigments is weakest in the visible green – orange (500–620 nm) and the far visible red (700 nm), therefore, these wavelengths are sensitive to stress and subtle changes in chlorophyll content [46,51]. Red edge indices are frequently used to detect stress at the boundary of the far red and NIR to detect changes in chlorophyll content. The red edge is a reflectance spike caused by steep transition from low reflectance in the far red to high reflectance in the NIR, which is used as reference because it is determined by the physical structure of plant leaves. Far red reflectance (near 700 nm) is known to increase as a plant exhibits a physiological stress response of reducing chlorophyll content to down regulate photosynthesis, thereby reflecting proportionally more potentially damaging photosynthetically active radiation compared to healthy vegetation, which can utilize all of that incoming energy. Vegetation stress caused by depleted oxygen concentrations around the roots from natural gas leaks and elevated soil CO_2 has been detected in the red edge region of the EM spectrum [23–24,33,45,52]. Water stress has also been found to cause reflectance changes related to chlorophyll pigment concentrations in the red edge [53–54].

Shifts of the red edge 5–10 nm towards shorter wavelengths, termed the "blue shift", has also been associated with a decline in chlorophyll absorption for stressed vegetation [32–33,45,52,55–58]. This phenomenon has been associated with the early detection of vegetation stress by hyperspectral remote sensing instruments. The sensitivity of hyperspectral instruments to detect vegetation stress early, and perhaps even before it is visible to the human eye, has distinct advantages over conventional monitoring methods, because land managers could then take quicker ameliorative action to minimize losses to valuable resources [37].

Vegetation reflectance in the green portion of the EM spectrum is associated with xanthophyll pigments in addition to chlorophyll. Xanthophyll pigments contained within plants' palisade mesophyll perform photoprotective roles that also influence photosynthetic radiation use efficiency and are sensitive to stress [42,59]. Vegetation reflectance within the green, therefore, will increase in response to stress to down regulate photosynthetic activity because of both xanthophyll pigment activity and decreases in chlorophyll content. The Photochemical Reflectance Index (PRI) is a narrow-band, hyperspectral index ratioing green reflectance and is known to be sensitive to xanthophyll pigment activity. The PRI has been used as an early indicator of water stress in plants caused by drought conditions [34–36,43].

Water stress responses in vegetation have also been detected in spectral water absorption features in the shortwave infrared (SWIR) portion of the EM spectrum. Reflectance within the 1300–2500 nm wavelengths has been associated with leaf water content and has been observed to increase as a primary response of plants to water stress [53]. Visible spectrum reflectance has also been shown to increase, but only as a secondary plant response to water stress, in which chlorophyll pigment concentration decreases as a mechanism of down regulating photosynthesis during prolonged dehydration in vegetation [53–54]. Spectral measurements of leaf water content in the SWIR, therefore, directly monitors for plant dehydration and would be a more immediate method of detecting a plant's response to water stress. The primary response of plants to drought within the water absorption bands also could possibly be spectrally distinct from other types of physiological stressors, including stress caused by elevated soil CO_2.

Our greenhouse experiment objectives were (1) to evaluate if elevated soil CO_2 stress can be detected in alfalfa plants using hyperspectral data, and if so how quickly; (2) to determine if CO_2 stress is spectrally distinguishable from water stress since water availability is likely to be spatially variable at GCS sites and water stressed vegetation will likely co-occur with CO_2 stressed vegetation during a CO_2 leak scenario; and (3) to determine if CO_2 stress is distinguishable during drought conditions when all vegetation is water stressed during a CO_2 leak scenario. It is still unknown whether the plant stress signal caused by elevated soil CO_2 is spectrally distinguishable from other forms of plant physiological stress. Water stress or drought conditions in particular, are common natural occurrences that might confound a CO_2 stress signal if both plant physiological stressors coincide. Remote CO_2 leak detection might therefore be problematic in the context of GCS monitoring. Although this experiment does not directly test the use of hyperspectral imaging for CO_2 leak detection in a real-world aerial monitoring context, understanding the spectral discernability between stress caused by elevated soil CO_2, water stress, and their interaction is important for determining the appropriate times and conditions in which remote sensing image acquisition should commence over GCS sites.

Methods

Data Collection

Alfalfa (*Medicago sativa*) seeds were planted on June 1, 2010, to obtain 40 mature plants for this greenhouse experiment. Each alfalfa seedling was transplanted into a 45 cm tall, 9.6-l tree pot (Stuewe & Sons, Inc.). These tree pots were modified in several ways prior to planting in order to facilitate CO_2 injection into each individual plant pot and to best emulate a CO_2 leak scenario from a GCS site. The drainage holes at the bottom of each pot were sealed using plastic wrap, silicone, and Gorilla tape to minimize injected CO_2 from readily escaping through the bottom of the pots. The pots were then filled with 1.0 l of pea-sized gravel placed on top of a layer of gauze cotton cloth that covered drain holes at the bottom of each pot. A 0.6-cm diameter tube was inserted into each plant pot above the bottom 5-cm layer of gravel. The tube terminated in a 5-cm long, ceramic air stone (Top Fin) intended to disperse the CO_2 gas radially, as opposed to being released from a single point source. The embedded air stone plumbing was topped with an additional l of gravel to assist in the dispersion of the CO_2 gas throughout each pot and to allow for adequate water drainage from the soil (Figure 1).

Soil was placed on top of another layer of gauze above the gravel to keep the media separate. The soil used in the pots was a mixture of equal parts loam, washed concrete sand, and Canadian Sphagnum peat moss. Additionally, AquaGro 2000 wetting agent was blended into the soil at a ratio of 0.5 kg/m^3 soil mix. The soil mix was pasteurized with aerated steam at 70°C for 60 minutes. A 10 cm long piece of 2.54 cm diameter PVC pipe was inserted 5 cm down into the surface of the soil of every plant to allow for immediate access for soil CO_2 concentration measurements with a Vaisala GMP221 probe to verify injection. The alfalfa plants were watered with approximately 300 ml of water at 0800 hours every Monday, Wednesday, and Friday. They were fertilized twice a week with a $^{1\ gm}/_{0.5\ l}$ dilution of NPK fertilizer (20-10-20) every Wednesday and Friday until CO_2 injection began. The greenhouse temperature averaged 21.25±1.60°C over the course of the experiment. Sixteen hours a day of supplemental artificial lighting was provided using 1000 watt metal halide growth bulbs (GE Multi-Vapor MVR 1000/C/U) from 0500 to 2100 hours. Greenhouse integral photosynthetic photon flux measured within each of the treatment blocks ranged from 9.41 to 15.79 mol/m^2/day on peak sunny days and 0.09 to 0.11 mol/m^2/day on cloudy days (Apogee Instruments Line quantum sensors).

Five treatment blocks were developed, containing eight plants each (two replicates of each of four treatment groups per block). The four treatment groups included: (1) CO_2 injection (I); (2) water stress (WS); (3) an interaction group that was subjected to both water stress and CO_2 injection (WSI); and (4) adequate water and no CO_2 injection (C). Each treatment block contained two plants from each treatment group. CO_2 gas was delivered to the plant pots through a plumbing manifold that used equal lengths of tubing between each of the injected plants and a 2-stage pressure reducing regulator (Smith) at the CO_2 tank, which maintained constant pressure and did not require adjustment as the pressure within the CO_2 storage cylinder decreased over time. The gas was delivered from cylinders containing 50 lbs. liquid CO_2 at a constant pressure of 15 psig.

Alfalfa plants were randomly assigned to treatment blocks and to discrete positions within each block using a random number generator. Treatments within the blocks were randomly assigned using a stratified randomization developed to maintain equal lengths tubing between the injected plants (Figure 2). The plants that received water continued on the same watering regiment of 300 $^{ml}/_{day}$, while the water-stressed plants were not given any water in order to mimic drought conditions once CO_2 injection began.

Each alfalfa plant was scanned three times a week for three weeks, and all 40 alfalfa plants were sampled twice to obtain two spectral measurements per plant on each scanning date using an ASD Field Spec Pro 350. The ASD is a 16-bit spectrometer that has a spectral range of 350–2500 nm. The spectral resolution is 3 nm at 700 nm, 8.5 nm at 1400 nm, and 6.5 nm at 2100 nm. The sampling interval for the instrument is 1.4 nm at 350–1000 nm and 2 nm at 1000–2500 nm. Prior to interpolation there are a total of 1512 channels that are used in scan acquisition. The instrument takes a scan every 100 ms (ASD Inc.). A single sample consisted of two clipped alfalfa leaves, each containing three leaflets, from a randomly selected stem, determined by the roll of a 4-sided die. Each selected stem was clipped off at the bottom of the sixth node and spectral scans were taken from the four largest leaves on the cut stem. This sampling design intended to measure leaves from both the lower and upper canopy of each alfalfa plant to ensure that stress was not exhibited in one more than the other. The first sample consisted of the upper two biggest leaves and the second was taken from the lower two largest leaves of each cut stem. The leaves were dissected, meaning each leaflet was clipped from each sampled leaf (the leaves were left entire for ease and maximum coverage of the viewing spot later in the experiment,

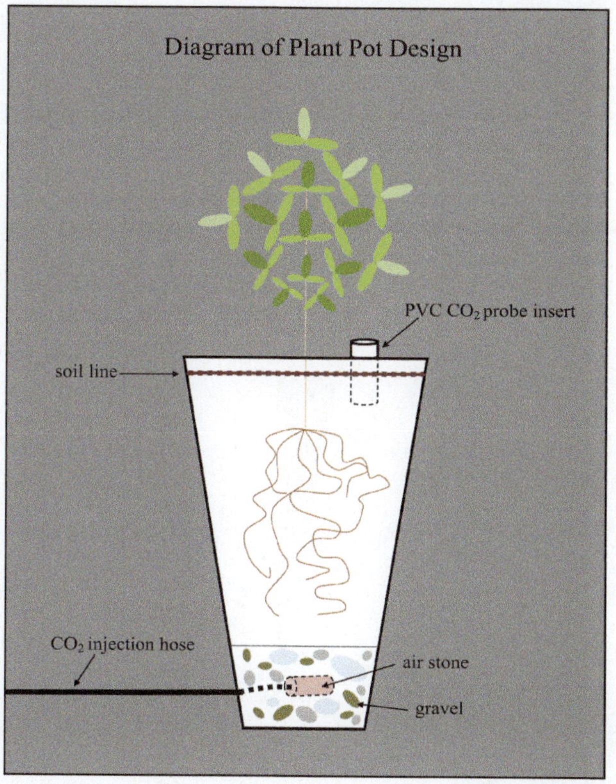

Figure 1. Diagram of a plant pot for the greenhouse experiment.

Figure 2. Treatment block design. Single treatment block shown with plumbing and labeled treatments shown in white for each plant (left). Four of five treatment blocks are shown within the greenhouse (right).

when samples had leaves that were too desiccated and brittle to dissect). The three leaflets from the bottom leaf of the top sample were placed in parallel on a spectralon target overlapped by the three leaflets of the upper leaf from the same sample to achieve total leaf surface coverage of the ASD fiber optic sensor viewing spot, the region from which spectral information is collected. The process was repeated for the lower set of leaves from the same plant for the second sample. These protocols were implemented both to preserve the spatial relationship among the leaf samples, as well as to keep the orientation of the leaves relative to the plant consistent to best simulate what a remote sensing instrument from an airborne platform would view (Figure 3).

The fiber optic cable of the ASD was equipped with a plant probe accessory for leaf-level spectral measurements. The plant probe was 10 in. long and contained an internal, low-intensity halogen bulb, which produces little heat, for collecting spectral scans from vegetation. The viewing geometry of the mounted fiber optic cable created a 10 mm by 13 mm oval viewing spot from which spectral information was collected.

Initial spectral readings were taken for all 40 plants at 1000 hours on February 7, 2011, as a baseline before treatments were applied. CO_2 injection began at 1400 hours on February 7, 2011. A Vaisala soil probe was used to monitor CO_2 concentrations over time and to confirm successful injection into the individual plant pots. A Swagelok G2 model variable area flowmeter also was used

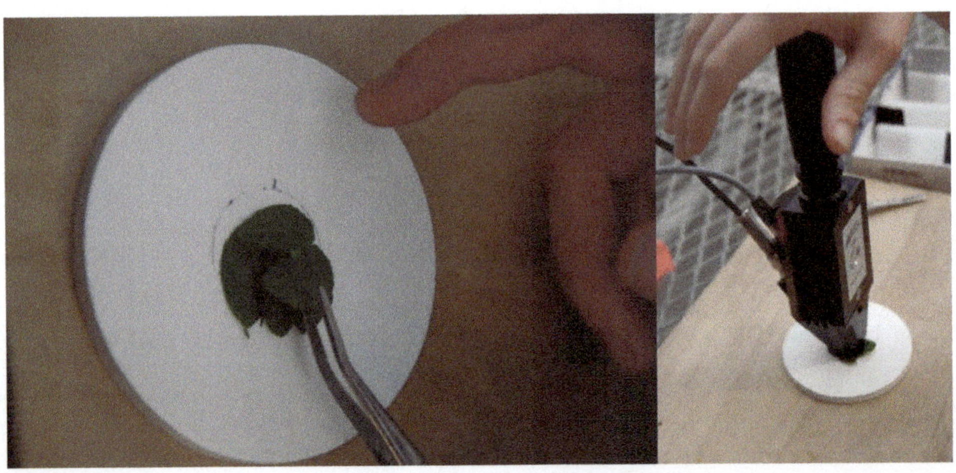

Figure 3. Sampling method. Two dissected alfalfa leaves on the spectralon target (left). The ASD fiber optic and plant probe assembly (right).

to measure CO_2 flux at the quick-connect union for each plant. Injection rates ranged from 0.25 to >5 l/hr for all injected plants. All 40 plants were spectrally scanned on Monday, Wednesday, and Friday for three weeks. Each plant was sampled and scanned on nine dates, except one plant, which was only scanned on eight dates because it was discovered on February 9 that a plant was not receiving CO_2 injection for some unknown reason. The plant therefore, was swapped out. A total of 18 scans were acquired for all plants, except for the exchanged plant, which had 16 scans. The injection was terminated on February 25, 2011, upon the completion of scanning all the plants earlier that day.

All spectral measurements were in reflectance and were a derived average of 25 individual scans. The instrument was calibrated to set gains and off-sets by an optimization process in the RS^3 software (ASD Inc.). A dark current was collected subsequently and the sensor was white referenced using a spectralon target. The ASD instrument was periodically recalibrated on each scanning date to ensure accurate and repeatable spectral measurements. A new white reference and dark current reading was made after the acquisition of both spectral scans for a single plant. The ASD was re-optimized upon the completion of acquiring data for an entire treatment block–eight plants.

Data Analysis

We developed a classification tree model for each individual scanning date to determine the timing and extent to which the different treatment groups were spectrally distinguishable. Classification tree analysis is a non-parametric statistical modeling method that has been successfully used to discriminate vegetation stress classes using hyperspectral data because it can utilize different band combinations to distinguish each class [37]. Classification trees use recursive, decision-based rules that can be interpreted by an analyst. This is important when trying to ascertain how and where the different plant physiological stressors are spectrally distinguished.

Although the ASD contained spectral data within the ultraviolet portion (350–400 nm) of the EM spectrum, these data were removed from further analysis because they were noisy and airborne hyperspectral sensors generally do not collect data at these wavelengths. The ASD measurements for the 400–2500 nm spectral range were output in an ascii format and exported into an Excel spreadsheet. TIBCO Spotfire S+ statistical software package was used to fit classification tree models to each scanning date's data using treatment group as the response variable and the ASD bands as explanatory (predictor) variables. Additionally, a factor variable indicating the relative position (upper or lower) of each leaf sample was included as a possible predictor variable in case sample location had an influence on treatment response. Cross-validation trees within S+ were used to prune the classification tree models for each date, so that they were unbiased and not over fit [60]. A standard 10-fold cross-validation was performed to determine the appropriate number of terminal nodes to lower deviance among samples, except that, given the small sample size, a script was written in S+ to take a random sample of scans stratified on treatment type (I, WS, WSI, or C) for each cross-validation tree to ensure that a balanced sample for each treatment group was withheld for validation purposes. This process was repeated five times, taking a different stratified random sample for each successive cross-validation tree and the plurality determined the appropriate size for pruning each classification tree. The spectral locations (wavelengths) used to distinguish between treatment groups for each of the binary nodes within each pruned classification tree model were examined to elucidate the spectral regions that best discerned between water stress and stress caused by elevated soil CO_2 in plants.

A random forest classification [61] also was performed on the individual scanning dates to evaluate the levels of prediction accuracy that could be achieved in discerning water and CO_2 stress agents. The algorithm is a "bagging" method that takes a bootstrap sample from the data observations to develop a classification tree by using a random subset of all possible explanatory variables (spectral bands) at each binary split [61]. This is done iteratively to form hundreds of classification trees (the forest) and then each observation is classified on the resultant plurality vote of the forest. Random forest models were derived with an ensemble of 500 classification trees with 45 randomly selected explanatory variables (of the possible 2100 spectral bands or the categorical variable indicating relative leaf position in the canopy) to be tried at each binary split within each tree. Variable importance plots from the random forest models were used to highlight the spectral regions used in distinguishing between the different treatment classes. Hyperspectral data have been modeled using random forest to greatly increase classification accuracies, as compared to other methods, as well as provide an unbiased, reliable internal estimate of accuracy for mapping land cover [62] and invasive plants [63]. A disadvantage of random forest is the inability to ascertain the relevance of individual explanatory variables and make meaningful interpretations of the model. Since hundreds of individual trees contribute to a random forest model, the classification results are essentially determined inside a "black box" that is effective for the purpose of prediction, but not for interpreting the decision-based rules that determine those predictions [64].

The lack of replicates in this study forced us to rely on internal measurements of predictive accuracy to evaluate spectral distinguishability. No data were withheld in the construction of the individual classification trees and the random forest classifier withheld approximately one third of each date's data as a bootstrap (out-of-bag) sample for internal validation within each of the 500 constructed trees. The random forest classifier has proved itself as a robust classification tool that contains an unbiased internal accuracy assessment that does not require a separately withheld dataset for validation [63]. Overall classification accuracies, Kappa statistics (a more robust estimate of classification accuracy because it takes into account chance agreement by differencing the observed accuracy from that of a total random assignment), and individual user's (mapping errors of commission) and producer's accuracies (mapping errors of omission) were used to assess predictive capabilities for the class distinctions through time [65].

Results

Examination of the individual sample spectral signatures for the different treatment groups illustrated the spectral regions that were critical in their distinction and accurate classification (Figure 4). Sample spectra for each of the treatment groups appeared to exhibit similar reflectance properties in the visible wavelengths, while subtle spectral deviations occurred in the near and SWIR peak reflectance features before treatments were applied, on February 07. The stressed sample spectra, in contrast, appeared to exhibit increased reflectance in the visible spectrum, especially within the visible green, as compared to the C class on February 21, at the height of spectral distinction among treatment groups. Water stressed samples exhibited increased reflectance in the SWIR as compared to the C and I classes. The WSI class predominantly exhibited higher reflectance in the visible and

SWIR regions compared to all other samples. The red edge spectral boundary shifted towards shorter wavelengths (the "blue shift") for stressed samples after four days of treatment application, however, visual symptoms of stress were also observed on this day. The blue shift was not uniquely expressed among stressed treatment groups.

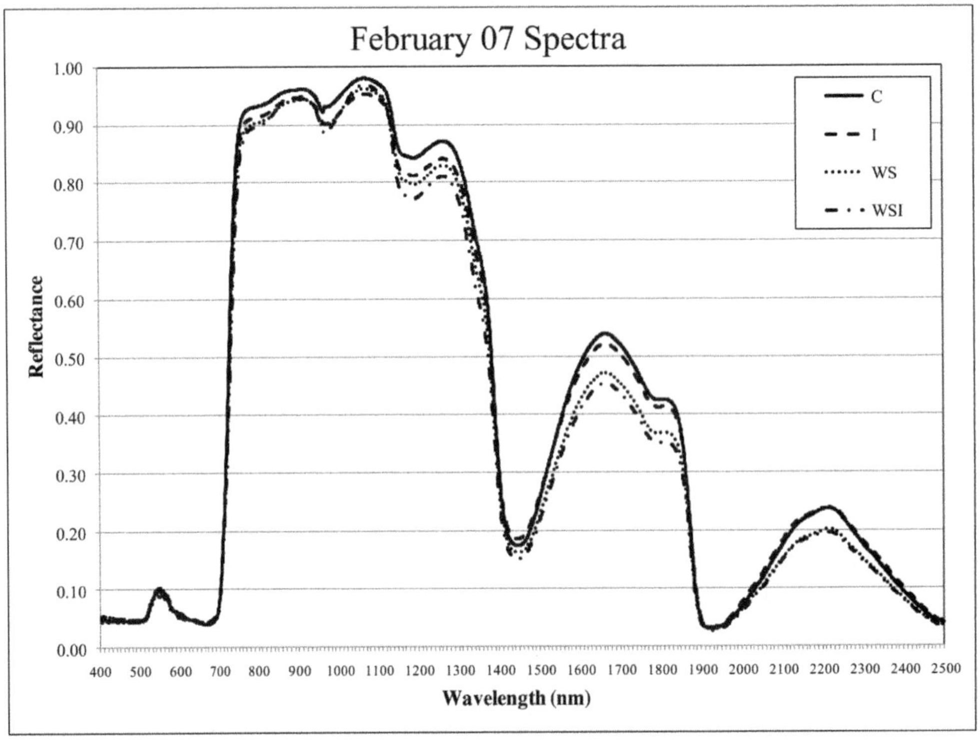

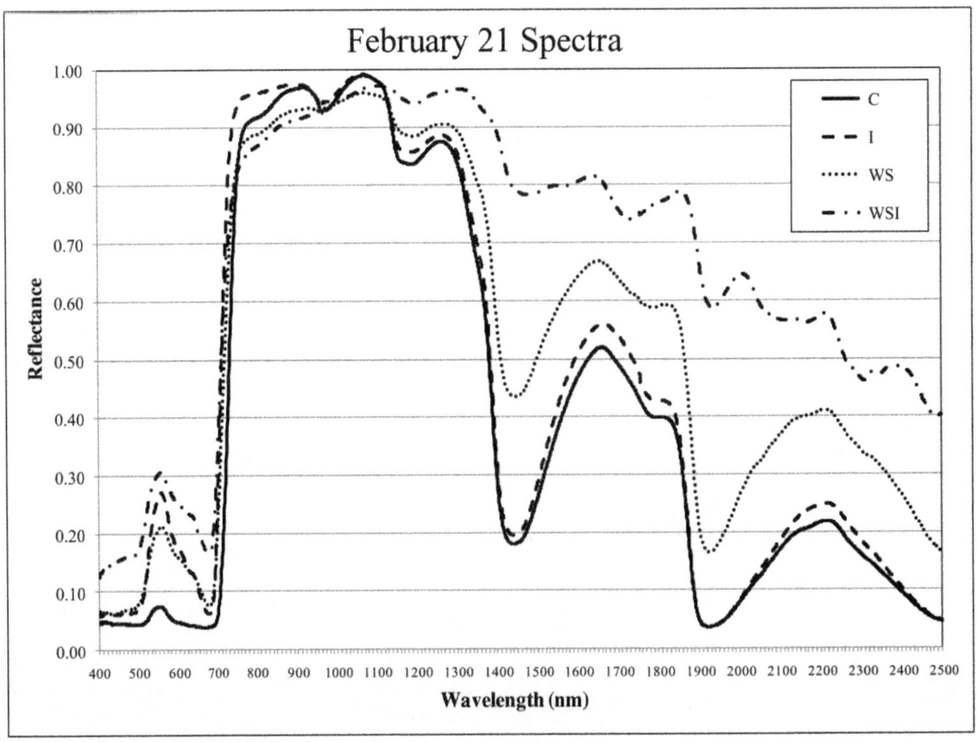

Figure 4. Sample reference spectra for each of the four treatment groups. Spectral signatures for February 07 (before treatment application) and February 21 (when spectral distinction was greatest between treatment groups).

Classification Tree Analysis

Cross-validation results determined that four terminal nodes were appropriate for February 21 and 23, otherwise the classification trees were pruned smaller to ensure they were not over fit (Table 1). Only the February 21 and 23 classification tree models, therefore, distinguished among all four treatment groups. A decision split or node was not justified for the classification tree models during the first week of the experiment. Not until February 14 were two or more terminal nodes warranted (Figure 5). The categorical variable indicating relative leaf sample position within each alfalfa canopy was never used as a splitting rule for any of the classification trees.

WS and WSI classes exhibited greater reflectance than the C and I classes within the water absorption bands near the 1400, 1900, and 2500 nm wavelengths. These distinctions in the SWIR infrared regions began on February 14 and persisted until the end of the experiment. Second level distinctions (decisions in trees with three or more terminal nodes) were made within the red edge at the 717 nm wavelength, where greater reflectance was exhibited by leaf samples from the I class as compared to the C class. The I class also exhibited greater reflectance in the visible green-visible orange at the 545, 621, and 541 nm wavelengths as compared to the C class. The WSI interaction group exhibited compound stress effects with greater reflectance as compared to the WS class in the water absorption bands at 1429 nm and 1901 nm.

Confusion early in the experiment, from February 14–18, occurred almost exclusively between (1) I and C classes, and (2) WSI and WS classes (Table 2). The C class was accurately distinguished by the second week of stress treatment application (user's and producer's accuracies ≥91%). Spectral differentiation between treatment groups was greatest on February 21 and February 23 (after two weeks of treatment application) with overall classification accuracies of 90% (Kappa of 0.87) and 81% (Kappa 0.75), respectively. Confusion during maximum spectral distinction primarily occurred (1) with I samples being classified as WS, and (2) WS being classified as WSI. Producer's accuracies for the I class were 80% and 55%, while the WSI class had perfect producer's accuracies (100%) on both February 21 and 23. Overall classification accuracy dropped to 66% (Kappa of 0.55) on the last scanning date, February 25.

Random Forest Analysis

The trend in accuracy of the random forest models throughout the time series was similar to that of the classification tree models (Figure 6). Out-of-bag accuracy predominantly exhibited a curvilinear trend with increasing out-of-bag accuracy from 25% (Kappa of 0) on February 7 (before treatments were applied) to peak classification accuracy of 83%, (Kappa of 0.77) on February 21. Subsequently, classification accuracy decreased. By February 25 out-of-bag accuracy had dropped to 65% (Kappa of 0.53). The C treatment group was best distinguished with user's and producer's accuracy ≥80% after one week of treatment application (Table 3). Most class confusion existed between WSI and WS treatment groups, once overall out-of-bag accuracy was ≥65% (February 14–25). I class producer's accuracy was ≥65% from February 14 onward. WSI class producer's accuracy was 90% for February 21 and 23, but was much lower on other dates.

The variable importance plots provided a visual display of the spectral regions that were the most influential in single date random forest model prediction (Figure 7). The variable importance plots were very noisy before there was spectral discernability among treatment groups. Spectral regions became more clearly defined as classification accuracies increased. The red edge region (700–750 nm) was shown to be the most important on February 14 and 16. On February 18, variable importance was predominantly placed on the spectral wavelengths centered on approximately 1400 nm. During maximum spectral distinction among treatment groups, variable importance was placed (1) in the visible green-yellow portion of the EM spectrum centered at approximately 550 nm; (2) the red edge (700–750 nm); and (3) the three water absorption features centered at approximately 1400, 1900, and 2500 nm. On February 25, the last scanning date, more emphasis was placed on the visible and NIR spectral regions rather than the water absorption features. The categorical variable indicating relative leaf sample position within each alfalfa canopy was never deemed important in any of the random forest classification models (variable importance<0.01).

Discussion

Our analyses demonstrated that hyperspectral spectrometry could distinguish between CO_2 stressed and healthy alfalfa leaves (I v. C). Furthermore, the spectral distinction of plant stress caused by elevated soil CO_2, water stress conditions, and their interaction was possible at certain times during the greenhouse experiment.

Table 1. Number of terminal nodes determined by the cross-validation results for the single date classification tree models.

Cross-Validation Tree Results

Date	x1	x2	x3	x4	x5	Plurality
2/7	1	1	1	1	1	1
2/9	1	1	1	1	1	1
2/11	1	1	2	2	1	1
2/14	2	1	2	2	1	2
2/16	3	3	3	3	2	3
2/18	2	2	5	2	2	2
2/21	4	4	4	4	4	4
2/23	4	4	4	4	4	4
2/25	3	2	2	3	3	3

The plurality of the five cross-validation trials determined the appropriate size of the final trees.

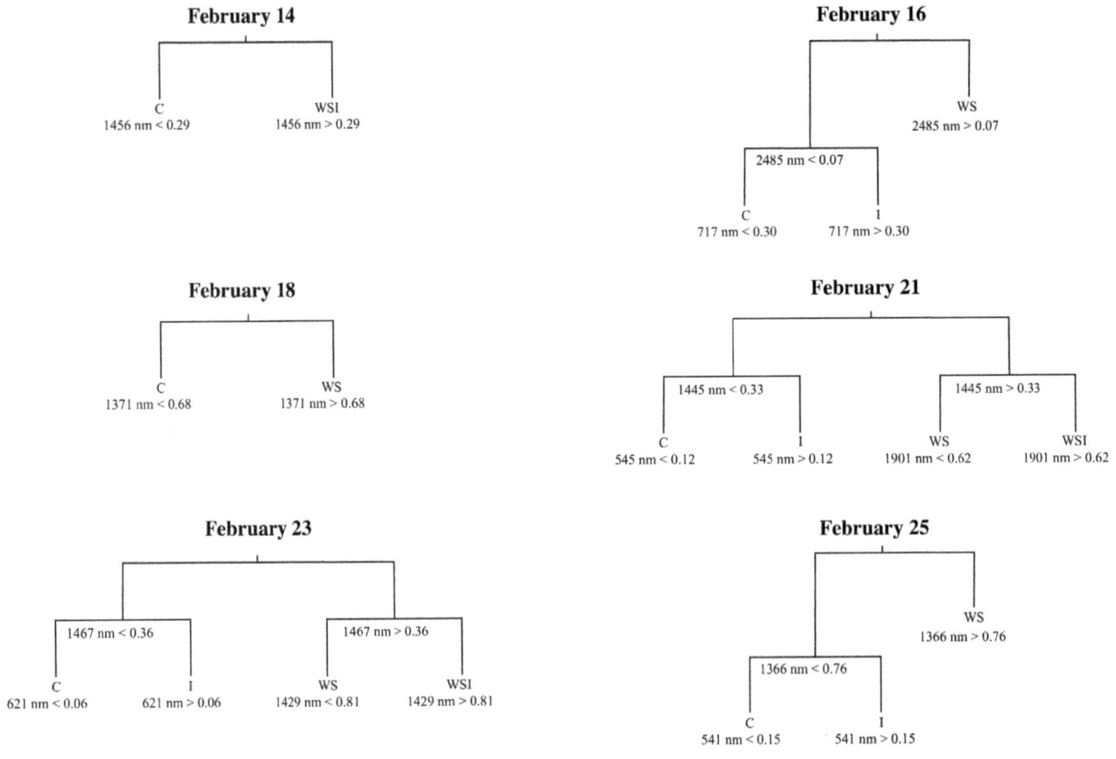

Figure 5. Pruned single date classification trees. Each tree shows the utilized spectral bands (by central band wavelength) and reflectance levels for each splitting rule. Splitting rules apply to the left branches of the tree. C = control class; I = CO_2 injection class; WS = water stress class; and WSI = water stress and CO_2 injection interaction class.

Those findings have significant implications for the use of hyperspectral imaging to monitor GCS sites for CO_2 leaks given that water stress is a prevalent environmental condition and could be a confounding factor. Remote sensing monitoring of CO_2 leaks amid a landscape of patchy, water stressed vegetation would most likely require differentiation of CO_2 and water stress (I v. WS), whereas during widespread drought, vegetation stressed by both CO_2 and water stress would need to be spectrally distinct from surrounding vegetation (WSI v. WS).

Confusion predominantly occurred between (1) I and C classes, (2) I and WS classes, and (3) WSI and WS classes, which could be problematic for CO_2 leak detection in a GCS monitoring context. Mapping errors of omission (low producer's accuracy) for CO_2 stressed vegetation (I and WSI) would be less acceptable from a hyperspectral monitoring standpoint, while mapping errors of commission (low user's accuracy) would be more tolerable, since a CO_2 leak would be less likely to go unidentified. High commission error and low omission error would result in the overestimation of CO_2 leaks, perhaps wasting labor resources for ground verification, however, CO_2 leaks would less likely be missed.

CO_2 Leak Detection When Water Is Not Limiting

Hyperspectral monitoring of GCS sites when water stressed vegetation is not present would consider the spectral discrimination between I and C classes. Spectral distinctions in the red edge, visible green, and visible orange regions were used to distinguish I from C classes. Rules discerned increased green reflectance within the I class multiple times near 550 nm, a spectral region sensitive to xanthophyll pigment activities associated with photosynthetic efficiency [42,59] and changes to chlorophyll content [46]. Additionally, increased reflectance at the 620 nm and 717 nm wavelengths distinguished I from C classes. These are also spectral regions that are known to be highly sensitive to chlorophyll *a* content and absorption [32]. The primary spectral response of the I class to elevated soil CO_2 was increased reflectance within the visible green–orange spectrum.

The I class was distinguished from all treatment groups with user's and producer's accuracies ≥70% beginning February 16, after ten days of CO_2 injection. Classification accuracies of the I class remained high until the end of the experiment, suggesting that a detectable stress signal caused by a CO_2 leak could persist for a substantial period of time. I and C classes were likely confounded in the experiment because some alfalfa plants received comparatively higher CO_2 injection rates due to minor differences in gas flow dynamics influenced by plant pot plumbing geometry and soil compaction. Severity and timing of stress caused by CO_2 injection therefore, varied. Healthy plants also appeared to be disproportionately targeted by a two-spotted spider mite infestation, which possibly caused a stress response that was most spectrally similar to stress caused by elevated soil CO_2 given that the insects caused a discoloration in the leaves most similar to the chlorosis observed in the alfalfa plants receiving CO_2 injection. The two-spotted spider mites were treated with numerous pesticides and biological controls, however, they probably remained an influential stress factor for healthier alfalfa plants. The results of the random forest classification demonstrated that stressed and non-stressed vegetation were accurately distinguished after one week of treatment application (C class accuracies ≥80%). Spectral differentiation among the stress classes was more difficult and time dependent, however.

Table 2. Internal classification accuracies and Kappa statistics for each single date classification tree model containing at least two terminal nodes.

February 14, 2011	C	I	WS	WSI	User's Accuracy	February 16, 2011	C	I	WS	WSI	User's Accuracy
C	20	20	6	3	41%	C	16	0	0	0	100%
I	0	0	0	0	na	I	4	19	0	0	83%
WS	0	0	0	0	na	WS	0	1	20	20	49%
WSI	0	0	14	17	55%	WSI	0	0	0	0	na
Producer's Accuracy	100%	0%	0%	85%	Overall Accuracy 46%	Producer's Accuracy	80%	95%	100%	0%	Overall Accuracy 69%
Kappa Statistic	0.28					Kappa Statistic	0.58				
February 18, 2011	**C**	**I**	**WS**	**WSI**	**User's Accuracy**	**February 21, 2011**	**C**	**I**	**WS**	**WSI**	**User's Accuracy**
C	20	20	0	0	50%	C	20	0	0	0	100%
I	0	0	0	0	na	I	0	16	0	0	100%
WS	0	0	20	20	50%	WS	0	4	16	0	80%
WSI	0	0	0	0	na	WSI	0	0	4	20	83%
Producer's Accuracy	100%	0%	100%	0%	Overall Accuracy 50%	Producer's Accuracy	100%	80%	80%	100%	Overall Accuracy 90%
Kappa Statistic	0.33					Kappa Statistic	0.87				
February 23, 2011	**C**	**I**	**WS**	**WSI**	**User's Accuracy**	**February 25, 2011**	**C**	**I**	**WS**	**WSI**	**User's Accuracy**
C	20	2	0	0	91%	C	20	1	0	0	95%
I	0	11	0	0	100%	I	0	13	0	0	100%
WS	0	6	14	0	70%	WS	0	6	20	20	43%
WSI	0	1	6	20	74%	WSI	0	0	0	0	na
Producer's Accuracy	100%	55%	70%	100%	Overall Accuracy 81%	Producer's Accuracy	100%	65%	100%	0%	Overall Accuracy 66%
Kappa Statistic	0.75					Kappa Statistic	0.55				

Treatment group columns represent the reference data while the rows represent the classified data.

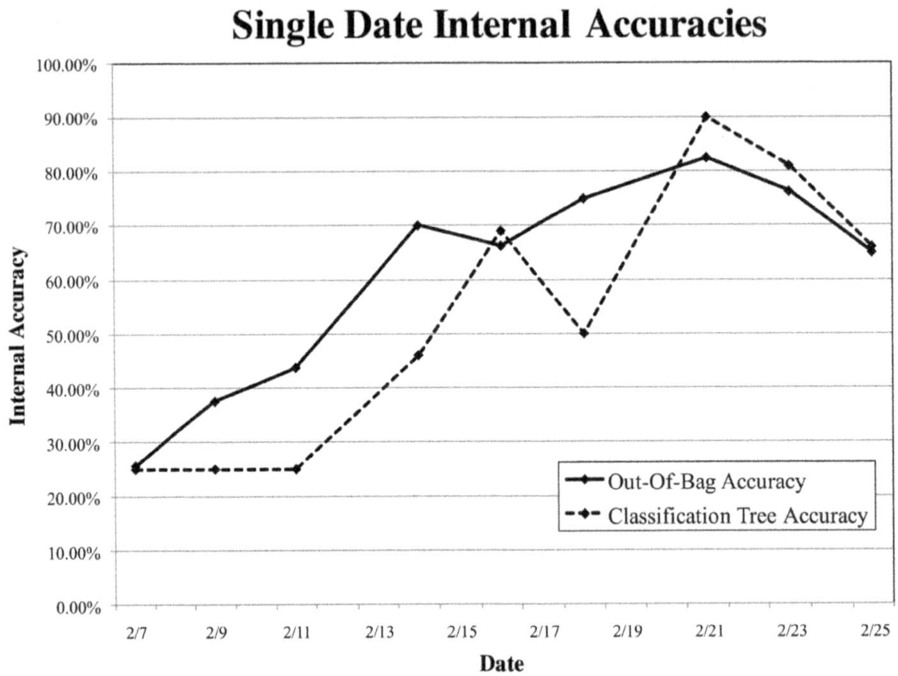

Figure 6. Overall internal accuracies for each random forest and classification tree model over the course of the experiment. Note that the February 7–11 classification trees did not contain ≥ two terminal nodes.

CO_2 Leak Detection When Soil Moisture Is Spatially Variable

Identification of a CO_2 leak would likely require water stress to be spectrally distinguished from stress caused by elevated soil CO_2 because soil moisture availability is highly variable across most landscapes [66–67]. Increased reflectance within the water absorption regions near 1400 and 2500 nm was the most important characteristic that distinguished water stressed classes (WSI and WS) from C and I classes. The primary spectral response of leaf dehydration occurs within the water absorption regions (near 1400, 1900, and 2500 nm) that are sensitive to leaf water content [53].

Patterns of variable importance for each of the random forest models identified discrete spectral regions that were particularly useful in discriminating among treatment groups. The red edge and the visible green-visible orange spectral regions were at least as important as the water absorption bands for distinguishing among the different classes. However, the "black box" nature of the random forest classifier did not allow for specific interpretation of the spectral distinctions made by the model [64]. It is possible that the variable importance within the visible and NIR wavelengths was related to distinguishing between the I and C classes given that these two treatment groups appeared to be spectrally similar in the SWIR region, while the water stress classes were likely separated by distinctions in the water absorption regions. Eventually increased reflectance in the visible, NIR, and SWIR regions can be exhibited by all stressed plants regardless of the cause, however [53].

Alfalfa plants of the I treatment group exhibited visual stress predominantly as chlorosis (yellowing of plant leaves due to lack of chlorophyll production), although some CO_2 injected plants' leaves did become desiccated by the end of the experiment, suggesting leaf water loss. Leaves of the WS treatment group generally exhibited a lack of vigor (leaves were droopy) early in the experiment, and eventually, they became dessicated and brittle as the water stress severity increased. Overall these classes were distinguished reasonably well. Only minor confusion occurred probably because of similar reflectance properties in the visible wavelengths due to eventual chlorophyll loss to water stressed samples and due to eventual leaf moisture loss, expressed as increased reflectance in the water absorption bands of I class samples [53].

CO_2 Leak Detection During Widespread Drought

Remote monitoring of GCS sites could require CO_2 leaks to be spectrally distinguished during drought conditions when all vegetation is subjected to water stress. Elevated soil CO_2 would thereby have to interact with water stress to cause a compound stress response in vegetation that was spectrally discernable. Greater reflectance in the water absorption bands near 1400 nm was used to distinguish WSI from WS classes, perhaps in response to diminished water-use-efficiency caused by comparatively faster leaf moisture loss or reduced water uptake at the root level. This was consistent with a compound stress interaction being exhibited by the WSI class. Treatment class accuracies indicated that alfalfa plants that went without water for one week exhibited water stress that became spectrally discernable on February 14. The confusion existed primarily *between* them and not with the other classes. Individual WS and WSI class accuracies, therefore, were low. This was consistent with a water stress signal that was more easily discerned because of the discrete water absorption features where leaf reflectance is directly related to leaf moisture content. Spectral distinction of the WSI class remained relatively convoluted with the WS class throughout the experiment. The WSI class was distinguished with high producer's accuracies (low omission error) on February 21 and 23, perhaps suggesting that vegetation could exhibit stress in response to a CO_2 leak that could be distinguished during drought conditions, albeit a narrow time window.

The plant leaves of the water stressed treatment groups became desiccated and brittle as the experiment progressed. These plants

Table 3. Out-of-bag accuracies and Kappa statistics for each single date random forest model.

February 7, 2011	C	I	WS	WSI	User's Accuracy	February 9, 2011	C	I	WS	WSI	User's Accuracy
C	5	2	3	5	33%	C	9	2	4	2	53%
I	3	4	6	2	27%	I	4	9	6	8	33%
WS	4	6	5	7	23%	WS	3	4	7	5	37%
WSI	8	6	6	6	23%	WSI	4	5	3	5	29%
Producer's Accuracy	25%	22%	25%	30%	Overall Accuracy 26%	Producer's Accuracy	45%	45%	35%	25%	Overall Accuracy 38%
Kappa Statistic	0					Kappa Statistic	0.17				
February 11, 2011	**C**	**I**	**WS**	**WSI**	**User's Accuracy**	**February 14, 2011**	**C**	**I**	**WS**	**WSI**	**User's Accuracy**
C	9	10	2	3	38%	C	17	2	2	0	81%
I	10	7	3	3	30%	I	2	13	1	2	72%
WS	0	2	10	5	59%	WS	1	2	12	4	63%
WSI	1	1	5	9	56%	WSI	0	3	5	14	64%
Producer's Accuracy	45%	35%	50%	45%	Overall Accuracy 44%	Producer's Accuracy	85%	65%	60%	70%	Overall Accuracy 70%
Kappa Statistic	0.25					Kappa Statistic	0.6				
February 16, 2011	**C**	**I**	**WS**	**WSI**	**User's Accuracy**	**February 18, 2011**	**C**	**I**	**WS**	**WSI**	**User's Accuracy**
C	16	4	0	0	80%	C	17	4	0	0	81%
I	4	15	1	1	71%	I	3	14	1	0	78%
WS	0	1	13	10	54%	WS	0	2	14	5	67%
WSI	0	0	6	9	60%	WSI	0	0	5	15	75%
Producer's Accuracy	80%	75%	65%	45%	Overall Accuracy 66%	Producer's Accuracy	85%	70%	70%	75%	Overall Accuracy 75%
Kappa Statistic	0.55					Kappa Statistic	0.67				
February 21, 2011	**C**	**I**	**WS**	**WSI**	**User's Accuracy**	**February 23, 2011**	**C**	**I**	**WS**	**WSI**	**User's Accuracy**
C	19	1	0	0	95%	C	19	2	0	0	90%
I	1	16	2	0	84%	I	1	14	3	0	78%
WS	0	3	13	2	72%	WS	0	4	10	2	63%
WSI	0	0	5	18	78%	WSI	0	0	7	18	72%
Producer's Accuracy	95%	80%	65%	90%	Overall Accuracy 83%	Producer's Accuracy	95%	70%	50%	90%	Overall Accuracy 76%
Kappa Statistic	0.77					Kappa Statistic	0.68				
February 25, 2011	**C**	**I**	**WS**	**WSI**	**User's Accuracy**						
C	19	2	1	0	86%						
I	1	15	2	0	83%						
WS	0	1	8	10	42%						

Table 3. Cont.

February 7, 2011	C	I	WS	WSI	User's Accuracy	February 9, 2011	C	I	WS	WSI	User's Accuracy
WSI	0	2	9	10	48%						
Producer's Accuracy	95%	75%	40%	50%	Overall Accuracy 65%						
Kappa Statistic	0.53										

Treatment group columns represent the reference data while the rows represent the classified data.

eventually began to die. Spectral distinction within the water absorption bands was probably difficult due to the extreme leaf moisture loss caused by severe drought by the end of the experiment. Little variable importance was placed on the water absorption features in the last classification date (February 25), where the water stressed treatment groups were poorly distinguished (accuracies≤50%), while the I and C treatment groups were classified with reasonable accuracy (accuracies ≥75%). This was evidence that the alfalfa plants' response to severe drought surpassed a stress threshold that rendered the CO_2 stress signal spectrally indistinguishable.

Pre-visual Plant Stress Detection

Pre-visual stress detection would be a critical attribute and an advantage to using hyperspectral remote sensing for early CO_2 leak detection at a GCS site. February 11 was the first day that subtle visual evidence of stress to some of the WSI and I treatment group plants was evident. Early visual symptoms of stress (i.e., chlorosis and languid leaves) were confined to older leaves located lower on the stems within the alfalfa plant canopies. Visible stress was most prominent within the WSI treatment group, which exhibited the compound effects of both CO_2 injection and lack of water. The accurate classification of the WSI samples did not occur until February 14, one week into the experiment, even though stress was noted visually three days before that time. The blue shift of the red edge towards shorter wavelengths has been associated with the pre-visual detection of plant stress [32,55,57], and was observed for some sample spectra by February 11. Blue shift spectral characteristics were similar in depth and slope for all of the stressed classes and, therefore, could not be used to distinguish among stress agents. This experiment provided no evidence that the pre-visual detection of plant stress was possible with hyperspectral data.

Summary

The primary conclusions that can be drawn from this greenhouse experiment are (1) that plant stress caused by elevated soil CO_2 was spectrally detected, probably in the visible spectrum, after approximately one week of CO_2 injection through the end of the experiment (random forest I class accuracies ≥65%). This is potentially, evidence that a spectral stress signal caused by elevated soil CO_2 could persist for a substantial period of time; (2) CO_2 and water stress were spectrally distinguishable; and (3) Elevated soil CO_2 appeared to cause a compound stress response detected in plants that were already water stressed, however, there was a relatively narrow time window when the WSI class was spectrally distinguished from the WS class. This was indicative of a time-dependent compound stress response caused by the interaction of elevated soil CO_2 and drought.

Although the results were variable between the individual classification tree and random forest classifier results, they both illustrated the same pattern of predictive success for plant stress detection. Differences between the two modeling strategies were likely due to the small sample size relative to the high dimensionality of the hyperspectral data. There were a total of 40 plants, 10 plants per class, as compared to the 1512 possible predictor variables. Depending on which samples were withheld for validation purposes during model development, variability did occur between outputs. Additionally, the random forest classifier uses a randomized node optimization which would introduce additional variability. Therefore, these differences were expected especially when using two distinctly different classification methods.

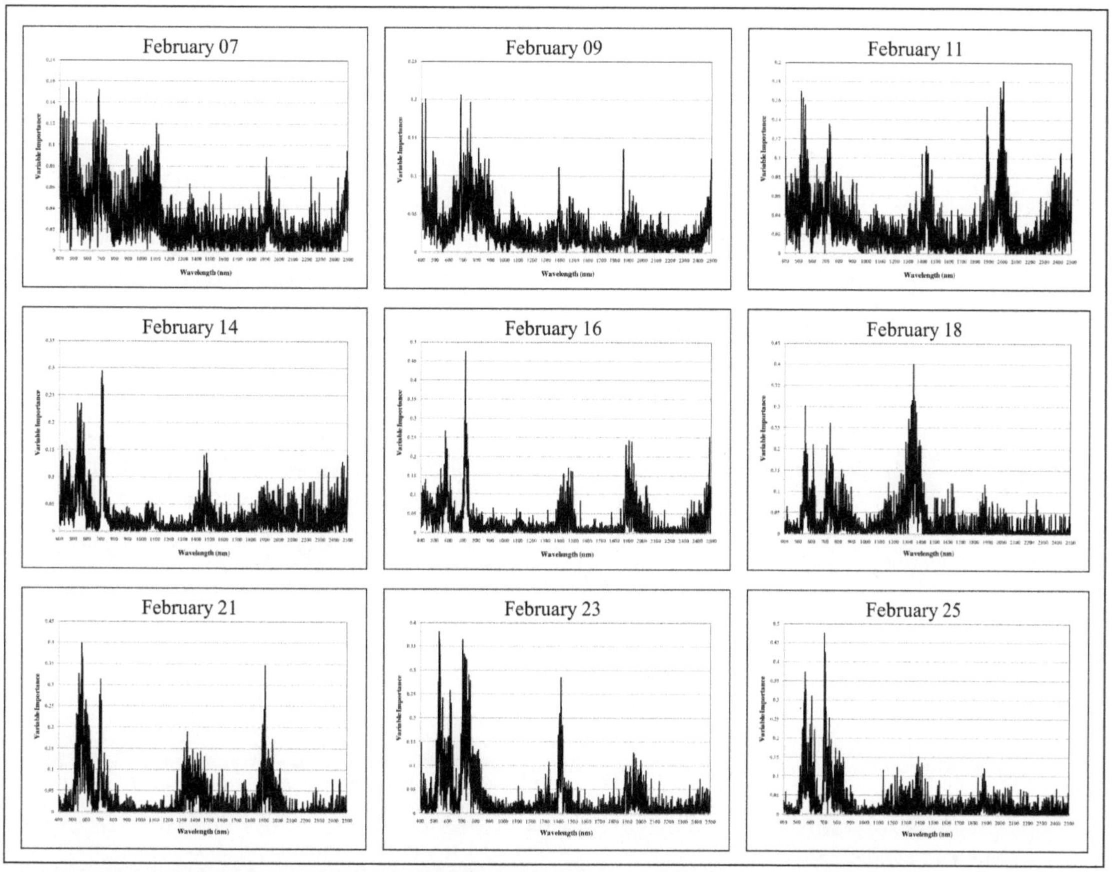

Figure 7. Variable importance plots for the single date random forest models. Note that the leaf position categorical variable is not included.

These results suggest that remote sensing may be used to monitor GCS sites for CO_2 leaks. Detection of a CO_2 leak when the availability of soil water is highly variable across space might be possible even if there are co-occurring patches of water stressed vegetation. Plants appear to hit a stress threshold, however, that renders spectral detection of a CO_2 leak unlikely during severe drought conditions. Regardless, this research demonstrated the necessity for remote sensing instruments to be spectrally sensitive to SWIR reflectance in order to accurately distinguish CO_2 and water stress.

Even though pre-visual stress detection was not possible at the leaf-scale, the early detection of stress caused by elevated soil CO_2 was achieved. Aerial hyperspectral imaging has been demonstrated to be effective for early stress detection at broader spatial scales [37] and the random forest classifier has been used to successfully map invasive plants in a natural setting [63]. Monitoring for CO_2 leaks at GCS sites will possibly require large area coverage, which will only be possible from an airborne platform. Hyperspectral instruments with high spatial resolution optics for canopy-level monitoring at altitude, therefore, might be important for the early detection of CO_2 leaks. The high dimensional nature of hyperspectral data will require robust data analysis methods that can detect a plant stress signature or signal within large datasets. Different data mining tools exist that could both reduce the dimensionality of hyperspectral data and discern the targeted CO_2 stress signature in an aerial monitoring context. For example, an orthogonal operator has been used to reduce the dimensionality of hyperspectral data while retaining the meaningful signature of interest [68–69]. Additionally, a hypergraph method that jointly relates the spatial context of image pixels to their associated spectral characteristics has been demonstrated to be effective in hyperspectral image classification and may be useful in detecting pockets of plant stress caused by elevated soil CO_2 [70]. Further exploration of different data analysis methods will be essential to optimize a workflow that would enable an analyst to efficiently assimilate large datasets and effectively detect a CO_2 leak in a GCS monitoring context.

Additional research is warranted to understand the extent to which elevated soil CO_2 stress could be detected at the canopy-level using airborne remote sensing, especially when water stress conditions exist. Spectral differentiation between CO_2 and water stressed vegetation would likely include spectral distinctions within the water absorption features. Spectral data acquired within the water absorption regions from an airborne remote sensing platform would likely be attenuated by atmospheric water vapor, perhaps further confounding CO_2 leak detection [47,54]. Snow cover and drought are going to limit the applicability of remote sensing data in temperate climates, therefore hyperspectral monitoring will need to be used in conjunction with other monitoring methods to obtain adequate temporal coverage of GCS sites. Improved understanding of vegetation spectral responses to stress caused by plant senescence will also be critical in determining the seasonal timing in which remote sensing data are appropriate for the detection of plant stress caused by an underground CO_2 leak.

Author Contributions

Conceived and designed the experiments: GB SP RL KR TD. Performed the experiments: GB SP RL KR TD. Analyzed the data: GB SP RL KR TD. Contributed reagents/materials/analysis tools: GB SP RL KR TD. Wrote the paper: GB SP RL KR TD.

References

1. Cuffey KM, Vimeux F (2001) Covariation of carbon dioxide and temperature from the Vostok ice core after deuterium-excess correction. Nature 412: 523–527.
2. Masarie K, Tans PT (1995) Extension and integration of atmosphere carbon dioxide data into a globally consistent measurement record. Journal of Geophysical Research 100: 11593–11610.
3. Monnin E, Indermühle A, Dällenbach A, Flückiger J, Stauffer B, et al. (2001) Atmospheric CO_2 concentrations over the last glacial termination. Science 291: 112–114.
4. Petit JR, Jouzel J, Raynaud D, Barkov NI, Barnola JM, et al. (1999) Climate and atmospheric history of the past 420,000 years from the Vostok ice core, Antarctica. Nature 399: 429–436.
5. Scripps Institute of Oceanography (2007) Monthly average carbon dioxide concentration. Available: http://scrippsco2.ucsd.edu/graphics_gallery/mauna_loa_record/mauna_loa_record.html. Accessed 2014 Sep 18.
6. Seigenthaler U, Stocker TF, Monnin E, Lüthi D, Schwander J, et al. (2005) Stable carbon cycle-climate relationship during the Late Pleistocene. Science 310: 1313–1317.
7. Intergovernmental Panel on Climate Change (IPCC) Fourth Assessment Report (2007) IPCC Fourth Assessment Report on Climate Change. Cambridge University Press, Cambridge, UK.
8. Parmeson C (2006) Ecological and evolutionary responses to recent climate change. Annual Review of Ecology, Evolution, and Systematics 37: 637–669.
9. Hansen J, Sato M, Kharecha P, Beerling D, Masson-Delmotte V (2008) Target atmospheric CO_2: where should humanity aim? Open Atmospheric Science Journal 2: 217–231.
10. Norby RJ, Luo Y (2006) Evaluating ecosystem responses to rising atmospheric CO_2 and global warming in a multi-factor world. New Phytologist 162: 281–293.
11. Shackleton NJ (2000) The 100,000-year ice-age cycle identified and found to lag temperature, carbon dioxide, and orbital eccentricity. Science 289: 1897–1902.
12. Vinnikov KY, Grody NC (2003) Global warming trend of mean tropospheric temperature observed by satellites. Science 302: 269–272.
13. Hepple RP (2002) Implications of surface seepage on the effectiveness of geologic storage of carbon dioxide as a climate change mitigation strategy. Lawrence Berkeley National Laboratory (LBNL), Paper LBNL-51267.
14. Hepple RP, Benson SM (2005) Geologic storage of carbon dioxide as a climate change mitigation strategy: performance requirements and the implications of surface seepage. Environmental Geology 47: 576–585.
15. Herzog HJ (2001) What future for carbon capture and sequestration? American Chemical Society 35: 148–153.
16. Intergovernmental Panel on Climate Change (IPCC) (2005) IPCC Special Report on Carbon Dioxide Capture and Storage. Cambridge University Press, Cambridge, UK.
17. Lawrence Berkeley National Laboratory (LBNL) (2000) An overview of geologic sequestration of CO_2. In: ENERGEX'2000: Proceedings of the 8th International Energy Forum, Las Vegas, NV.
18. Pacala S, Socolow R (2004) Stabilization wedges: solving the climate problem for the next 50 years with current technologies. Science 305: 968–972.
19. Wilson EJ, Friedmann SJ, Pollak MF (2007) Research for deployment: Incorporating risk, regulation, and liability for carbon capture and sequestration. Environmental Science & Technology 41: 5945–5952.
20. Cortis A, Oldenburg CM, Benson SM (2008) The role of optimality in characterizing CO_2 seepage from geologic carbon sequestration sites. International Journal of Greenhouse Gas Control 2: 640–652.
21. Oldenburg CM, Bryant SL, Nicot JP (2009) Certification framework based on effective trapping for geologic carbon sequestration. International Journal of Greenhouse Gas Control 3: 444–457.
22. Schuerger AC, Capelle GA, Di Benedetto JA, Mao CY, Thai CN, et al. (2003) Comparison of two hyperspectral imaging and two laser-induced fluorescence instruments for the detection of zinc stress and chlorophyll concentration in bahia grass (Paspalum notatum Flugge.). Remote Sensing of Environment 84: 572–588.
23. Keith CJ, Repasky KS, Lawrence RL, Jay SC, Carlsten JL (2009) Monitoring effects of a controlled subsurface carbon dioxide release on vegetation using a hyperspectral imager. International Journal of Greenhouse Gas Control 3: 626–632.
24. Male EJ, Pickles WL, Silver EA, Hoffmann GD, Lewicki JL, et al. (2010) Using hyperspectral plant signatures for CO_2 leak detection during the 2008 ZERT CO_2 sequestration field experiment in Bozeman, Montana. Environmental Earth Sciences 60: 251–261.
25. Kling GW, Evans WC, Tanyileke G, Kusakabe M, Ohba T, et al. (2005) Degassing Lakes Nyos and Monoun: Defusing certain disaster. Proceedings of the National Academy of Sciences 102: 14185–14190.
26. Bergfeld D, Evans WC, Howle JF, Farrar CD (2006) Carbon dioxide emissions from vegetation-kill zones around the resurgent dome of Long Valley caldera, eastern California, USA. Journal of Volcanology and Geothermal Research 152: 140–156.
27. Jong SM (1998) Imaging spectrometry for monitoring tree damage caused by volcanic activity in the Long Valley caldera, California. ITC Journal 1–10.
28. Pickles WL, Kasameyer PW, Martini BA, Potts DC, Silver EA (2001) Geobotanical remote sensing for geothermal exploration. Geothermal Resources Council Transactions 25: 307–312.
29. Maček I, Pfanz H, Francetič V, Batič F, Vodnik D (2005) Root respiration response to high CO2 concentrations in plants from natural CO2 springs. Environmental and Experimental Botany 54: 90–99.
30. Spangler LH, Dobeck LM, Repasky KS, Nehrir AR, Humphries SD, et al. (2009) A shallow subsurface controlled release facility in Bozeman, Montana, USA, for testing near surface CO2 detection techniques and transport models. Environmental Earth Science, http://dx.doi.org/10.1007/s12665-009-0400-402.
31. Goetz AFH, Vane G, Solomon JE, Rock BN (1985) Imaging spectrometry for Earth Remote Sensing. Science 228: 1147–1153.
32. Carter GA (1993) Responses of leaf reflectance to plant stress. American Journal of Botany 80: 239–243.
33. Noomen MF, Skidmore AK (2009) The effects of high soil CO_2 concentrations on leaf reflectance of maize plants. International Journal of Remote Sensing 30: 481–497.
34. Peguero-Pina JJ, Morales F, Flexas J, Gil-Pelegrin E, Moya I (2008) Photochemistry, remotely sensed physiological reflectance index and de-epoxidation state of the xanthophyll cycle in Quercus coccifera under intense drought. Oecologia 156: 1–11.
35. Penuelas J, Llusia J, Pinol J, Filella I (1997) Photochemical reflectance index and leaf photosynthetic radiation-use-efficiency assessment in Mediterranean trees. International Journal of Remote Sensing 18: 2863–2868.
36. Suarez L, Zarco-Tejada PJ, Sepulcre-Canto G, Perez-Priego O, Miller JR, et al. (2008) Assessing canopy PRI for water stress detection with diurnal airborne imagery. Remote Sensing of Environment 112: 560–575.
37. Lawrence RL, Labus M (2003) Early detection of Douglas-Fir Beetle infestation sith subcanopy resolution hyperspectral imagery. Westeren Journal of Applied Forestry 18: 202–206.
38. Pu R, Kelly M, Anderson GL, Gong P (2008) Using CASI hyperspectral imagery to detect mortality and vegetation stress associated with a new hardwood forest disease. Photogrammetric Engineering & Remote Sensing 74: 65–75.
39. Fitzgerald GJ, Maas SJ, Detar WR (2004) Spider mite detection and canopy component mapping in cotton using hyperspectral imagery and spectral mixture analysis. Precision Agriculture 5: 275–289.
40. Muhammed HH (2005) Hyperspectral crop reflectance data for characterizing and estimating fungal disease severity in wheat. Biosystems Engineering 91: 9–20.
41. Naumann JC, Young DR, Anderson JE (2009) Spatial variations in salinity stress across a coastal landscape using vegetation indices derived from hyperspectral imagery. Plant Ecology 202: 285–297.
42. Gamon JA, Serrano L, Surfus JS (1997) The photochemical reflectance index: an optical indicator of photosynthetic radiation use efficiency across species, functional types, and nutrient levels. Oecologia 112: 492–501.
43. Strachan IB, Pattey E, Boisvert JB (2002) Impact of nitrogen and environmental conditions on corn as detected by hyperspectral reflectance. Remote Sensing of Environment 80: 213–224.
44. Meroni M, Rossini M, Picchi V, Panigada C, Cogliati S, et al. (2008) Assessing steady-state fluorescence and PRI from hyperspectral proximal sensing as early indicators of plant stress: The case of ozone exposure. Sensors 8: 1740–1754.
45. Noomen MF, Smith KL, Colls JJ, Steven MD, Skidmore AK, et al. (2008) Hyperspectral indices for detecting changes in canopy reflectance as a result of underground natural gas leakage. International Journal of Remote Sensing 29: 5987–6008.
46. Carter GA, Knapp AK (2001) Leaf optical properties in higher plants: linking spectral characteristics to stress and chlorophyll concentration. American Journal of Botany 88: 677–684.
47. Knipling EB (1970) Physical and physiological basis for the reflectance of visible and near-infrared radiation from vegetation. Remote Sensing of Environment 1: 155–159.
48. Smith KW, Vogelmann TC, DeLucia EH, Bell DT, Shepherd KA (1997) Leaf form and photosynthesis. BioScience 47: 785–793.
49. Zarco-Tejada PJ, Miller JR, Mohammed GH, Noland TL (2000) Chlorophyll fluorescence effects on vegetation apparent reflectance: I. leaf-level measurements and model simulation. Remote Sensing of Environment 74: 582–595.
50. Lichtenhaler HK (1987) Chlorophylls and carotenoids: pigments of photosynthetic biomembranes. Methods in Enzymology 148: 350–382.

51. Buschmann C, Nagel E (1993) In vivo spectroscopy and internal optics of leaves as basis for remotes sensing of vegetation. International Journal of Remote Sensing 14: 711–722.
52. Smith KL, Steven MD, Colls JJ (2004) Use of hyperspectral derivative ratios in the red-edge region to identify plant stress responses to gas leaks. Remote Sensing of Environment 92: 207–217.
53. Carter GA (1991) Primary and secondary effects of water content on the spectral reflectance of leaves. American Journal of Botany 78: 916–924.
54. Tucker CJ, Elgin JH, McMurtrey JM (1980) Relationship of red and photographic infra-red spectral data to alfalfa agronomic values. International Journal of Remote Sensing 1: 69–75.
55. Carter GA, Miller RL (1994) Early detection of plant stress by digital imaging within narrow stress-sensitive wavebands. Remote Sensing of Environment 50: 295–302.
56. Smith KL, Steven MD, Colls JJ (2004) Spectral responses of pot-grown plants to displacement of soil oxygen. International Journal of Remote Sensing 25: 4395–4410.
57. Rock BN, Hoshizaki T, Miller JR (1988) Comparison of in situ and airborne spectral measurements of the blue shift associated with forest decline. Remote Sensing of Environment 24: 109–127.
58. Vogelmann JE, Rock BN, Moss DM (1993) Red edge spectral measurements from sugar maple leaves. International Journal of Remote Sensing 14: 1563–1575.
59. Gamon JA, Surfus JS (1999) Assessing leaf pigment content and activity with a reflectometer. New Phytologist 143: 105–117.
60. Venables WN, Ripley BD (1997) Modern Applied Statistics with S-Plus, Second Edition. Springer-Verlag, New York, NY 548 p.
61. Breiman L (2001) Random Forests. Machine Learning 45: 5–32.
62. Gislason PO, Benediktsson JA, Sveinsson JR (2006) Random Forest for land cover classification. Pattern Recognition Letters 27: 294–300.
63. Lawrence RL, Wood SD, Sheley RL (2006) Mapping invasive plants using hyperspectral imagery and Breiman Cutler classifications (RandomForest) Remote Sensing of Environment 100: 356–362.
64. Prasad AM, Iverson LR, Liaw A (2006) Newer classification and regression tree techniques: Bagging and random forests for ecological prediction. Ecosystems 9: 181–199.
65. Congalton RG, Green K (2009) Assessing the accuracy of remotely sensed data: principles and practices Second edition. Taylor and Francis Group.
66. Baier W (1969) Concepts of soil moisture availability and their effect on soil moisture estimates from a meteorological budget. Agricultural Meteorology 6: 165–178.
67. Denmead OT, Shaw RH (1962) Availability of soil water to plants as affected by soil moisture content and meteorological conditions. Agronomy Journal 54: 385–390.
68. Harsanyi JC, Chang C (1994) Hyperspectral image classification and dimensionality reduction: an orthogonal subspace projection approach. IEEE Transactions On Geoscience and Remote Sensing 32: 779–785.
69. Rodarmel C, Shan J (2002) Principal component analysis for hyperspectral image classification. Surveying and Land Information Science 62: 115–123.
70. Ji R, Gao Y, Hong R, Liu Q, Tao D, et al. (2014) Spectral-spatial constraint hyperspectral image classification. IEEE Transactions On Geoscience and Remote Sensing 52: 1811–1824.

3

Streamflow Impacts of Biofuel Policy-Driven Landscape Change

Sami Khanal[1]*, Robert P. Anex[2]*, Christopher J. Anderson[3], Daryl E. Herzmann[3]

1 School of Environment and Natural Resources, Ohio State University, Wooster, OH, United States of America, 2 Dept. of Biological Systems Engineering, University of Wisconsin-Madison, Madison, Wisconsin, United States of America, 3 Dept. of Agronomy, Iowa State University, Ames, Iowa, United States of America

Abstract

Likely changes in precipitation (P) and potential evapotranspiration (PET) resulting from policy-driven expansion of bioenergy crops in the United States are shown to create significant changes in streamflow volumes and increase water stress in the High Plains. Regional climate simulations for current and biofuel cropping system scenarios are evaluated using the same atmospheric forcing data over the period 1979–2004 using the Weather Research Forecast (WRF) model coupled to the NOAH land surface model. PET is projected to increase under the biofuel crop production scenario. The magnitude of the mean annual increase in PET is larger than the inter-annual variability of change in PET, indicating that PET increase is a forced response to the biofuel cropping system land use. Across the conterminous U.S., the change in mean streamflow volume under the biofuel scenario is estimated to range from negative 56% to positive 20% relative to a business-as-usual baseline scenario. In Kansas and Oklahoma, annual streamflow volume is reduced by an average of 20%, and this reduction in streamflow volume is due primarily to increased PET. Predicted increase in mean annual P under the biofuel crop production scenario is lower than its inter-annual variability, indicating that additional simulations would be necessary to determine conclusively whether predicted change in P is a response to biofuel crop production. Although estimated changes in streamflow volume include the influence of P change, sensitivity results show that PET change is the significantly dominant factor causing streamflow change. Higher PET and lower streamflow due to biofuel feedstock production are likely to increase water stress in the High Plains. When pursuing sustainable biofuels policy, decision-makers should consider the impacts of feedstock production on water scarcity.

Editor: Ben Bond-Lamberty, DOE Pacific Northwest National Laboratory, United States of America

Funding: This work is supported by the National Science Foundation Grant CBET-1137677. The funders had no role in study design, data collection and analysis, decision to publish, or preparation of the manuscript.

Competing Interests: The authors have declared that no competing interests exist.

* Email: khanal.3@osu.edu (SK); anex@wisc.edu (RA)

Introduction

As demand for renewable fuels grows, biofuels from lignocellulosic feedstock are considered a promising alternative to corn-based ethanol [1–2]. Cellulosic biofuels are expected to be both environmentally and energetically superior to grain-based biofuels [3–6]. The mandate set by the Renewable Fuel Standard [7] to use 16 billion gallons of cellulosic biofuel per year by 2022 is projected to have significant impact on agricultural land use in the U.S. as lands are converted for the production of bioenergy crops [8]. Prior studies [6,9] have investigated yields, land use, economics and greenhouse gas emissions of bioenergy crops, but one key factor often overlooked is the hydrologic balance associated with bioenergy crop production.

There is strong coupling between the land surface and atmosphere that is heavily influenced by the vegetative land cover [10–13]. Change in land cover thus has the potential to impact local and regional climate through alteration of the energy and moisture balances of the land surface [14–18]. The longer growing season and greener vegetative cover of biofuel crops result in higher water loss to the atmosphere through evapotranspiration (ET), decline in soil water depth [17,19] and reduced surface runoff [20] relative to annual cropping systems. Changes in soil moisture and runoff determine streamflow, groundwater recharge and influence water quality.

Bioenergy crops, e.g., switchgrass and miscanthus, can transpire as much as 38% more than corn over a growing season [20]. Replacing traditional annual cropping systems with switchgrass in the Midwest and High Plains may cause additional stress to water resources because the agricultural crop production in large portions of these areas (e.g., Kansas and Nebraska) is dependent upon irrigation water from already stressed local resources [21]. Streamflow volume (Q) is responsive to changes in both climate and land cover [22–23], and changes in Q have important biological and socioeconomic implications [24–26]. Anthropogenic alteration of Q has been shown to impair aquatic communities and ecosystems, and the likelihood of impairment rises rapidly with increasing severity of reduced Q [25].

Water may be a significant limiting factor for biofuel crop production in many agricultural regions. It is important that we develop projections of future water use for agricultural crop production under climate change induced by land use change and to account for the impact of that water use on critical water resources. Prior studies have examined the potential for biofuel crops to affect regional climate [15]. The climate feedback of biofuel crops examined in these studies, however, is based on hypothetical scenarios that do not account for the socio-economic responses of land managers and thus do not represent plausible land use patterns that might result from current biofuel policies. To our knowledge, no prior studies have explored the changes in streamflow volume in response to climate change induced by land use/land cover (LULC) change; certainly none have examined this under the constraints of enacted legislation. Future projections of climate and climate-driven streamflow under plausible landscape scenarios will aid state and federal agencies in assessing the local cost of adaptation, increase public awareness, and guide the development of new mitigation programs related to water resources.

In this study, we examine changes in hydrologic processes including precipitation (P), ET, PET, runoff and Q that result from modification of local/regional climate driven by switchgrass cropping systems predicted to replace current cropping systems in the High Plains (hereafter referred to as the "biofuel scenario"). A regional climate model coupled to a land surface model is used to capture feedback between changes in the vegetation canopy due to switchgrass planting and regional climate processes. The change in Q to climate change under the biofuel scenario compared to the current cropping system scenario (hereafter referred to as the "baseline scenario") is estimated based on widely used non-parametric approaches [22–23,27]. These non-parametric approaches utilize the concept of elasticity of Q that is usually derived using the historic relationship between Q, P and PET. Following the similar approach, we first derived the elasticity of Q to climate, and later used it in combination with projected changes in P and PET to derive changes in Q under the biofuel scenario relative to baseline across the conterminous U.S.

Materials and Methods

Regional climate modeling framework

The Weather Research Forecast (WRF) model version 3.1.1 [28–29] and NOAH land surface model (NOAH LSM) are used for regional climate simulations. The simulation domain covers the continental U.S. at a resolution of 0.25 degree (i.e., 24 km). The NOAH LSM coupled to the WRF model is used to represent the interaction of soil and vegetation with the atmosphere [30]. Regional climate simulations were produced for the period of 1979–2004 under the two land use scenarios. The choice of the simulation period is constrained by 1) availability of data describing the baseline scenario; 2) the accuracy of extrapolating land use categories derived from 1991–1995 satellite data further into the future; and 3) the large computational burden associated with longer simulations. The baseline scenario represents land use categories and monthly phenology based on satellite derived data from 1991–1995, and the biofuel scenario represents projected alternative (i.e., switchgrass) land use categories (Figure 1) with identical atmospheric forcing data. Following the recent studies [16,31–32] that have used 2-years as a minimum length for spin-up, we discarded the first two years (i.e., 1979 and 1980) of each simulation to allow for adjustment of the land surface with the atmosphere. Details of the model configuration are provided in the Anderson et al [19], and thus are not described here.

Land use scenarios

The baseline scenario uses the NOAH LSM default settings of land use and vegetation parameters, including 24 vegetation classes, a vegetation parameter table and satellite-based (1991–1995) monthly vegetation fraction from which leaf area index (LAI) and albedo are derived (Table S1). The projection of LULC change produced by the Policy Analysis System (POLYSYS) model [33–34] in support of the DOE study report "U.S. Billion-Ton Update" [7] is used to create the biofuel land use scenario. The Billion Ton Update study examines the feasibility of attaining annual production of one billion dry tons of biomass feedstock by 2030 based on projections of future biomass demand, inventory, production capacity, availability, and technology. In the 2011 update, the greatest conversion of traditional cropping system to switchgrass occurs in the Great Plains, with 20–30% in Kansas and 30–45% in northern and northwest Oklahoma [35]. The POLYSYS simulation contained county level switchgrass and crop production for 2022, the first year Renewable Fuel Standard (RFS) goals reach maximum levels, at a $60/dry ton farm gate switchgrass price with the Billion-Ton Study baseline assumptions, including an extension of the USDA 10-year yield forecast for major food and forage crops to 2022. An area weighted method was used to resample county-level POLYSIS estimates of switchgrass conversion to the WRF grid [19]. The NOAH LSM uses a single vegetation category for each grid cell, and vegetation parameters are homogenous within each grid cell. Thus, a grid cell that contains a mixture of vegetation types does not explicitly account for each type, but represents an average of vegetation parameters over all vegetation types present in the grid cell.

For the biofuel scenario, four new vegetation classes, their related vegetation parameters and monthly vegetation fraction are introduced in the WRF model based on default land use categories (Figure 1A). The new vegetation classes are modified to represent a mixture of the default land use categories and switchgrass (Figure 1B). This reflects that a mixture of biofuel and conventional crops are expected in regions where biofuel crops are adopted rather than complete replacement of conventional crops with switchgrass. Two of the new classes (Switchgrass/Grassland Mosaic and Switchgrass/Cropland Mosaic) are used for grid cells in which the switchgrass fraction exceeds 30%, and two (Grassland/Switchgrass Mosaic and Cropland/Switchgrass Mosaic) are used for grid cells in which the switchgrass fraction is below 30% [36] (Figures 1B and 1C). We prevented land use change to switchgrass in regions beyond that projected by the Billion ton update by only changing the parameters based on latitude/latitude. For example, land cover changes were mostly in Oklahoma and Kansas (Figure 1B); we did not change parameters in California.

As the new land use categories under the biofuel scenario reflect a mixture of switchgrass and conventional crops rather than complete replacement of switchgrass, the phenology for new land use categories is characterized in our simulations by adjusting the satellite-based monthly vegetation fraction based on discussion with scientists working on field trials of switchgrass. The monthly greenness fraction is the basis for LAI and albedo calculations in NOAH LSM, and both LAI and albedo increase with vegetation fraction. During the growing season, spectrally weighted albedo (which is required in WRF) increases as a crop is greening up. To represent change in phenology consistent with managed plots of switchgrass stands, vegetation fraction is increased during February–October to simulate earlier greening, denser foliage at peak LAI, and later senesce of switchgrass (a perennial grass) compared to annual crops and rangeland. Increase of vegetation fraction ranges 10–20%, except north central Oklahoma where it is increased by 90% to offset low LAI due to winter wheat harvest [19].

The maximum and minimum values of LAI and albedo are adjusted to reflect changes in vegetation under the biofuel scenario (Table S1). And, the same monthly phenology is imposed for each simulated year. When the vegetation fraction is at its peak, the LAI and albedo are as well. Maximum LAI is based upon observations of field stands of managed perennial grass that grows a denser

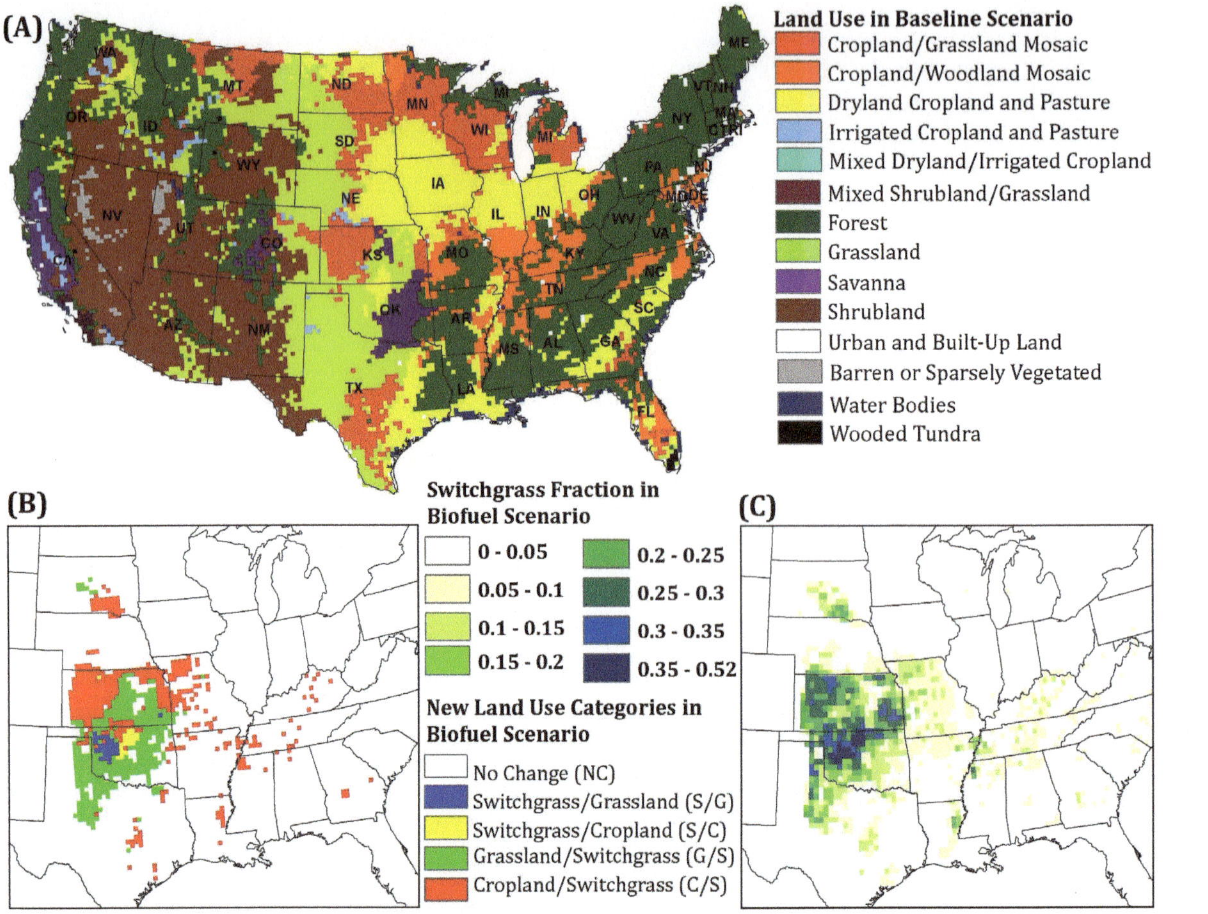

Figure 1. Default land use categories in the WRF model (A); new land use categories defined for the biofuel scenario (B); and fraction of land use that is switchgrass in the biofuel scenario (C).

canopy than prairie grass. Although LAI>6 as measured in field trials of managed switchgrass is used in simulations of switchgrass production [37–38], maximum LAI<6 is set to reflect a regional vegetation mixture. This approach is consistent in simulations with Van Loocke *et al* [17].

Climate elasticity of streamflow and changes in streamflow

In this version of WRF model, surface runoff is computed as the excess of precipitation that does not infiltrate into the soil [39]. Although NOAH-LSM describes the canopy and root zone in detail, the interactions between groundwater, the root zone, and surface water were not yet included at the time our project was undertaken and completed. This version parameterizes surface runoff with a simple infiltration-excess scheme rather than terrain slope channel routing, and it treats baseflow as a linear function of bottom soil-layer drainage [40]. Thus, runoff estimates from WRF are not representative of the changes in streamflow. For the purpose, we used non-parametric approaches that are demonstrated as or more robust than complex and detailed hydrologic models for evaluating the sensitivity of streamflow to climate [22,27,41]. These non-parametric approaches use the concept of climate elasticity of streamflow (ε_x), computed based on historic P, PET and Q data.The climate elasticity of streamflow is defined by the proportional change in Q to the change in a climate variable (x), such as P or PET [42]. It is an index commonly used to quantify the sensitivity of Q to changes in climate. Often this index (i.e., ε_x) is derived from the historic climate and hydrologic data (i.e., P, PET and Q) [23,42]. Streamflow in unimpaired watersheds (i.e., watersheds in which streamflow are not subject to regulation or diversion, and defined as reference watershed in this study) can be modeled as a function of P and PET [23]. The changes in Q due to changes in P and PET can be approximated as:

$$\Delta Q/Q = \varepsilon_p \Delta P/P + \varepsilon_{pet}\ \Delta PET/PET \quad (1)$$

In equation 1, ΔQ, ΔP and ΔPET are changes in Q, P and PET, respectively; ε_p and ε_{pet} are the elasticities of streamflow with respect to P and PET. Prior studies [23,27,43] have proposed non-parametric approaches to estimate ε_x from observed climatic data. Of the various approaches (Table S2, Figure S2) found in the literature, we have no reason to favor one over the others, and thus use the average of ε_p estimated from all available non-parametric approaches to predict average changes in Q under the alternative LULC scenario. Details about the non-parametric approaches used in this study and hydro-climatology of the conterminous U.S are discussed in Figure S3.

To compute elasticity estimates, we used historical annual streamflow, precipitation and PET information for 1,845 reference watersheds across the conterminous United States (see Figure S1).

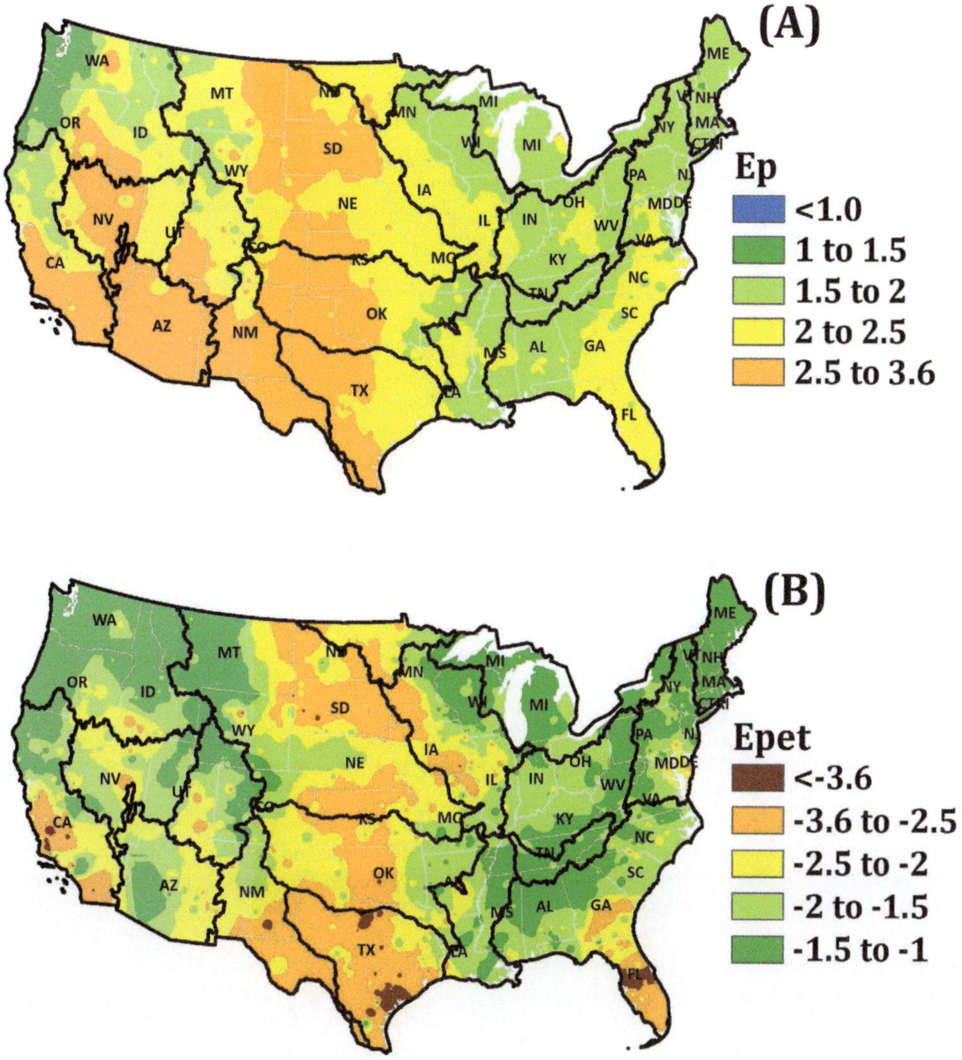

Figure 2. Hydro-climatology of the conterminous US; (A) Precipitation elasticity of streamflow (ε_p) and (B) Evapotranspiration elasticity of streamflow (ε_{pet}).

To compute ε_x estimates from the historic climate record, we used PET instead of ET similar to prior studies [22–23] due to data limitations associated with the estimation of actual ET from 1950–2009. Further, as ε_{pet} was computed using historic PET estimates, we used PET instead of ET from the regional climate model simulations to estimate changes in Q to maintain consistency in the methodology. These climate elasticity values were combined with the differences in mean annual P and PET between the biofuel and baseline scenarios, expressed as a percentage of the baseline, to compute the relative change in Q across the nation under the biofuel scenario (equation 1).

Sensitivity of streamflow change

Uncertainty in the estimated change in Q under the biofuel scenario is evaluated based on: 1) difference in elasticity estimates computed from various non-parametric approaches (discussed in the supporting material), and 2) year to year changes in simulated P and PET for the period 1981–2004. To evaluate the sensitivity of changes in streamflow to changes in both climate and elasticity estimates, we calculated the standard deviation (std) of percent annual change in P and PET between the biofuel and baseline scenarios, and the std of elasticity estimates from the seven different non-parametric methods as shown in equations 2 and 3.

$$\Delta Q = (P \pm \text{std } P) \times \varepsilon_p + (PET \pm \text{std } PET) \times \varepsilon_{pet} \qquad (2)$$

$$\Delta Q = P \times (\varepsilon_p \pm \text{std } \varepsilon_p) + PET \times (\varepsilon_{pet} \pm \text{std } \varepsilon_{pet}) \qquad (3)$$

$$\Delta Q = (P \pm \text{std } P) \times \varepsilon_p + PET \times \varepsilon_{pet} \qquad (4)$$

In equations 2, 3 and 4, P and PET indicate the percent change in precipitation and evapotranspiration under the biofuel scenario relative to the baseline. ε_p and ε_{pet} indicate the mean values of ε_p and ε_{pet} from the seven empirical methods.

The sensitivity of estimated change in Q is evaluated by varying the percent change in mean annual P and PET under the biofuel scenario (equation 2; Figure S6A and S6B), and the mean estimates of ε_{pet} and ε_p (equation 3; Figures S6C and S6D) by ±

their standard deviation. We estimated the sensitivity of predicted change in Q by varying the percent change in mean annual P and PET simultaneously because P observed in many regions including High Plains are correlated with PET [44] and ET [45]. Sensitivity measures are also computed varying only P (equation 4; Figure S7) because the impacts of varying P and PET simultaneously will tend to cancel each other in the regions where impacts are inversely correlated, thus not reflecting the full contribution of P or PET to Q (equation 4).

Results

Climate elasticity of streamflow

Precipitation elasticity of streamflow is estimated in the range of 1–3.6 with a mean of 2.2 for watersheds across the U.S., implying that a 1% change in P will result in more than a 1% change in Q. The relationship between P and Q is generally non-linear, and this non-linearity is influenced by catchment properties including storage processes, ET and vegetation properties, and these factors are implicitly factored into elasticity estimates. Approximately 46% of all watersheds examined have ε_p higher than 2, and these watersheds are clustered in the Southwest, Midwest and Southeastern parts of the nation (Figure 2A) where PET usually exceeds P. Only a few basins in the Northwest have ε_p less than 1.5.

Evapotranspiration elasticity of streamflow for watersheds across the U.S. is estimated to be in the range of negative 3.6 to negative 1, with a mean of negative 1.9, indicating that a 1% reduction in PET would result in about a 1.9% increase in Q. The geographical distribution of ε_{pet} (Figure 2B) in the conterminous U.S. is similar to the distribution of ε_p with lower values ε_{pet} in the arid and semiarid regions of the Southwest and Midwest.

Variability in climate elasticity of streamflow

The variability in elasticity estimates computed using seven different non-parametric approaches (expressed as a standard deviation) is high in the arid and semi-arid regions of the Midwest and Southwest, and lower in the humid and semi-humid regions (Figures S4A and S4B). Also, the standard deviation of ε_{pet} estimates is higher than the standard deviation of ε_p estimates. This is due to differences among the seven different approaches; five of seven approaches [46–51] depend upon aridity index (PET/P) to estimate elasticity estimates, while the other two depend on Q and P or PET [23,27] (see Table S2 for details).

Projected climate change under biofuel scenario

Change in annual precipitation and evapotranspiration. Under the biofuel scenario, the mean annual change in P relative to the baseline is projected to be in the range of negative 10% to positive 10% (Figure 3A). About 84% of the conterminous U.S. is predicted to experience changes in P in the range of negative 5% to positive 5% under the biofuel scenario; and a general decrease in P is predicted over 52% of the area. The magnitude of projected change in mean annual P in the main biofuel crop producing region (i.e., Kansas and Oklahoma) is between 2.5% to 5% relative to the baseline scenario. Under the biofuel scenario, western Kansas and Oklahoma show 5 to 15 mm higher annual P than under the baseline scenario.

The projected change in mean annual PET under the biofuel scenario is estimated to be in the range of negative 6% to positive 16% relative to the baseline (Figure 3C). Higher PET in the switchgrass planted region is consistent with lower mean temperatures due to earlier green-up and the higher LAI of switchgrass compared to current vegetation, and higher net radiation under the biofuel scenario.

In addition to the observed changes in climate in switchgrass planted region, we observed changes in climate patterns in areas away from the switchgrass concentrated area. Under the biofuel scenario, southern regions including parts of Arizona, New Mexico and Texas, the High Plains including western Kansas and Oklahoma, the Midwest including eastern part of Nebraska and Iowa, and a large region of the eastern states show an increase in annual P of between 2.5% to 10% relative to the baseline scenario. A large decline in annual P (i.e., between 2.5% to 10% relative to the baseline) is predicted across much of the northern (i.e., northern Minnesota, South Dakota, Wisconsin), Midwestern (i.e., Wyoming, Idaho, northern Colorado) and the High Plains (i.e., Missouri) regions. Due to the internal non-linear climate dynamics, a single simulation is insufficient to conclude that they are systematically caused by land use change in the Great Plaines. They could, in fact, be an artifact of the initial atmospheric conditions.

PET is usually a good representation of actual ET when there is no plant water stress, and is thus commonly used in precipitation-runoff modeling applications [41]. We observed similar trends in PET and ET in the switchgrass perturbed region (Figures 4B and 4C) although they differed in magnitude. Increases in ET increase low-level humidity and the potential for more P [19]. Change in mean annual P, PET, ET and runoff when examined by land use (i.e., switchgrass altered and unaltered) categories in the High Plains, suggests that the difference in ET, PET and runoff represent a change induced by land cover perturbation (Figures 4 and S5).

Differences in climate between the baseline and biofuel projections vary annually in both magnitude and direction between 1981 and 2004. The magnitude of mean annual change in PET and ET is higher than the year to year changes in PET and ET in switchgrass planted regions under the biofuel scenario. Also, the mean annual runoff in the switchgrass dominated region is lower than other regions in Kansas and Oklahoma. Compared to land cover with a lower fraction of switchgrass (i.e., cropland/switchgrass and grassland/switchgrass), the land cover with a large fraction (i.e., >30%) of switchgrass (i.e., switchgrass/grassland and switchgrass/cropland) demonstrated a higher magnitude of change in PET, ET and runoff in the biofuel scenario (Figure 4). This indicates that the observed changes in PET, ET and runoff associated with biofuel feedstock production are large and significant. Decrease in runoff results from lower soil moisture levels due to higher evapotranspiration of switchgrass during the growing season [19]. Conversely, the inter-annual variability of change in P is as large as or larger than the magnitude of mean annual change in P (Figure 4). Thus it is hard to conclude that P is changed under the biofuel scenario.

Streamflow response to projected climate change. Mean annual change in Q in response to changes in P and PET under the biofuel scenario is shown in Figure 5. Across the conterminous U.S., the change in mean Q under the biofuel scenario is estimated to be in the range of negative 56% to positive 20% relative to the baseline scenario. An increase in Q with magnitude greater than 5% is predicted over 12% of the area. Lower PET but higher P in New Mexico and Arizona are estimated to increase Q under the biofuel scenario. However, a decrease in Q with magnitude greater than 5% is predicted over 30% of the area. The increase in P is smaller than the increase in PET under the biofuel scenario, and this causes a net decline in Q in the High Plains. In the High Plains, Q is predicted to be about 20% lower than the baseline. Streamflow in the biofuel crop region within the High Plains is 18% lower relative to the baseline (Figure 5). The switchgrass areas in the biofuel crop region show decreases in

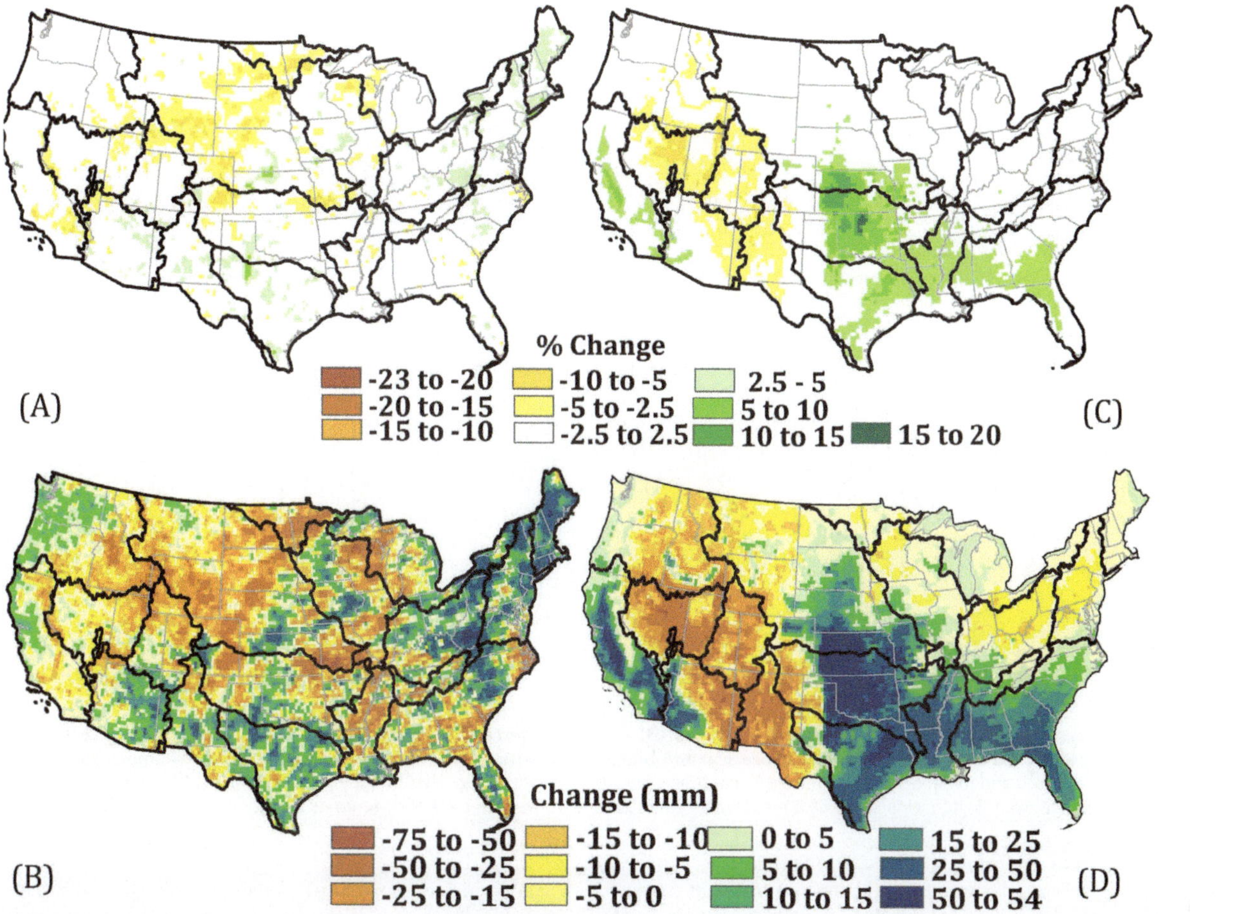

Figure 3. Change in mean annual precipitation and potential evapotranspiration for 1981–2004 expressed in A) percentage change in mean annual precipitation; B) change in mean annual precipitation (millimeters); C) percentage change in mean annual PET; and D) change in PET (millimeters) under the biofuel scenario. Percent change under the biofuel scenario relative to the baseline scenario for a given location is estimated as the mean of: $\frac{Baseline_{xi} - Biofuel_{xi}}{Baseline_{xi}} \times 100$ where x is either P or PET and i is year between 1981 and 2004.

streamflow that are twice as large as the decrease in the unperturbed area.

Sensitivity of streamflow. The change in Q under the biofuel scenario is observed to be highly sensitive to mean annual P and PET for large parts of the conterminous U.S., however, it is less sensitive in the switchgrass planted region where change in PET is larger than P (Figure S6). As parameters associated with precipitation and evapotranspiration are very likely to be correlated with each other, PET and P were changed independently in the sensitivity test to examine the relative significance of changes in P and PET on streamflow estimates (see the section 'Materials and Method' for details). Under these sensitivity analyses, Q always decreases under the biofuel scenario, even in the case when PET is held constant and P is increased by one standard deviation (Figure S7). Sensitivity analysis also showed that predicted changes in streamflow are robust to differences in ε_p and ε_{pet} stimation approaches. Change in Q across the range of ε_p and ε_{pet} estimation approach (Figure S6C and S6D) is less than the change in Q resulting from changes in P and PET equal to their inter-annual variability (Figure S6A and S6B).

Discussion

Climate change is a likely result of a policy that encourages expansion of energy crop production in the High Plains and Midwest U.S [19]. This climate change is caused by alteration of the surface energy and moisture balances induced by changes in land cover when current cropping systems are replaced by energy crops. Notable changes include lower temperature, higher PET, lower runoff and a decline in Q in the region where switchgrass is predicted to replace current vegetation. Our analyses show changes in the mean annual PET, ET and streamflow to be a stronger climate change impact of biofuel crop production than changes in mean annual P.

The main goal of this study is to use a simple and robust model to communicate to policy makers and analysts the potential implications of biofuels policy-induced climate change on streamflow to inform the search for a sustainable renewable fuel production system. While adopting the simple but robust model for estimating streamflow changes, we made several assumptions, and these assumptions are likely to introduce uncertainties in our streamflow estimates. Despite some of the limitations (as discussed below), our conclusion that increase cellulosic feedstock production are likely to reduce water yield is found to be in agreement with other studies conducted [52,20] in other parts of the nation.

Weather conditions simulated with regional and global climate models over short periods are sensitive to their initial conditions. A change in the vegetation type at the initial condition will result in a different sequence of weather conditions. Thus, differences

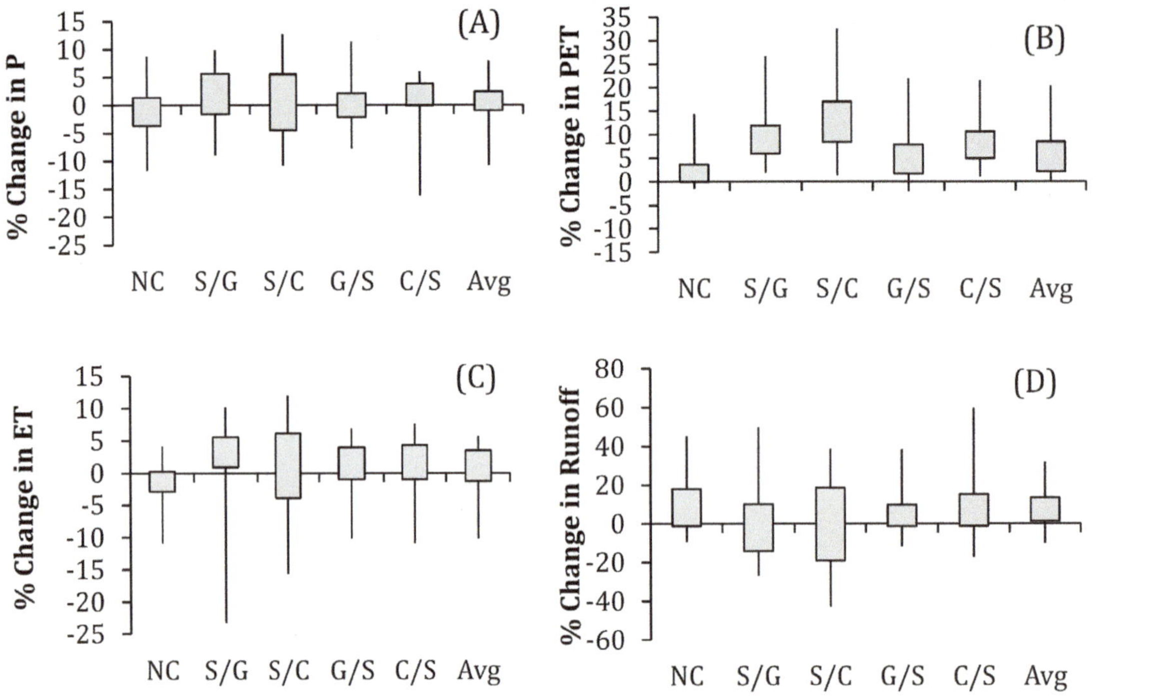

Figure 4. Percent difference in (A) annual precipitation; (B) PET averaged by land use categories in switchgrass altered regions in Kansas and Oklahoma as shown in Figure 1. Box top and bottom edges are the interquartile range of percent difference for each year, and whiskers are maximum and minimum annual values. X-axis labels are land use categories: No Change (NC), Switchgrass/Grassland (S/G), Switchgrass/Cropland (S/C), Grassland/Switchgrass (G/S), Cropland/Switchgrass (C/S), and average over all categories (Avg).

between simulations may be a combination of a transient response from unpredictable nonlinear dynamics acting upon a different initial state as well as a systematic response from a structural change in forcing of local climate, in this case land use change [53–54]. The average change for a transient response is expected to be zero given either a long data record or multiple simulations

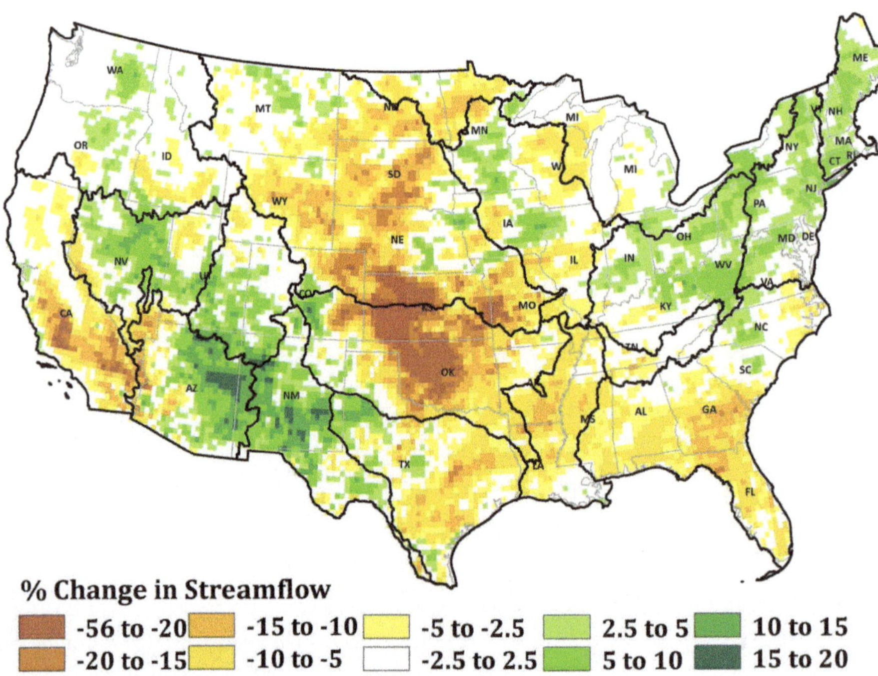

Figure 5. Percent change in mean annual streamflow as a function of change in annual precipitation and potential evapotranspiration under the biofuel scenario.

from alternative initial conditions. Changes in Q estimated in this study thus include uncertainties inherent in P and PET estimated over a relatively short time series (in this case 24 years). Anderson et al [19], using one-way ANOVA analysis, found statistically significant change of monthly values for ET and P. However, the small change in annual ET resulted from offsetting statistically significant monthly changes. To examine further whether the ET and P responses were possibly transient, they performed a second simulation of a single year of the control design beginning from a different initial condition. The difference of ET in the two control simulations was much smaller than the difference between the control and LULC scenario. Precipitation, however, showed substantial sensitivity to the initial condition. This sensitivity of P to initial conditions indicates a larger number of simulations would be needed to identify a forced response in P, if one exists at all. Therefore, we are unable to state that a forced change in P exists anywhere in our simulation domain. Since P, ET and PET are correlated, we conclude that outside of the switchgrass planted region, change in PET and Q should not be considered a forced response. Within the switchgrass planted region, however, change of PET and Q is a forced response. Another source of uncertainty in our estimates can be driven by the use of temperature based approach to estimate PET instead of approach like Penman-Monteith (see the supporting material). As temperature-based PET estimates are likely to overestimate the impacts of changes in temperature, our streamflow estimates might have been overestimated to some extent.

Another possible area of limitation includes elasticity estimates. In the study, we interpolated the elasticity estimates based on reference watershed for the conterminous US. This means, our elasticity estimates may not accurately reflect the connection of climate to Q in non-reference watersheds where streamflow is heavily influenced by land use practices and ground water pumping. Predicted change in Q in non-reference watersheds thus might be over or under estimated in the biofuel scenario. Despite this limitation, this study is useful in providing the direction of change in streamflow under the biofuel scenario without requiring the use of detailed hydrologic models that are computationally complex and often provide ambiguous results when compared [27].In the study we show a potential change in annual streamflow volume as an outcome of a landscape influenced climatic system. Our analyses also suggest that under the biofuel scenario, there is a change in seasonal P and ET. In the growing seasons (i.e., April–June), P decreases and ET increases. Evapotranspiration increases until soil moisture nears wilting point, eliminating transpiration and inhibiting further decline in soil moisture [19]. Decrease in P and increase in ET suggests the possibility for a higher magnitude of change in Q seasonally than annually, and we recommend that this possibility should be explored further with detailed hydrologic models.

This study examines landscape induced climate change and ignores projected climate change due to atmospheric concentration of greenhouse gases (GHGs). There is a strong coupling among landscape processes, atmospheric GHGs and climate, since landscape acts as a sink or source to GHGs like CO_2 which affect the distribution of heat over land and in the atmosphere and feedback to the climate [55]. However, these processes are often uncoupled when we make future projections, and this is likely to introduce biases in the projection of future changes in climate. Thus, we recommend the direct coupling of landscape and other climate forcing factors (e.g., GHGs) to better predict the range of future climate change and its impacts on environmental resources [19]. In this study, we examine landscape induced climate change and its impact on Q on an annual basis. Changing landscape and climate alter not only Q, but also stream quality due to changes in the nutrient and sediment content of runoff, particularly from agricultural fields. Evaluation of stream quality under landscape driven climate change is thus recommended as a topic for future research.

In this study, we find interaction between land use change and climate change in Q that is not considered in previous studies. Although processes related to landscape characteristics, such as soil moisture, infiltration, and surface roughness could affect Q, but are not considered in estimation in Q in this study, we assume that the effect on streamflow of changing climate is larger than that resulting from change in landscape characteristics [56–58]. For example, Tu [56] examined change in streamflow in eastern Massachusetts under three different climate change scenarios under IPCC and land use scenarios relative to current conditions, individually and in combination. Tu [56] found that the change in streamflow under both climate and land use scenarios is similar to the streamflow changes examined under a climate change scenario only. Although the process-based hydrologic models (e.g., Soil and Water Assessment Tool (SWAT) and the Variable Infiltration Capacity (VIC)) have the ability to integrate the effects of land use change and climate change on Q and streamflow quality, they are computationally expensive given that they operate at the field- or sub-field scale and need to be calibrated with large amounts of empirical data. Using the method developed here, areas of concern can be identified and then detailed hydrologic models can be used to examine in more detail the specific trade-offs between land use and management options and streamflow.

The results here suggest that changes induced in the climate system by biofuel crop production may increase stress on the water resources of the High Plains. Under the biofuel scenario, increased ET reduces soil moisture [19], and lower soil moisture during the growing season can cause plant water stress and reduce crop yield. Irrigation can reduce water stress but additional irrigation will increase pressure on already strained water resources in arid agricultural regions such as the High Plains. The predicted 20% decline in Q in the biofuel crop producing region under the biofuel scenario would exacerbate on-going conflicts over water allocation between agriculture and other uses. As we develop and implement policies to pursue more sustainable cellulosic biofuel production, we should carefully consider potential water limitations and other impacts to the hydrologic cycle.

Supporting Information

Figure S1 Location of stream gauges used in this study and major water resource regions in the U.S.

Figure S2 Relationship between climate elasticity of streamflow and aridity index in the U.S. watersheds. Precipitation elasticity of streamflow versus aridity index (left); Evapotranspiration elasticity of streamflow versus aridity index (right).

Figure S3 Hydro-climatology of the conterminous US; (A) Aridity Index (Ø); (B) Runoff Coefficient; (C) Correlation coefficient between precipitation and streamflow; and (D) Dominant land cover of the watersheds in year 2006.

Figure S4 Standard deviation of (A) ε_p; (B) ε_{pet}; (C) mean annual change (%) in precipitation; and (D) mean annual change (%) in PET under the biofuel scenario relative to baseline scenarios. Standard deviation of elasticity estimates here reflects the variability in elasticity estimates among seven non-parametric approaches from their mean estimate, computed as $\sqrt{\frac{1}{N-1}\sum_{i=1}^{N}(x_i-\bar{x})^2}$ where N = 7, x_i is $\varepsilon_{pet(i)}$ or $\varepsilon_{p(i)}$ where i represents a non-parametric approach and $\bar{x}$ is the mean of $\varepsilon_{pet\,(i)}$ or $\varepsilon_{p\,(i)}$ from seven non-parametric approaches for a given location. Standard deviation of precipitation and PET change under the biofuel scenario here reflects the inter-annual variability in changes in P or PET, and is computed as $\sqrt{\frac{1}{N-1}\sum_{i=1}^{N}(x_i-\bar{x})^2}$ where N = 24, the number of years between 1981 and 2004, x_i is $\frac{Baseline_{x_i}-Biofuel_{x_i}}{Baseline_{x_i}}\times 100$ where $Baseline_{xi}$ and $Biofuel_{xi}$ represent P or PET in a year i in a given location under the baseline and biofuel scenarios, respectively, and $\bar{x}$ is the mean of x_i.

Figure S5 Change in mean annual runoff over 1981–2004 expressed in A) percentage change; B) absolute change (millimeters).

Figure S6 Change in streamflow volume when mean annual P and PET are varied by (A) adding and (B) subtracting a standard deviation of P and PET; (C) adding a standard deviation of elasticity (ε_p and ε_{pet}) estimates; and (D) subtracting a standard deviation of elasticity estimates among seven non-parametric approaches.

Figure S7 Streamflow prediction when mean annual P and PET are varied by (A) adding, and (B) subtracting a standard deviation of P holding PET constant at its mean annual value.

Table S1 Values of land use parameters in the NOAH land surface model as coupled within the WRF regional climate model. Text in bold face are the new land use categories and the associated parameterization of these categories used in the switchgrass scenario. Meanings of the parameters are listed below the table (from Anderson et al. [s19]).

Table S2 Mathematical expression of f(∅) and f'(∅) across the studies.

Text S1 Climate Elasticity of Streamflow.

Acknowledgments

We would like to acknowledge the valuable suggestions of Drs. Chris Kucharik and Anita Thompson. This material is based upon work supported by the National Science Foundation Grant CBET-1137677. Any opinions, findings, and conclusions or recommendations expressed in this material are those of the author(s) and do not necessarily reflects the views of the National Science Foundation.

Author Contributions

Conceived and designed the experiments: RA CA SK. Performed the experiments: CA SK. Analyzed the data: SK DH. Contributed reagents/materials/analysis tools: SK. Wrote the paper: SK RA CA.

References

1. Hill J (2007) Environmental costs and benefits of transportation biofuel production from food-and lignocellulose-based energy crops. A review. Agronomy for Sustainable Development 27: 1–12.
2. Hess JR, Foust TD, Hoskinson R, Thompson D (2003) Roadmap for agriculture biomass feedstock supply in the United States. DOE/NE-ID-11129, US Department of Energy, Washington DC, USA.
3. Costello C, Griffin WM, Landis AE, Matthews HS (2009) Impact of biofuel crop production on the formation of hypoxia in the Gulf of Mexico. Environ Sci Technol 43: 7985–7991.
4. Donner SD (2007) Surf or turf: A shift from feed to food cultivation could reduce nutrient flux to the Gulf of Mexico. Global Environ Change 17: 105–113.
5. Schnoor J, Doering III OC, Entekhabi D, Hiler EA, Hullar TL, et al. (2008) Water implications of biofuels production in the United States. National Academy of Sciences, Washington DC, USA.
6. Tilman D, Hill J, Lehman C (2006) Carbon-negative biofuels from low-input high-diversity grassland biomass. Science 314: 1598–1600.
7. U.S. Department of Energy (2011) U.S. Billion-Ton Update: Biomass Supply for a Bioenergy and Bioproducts Industry. Perlack, R.D., Stokes, B.J.; Leads, ORNL/TM-2011/224. Oak Ridge National Laboratory, Oak Ridge, TN. 227.
8. Marshall E, Caswell M, Malcom S, Motamed M, Hrubovcak J, et al. (2011) Measuring the indirect land-use change associated with increased biofuel feedstock production. A review of modeling efforts. US Department of Agriculture-ERS, AP054. Washington DC, USA.
9. Schmer MR, Vogel KP, Mitchell RB, Perrin RK (2008) Net energy of cellulosic ethanol from switchgrass. PNAS 105: 464–469.
10. Diffenbaugh N (2009) Influence of modern land cover on the climate of the United States. Climate Dynamics 33: 945–958.
11. Pielke Sr RA, Adegoke J, Beltran-Przekurat A, Hiemstra CA, Lin J, et al. (2007) An overview of regional land-use and land-cover impacts on precipitation. Tellus B 59: 587–601.
12. Twine TE, Kucharik CJ, Foley JA (2004) Effects of land cover change on the energy and water balance of the mississippi river basin. J Hydrometeor 5: 640–655.
13. Bonan GB (2001) Observational evidence for reduction of daily maximum temperature by croplands in the Midwest United States. J Clim 14: 2430–2442.
14. Halgreen W, Schlosser CA, Monier E, Kicklighter D, Sokolov A, et al. (2013) Climate impacts of a large-scale biofuels expansion. Geophys Res Lett 40: 1624–1630.
15. Georgescu M, Lobell D, Field C (2011) Direct climate effects of perennial bioenergy crops in the United States. PNAS 108(11): 4307–4312.
16. Mishra V, Cherkauer KA, Niyogi D, Lei M, Pijanowski BC, et al. (2010) A regional scale assessment of land use/land cover and climatic changes on water and energy cycle in the upper Midwest United States. Int. J. of Climatology 30(13): 2025–2044.
17. Vanloocke A, Bernacchi CJ. Twine TE (2010) The impacts of Miscanthus× giganteus production on the Midwest US hydrologic cycle. GCB Bioenergy, 2, 180–191. Skamarock WC, Klemp JB (2008) A time-split nonhydrostatic atmospheric model for weather research and forecasting applications. Journal of Computational Physics 227: 3465–3485.
18. Pielke Sr RA (2005) Land use and climate change. Science 310: 1625–1626.
19. Anderson CJ, Anex RP, Arritt RW, Gelder BK, Khanal S, et al. (2013) Regional climate impact of a biofuels policy projection. Geophys Res Lett 40: 1217–1222. doi:10.1002/grl.50179.
20. Le PVV, Kumar P, Drewry DT (2011) Implications for the hydrologic cycle under climate change due to the expansion of bioenergy crops in the Midwestern United States. PNAS 108: 15085–15090.
21. Stone K, Hunt P, Cantrell K, Ro K (2010) The potential impacts of biomass feedstock production on water resource availability. Bioresource Technology 101: 2014–2025.
22. Chiew F, Teng J, Vaze J, Kirono D (2009) Influence of global climate model selection on runoff impact assessment. Journal of Hydrology 379: 172–180.
23. Zheng H, Zhang L, Zhu R, Liu C, Sato Y, et al. (2009) Responses of streamflow to climate and land surface change in the headwaters of the Yellow River Basin. Water Resources Research 45 W00A19.
24. Anderson CJ (2012) Local adaption to changing flood vulnerability in the Midwest. Climate change in the Midwest; Pryor, S.C., Eds.; Indiana University Press, ISBN: 978-0-253-00682-0.
25. Carlisle DM, Wolock DM, Meador MR (2011) Alteration of streamflow magnitudes and potential ecological consequences: a multiregional assessment. Front Ecol Environ 9(5): 264–270.
26. Vogel RM, Wilson I, Daly C (1999) Regional regression models of annual streamflow for the United States. Journal of Irrigation and Drainage Engineering 125: 148–157.
27. Sankarasubramanian A, Vogel RM, Limbrunner JF (2001) Climate elasticity of streamflow in the United States. Water Resources Research 37: 1771–1781.
28. Skamarock W, Klemp JB (2008) A time-split nonhydrostatic atmospheric model for weather research and forecasting applications. J. of Computational Physics 227(7): 3465–3485.

29. Skamarock W, Klemp JB, Dudhia J, Gill DO, Barker DM, et al. (2008) A description of the Advanced Research WRF version 3. NCAR Technical Note NCAR/TN-475 STR.
30. Ek M, Mitchell K, Lin Y, Rogers E, Grunmann P, et al. (2003) Implementation of NOAH land surface model advances in the National Centers for Environmental Prediction operational mesoscale Eta model. J Geophys Res 108: doi:10.1029/2002JD003296.
31. Steiner AL, Pal JS, Rauscher SA, Bell JL, Diffenbaugh NS, et al. (2009) Land surface coupling in regional climate simulations of the West African monsoon, Climate Dynamics, 33(6):869–892.
32. Ela R, Caya D, Côté H, Frigon A, Biner S, et al. (2008) Evaluation of uncertainties in the CRCM-simulated North American climate. Climate Dynamics 30: 113–132.
33. Ugarte DG De La Torre, Ray DE (2000) Biomass and bioenergy applications of the POLYSYS modeling framework. Biomass Bioenergy 18: 291–308.
34. Ray DE, Moriak TF (1976) POLYSIM: A National Agricultural Policy Simulator. Agricultural Sector Models for the United States 28: 14–21.
35. Dicks MR, Campiche J, Ugarte DDLT, Hellwinckel C, Bryant HL, et al. (2009) Land use implications of expanding biofuel demand. J. of Agric. And App. Econ. 41(2): 435–453.
36. Khanal S, Anex RP, Anderson CJ, Herzmann DE, Jha MK (2013) Implications of biofuel policy-driven land cover change for rainfall erosivity and soil erosion in the United States. GCB Bioenergy: doi:10.1111/gcbb.12050.
37. Miguez F, Maughan M, Bolero GA, Long SL (2012) Modeling spatial and dynamic variation in growth, yield, and yield stability of the bioenergy crops Mixcanthus x giganteus and Panicum virgatum across the conterminous United States, Global Change Biology-Bioenergy doi:10.1111/j.1757-1707.2011.01150.x.
38. Mitchell R, Schmer MR (2012) Switchgrass harvest and storage. In A. Monti (ed.) Switchgrass: A valuable biomass crop for energy (Green Energy and Technology), pp. 113–127.
39. Chen F, Dudhia J (2001) Coupling an Advanced Land Surface-Hydrology Model with the Pen State-NCAR MM5 Modeling System. Part I: Model Implementation and Sensitivity. Monthly Weather Review1 29: 569–585.
40. Rosero E, Gulden LE, Yang Z, Goncalves L, Nui G, et al. (2011) Ensemble Evaluation of Hydrologically Enhanced Noah-LSM: Partitioning of the Water Balance in High-Resolution Simulations over the Little Washita River Experimental Watershed. J. Hydrometeor 12: 45–64.
41. Chiew FHS (2006) Estimation of rainfall elasticity of streamflow in Australia. Hydrological Sciences 51: 613–625.
42. Schaake JC (1990) From climate to flow. In: Climate change and US water resources, Waggoner, P.E., Eds.; J. Wiley and Sons, pp. 177–206.
43. Arora VK (2002) The use of the aridity index to assess climate change effect on annual runoff. Journal of Hydrology 265: 164–177.
44. Zhang L, Dawes W, Walker G (2001) Response of mean annual evapotranspiration to vegetation changes at catchment scale. Water Resources Research 37: 701–708.
45. Koster RD, Dirmeyer PA, Guo Z, Bonan G, Chan E, et al. (2004) Regions of strong coupling between soil moisture and precipitation. Science 305: 1138–1140.
46. Zhang L, Dawes W, Walker G (2001) Response of mean annual evapotranspiration to vegetation changes at catchment scale. Water Resources Research 37: 701–708.
47. Pike J (1964) The estimation of annual run-off from meteorological data in a tropical climate. Journal of Hydrology 2: 116–123.
48. Budyko MI (1963) Evaporation under natural conditions, Israel Program for Scientific Translations; [available from the Office of Technical Services, US Dept. of Commerce, Washington].
49. Turc L (1954) The water balance of soils. Relation between precipitation evaporation and flow. Ann Agron. 5: 491–569.
50. Ol'Dekop E (1911) On evaporation from the surface of river basins. Trans. Met. Obs. lurevskogo, Univ. Tartu, 4.
51. Schreiber P (1904) Über die Beziehungen zwischen dem Niederschlag und der Wasserführung der Flüsse in Mitteleuropa. Meteorologische Zeitschrift 21: 441–452.
52. Wu Y, Liu S (2012) Impacts of biofuels production alternatives on water quantity and quality in the Iowa River Basin. Biomass and Bioenergy 36: 182–191.
53. Braun M, Caya D, Frigon A, Slivitzky M (2011) Internal variability of Canadian RCM's hydrological variables at the basin scale in Quebec and Labrador. Journal of Hydrometeorology 13: 443–462.
54. Solomon S, Qin D, Manning M, Chen Z, Marquis M, et al. (2007) Climate Change 2007: The Physical Science Basis, IPCC Fourth Assessment Report, Miller, H.L., Eds.; Cambridge University Press, United Kingdom and New York, NY, USA, pp. 996.
55. Marland G, Pielke RA Sr, Apps M, Avissar R, Betts RA, et al. (2003) The climatic impacts of land surface change and carbon management, and the implications for climate-change mitigation policy. Climate Policy 3: 149–157.
56. Tu J (2009) Combined impact of climate and land use changes on streamflow and water quality in eastern Massachusetts, USA. Journal of Hydrology 379: 268–283.
57. Guo G, Hu Q, Jiang T (2007) Annual and seasonal streamflow responses to climate and land-cover changes in the Poyang Lake Basin China. Journal of Hydrology 355: 106–122.
58. Hu Q, Willson GD, Chen X, Akyuz A (2004) Effects of climate and landcover change on stream discharge in the Ozark highlands, USA. Environ. Model Assess 10: 9–19.

4

Improved Health among People Living with HIV/AIDS Who Received Packages of Proven Preventive Health Interventions, Amhara, Ethiopia

Ciara E. O'Reilly[1]*, Ethel V. Taylor[1,2], Tracy Ayers[1], Ribka Fantu[3], Sisay Alemayehu Abayneh[3], Barbara Marston[4], Yordanos B. Molla[5], Tegene Sewnet[5], Fitsum Abebe[5], Robert M. Hoekstra[1], Robert Quick[1]

1 Division of Foodborne, Waterborne, and Environmental Diseases, Centers for Disease Control and Prevention, Atlanta, Georgia, United States of America, **2** Epidemic Intelligence Service, Centers for Disease Control and Prevention, Atlanta, Georgia, United States of America, **3** Centers for Disease Control and Prevention, Addis Ababa, Ethiopia, **4** Division of Global HIV/AIDS, Centers for Disease Control and Prevention, Atlanta, Georgia, United States of America, **5** Michael Dejene Public Health Consultancy Services, Addis Ababa, Ethiopia

Abstract

In 2009, basic care packages (BCP) containing health products were distributed to HIV-infected persons in Ethiopia who were clients of antiretroviral therapy clinics. To measure health impact, we enrolled clients from an intervention hospital and comparison hospital, and then conducted a baseline survey, and 7 bi-weekly home visits. We enrolled 405 intervention group clients and 344 comparison clients. Intervention clients were more likely than comparison clients to have detectable chlorine in stored water (40% vs. 1%, $p<0.001$), soap (51% vs. 36%, $p<0.001$), and a BCP water container (65% vs. 0%, $p<0.001$) at every home visit. Intervention clients were less likely than comparison clients to report illness (44% vs. 67%, $p<0.001$) or health facility visits for illness (74% vs. 95%, $p<0.001$), and had lower median illness scores (1.0 vs. 3.0, $p<0.05$). Participation in the BCP program appeared to improve reported health outcomes.

Editor: Hong Shang, China Medical University, China

Funding: Funding for the study was provided by the US President's Emergency Plan for AIDS Relief, Global AIDS Program, and Centers for Disease Control and Prevention, Ethiopia. Additional support for technical assistance was provided by the U.S. Agency for International Development through an Inter-Agency Agreement with the U.S. Centers for Disease Control and Prevention. Michael Dejene Public Health Consultancy Services was formally contracted by the Centers for Disease Control and Prevention, Ethiopia to carry out the data collection component of this study only. Co-authors YM, TS, and FA were employed by Michael Dejene Public Health Consultancy Services, Addis Ababa, Ethiopia.

Competing Interests: Co-authors YM, TS and FA are employed by Michael Dejene Public Health Consultancy Services, Addis Ababa, Ethiopia.

* Email: bwf1@cdc.gov

Introduction

Diarrhea and other opportunistic infections are important causes of morbidity and mortality among people living with HIV/AIDS (PLHIV) [1–5]. Inexpensive interventions to prevent opportunistic infections in resource-poor settings can improve clinical outcomes in PLHIV [5–11] and may help reduce viral loads and maintain CD4+ cell counts (CD4 counts) [12–15]. Consequently, the US President's Emergency Plan for AIDS Relief (PEPFAR) now authorizes use of program funds to procure and distribute proven health interventions (e.g., insecticide treated bednets, condoms, household water treatment products, soap) as part of HIV care and support programs [16].

Among the interventions shown to reduce the risk of diarrhea in PLHIV is household water treatment using a locally produced sodium hypochlorite solution and safe water storage using a narrow-mouth container with tap [5,7]. In Ethiopia, two water treatment products for home chlorination are available: sodium hypochlorite solution with the brand name Wuha Agar, which is marketed at a price of USD 0.24 for a bottle that treats 1,000 liters of water, and a flocculent-disinfectant product with the brand name PuR that clarifies and disinfects water [17], and is available in Ethiopia at a price of USD 0.06 for a sachet that treats 10 liters.

In June 2009, with PEPFAR funding provided though the United States Agency for International Development (USAID) program, the Ministry of Health of Ethiopia and Population Services International (PSI) began a program to distribute basic preventive care packages (BCPs) [18,19] to HIV-infected persons receiving antiretroviral therapy (ART). The BCPs were distributed through a phased roll-out to clinics in five regions of Ethiopia: Tigray, Amhara, Oromia, Southern Nations, Nationalities and Peoples (SNNP), and Addis Ababa. Each BCP included: 6 bottles of Wuha Agar (enough to treat 20 liters of water/day for 6 months), 24 sachets of PuR (enough to treat 10 liters of water/day for 24 days), a 20 L jerrican with a lid and tap for safe water storage, 100 condoms, 4 bars of soap, 4 doses of albendazole (enough for one dose per quarter), and a low literacy illustrated information booklet in Amharic promoting appropriate water treatment, sanitation and hygiene practices. Cotrimoxazole prophylaxis was provided free to clients at ART clinics when indicated, as per Ethiopian Federal Ministry of Health guidelines

[20]. Distribution of the BCP to clients was accompanied by demonstration of product use by trained hospital staff. All products contained in the BCP, except for the water storage container, were available on the market in Ethiopia before the BCP program began.

During the last quarter of 2009, 75,000 BCPs were distributed, with another 125,000 BCPs distributed in 2010 and 2011. The expense and effort to implement this program warrants an evaluation of impact. From September 2009 through January 2010, we assessed the use, acceptability, and health impact of the BCP on ART clients in Amhara, the region with the highest proportion of patients receiving ART in Ethiopia as a consequence of its large population size, and the high adult HIV prevalence rates (approximately 4.5%: 3.2% in rural and 13.5% in urban settings) [21,22].

Materials and Methods

Study Population and Sample Size Calculations

Using previously published methods [5,7] we conducted a longitudinal evaluation of the BCP program among PLHIV (hereafter referred to as clients) enrolled in ART programs. Because of resource constraints, this evaluation was limited to a convenience sample of two health facilities in Amhara Region, Ethiopia. We selected the health facilities based on similar numbers of clients currently receiving ART, demographic characteristics, estimated risk for diarrheal disease, and, for logistical reasons, ease of access by road. The Amhara Regional Health Bureau designated Gonder University Hospital, a large teaching hospital under the administration of the Department of Education that provides treatment for 3,175 ART clients, as the intervention site and Debre Markos Zonal Hospital, a large regional hospital under the auspices of the Ethiopian Ministry of Health that provides treatment for 2,056 ART clients, as the comparison site. The ART programs in both of these facilities were supported by International Training & Education Center for Health (ITEC, Seattle, Washington). Even though Gonder Hospital is an academic hospital, the HIV clinics at both hospitals are run by ITEC as the PEPFAR contracted partner organization. The two sites were similar in terms of clinical oversight, located in a town, access to the health facility (all participants lived within 10 km of the health facility), geography, climate, and ethnic group representation (>95% of participants at both sites were of Amhara ethnicity). Both communities are at altitudes >2,000 m (i.e. not at high risk of malaria transmission [Gonder 2,133 m; Debre Markos 2,446 m]) and have similar climates.

A sample size of approximately 400 persons each in the intervention and comparison groups was estimated to detect a 10% reduction in illnesses in the intervention group relative to the comparison group, with 80% power, 95% confidence, and a 25% drop out rate. We limited participation to one client per household.

Adult clients (≥18 years old) enrolled in the HIV care and treatment program and receiving ART treatment who lived within a 10 km radius of the selected health facilities were eligible for inclusion. Because records lacked sufficient information to locate patients and daily patient load was relatively low, random selection of participants was not feasible. Clients who consented to participate were consecutively enrolled during clinic visits until the target sample size was reached or the eligible client population was exhausted.

Baseline Survey

Field workers administered a baseline questionnaire in Amharic from September 24–October 23, 2009 to participants at both health facilities to gather information about demographic and socioeconomic characteristics; water, sanitation and hygiene knowledge and practices; and utilization rates of BCP components. Clinic nurses at both health facilities also abstracted medical information from clients' hospital charts including the most recently recorded CD4 count, functional status, prescription of cotrimoxazole prophylaxis, and duration of ART. Intervention group clients received the BCP after completion of the baseline questionnaire. Fieldworkers accompanied clients to their homes after the baseline interview to make observations on water treatment, storage, and hygiene and sanitation facilities, and to test stored drinking water for free residual chlorine using the N, N-diethyl-p-phenylenediamine (DPD) colorimetric method (LaMotte Co., Baltimore, MD) as an objective measure of water treatment.

Biweekly Home Visits

Following the baseline survey, field teams began biweekly home visits to study participants from both groups on October 8, 2010 to ask about water, sanitation and hygiene practices; ongoing use of BCP components and cotrimoxazole prophylaxis; illnesses in the 48 hours preceding the visit; health facility visits or hospitalizations in the previous 14 days; and perceived BCP-associated stigma among clients in the intervention group. Fieldworkers tested stored drinking water for free residual chlorine and made observations of BCP components and cotrimoxazole tablets in the home. Deaths among study participants were also recorded. Biweekly home visits continued for seven rounds. Comparison group clients received the BCP at the end of the study, in January 2010.

Definitions and Statistical Analysis

We defined diarrhea as ≥3 loose or watery stools in a 24 hour period; respiratory infection as reported fever and ≥2 of the following signs and symptoms: cough, difficulty breathing, or respiratory rate >20 breaths per minute; and febrile illness as reported subjective fever without concurrent signs and symptoms of diarrhea or respiratory illness. Additionally, we calculated a total illness score for each client that summed all illnesses reported for the previous 48 hours at each home visit and all reported health facility visits or hospitalizations for any illness. Illnesses spanning multiple home visits were counted each time they were reported.

Data were double-entered in Ethiopia into EPI Info (CDC, Atlanta, GA, USA). Data analysis was performed using SAS version 9.2 (SAS Institute, Cary, NC, USA). Product use and illness data between groups were compared using Chi-Square tests for categorical variables. We performed categorical data analysis on continuous variables based on natural breakpoints within the data and compared differences between the two groups using Chi-Square tests. The total illness score was weighted to account for severity of illness, with self-reported health facility visits for illness or hospitalizations contributing double the value of self-reported acute illnesses in the previous 48 hours for which clients did not seek care. Additional weighting factors for health facility visits and hospitalizations were also evaluated, with no impact on outcomes, therefore the most conservative weighting factors were chosen. Health facility visits for HIV follow-up care (e.g., ART refills, blood testing for CD4 counts) were not included in the total illness score. Because total illness scores were not normally distributed, we used the Wilcoxon rank-sum test to test for statistical significance. Estimation of correlation structures within individuals over time and across all measured variables was not appropriate in this study due to the lack of stability in such estimates that can arise from the relatively small sample size, weak correlation, and highly skewed distributions. In collapsing the data over time, we lost power, but gained simplicity and robustness. We choose to present a common analysis for all variables that facilitates comparisons, although for some variables the data were sufficient to incorporate correlation structures in the analysis; the scientific findings were not altered by this approach.

Ethical Review

The study protocol was approved by an Institutional Review Board of the Centers for Disease Control and Prevention, Atlanta, GA, USA (Protocol Number 5543), and the Ethiopian Public Health Association (Protocol Number IERC 0030). Written informed consent was obtained at the beginning of the study from all participants.

Results

We enrolled 405 clients in the intervention group and 344 clients in the comparison group. Overall, 26 (0.3%) clients refused to participate. We excluded from analysis 29 (7%) intervention group clients and 36 (10%) comparison group clients who did not complete all rounds of biweekly surveillance, including 8 who died (3 in the intervention and 5 in the comparison group). The reported cause of death as per the medical chart in the intervention group clients was suicide (1), pneumonia (1), unknown cause (1), and in the control group was car accident (1), chronic liver disease and pulmonary tuberculosis (1), diarrhea/vomiting (2), pulmonary tuberculosis and pneumonia (1).

Baseline Survey

The mean age of study participants was 36 years (range 19–69 years); 75% were female (Table 1). Over a third of participants had no formal schooling, and more than 65% earned no income. Nearly 70% of clients lived in a rented home and over 80% lived in households with 5 or fewer members. There were no significant differences in demographic characteristics between intervention and comparison group clients.

At baseline, more than 95% of clients in both groups were assessed by ART clinical staff as able to work (Table 2). At the time of enrollment, 81% of clients in the intervention group and 77% in the comparison group had been on ART for >12 months. Based on abstracted medical records, >99% of clients in both groups had good adherence to ART (<5% of doses missed). Most recently recorded median CD4 counts were similar at baseline for both intervention and comparison group clients (284 and 274 cells/μL, respectively). The range of CD4 counts in the intervention and control groups respectively were 26% vs. 33% <200 cells/μL, 38% vs. 35% 200–<350 cells/μL, 20% vs. 19% 350–>500 cells/μL, and 17% vs. 12% ≥500 cells/μL. A higher percentage of clients in the comparison group than in the intervention group were receiving cotrimoxazole prophylaxis (79% vs. 68%, $p<0.05$) at baseline.

The main source of drinking water was either public or private piped water for 98% of intervention group participants and 88% of comparison group participants. All households stored drinking water in the home. At baseline, 48% of clients reported that they did not treat their drinking water. At baseline, before receiving the BCP, 29% of intervention group clients reported currently treating their drinking water with Wuha Agar compared to 13% of comparison group clients ($p<0.05$); 20% of stored water samples in the intervention group and 8% in the comparison group had detectable free chlorine residuals ($p<0.05$); baseline differences were likely a result of intermittent distributions of products in the community by non-governmental organizations.

At baseline, similar percentages of clients in the intervention and comparison groups reported having soap in the home (76% vs. 78%). A significantly higher percentage of clients in the intervention than in the comparison group reported presence of PuR (21% vs. 0%, $p<0.05$), condoms (24% vs. 12%, $p<0.05$), and albendazole tablets (16% vs. 2%, $p<0.05$) in the home, while significantly fewer intervention group clients than comparison group clients reported having Wuha Agar in the home (24% vs. 42%, $p<0.05$).

Biweekly Home Visits

Over seven rounds of home visits (16 weeks), a significantly higher percentage of intervention group clients than comparison clients were observed to have Wuha Agar (91% vs 19%, $p<0.001$), PuR (73% vs 0%, $p<0.001$), condoms (60% vs 3%, $p<0.001$), soap (51% vs 36%, $p<0.001$), albendazole (69% vs 0%, $p>0.001$), and a BCP water storage container (65% vs 0%, $p<0.001$) in the home at all visits (Table 3). A similar percentage of intervention and comparison group clients had cotrimoxazole tablets in the home at all visits (53% vs 57%).

During home visits, the percent of intervention group clients reporting water treatment with Wuha Agar ranged from 71% to 82% and the percentage with free chlorine residuals detected in drinking water ranged from 67% to 78% (Figure 1). In the comparison group, neither reported Wuha Agar use, nor observed free chlorine residuals exceeded 10% at any home visit. Observed PuR in the home among the intervention group ranged from 83% to 90% during the home visits, with a median count of 24 sachets (range 3, 37) at all 7 home visits. Over 7 rounds of home visits, a significantly higher percentage of intervention group clients than comparison clients reported treating their water with Wuha Agar (45% vs 2%, $p<0.001$) and were observed to have free residual chlorine in stored water at all visits (40% vs 1%, $p<0.001$) (Table 3).

A similar percentage of intervention and comparison group clients reported during 50% or more of home visits that they had purchased soap in the preceding 2 weeks (Table 3). More comparison than intervention clients reported during at least 50% of home visits that they had purchased cotrimoxazole (18% vs 4%, $p<0.001$) or received free cotrimoxazole (13% vs 2%, $p<0.001$) in the preceding 2 weeks.

During the study, 75% of intervention group clients reported they were concerned that their friends or neighbors might learn they had received the BCP, which could signal their HIV status. However, only 8% reported feeling discriminated against because of the BCP, and 67% told friends or family about components of the BCP that they had received.

Intervention group clients were less likely than comparison clients to report illness from any cause (44% vs 67%, $p<0.05$) or at least one episode of febrile illness (28% vs. 38%, $p<0.05$) in the 48 hours preceding home visits (Table 4). Over the 14 days preceding home visits, intervention clients were less likely than comparison clients to report health facility visits for any illness (74% vs. 95%, $p<0.05$), respiratory illness (5% vs. 10%, $p<0.05$), or fever (9% vs. 15%, $p<0.05$). Intervention group clients reported fewer hospitalizations for any cause (1% vs. 3%, $p=0.12$) and health facility visits for diarrhea (4% vs. 7%, $p=0.11$) than comparison clients, but the differences were not statistically significant. The median illness score was 1.0 (range 0, 22) for intervention group clients and 3.0 (range 0, 33) for comparison group clients ($p<0.05$, Table 5).

Table 1. Baseline demographic characteristics of study participants in intervention and comparison groups, basic care package evaluation, Gonder and Debre Markos, Ethiopia, October 2009.

Characteristic		Intervention Group (n = 376)	Comparison Group (n = 308)	p-value	Overall (n = 684)
Mean age ± SD[a]		35.6±9.5	37.0±10.0	0.05	36.2±9.7
Age group, n (%)[b]	18–25 years	49 (12.1)	43 (12.5)	0.7	92 (12.3)
	26–35 years	193 (47.7)	145 (42.2)	0.3	338 (45.1)
	36–45 years	101 (24.9)	88 (25.6)	0.6	189 (25.2)
	>45 years	62 (15.3)	68 (19.8)	0.06	130 (17.4)
Female, n (%)[b]		278 (73.9)	234 (75.9)	0.6	512 (74.9)
Ethnicity, n (%)	Amhara[b]	360 (95.7)	302 (98.1)	0.1	662 (96.8)
	Tigray[c]	10 (1.5)	2 (0.7)	0.08	12 (1.8)
	Oromo[c]	2 (0.5)	3 (0.8)	0.7	5 (0.7)
No formal schooling, n (%)[b]		121 (32.2)	109 (35.4)	0.4	230 (33.6)
Do not earn income, n (%)[b]		254 (67.6)	200 (64.9)	0.5	454 (66.4)
Living in rented house, n (%)[b]		260 (69.2)	212 (68.8)	0.9	472 (69.0)
Number of persons living in household in past six months, n (%)[b]	1 to 5	298 (79.3)	269 (87.3)	<0.05	567 (82.9)
	>5	78 (20.7)	39 (12.7)	<0.05	117 (17.1)

Comparison between intervention and comparison groups using Kruskal-Wallis test[a], Chi-Square test[b], or Fisher exact test[c].

Table 2. Clinical characteristics and treatment of ART clients at baseline, by study group, basic care package evaluation, Gonder and Debre Markos, Ethiopia, October 2009.

Characteristic		Intervention Group (n = 376)	Comparison Group (n = 308)
Functional status, n (%)			
	Ambulatory	11 (3.0)	13 (4.2)
	Working	361(97.0)	295 (95.8)
CD4 count in cells/μL, median (range)		284 (3–996)	274 (31–1088)
Time on ART treatment in months, median (range)		28.5 (0.1–67.3)	24.6 (0.2–73.2)
>12 months on ART treatment, n (%)		304 (81.1)	238 (77.3)
Adherence to ART drugs[a], n (%)			
	Good	372 (99.2)	306 (99.4)
	Fair	3 (0.8)	1 (0.3)
	Poor	0	1 (0.3)
Currently on cotrimoxazole[b], n (%)		256 (68.3)	243 (78.9)
Adherence to cotrimoxazole[a], n (%)			
	Good	253 (98.8)	240 (98.8)
	Fair	1 (0.3)	2 (0.6)
	Poor	3 (0.8)	1 (0.3)
Client reported following conditions at visit, n (%)			
	Diarrhea	2 (0.5)	0
	Coughing[a]	2 (0.5)	6 (2.0)
	Tuberculosis positive[c]	47 (22.6)	27 (6.5)

[a]Self reported adherence obtained from medical records defined as follows: good (<5% of doses missed); fair (5–15% of doses missed); poor (>15% of doses missed).
[b]Significant difference observed between intervention and comparison groups at $p<0.05$ using Chi-square test.
[c]Tuberculosis diagnosis and staging differed between the health care facilities; for this reason no further analysis of tuberculosis was conducted.

Table 3. Percentage of clients observed to have interventions in the home; reported and observed water treatment behaviors; and reported receipt of interventions, by intervention and comparison group, over 16 weeks of basic care package evaluation, Gonder and Debre Markos, Ethiopia, October 2009–January 2010.

Characteristic	Intervention Group (n = 376)	Comparison Group (n = 308)	p-value
Interventions observed in home at every visit			
Wuha Agar	341 (90.7)	59 (19.2)	<0.001[a]
PuR	276 (73.4)	0 (0.0)	<0.001[a]
Condoms	224 (59.6)	10 (3.3)	<0.001[a]
Soap	192 (51.0)	111 (36.0)	<0.001[a]
Cotrimoxazole	198 (52.7)	174 (56.5)	0.35
Albendazole	258 (68.6)	0 (0.0)	<0.001[a]
BCP container	243 (64.6)	0 (0.0)	<0.001[a]
Water treatment			
Reported treating water at every visit	177 (47.1)	5 (1.6)	<0.00[a]
Reported treating water with Wuha Agar at every visit	170 (45.2)	5 (1.6)	<0.001[a]
Water tested positive for free chlorine residual at every visit	152 (40.4)	3 (1.0)	<0.001[a]
Reported buying water treatment products at ≥5% of visits	5 (1.4)	2 (5.0)	0.15
Received free Wuha Agar at ≥5% of visits	19 (5.1)	51 (16.6)	<0.001[a]
Reports of interventions purchased or received			
Bought soap at ≥50% of visits	316 (84.0)	267 (86.7)	0.38
Bought soap at every visit	149 (39.6)	117 (38.0)	0.69
Received free soap at ≥10% of visits	26 (6.9)	35 (11.4)	0.04
Bought cotrimoxazole at ≥10% of visits	13 (3.5)	55 (17.9)	<0.001[a]
Received free cotrimoxazole at ≥50% of visits	6 (1.6)	39 (12.7)	<0.001[a]
Received free condoms at ≥15% of visits	3 (0.8)	16 (5.2)	<0.001[a]
Bought condoms at ≥10% of visits	20 (5.3)	12 (3.9)	0.47

[a]Remains significant at <0.001 after controlling for False Discover Rate [30].

Discussion

Findings of this study, which was the first to attempt to ascertain the health impact of a program to provide BCPs to PLHIV, suggested that recipients of the BCP were likely to have a lower illness score derived from self-reported illnesses, health facility visits, and hospitalizations than PLHIV who did not receive the BCP. Rates of reported diarrheal illness, and health facility visits or hospitalizations for diarrhea among intervention clients were also lower than in comparison clients, but disease rates were too low to provide the power to conclude that the differences were statistically significant. Nevertheless, the results are consistent with evidence from five high-quality studies, including four randomized controlled trials and one systematic review and meta-analysis which support the beneficial impact of interventions to improve household drinking water quality on the health of PLHIV and their families [5,23–26].

The differences between the two groups in reported illness and health facility visit rates occurred following documentation of similar functional status, CD4 cell counts, duration on ART, adherence to ART, and adherence to cotrimoxazole prophylactic treatment among intervention and comparison clients at baseline, and in spite of the finding that a lower percentage of intervention clients than comparison clients were on cotrimoxazole prophylaxis at baseline. Furthermore, a higher percentage of clients in the comparison than the intervention group reported during home visits that they had received free Wuha Agar, soap, condoms, and cotrimoxazole from other health facility-based programs and had bought cotrimoxazole, all of which would likely diminish the differences in health outcomes between the two groups. Finally, these findings were observed despite relatively low rates of illness, which were likely a result of ART and cotrimoxazole prophylaxis [5,6], and suggested that, even in a setting of low illness rates, the BCP may provide incremental clinical benefits to patients receiving these treatments. An alternative interpretation is that differences between the two communities may have accounted for the findings. Health findings were congruent with reported and confirmed use of BCP components documented in this study. Specifically, intervention clients were more likely than comparison clients to be observed with BCP components in their homes, including Wuha Agar, PuR, BCP water containers, condoms, soap, and albendazole. Similarly, clients who received the BCP were significantly more likely than comparison clients to report treating their stored drinking water with Wuha Agar and have free chlorine residuals in stored drinking water. By removing the economic barriers to access to these products through free distribution, [27–29] the program created an opportunity for participants to use them, and participants took advantage of this opportunity, as has been observed in at least one other evaluation [17]. Without free distribution, it is likely that use of different BCP components would have been lower in this population because two-thirds earned no income.

Economic barriers to sustained use of the BCP components could diminish or eliminate health gains observed in this study

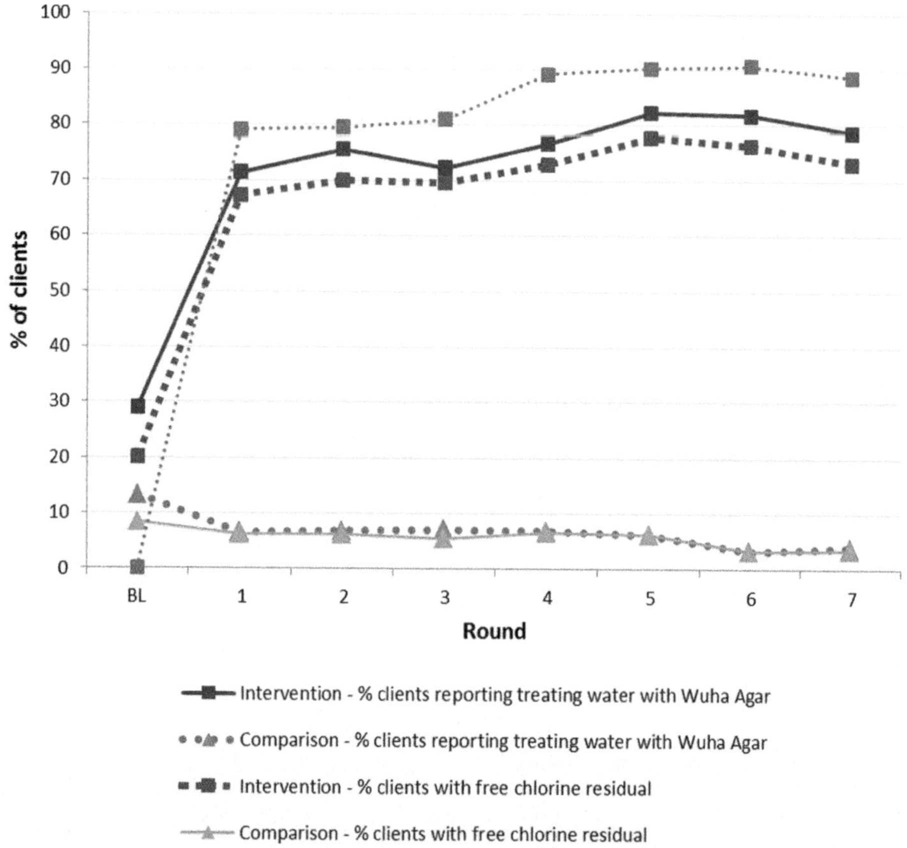

Figure 1. Percent of clients with reported water treatment with Wuha Agar, free chlorine residual in stored water, and observed BCP container, at baseline (before receiving the BCP) and by home visit round, by intervention and comparison group, basic care package evaluation, Gonder and Debre Markos, Ethiopia, October–January, 2010.

because only one BCP per client is provided and clients may not be able or willing to replenish consumable commodities in the BCP such as Wuha Agar, condoms, and albendazole. Research in other settings has shown that repeated purchase of inexpensive commodities in low income populations often does not take place, [27–29] even following a free trial [27,30,31]. To our knowledge, only one other evaluation, which took place in Uganda, has examined use of BCP components among PLHIV, and this study was conducted before depletion of the original supply of intervention components [18]. An evaluation of use of BCP components over time in both the Ethiopian and Ugandan programs is warranted and could help guide programming. In addition, an assessment of the costs and benefits of a program of free resupply of BCP commodities would be useful for guiding policy, [32,33] particularly in view of evidence that integration of free health interventions into health services can serve as an incentive to increase use of services [34].

Some of the components of the BCP distributed in Ethiopia may be unnecessary. For example, soap was observed in a similar percentage of homes in both evaluation groups throughout the evaluation, and similar percentages of clients reported purchasing soap, suggesting that the BCP did not likely change access to or use of soap. In contrast, despite the baseline observation that a higher percentage of intervention than comparison households had PuR in the home, possibly as a result of previous distribution efforts by local organizations, use of PuR in this evaluation appeared to be low, as suggested by the observation that the median number of PuR sachets in intervention client homes remained at 24, the total number provided in the BCP, throughout the 7 rounds of home visits. One possible explanation for this observation is that most intervention clients had access to piped water, which tended to be clear and not in need of flocculation, and access to Wuha Agar for in-home chlorination. By eliminating items from the BCP that are already present or infrequently used in most households, resources could be more efficiently directed toward alternative interventions.

The finding that about half of clients in both groups had cotrimoxazole in the home at every visit, and that relatively few patients reported buying or receiving it for free, was of some concern. Prophylactic treatment with cotrimoxazole has been shown to reduce morbidity and mortality from opportunistic infections among PLHIV [6], including those on ART and has been recommended for inclusion in BCPs [35]. High adherence to recommended prophylactic treatment has been achieved in other programs [6,11] and assuring regular access and use of the drug should be a priority.

This study had several important limitations. First, BCPs were not randomly distributed because that would have caused potentially disruptive, inequitable treatment of clients, logistical complications for nursing staff, and ethical concerns. We therefore chose to compare health outcomes in the context of a phased BCP distribution to intervention and comparison sites that were chosen on the basis of similar population characteristics. The resultant quasi-experimental design with just two study sites was subject to potential confounding factors, such as differences between the two study sites in geography; climate; demographic and socio-economic status; household water management habits; and health

Table 4. Percent of clients self-reporting one or more episodes of illness, health facility visit, and hospitalization by intervention and comparison group, over 16 weeks of basic care package evaluation, Gonder and Debre Markos, Ethiopia, October 2009–January 2010.

Characteristic[a]	Intervention Group (n = 376)	Comparison Group (n = 308)	p-value
Illness in previous 48 hours	165 (43.9)	206 (66.9)	<0.001[c]
Diarrhea in previous 48 hours	37 (9.8)	39 (12.7)	0.27
Respiratory illness in previous 48 hours	53 (14.1)	50 (16.2)	0.55
Fever in previous 48 hours	105 (27.9)	117 (38.0)	<0.005[c]
Health facility visits in previous 14 days	277 (73.7)	292 (94.8)	<0.001[c]
Health facility visit for diarrhea in previous 14 days	14 (3.7)	20 (6.5)	0.11
Health facility visit for respiratory illness in previous 14 days	20 (5.3)	32 (10.4)	0.01[c]
Health facility visit for malaria in previous 14 days	5 (1.3)	3 (0.1)	0.74
Health facility visit for fever in previous 14 days	35 (9.3)	45 (14.6)	0.04
Health facility visit for STI in previous 14 days	3 (0.8)	1 (0.3)	–
Health facility visit for HIV care in previous 14 days[b]	30 (8.0)	67 (21.8)	<0.001[c]
Health facility visit for other illness in previous 14 days	93 (24.7)	111 (36.0)	0.001[c]
Hospitalization in previous 14 days	5 (1.3)	10 (3.3)	0.12
Hospitalization for diarrhea in previous 14 days	0 (0)	2 (0.7)	0.2
Hospitalization for respiratory illness in previous 14 days	0 (0)	1 (0.3)	0.45
Hospitalization for malaria in previous 14 days	0 (0)	0 (0)	–
Hospitalization for fever in previous 14 days	0 (0)	2 (0.7)	0.2
Hospitalization for STI in previous 14 days	0 (0)	0 (0)	–
Hospitalization for HIV care in previous 14 days[b]	0 (0)	2 (0.7)	0.2
Hospitalization for other illness in previous 14 days	5 (1.3)	10 (3.3)	0.12

[a]Illnesses spanning multiple home visits were counted each time they were reported over 16 weeks of basic care package evaluation.
[b]Health facility visits and hospitalizations for HIV care, including changes in ART medications, CD4 counts, etc. do not contribute towards the illness score.
[c]Remains significant at <0.05 after controlling for False Discover Rate [38].

care delivery practices. Thus, we cannot rule out that the findings observed could have been due to differences between the two communities. However, while health care delivery and water treatment practices differed at baseline between the two groups, the large magnitude of difference observed between the two populations during the study would have been highly unlikely without the relatively intensive intervention that took place. Second, because of logistical and financial considerations, we were limited to two sites in one region of Ethiopia, so the evaluation population was neither representative of Amhara state nor the country as a whole and, therefore, different results might be observed in other populations. Donors providing major investments into interventions such as those described here should budget adequately for an evaluation component to avoid similar situations. Third, because all clients enrolled in the study lived within a 10 km radius of the town, information about clients living in more remote settings was unavailable. Fourth, biweekly home visits may have influenced responses to questions and observable

Table 5. Total number of illnesses, health facility visits, and hospitalizations reported over 16 weeks of basic care package evaluation, and illness scores derived from these events, by intervention and comparison group, Gonder and Debre Markos, Ethiopia, October 2009–January 2010.

Characteristic	Intervention Group (n = 2,632)	Comparison Group (n = 2,156)	Weighting factor	Illness Score for Intervention Group	Illness Score for Comparison Group
Illness in previous 48 hours	653	946	1	653	946
Health facility visits in previous 14 days	220	310	2	440	620
Hospitalization in previous 14 days	5	15	2	10	30
Total illness score				1,103	1,596
Median illness score per patient (range)				1.0 (0–22)	3.0 (0–33)

household practices, resulting in a Hawthorne effect [36], but this effect would likely have impacted both the intervention and comparison groups. Fifth, because the evaluation was designed to evaluate the BCP as a whole we had limited ability to tease out the effect of specific interventions in the package on specific diseases, and could not elucidate which interventions in the package drove the findings. Finally, because of resource limitations, active surveillance was limited to 14 weeks; therefore we were not able to follow clients until consumable BCP components were exhausted, limiting our ability to determine whether their use was sustained over time. In addition, the short duration of the study made it difficult to assess whether illness patterns varied seasonally and over time, or whether there was attrition in the health impact of water treatment interventions over time, which has been observed elsewhere [37].

Because of these limitations, the findings of this evaluation must be interpreted with caution. However, results of this study suggest that, taken as a whole, the BCP is a promising method of increasing access to inexpensive interventions that can reduce the risk of disease and improve the quality of life of PLHIV. This evaluation also highlighted the importance of evaluating BCP programs in order to assure maximum health impact and optimal use of scarce resources.

Acknowledgments

We would like to thank the management and staff of Gonder University Hospital, and Debre Markos Zonal Hospital, Amhara, Ethiopia, the International Training and Education Center for Health (I-TECH), Ethiopia, the Amhara Regional Health Bureau, Ethiopia, the Federal HIV/AIDS Prevention and Control Office, Ethiopia, the United States Agency for International Development (USAID), Ethiopia, and the following individuals for their assistance: Achuyt Bhattarai and Victoria Cuellar, Centers for Disease Control and Prevention, Atlanta, GA, USA; Beniam Feleke, Seymour Williams, Elias Mekonnen, Merid Worku, Centers for Disease Control and Prevention, Addis Ababa, Ethiopia; Michael Dejene, Ermias Mekuria Lewte, and Teshome Mulugeta Kifetew, Michael Dejene Public Health Consultancy Services, Addis Ababa, Ethiopia; Marty Bell, Abdu Dawud, and Meskerem Abera, Population Services International, Addis Ababa, Ethiopia; the enumerators, data entry assistants, and participants in this study. Disclaimer: The findings and conclusions in this report are those of the authors and do not necessarily represent the official position of the U.S. Centers for Disease Control and Prevention.

Author Contributions

Conceived and designed the experiments: COR ET TA BM RH RQ. Analyzed the data: ET COR TA. Contributed reagents/materials/analysis tools: COR ET TA RF SA BM RH RQ. Contributed to the writing of the manuscript: COR ET TA RF SA BM RH RQ. Carried out data collection and management: FA TS YM.

References

1. Brink A, Mahe C, Watera C, Lugada E, Gilks C, et al. (2002) Diarrhoea, CD4 counts and enteric infections in a community-based cohort of HIV-infected adults in Uganda. J Infection 45: 99–106.
2. Colebunders R, Francis H, Mann J, Bila K, Izaley L, et al. (1987) Persistent diarrhea, strongly associated with HIV infection in Kinshasa, Zaire. Am J of Gastroenterol 82: 859–864.
3. Grant A, Djomand G, De Cock K (1997) Natural history and spectrum of disease in adults with HIV/AIDS in Africa. AIDS 11: S43.
4. Kaplan JE, Hu DJ, Holmes KK, Jaffe HW, Masur H, et al. (1996) Preventing opportunistic infections in human immunodeficiency virus-infected persons: implications for the developing world. Am J Trop Med Hyg 55: 1–11.
5. Lule JR, Mermin J, Ekwaru JP, Malamba S, Downing R, et al. (2005) Effect of home-based water chlorination and safe storage on diarrhea among persons with human immunodeficiency virus in Uganda. Am J Trop Med Hyg 73: 926.
6. Mermin J, Lule JR, Ekwaru JP, Malamba S, Downing R, et al. (2004) Effect of co-trimoxazole prophylaxis on morbidity, mortality, CD4-cell count, and viral load in HIV infection in rural Uganda. Lancet 364: 1428–1434.
7. Barzilay EJ, Aghoghovbia TS, Blanton EM, Akinpelumi AA, Coldiron ME, et al. (2005) Diarrhea prevention in people living with HIV/AIDS: an evaluation of a point of use water quality intervention in Lagos, Nigeria. AIDS Care 23: 330–339.
8. World Health Organization (2007) Essential prevention and care interventions for adults and adolescents living with HIV in resouce-limited settings. Geneva: WHO. 1–126.
9. Walker AS, Ford D, Gilks CF, Munderi P, Ssali F, et al. (2010) Daily co-trimoxazole prophylaxis in severely immunosuppressed HIV-infected adults in Africa started on combination antiretroviral therapy: an observational analysis of the DART cohort. Lancet 375: 1278–1286.
10. Hoffmann CJ, Fielding KL, Charalambous S, Innes C, Chaisson RE, et al. (2010) Reducing mortality with cotrimoxazole preventive therapy at initiation of antiretroviral therapy in South Africa. AIDS 24: 1709–1716.
11. Lowrance D, Makombe S, Harries A, Yu J, Aberle-Grasse J, et al. (2007) Lower early mortality rates among patients receiving antiretroviral treatment at clinics offering cotrimoxazole prophylaxis in Malawi. J of Acq Immun Def Synd 46: 56–61.
12. Barnabas RV, Webb EL, Weiss HA, Wasserheit JN (2011) The role of coinfections in HIV epidemic trajectory and positive prevention: a systematic review and meta-analysis. AIDS 25: 1559–1573.
13. Modjarrad K, Chamot E, Vermund SH (2008) Impact of small reductions in plasma HIV RNA levels on the risk of heterosexual transmission and disease progression. AIDS 22: 2179–2185.
14. Modjarrad K, Vermund SH (2010) Effect of treating co-infections on HIV-1 viral load: a systematic review. Lancet Infect Dis 10: 455–463.
15. Modjarrad K, Vermund SH (2011) An addition to the effect of treating co-infections on HIV-1 viral load. Lancet Infect Dis 11: 81.
16. The US President's Emergency Plan for AIDS Relief (2009) Annex: PEPFAR and prevention, care and treatment. In: Five Year Strategy. Washington, DC 1–32.
17. Doocy S, Burnham G (2006) Point-of-use water treatment and diarrhoea reduction in the emergency context: an effectiveness trial in Liberia. Trop Med Int Health 11: 1542–1552.
18. Colindres R, Mermin J, Ezati E, Kambabazi S, Buyungo P, et al. (2008) Utilization of a basic care and prevention package by HIV-infected persons in Uganda. AIDS Care 20: 139–145.
19. The US President's Emergency Plan for AIDS Relief (2009) Ethiopia 2009 Country Operational Plan. Washington, DC 1–801.
20. Federal Ministry of Health, HIV/AIDS Prevention and Control Office (2006) Guideline for Cotrimoxazole Prophylaxis in HIV/AIDS Care and Treatment. Addis Ababa, Ethiopia: Federal Ministry of Health of Ethiopia.
21. Federal Ministry of Health, HIV/AIDS Prevention and Control Office (2006) AIDS in Ethiopia. Addis Ababa, Ethiopia: Federal Ministry of Health of Ethiopia.
22. Federal Ministry of Health, HIV/AIDS Prevention and Control Office (2010) ART update as of October 8, 2009 per site. Addis Ababa, Ethiopia: Federal Ministry of Health of Ethiopia.
23. Abebe LS, Smith JA, Narkiewicz S, Oyanedel-Craver V, Conaway M, et al. (2013) Ceramic water filters impregnated with silver nanoparticles as a point-of-use water-treatment intervention for HIV-positive individuals in Limpopo Province, South Africa: a pilot study of technological performance and human health benefits. J Water Health, doi:10.2166/wh.2013.185.
24. Colford JM, Jr., Saha SR, Wade TJ, Wright CC, Vu M, et al. (2005) A pilot randomized, controlled trial of an in-home drinking water intervention among HIV + persons. J Water Health 3: 173–184.
25. Peletz R, Mahin T, Elliott M, Harris MS, Chan KS, et al. (2013) Water, sanitation, and hygiene interventions to improve health among people living with HIV/AIDS: a systematic review. AIDS 27: 2593–2601.
26. Peletz R, Simunyama M, Sarenje K, Baisley K, Filteau S, et al. (2012) Assessing Water Filtration and Safe Storage in Households with Young Children of HIV-Positive Mothers: A Randomized, Controlled Trial in Zambia. PLoS One 7(10): e46548. doi:10.1371.
27. Stockman LJ, Fischer TK, Deming M, Ngwira B, Bowie C, et al. (2007) Point-of-use water treatment and use among mothers in Malawi. Emerg Infect Diseases 13: 1077–1080.
28. Olembo L, Kaona F, Tuba M, Burnham G (2004) Safe water systems: an evaluation of the Zambia CLORIN program. (Environmental Health at USAID website. Available: http://www.ehproject.org/PDF/Others/Zambia%20Report%20Format.pdf. Accessed 2014 Aug 25).
29. Freeman M, Quick R, Abbott D, Ogutu P, Rheingans R (2009) Increasing equity of access to point-of-use water treatment products through social marketing and entrepreneurship: a case study in western Kenya. J Water Health 7: 527–534.
30. Luby SP, Mendoza C, Keswick BH, Chiller TM, Hoekstra RM (2008) Difficulties in bringing point-of-use water treatment to scale in rural Guatemala. Am J Trop Med Hyg 78: 382–387.

31. Arnold B, Arana B, Mausezahl D, Hubbard A, Colford JM (2009) Evaulation of a pre-existing, 3-year household water treatment and handwashing intervention in rural Guatemala. Int J Epidemiol 38: 1651–1661.
32. Shrestha RK, Marseille E, Kahn JG, Lule JR, Pitter C, et al. (2006) Cost-effectiveness of home-based chlorination and safe water storage in reducing diarrhea among HIV-affected households in rural Uganda. Am J Trop Med and Hyg 74: 884–890.
33. Pitter C, Kahn JG, Marseille E, Lule JR, McFarland DA, et al. (2007) Cost-effectiveness of cotrimoxazole prophylaxis among persons with HIV in Uganda. J Acq Immun Def Synd 44: 336–343.
34. Sheth AN, Russo ET, Menon M, Wannemuehler K, Weinger M, et al. (2010) Impact of the integration of water treatment and handwasing incentives with antenatal services on hygiene practices of pregant women in Malawi. Am J Trop Med Hyg 83: 1315–1321.
35. Mermin J, Bunnell R, Lule JR, Opio A, Gibbons A, et al. (2005) Developing an evidence based, preventive care package for persons with HIV in Africa. Trop Med Int Health 10: 961–970.
36. Fletcher RH, Fletcher SW (2005) Randomized control trials. Clinical epidemiology: the essentials. 4th ed. Baltimore: Lippincott, Williams & Wilkins. 128–136.
37. Arnold BF, Colford JM (2007) Treating water with chlorine at point-of-use to improve water quality and reduce child diarrhea in developing countries: a systematic review and meta-analysis. Am J Trop Med Hyg 76: 354–364.
38. Benjamini Y, Hochbert Y (1995) Controlling the false discovery rate: a practical and powerful approach to multiple testing. J R Stat Soc 57: 289–300.

5

The Role of Water Channel Proteins in Facilitating Recovery of Leaf Hydraulic Conductance from Water Stress in *Populus trichocarpa*

Joan Laur*, Uwe G. Hacke

Department of Renewable Resources, University of Alberta, 442 Earth Sciences Building, Edmonton, Alberta, Canada

Abstract

Gas exchange is constrained by the whole-plant hydraulic conductance (K_{plant}). Leaves account for an important fraction of K_{plant} and may therefore represent a major determinant of plant productivity. Leaf hydraulic conductance (K_{leaf}) decreases with increasing water stress, which is due to xylem embolism in leaf veins and/or the properties of the extra-xylary pathway. Water flow through living tissues is facilitated and regulated by water channel proteins called aquaporins (AQPs). Here we assessed changes in the hydraulic conductance of *Populus trichocarpa* leaves during a dehydration-rewatering episode. While leaves were highly sensitive to drought, K_{leaf} recovered only 2 hours after plants were rewatered. Recovery of K_{leaf} was absent when excised leaves were bench-dried and subsequently xylem-perfused with a solution containing AQP inhibitors. We examined the expression patterns of 12 highly expressed AQP genes during a dehydration-rehydration episode to identify isoforms that may be involved in leaf hydraulic adjustments. Among the AQPs tested, several genes encoding tonoplast intrinsic proteins (TIPs) showed large increases in expression in rehydrated leaves, suggesting that TIPs contribute to reversing drought-induced reductions in K_{leaf}. TIPs were localized in xylem parenchyma, consistent with a role in facilitating water exchange between xylem vessels and adjacent living cells. Dye uptake experiments suggested that reversible embolism formation in minor leaf veins contributed to the observed changes in K_{leaf}.

Editor: Silvia Mazzuca, Università della Calabria, Italy

Funding: U. G. H. acknowledges funding from a Natural Sciences and Engineering Research Council (NSERC) Discovery grant, the Canada Foundation for Innovation and the Canada Research Chair program. The funders had no role in study design, data collection and analysis, decision to publish, or preparation of the manuscript.

Competing Interests: The authors have declared that no competing interests exist.

* Email: laur@ualberta.ca

Introduction

High gas exchange rates can only be sustained when leaves are kept well hydrated. This, in turn, depends on the properties of the xylem pipeline and on the way in which water moves through living cells in roots and leaves [1,2]. Leaf hydraulic conductance is emerging as an important component of whole-plant hydraulic conductance [3–7]. Like in roots and stems, the hydraulic conductance of leaves declines as the water potential becomes more negative. This loss of hydraulic conductance is due to embolism formation in leaf veins [8,9], cell shrinkage [10], collapse of xylem conduits [11], and/or to decline in the permeability of extra-xylary tissues [12]. Compared with stems, leaves [13] and roots [14] are often more vulnerable to hydraulic dysfunction. In some cases, however, the hydraulic conductance of these plant organs may also be able to quickly recover from the effects of drought [3,15].

This recovery of hydraulic function may be facilitated by the activity of aquaporin (AQP) water channels [16–20]. AQPs belong to the major intrinsic protein (MIP) superfamily, a family of protein pores present in the membranes of almost all biological cells to facilitate the diffusion of a wide range of small uncharged solutes. Plant MIPs form a particularly large family of proteins, with 28 members in *Vitis vinifera* [21], ≥30 members in *Arabidopsis thaliana*, *Picea glauca* and *Oryza sativa* [19,22,23], and >50 members in *Populus trichocarpa* [24]. The plant-specific plasma membrane intrinsic proteins (PIPs), with their highly conserved phylogenetic subgroups PIP1 and PIP2, and tonoplast intrinsic proteins (TIPs) show significant water transport activity *in vitro* and *in planta* [25–27]. Regulation of AQPs via transcription, translation, post-translational modifications or trafficking allows plant cells and organs to respond to hydraulic changes in their surrounding environment [28].

In this present study, *Populus trichocarpa* plants were exposed to moderate drought and then rewatered. The objective was to study the recovery of K_{leaf} from water stress at both physiological and molecular levels. We hypothesized that leaves would quickly (i.e., within hours) recover from water stress, and that this would be associated with modulation of AQP activity. To test this hypothesis, we monitored K_{leaf} and Ψ_{leaf} during a dehydration-rehydration episode. We also explored the regulation of 12 leaf-expressed *AQP* isoforms as well as the tissue-specific location of PIP1, PIP2 and TIP2 proteins. Recovery of K_{leaf} was assessed in two ways: (i) intact plants were taken through a drying-rewatering cycle, and (ii) detached leaves were bench-dried and subsequently xylem-perfused with AQPs inhibitors.

Materials and Methods

Plant material and growing conditions

All experiments were carried out with *P. trichocarpa* clone 664042 cuttings (mother tree planted in Lotbinière, Québec, Canada from a 1973 IUFRO progeny collection in Washougal, Oregon, USA). Rooted cuttings were produced and established in a greenhouse at the University of Alberta for 2 months in 3.8 L containers with sunshine mix 4 (Sun Gro Horticulture Canada Ltd.) under semi-controlled conditions (22/20°C day: night cycle, 18/6 h light: dark, watered daily, and fertilized (2g L-1 NPK15-30-15) once a week).

Leaf hydraulic conductance measurements

Leaf hydraulic conductance was measured using the evaporative flux method (29) on six plants per treatment. A filtered (0.2 μm) 20 mM KCl+1 mM $CaCl_2$ solution (subsequently referred to as 'artificial xylem sap', AXS) was used for these measurements. Flow rate through leaves was measured with a balance (model CP 224S, Sartorius, Göttingen, Germany), which logged data every 30 s to a computer. The air was well stirred by a fan as explained by Sack & Scoffoni [29]. Leaves were illuminated with ~1000 μmol m^{-2} s^{-1} photosynthetically active radiation (PAR) at the leaf surface by an LED worklight (Husky, distributed by Home Depot, Atlanta, GA, USA). Leaf temperature was monitored by a thermocouple. Leaf water potential (Ψ_{leaf}) was measured using a pressure chamber (PMS Instruments, Albany, OR, USA). For hydrated leaves, the K_{leaf} was calculated as described previously [29] using the final Ψ_{leaf} (Ψ_{final}), which was determined at the end of the measurement of E, immediately after removing the leaf from the tubing system. The K_{leaf} was normalized by leaf area, which was determined with a scanner.

A leaf vulnerability curve was generated with plants experiencing different levels of water stress following methods of Sack & Scoffoni [29]. K_{leaf} was measured on leaves corresponding to leaf plastochron index (LPI) 9 [30]. The initial Ψ_{leaf} (Ψ_O) was measured using a leaf immediately above or below the leaf that was subsequently connected to the tubing system. Once a leaf was connected to the tubing system, a stable flow rate (E) was usually reached in less than 20 minutes; this value of E was subsequently used for calculating K_{leaf}. Leaves that did not provide stable E within 20 min were discarded. When dehydrated leaves are measured with the EFM, their Ψ_{leaf} may change because the petiole is connected to water at atmospheric pressure [3]. To test for this, Ψ_{final} and Ψ_O were compared. In most cases, Ψ_O and Ψ_{final} were similar and the more negative of these two values was used to calculate K_{leaf}. When Ψ_O and Ψ_{final} differed by more than 0.2 MPa, the leaf was discarded and K_{leaf} was not calculated. The leaf vulnerability curve (shown in Fig. 1) was fitted with a 3 parameter logistic function: $K_{leaf} = a/[1+(\Psi_{leaf}/x_0)^b]$. The curve fit was done using SigmaPlot v. 13 (Systat Software, San Jose, CA, USA). The curve fit was used to calculate the maximum leaf hydraulic conductance as well as the leaf water potentials at 50% and 80% loss of hydraulic conductance.

Recovery of leaf hydraulic conductance after dehydration

To study the recovery of K_{leaf} in intact plants, plants were randomly assigned to different watering regimes in the greenhouse. One group of plants was kept well watered (control). Another group of plants was subjected to a drought treatment. Water was withheld for several days until plants reached a Ψ_{leaf} of -0.77 ± 0.05 MPa (mean ± SE, n = 6). This Ψ_{leaf} was associated with a substantial reduction in K_{leaf}. A subset of drought-stressed plants was then rewatered, and Ψ_{leaf} and K_{leaf} were re-measured 2 h and 26 h after rewatering.

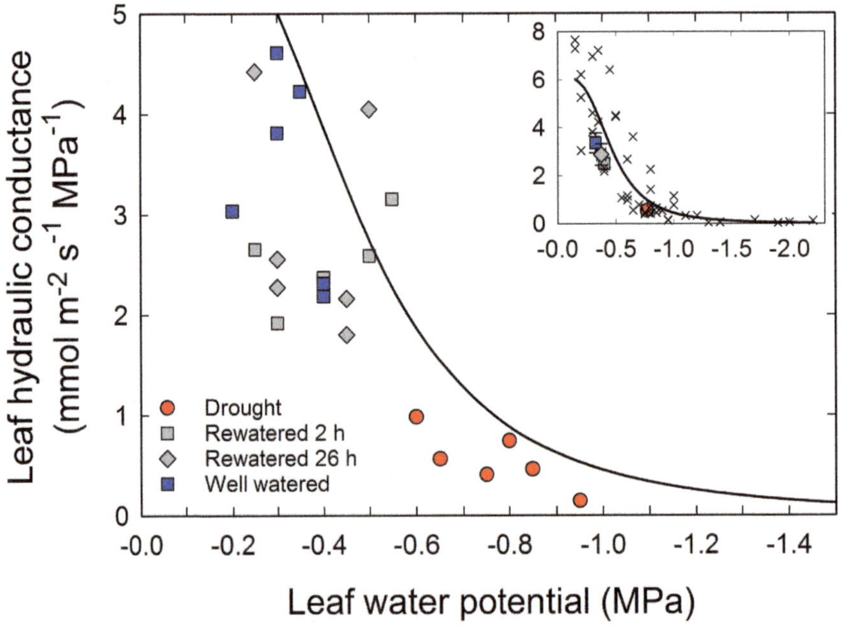

Figure 1. Effect of a change in water availability on leaf hydraulic conductance (K_{leaf}) in *Populus trichocarpa* saplings. K_{leaf} and the associated leaf water potential (Ψ_{leaf}) were measured in 6 well-watered control plants (blue squares), 6 drought-stressed plants (red circles), and drought-stressed plants 2 and 26 h after rewatering (grey squares and diamonds, respectively). Each data point represents a single measurement of K_{leaf}. The solid line shows the previously established vulnerability curve for K_{leaf}. A 3 parameter logistic function was used for the curve fit: $K_{leaf} = 6.154/[1+(\Psi_{leaf}/0.469)^{3.337}]$. An overview of the complete vulnerability curve is shown in the upper right corner of the figure. Individual measurements are shown as crosses; the mean values for each group (±SE, $n = 6$) are shown using the same symbols as explained above.

Table 1. Transcript abundance of 12 aquaporin genes expressed in leaves of well-watered control plants.

Aquaporin name	Expression (copies μg^{-1} of total RNA)
PtPIP1;1	112,960±9,067
PtPIP1;2	272,111±32,575
PtPIP1;3	229,960±44,252
PtPIP2;3	85,667±15,402
PtPIP2;4	273,655±33,728
PtPIP2;5	11,536±1,738
PtTIP1;3	23,105±2,540
PtTIP1;5	11,840±1,675
PtTIP1;6	2,330±121
PtTIP2;1	153,689±19,669
PtTIP2;2	24,863±3,451
PtTIP4;1	517±9

Values are the means ± SE from three biological samples which were tested in triplicate.

To assess the effect of AQP inhibitors and abscisic acid (ABA) on the recovery of K_{leaf}, excised leaves were bench-dried for 1 h (Ψ_{leaf} reached -1.00 ± 0.09 MPa (mean ± SE, n = 6)) and then perfused for 2 h with AXS, AXS+0.2 mM $HgCl_2$, AXS+50 mM H_2O_2 or AXS+50 µM ABA. Solutions were introduced into the transpiring leaf by immersing the petiole inside 50 mL containers. Leaves were placed near a fan; light was provided at a light level of ~1,000 µmol m^{-2} s^{-1} PAR. Mercury chloride and H_2O_2 have

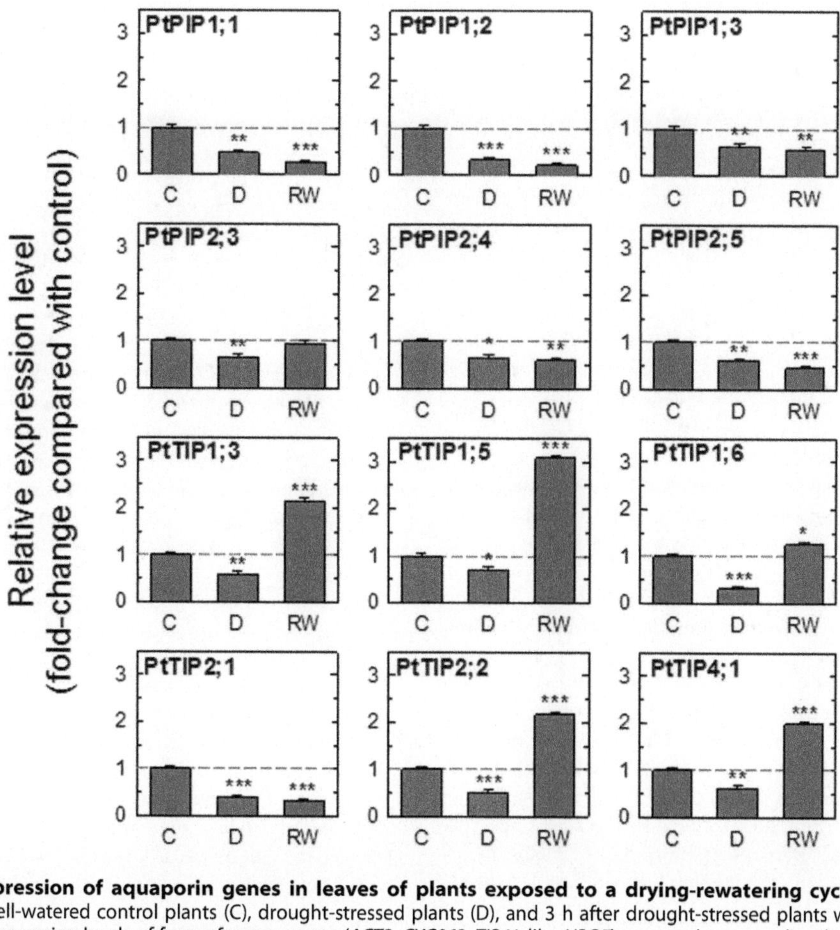

Figure 2. Relative expression of aquaporin genes in leaves of plants exposed to a drying-rewatering cycle. Gene expression was measured in leaves of well-watered control plants (C), drought-stressed plants (D), and 3 h after drought-stressed plants were rewatered (RW). The geometric mean of the expression levels of four reference genes (*ACT2, CYC063, TIP41-like, UBQ7*) was used to normalize the results. Asterisks denote significant differences in expression level compared to control levels (one-way ANOVA, followed by Bonferroni's post test, *$P\leq0.05$; **$P\leq0.01$***$P\leq0.001$). Data are means ± SE of three biological replicates.

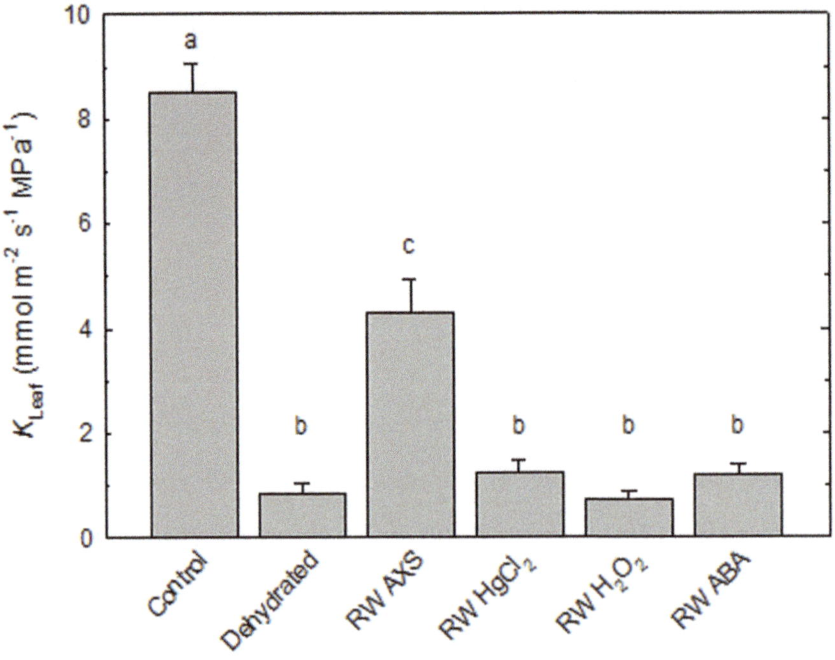

Figure 3. Response of leaf hydraulic conductance (A) and stomatal aperture (B) to different perfusion solutions. Control conditions refer to the K_{leaf} that was measured after leaves were xylem perfused with filtered (0.2 μm) 20 mM KCl+1 mM $CaCl_2$ solution (subsequently referred to as 'artificial xylem sap', AXS) for 2 h. K_{leaf} was also measured on leaves that were bench-dried for 1 h (Dehydrated) and on leaves that were bench-dried for 1 h and subsequently perfused for 2 h with AXS (RW AXS), AXS+0.2 mM $HgCl_2$ (RW $HgCl_2$), AXS+50 mM H_2O_2 (RW H_2O_2) or AXS+50 μM ABA (RW ABA). Values are means ± SE (n = 6). Different letters denote statistically significant differences by one-way ANOVA with Tukey's test.

been widely used as AQP inhibitors; ABA may also reduce AQP activity in leaves (reviewed in 28; 12). Control leaves were always kept hydrated and were perfused with pure AXS for 2 h. Immediately after perfusion with these solutions, K_{leaf} was determined as described above. All measurements were conducted at the same time of day (10:00–11:30 h).

After perfusion with the different solutions, stomatal pore aperture of leaves was measured as described in Laur & Hacke [31]. Images were recorded in six randomly selected fields of view of each leaf. Fields of view were located near the point of maximum leaf width on the abaxial leaf surface.

Dye uptake experiments

The extent of dye uptake in excised leaves was used as an additional method to assess xylem refilling in leaf veins during the rehydration phase. We also used the dye uptake experiments in an attempt to study how embolism reversal in leaf veins is impacted by mercury. Excised leaves were bench-dried for 1 h and rehydrated for 2 h by immersion of the petioles in filtered safranin solutions. Transpiration during dye uptake was promoted by placing leaves near a fan at a light level of ~1,000 μmol m^{-2} s^{-1} PAR (i.e., conditions similar to the protocol used to measure K_{leaf}). Dye (0.1% (w/v) safranin) was dissolved in pure AXS or AXS+ 0.2 mM $HgCl_2$. Control leaves were excised from well-watered plants and then perfused for 2 h with 0.1% safranin-containing AXS without prior dehydration treatment. Images were recorded in six randomly selected fields of view of each leaf. Fields of view were located near the point of maximum leaf width on the abaxial leaf surface.

Gene transcript measurements by quantitative real-time PCR

Fully expanded leaves corresponding to LPI 7–10 were collected, immediately frozen in liquid nitrogen and stored at −80°C until analyzed. Samples were always collected between 10:00 h and 11:30 h to minimize any diurnal effect on AQP expression. Total RNA was extracted from 3 plants per treatment following the CTAB method of Pavy et al. [32]. RNA quality was assessed on an agarose gel and quantified with a spectrophotometer (Nanodrop ND-1000, Thermo Scientific, Wilmington, DE, USA). RNA was treated as previously described [19]. cDNA quality was checked by PCR with intron-spanning actin primers. Putative leaf-expressed AQP genes were selected [33–35], specific primers (Table S1) were designed according to Rutledge & Stewart [36] using the QuantPrime online tool [37]. PCR efficiency was 100±7% for all primer pairs and specificity was checked using melting curves. Real-time qPCR was performed on a 7900 HT Fast Real-Time PCR system (Applied Biosystems, Foster City, CA, USA) as described previously [31]. Relative gene expression was measured according to Livak & Schmittgen [38] using the 2ΔΔC(t) method. The expression values were normalized to the geometric mean of four housekeeping genes (actin (POPTR_0001s31700), cyclophilin (POPTR_0005s26170), TIP4-like (POPTR_0009s09620.1) and ubiquitin (POPTR_0005s09940)). Relative gene expression was determined as the fold change of an AQP isoform at a given condition relative to its expression under control conditions. Real-time PCR was carried out using three biological replicates each with three technical replicates.

Immunolocalization

Samples were fixed in formaldehyde-acetic acid and embedded in paraffin as described previously [39]. Transverse sections,

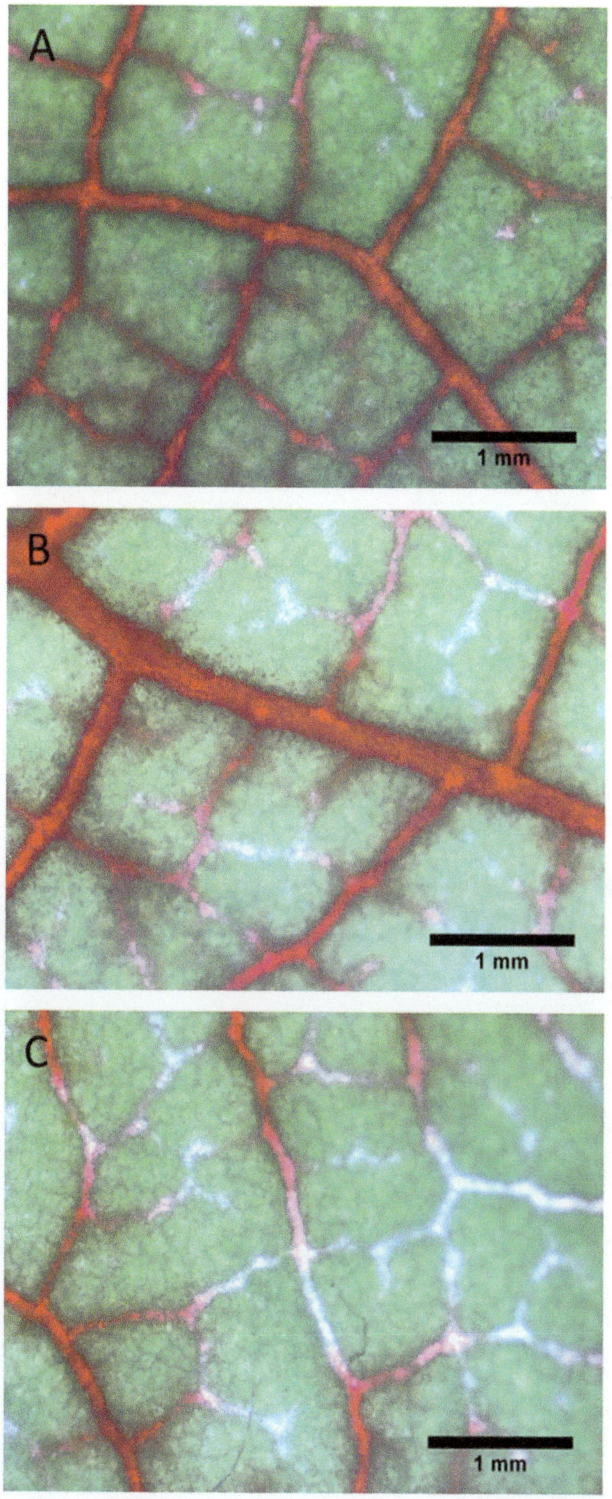

Figure 4. Typical images of transpiring *P. trichocarpa* leaves that were allowed to take up safranin solution. (A) A control leaf was excised from a well-watered plant, and the petiole was immersed for 2 h in safranin solution. Transpiration during dye uptake was promoted by placing the leaf near a fan at ~1,000 µmol m^{-2} s^{-1} photosynthetic active radiation. Most leaf veins were stained indicating minimal xylem embolism. (B) Dye uptake in a bench-dried leaf that was subsequently perfused with safranin solution for 2 h. Minor veins exhibited incomplete staining indicating the presence of embolized xylem conduits in minor veins. (C) Dye uptake of a bench-dried leaf subsequently perfused with safranin + $HgCl_2$ solution for 2 h. Mercury is an aquaporin inhibitor. Staining remained even more incomplete than in (B).

10 µm thick, were prepared with a microtome. Immunoreactions were performed following the protocol of Gong et al. [40]. Primary antibodies directed against the 42 N-terminal amino acids of AtPIP1;3 [41] and the conserved 10 amino acids of the C-terminal of PIP2s [19] were used. In addition, we applied a commercially available anti-TIP2 antibody (Sakurai et al. [42]); Agrisera AB, Sweden; alignment shown in Figure S1). AlexaFluo 488-conjugated goat anti-chicken, anti-mouse and anti-rabbit secondary antibodies (Life Technologies Inc., Burlington, ON, Canada) were applied respectively for 2 h at 37°C. Slides were mounted with Permount. Images were taken with a Zeiss LSM 700 confocal microscope (Carl Zeiss, Oberkochen, Germany).

Statistical analysis

All statistical analyses were carried out using SigmaPlot software. Differences due to the effect of physiological treatments were analyzed after testing for normality and equal variance by using a one-way ANOVA followed by a Tukey's test. A one-way ANOVA followed by Bonferroni's post test was used for the gene expression analysis. Differences were considered significant at $P \leq 0.05$.

Results and Discussion

Leaf hydraulic conductance is highly sensitive to drought

To assess how K_{leaf} declines as a function of Ψ_{leaf}, we first constructed a vulnerability curve. Water was withheld from plants in the greenhouse until plants reached different levels of water stress. Leaves were highly vulnerable with 50% and 80% loss of hydraulic conductance occurring at $\Psi_{leaf} = -0.47$ MPa and -0.71 MPa, respectively (Figure 1, insert). According to the curve fit in Figure 1 (insert), the maximum K_{leaf} measured during this experiment was 6.15 mmol m^{-2} s^{-1} MPa^{-1}. The drought-induced loss in K_{leaf} shown in Figure 1 may have been due to xylem cavitation, reduced water permeability of cell membranes and/or other factors [6,7]. The water potentials at 50% and 80% loss of hydraulic conductance (P_{50} and P_{80}, respectively) are well within the range of water potentials that trees experience under natural conditions [43,44]. It therefore appears that K_{leaf} is subject to substantial diurnal changes under natural conditions, similar to what has been observed in rice and other species [3,15,45,46]. Our data also indicates that leaf hydraulic conductance is more sensitive to decreasing water potentials than the hydraulic conductance of stems [43]. However, since we only worked with young greenhouse-grown plants, it remains to be seen whether leaves of field-grown trees are similar in their response to water stress.

Leaves of intact plants quickly recover from drought

We next tested whether K_{leaf} would recover after a drought treatment when plants were left intact during the dehydration-rehydration episode. In this experiment, leaves of well-watered control plants had a Ψ_{leaf} of -0.33 ± 0.03 MPa (±SE, n = 6), which was associated with a K_{leaf} of 3.37 ± 0.41 mmol m^{-2} s^{-1} MPa^{-1} (±SE, n = 6) (Figure 1, blue squares). The drought treatment resulted in a drop of Ψ_{leaf} to -0.77 ± 0.05 MPa (± SE, n = 6) and a six-fold drop of K_{leaf} to 0.55 ± 0.12 mmol m^{-2} s^{-1} MPa^{-1} (±SE, n = 6) (Figure 1, red circles). These values were in good agreement with the previously established vulnerability curve (Figure 1, insert). Only 2 h after rewatering (Figure 1, grey

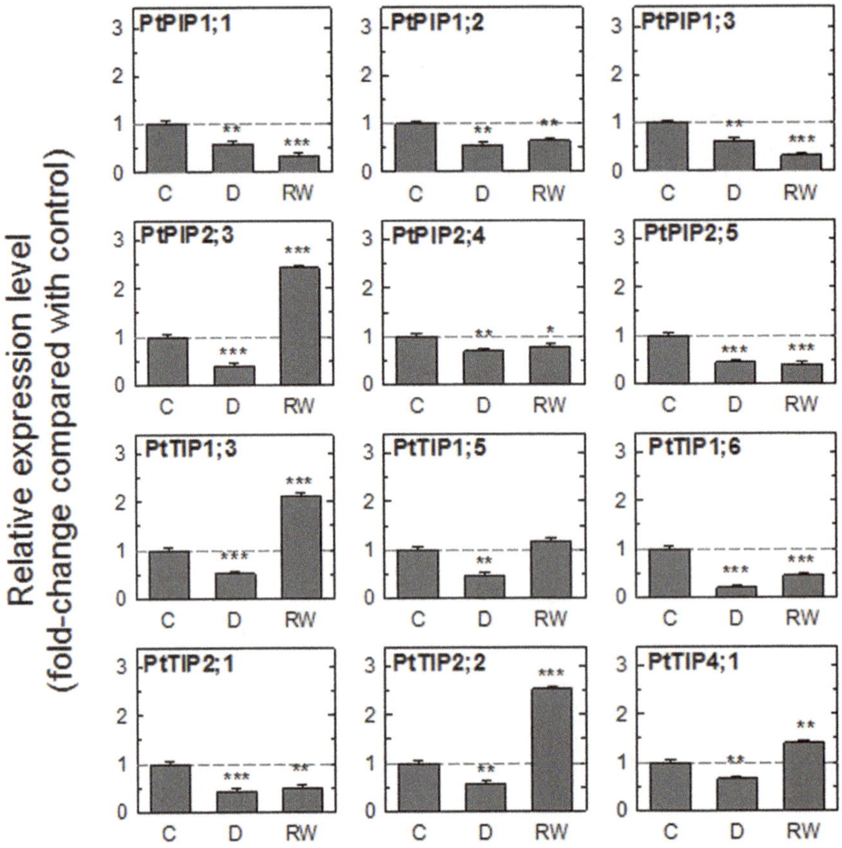

Figure 5. Relative expression of aquaporin genes in detached leaves during a dehydration-rehydration experiment. Data are from control leaves (C) after they were perfused with artificial xylem sap (AXS) for 2 h, leaves that were dehydrated on the bench top for 1 h (D), and leaves that were dehydrated on the bench top for 1 h and then perfused for 2 h with AXS (RW). The geometric mean of the expression levels of four reference genes (*ACT2, CYC063, TIP41-like, UBQ7*) was used to normalize the results. Asterisks denote significant differences in expression level compared to control levels (one-way ANOVA, followed by Bonferroni's post test, $^{*}P\leq0.05$; $^{**}P\leq0.01$$^{***}P\leq0.001$). Data are means ± SE of three biological replicates.

squares), both Ψ_{leaf} and K_{leaf} reached values that were not statistically different from well-watered control plants (*t* test, $P=0.083$ for K_{leaf}), indicating that leaves completely recovered their hydraulic function.

AQP expression in leaves collected from intact plants

To study the role of water channels in the recovery of K_{leaf}, AQP expression was measured in leaves at different stages during the dehydration-rehydration experiment. Three *PIP1*, three *PIP2*, and six *TIP* candidate genes were selected for analysis. Among them, *PtPIP1;1*, *PtPIP1;2*, *PtPIP1;3*; *PtPIP2;4* and *PtTIP2;1* exhibited the highest total number of mRNA molecules in leaves of control plants (Table 1).

The drought treatment resulted in a significant reduction in the expression of all tested genes (Fig. 2). In leaves collected 2 h after rewatering, there were two patterns of expression between the 12 isoforms. One group of genes (among them all *PIP1*s) remained down-regulated while the expression of a second group of genes matched or exceeded the transcript levels measured in control leaves. With the exception of *PtTIP2;1*, all tested *TIP*s were significantly up-regulated after 2 h. Among the *PIP*s, only the expression level of *PtPIP2;3* increased to match the control level.

Recovery of K_{leaf} in detached leaves is impaired by inhibitors

Another set of experiments was conducted on leaves that were excised from the plant prior to the dehydration-rehydration treatment. Working with detached leaves allowed us to study the effect of AQP inhibitors and ABA on the recovery of K_{leaf}. Control leaves exhibited a K_{leaf} of 8.49±0.57 mmol m^{-2} s^{-1} MPa^{-1} (±SE, n = 6), which is higher than the values shown in Figure 1. One difference between the data shown in Figures 1 and 3 is that all data in Figure 1 was derived from leaves that were excised (petioles were cut under water) from transpiring plants immediately before K_{leaf} was measured while the control leaves in Figure 3 were perfused with AXS for 2 h prior to measuring K_{leaf}. Hence, the absolute K_{leaf} values shown in Figures 1 and 3 are not readily comparable.

Bench-drying of leaves caused a ~10-fold decline in K_{leaf} relative to fully hydrated control leaves (Figure 3A). Dehydrated leaves that were subsequently xylem-perfused for 2 h with AXS exhibited a significant recovery to 50% of the hydraulic conductance measured in control leaves. The fact that recovery remained incomplete in detached leaves is consistent with an involvement of phloem transport in embolism repair [47,48]. Application of commonly used inhibitors allowed us to assess the impact of AQPs on K_{leaf} during leaf rehydration. Leaves fed with $HgCl_2$ and H_2O_2 did not exhibit any recovery of hydraulic

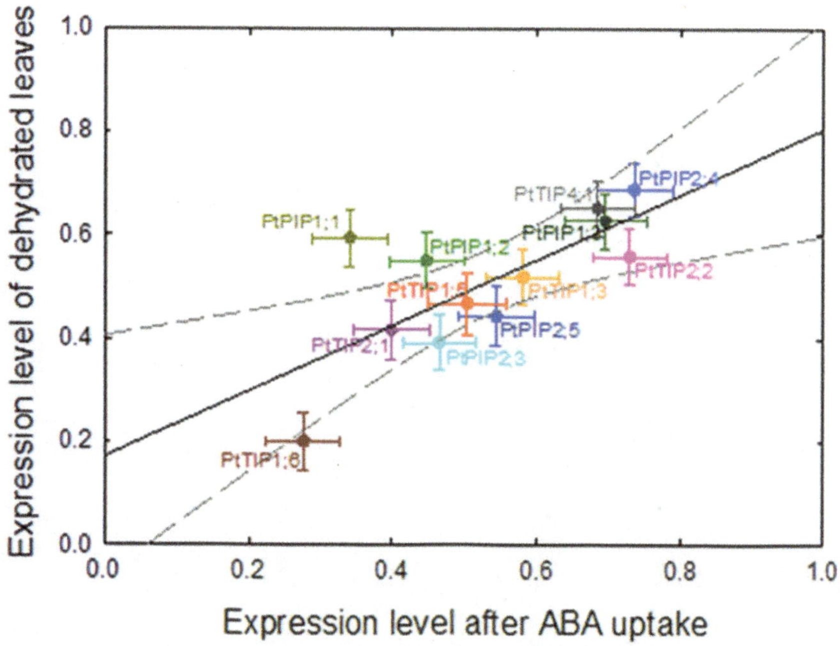

Figure 6. Relative expression of 12 aquaporin genes in response to dehydration (y-axis) and dehydration + perfusion with abscisic acid (x-axis). Detached leaves were either dehydrated on the bench top for 1 h or dehydrated for 1 h and subsequently perfused for 1 h with 50 µM abscisic solution (ABA). Data from fully hydrated detached leaves (perfused for 3 h with 20 mM KCl+1 mM $CaCl_2$ solution) were used as the control group, and their expression refers to a value of 1. Pearson's r = 0.725; $P \leq 0.01$. Data are means ± SE of three biological replicates.

conductance, indicating that AQPs were involved in the recovery of K_{leaf} after dehydration. A role of AQPs in embolism repair has also been proposed for other species and plant organs [18,19,49–51].

We next asked whether differences in K_{leaf} were associated with different degrees of stomatal closure. Stomatal apertures in fully hydrated control leaves were 9.5±0.1 µm (±SE, n = 6), similar to the value of ~10 µm previously reported for *P. trichocarpa* leaves [52]. Schulte and Hinckley [52] found that stomatal aperture in this species was not affected by a wide range of epidermal water potentials. Our data supports these findings as we also did not observe complete stomatal closure in any of our experimental treatments (Figure 3B). Even in dehydrated leaves and in leaves that were perfused with AQP inhibitors and ABA, stomatal aperture remained at ~6 µm. This value is similar to the maximum apertures found in *P. deltoides* and in *P. trichocarpa* x *deltoides* hybrids [31,52]. We conclude that the magnitude of the decline in K_{leaf} in dehydrated leaves and in leaves that were perfused with AQP inhibitors is greater than that of changes in stomatal aperture. Guyot et al. [53] also found a discrepancy between patterns of stomatal conductance and K_{leaf}, and they discuss possible reasons for a mechanistic independence of stomatal and leaf hydraulic conductance.

We used the dye uptake experiments in an attempt to study how embolism reversal in leaf veins is impacted by mercury. Nearly all veins of well-watered control leaves were stained and functional (Figure 4A). In leaves that were bench-dried and subsequently supplied with ASX + safranin for 2 h, many minor veins exhibited incomplete staining (Figure 4B). Staining was even less complete in leaves that were bench-dried and subsequently perfused with ASX + safranin + $HgCl_2$ (Figure 4C).

These findings suggest that embolism formation in minor veins had a substantial impact on the dynamics of K_{leaf}. Studying water transport in rice leaves, Stiller et al. [8] reported that the leaf xylem experienced high embolism levels, even in watered controls. Nardini et al. [54] found that minor veins of *Cercis siliquastrum* leaves underwent extensive embolism at leaf water potentials <−1.5 MPa, indicating that leaf vein embolism was closely related to K_{leaf} changes. Recently, Johnson et al. [55] suggested that reductions in K_{leaf} are directly related to vein embolism. On the other hand, a recent study found that hydraulic decline during mild dehydration was associated with leaf shrinkage [10]. The changes in K_{leaf} we observed in this present study were likely caused by xylem and extra-xylem components; it is difficult to determine the relative importance of either component. In addition, both xylem refilling in leaf veins and the permeability of extra-xylem tissues may be impacted by AQP function.

AQP expression in detached leaves

Aquaporin expression was measured in detached leaves undergoing a dehydration-rehydration cycle (Figure 5). Control leaves were perfused with AXS for 2 h before leaf tissue was sampled for the gene expression analysis. As previously seen in intact plants (Figure 2), water stress caused down-regulation of all tested *AQP*s (Figure 5). This agrees with several previous studies [19,56,57].

Notably, very similar degrees of down-regulation were found in bench-dried leaves and in dried leaves that were subsequently xylem-perfused with AXS + ABA (Figure 6, r = 0.725, $P<0.01$). Genes that were strongly down-regulated by dehydration, such as *PtTIP1;6* also exhibited strong down-regulation after perfusion with ABA solution while the expression of other genes, such as *PtPIP2;4*, changed less in response to either of these factors (Figure 6). Excluding *PtPIP1;1* from the analysis shown in Figure 6 further increased the strength of the linear relationship (r = 0.89, $P<0.001$).

The lack of recovery in ABA-perfused leaves and down-regulation of *AQP*s in leaves supplied with AXS + ABA is

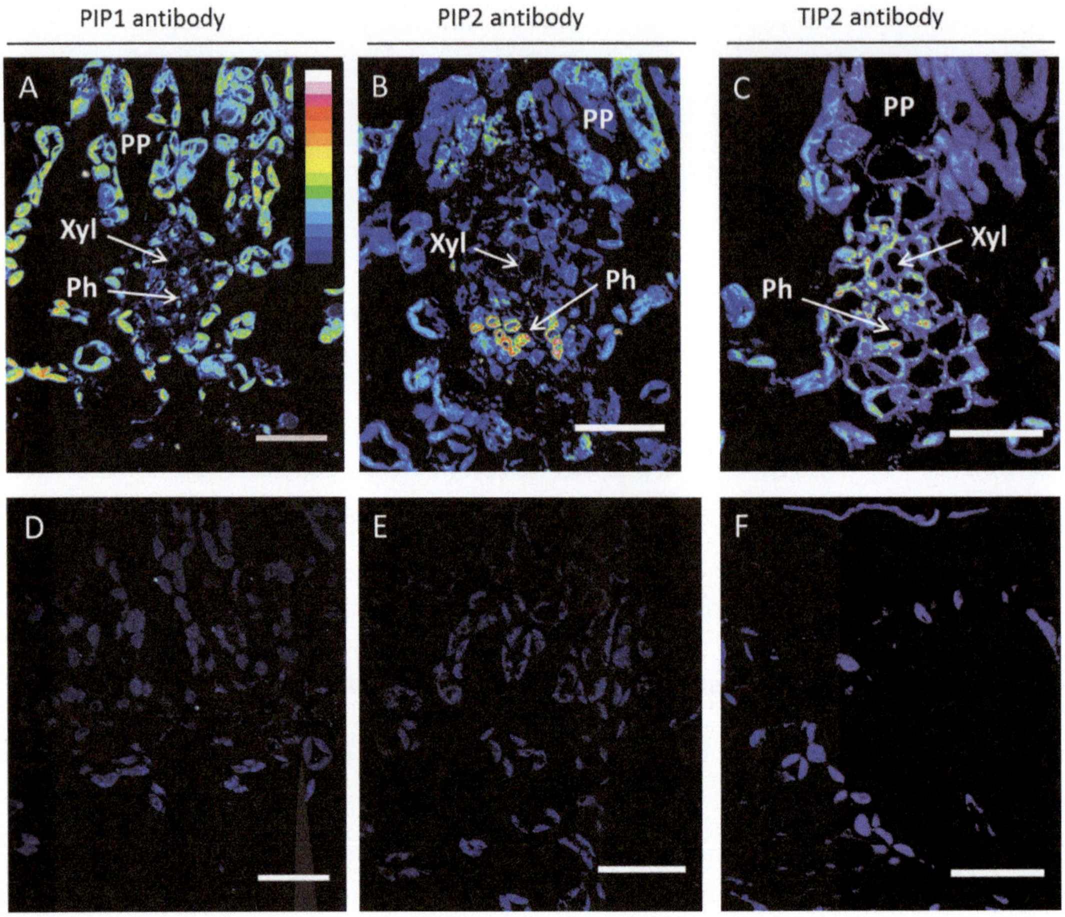

Figure 7. Immunolocalization of AQP proteins in leaves of *P. trichocarpa* saplings. Confocal laser scanning micrographs showing the localization of PIP1, PIP2, TIP2 proteins in minor veins of leaf transverse sections (A, B, C respectively). Controls with no primary antibody indicate minimal background fluorescence (D, E, F respectively). Images were taken at an identical setting and were color-coded with an intensity look-up-table (LUT; displayed in A), in which black was used to encode background, and blue, green, yellow, red and white to encode increasing signal intensities. Ph, phloem; PP, palisade parenchyma; Xyl, xylem. Scale bars = 20 µm.

consistent with the model of Shatil-Cohen et al. [12]. Working with *Arabidopsis*, these authors also used a 'detached leaf' approach to feed ABA to the xylem via the petiole. Feeding the leaf with ABA decreased K_{leaf} by nearly 50%. In contrast, smearing ABA on the leaf surface, while reducing transpiration, had no effect on K_{leaf}. Shatil-Cohen et al. [12] proposed that the membrane water permeability of bundle sheath cells is controlled by AQPs, and that the bundle sheath would act like a control center regulating K_{leaf} in response to signals from the xylem. As the concentration of ABA increases in the xylem, AQP activity in the bundle sheath would be down-regulated, reducing water flow into the leaf mesophyll. Bundle sheath cells, and perhaps xylem parenchyma cells, seem to have a specific responsiveness to ABA, which likely explains the negative effects of this hormone on K_{leaf} (for a recent review see 7). While our data is consistent with these observations, it is not clear yet which cells may perform the role of a 'control center' in *P. trichocarpa* leaves (Figure S2). While we previously observed prominent PIP1 and PIP2 labeling of the endodermis-like bundle sheath in *Picea glauca* needles [19], no such pattern was found in this present study.

In rehydrated leaves, four genes showed increased expression levels relative to control leaves (Figure 5). Three of these AQPs (*PtTIP1;3*, *PtTIP2;2*, and *PtTIP4;1*) were TIPs and were also found to be up-regulated when intact plants were rewatered after a drought (compare Figures 2 and 5). While TIPs have rarely been studied in the context of water flow through tissues and embolism repair, a recent study on grapevine plants found a striking positive correlation between K_{leaf} and the transcript abundance of *VvTIP2;1* [58]. Our immunolocalization experiments indicate that TIP2 protein was present in xylem parenchyma cells (Figure 7). This agrees with the expression pattern of *ZmTIP1* in leaves and stems of maize. *In situ* localization revealed that this tonoplast AQP was highly expressed in parenchyma cells surrounding xylem vessels, in phloem companion cells, and between the phloem and the xylem strands [59]. Barrieu et al. [59] hypothesized that the high expression of the ZmTIP1 tonoplast AQP in xylem parenchyma cells would allow these cells to control water movement in and out of the xylem vessels. Daniels et al. [60] found that AtTIP2 expression in mature leaves was generally restricted to vascular tissues. In stem xylem of hybrid poplar, a TIP2 AQP was highly expressed in contact cells, suggesting a role in increasing water exchange between vessels and xylem rays [61].

In this present study, we also determined the cell- and tissue-level localization of PIP1 and PIP2 proteins (Figure 7). All sections were taken from leaves of well-watered plants. Strong PIP1 signals were present in the palisade parenchyma (Figure 7A). PIP1 antibody was also detected in vein cells, including phloem and xylem parenchyma. This labeling pattern is consistent with a dual role of PIP1s in influencing permeability to water and CO_2 [62]. PIP2 was mostly localized in the phloem, which agrees with previous studies [19,27,61,63,64]. Weaker PIP2 labelling was evident in palisade parenchyma cells (Figure 7B).

Conclusions

We studied how AQPs may be involved in the recovery of water stress-induced declines in K_{leaf}. We examined how K_{leaf} responds to known AQP inhibitors and xylem-fed ABA. We also examined the expression of 12 highly expressed AQP genes during dehydration-rehydration experiments. Hydraulic measurements and gene expression assays were complemented by dye uptake and immunolocalization experiments. This has revealed that, while *P. trichocarpa* leaves are highly sensitive to dehydration, leaf hydraulic conductance can quickly recover when water becomes available again. Recovery of K_{leaf} was absent when excised leaves were xylem-perfused with AQP inhibitors, suggesting that the recovery of leaf hydraulic function is associated with AQP activity. Among the AQPs tested, several *TIP*s showed large increases in expression in rehydrated leaves, suggesting that TIPs play an important role in reversing drought-induced reductions in K_{leaf}.

Supporting Information

Figure S1 (a) Amino acid multiple sequence alignment of the N-terminal region of the *Arabidopsis thaliana* AtPIP1;3 and the *Populus trichocarpa* PtPIP1s; (b) of the conserved the C-terminal region of PIP2s, and (c) TIP2s. Consensus amino acids are underlined in black.

Figure S2 Transverse section of a *Populus trichocarpa* leaf showing minor veins with (left) and without (right) bundle sheath cell extensions. Scale bar = 20 μm.

Table S1 Primer sequences used for the gene expression study.

Acknowledgments

We are grateful to Dr. Barb R. Thomas (University of Alberta - Alberta-Pacific Forest Industries, Canada) and Pierre Périnet (Québec Ministry of Forests) for providing the plant material.

Author Contributions

Conceived and designed the experiments: JL UGH. Performed the experiments: JL UGH. Analyzed the data: JL UGH. Contributed reagents/materials/analysis tools: JL UGH. Contributed to the writing of the manuscript: JL UGH.

References

1. Sperry JS, Hacke UG, Oren R, Comstock JP (2002) Water deficits and hydraulic limits to leaf water supply. Plant, Cell Environ 25(2): 251–63.
2. Tyree MT, Sperry JS (1988) Do woody plants operate near the point of catastrophic xylem dysfunction caused by dynamic water stress? answers from a model. Plant Physiol (88): 574–80.
3. Scoffoni C, McKown AD, Rawls M, Sack L (2012) Dynamics of leaf hydraulic conductance with water status: quantification and analysis of species differences under steady state. J Exp Bot 63: 643–58.
4. Nardini A, Luglio J (2014) Leaf hydraulic capacity and drought vulnerability: possible trade-offs and correlations with climate across three major biomes. Funct Ecol 28(4): 810–818.
5. Brodribb TJ, Holbrook NM (2004) Stomatal protection against hydraulic failure: a comparison of coexisting ferns and angiosperms. New Phytol 162(3): 663–70.
6. Heinen RB, Ye Q, Chaumont F (2009) Role of aquaporins in leaf physiology. J Exp Bot 60(11): 2971–85.
7. Prado K, Maurel C (2013) Regulation of leaf hydraulics: from molecular to whole plant levels. Front Plant Sci 4: 255.
8. Stiller V, Lafitte HR, Sperry JS (2003) Hydraulic properties of rice and the response of gas exchange to water stress. Plant Physiol 132: 1698–706.
9. Johnson DM, McCulloh KA, Meinzer FC, Woodruff DR, Eissenstat DM (2011) Hydraulic patterns and safety margins, from stem to stomata, in three eastern U.S. tree species. Tree Physiol 31(6): 659–68.
10. Scoffoni C, Vuong C, Diep S, Cochard H, Sack L (2014) Leaf shrinkage with dehydration: coordination with hydraulic vulnerability and drought tolerance. Plant Physiol 164(4): 1772–88.
11. Brodribb TJ, Holbrook NM (2005) Water stress deforms tracheids peripheral to the leaf vein of a tropical conifer. Plant Physiol 137: 1139–46.
12. Shatil-Cohen A, Attia Z, Moshelion M (2011) Bundle-sheath cell regulation of xylem-mesophyll water transport via aquaporins under drought stress: a target of xylem-borne ABA? Plant J 67(1): 72–80.
13. Brodribb TJ, Holbrook NM, Edwards EJ, Gutiérrez MV (2003) Relations between stomatal closure, leaf turgor and xylem vulnerability in eight tropical dry forest trees. Plant, Cell Environ 26: 443–50.
14. Hacke UG, Sperry JS, Ewers BE, Ellsworth DS, Schäfer KVR, et al. (2000) Influence of soil porosity on water use in *Pinus taeda*. Oecologia 124(4): 495–505.
15. Stiller V, Sperry JS, Lafitte R (2005) Embolized conduits of rice (*Oryza sativa*, Poaceae) refill despite negative xylem pressure. Am J Bot 92(12): 1970–4.
16. Galmés J, Pou A, Alsina MM, Tomàs M, Medrano H, et al. (2007) Aquaporin expression in response to different water stress intensities and recovery in Richter-110 (*Vitis* sp.): relationship with ecophysiological status. Planta 226(3): 671–81.
17. Jang HY, Yang SW, Carlson JE, Ku YG, Ahn SJ (2013) Two aquaporins of *Jatropha* are regulated differentially during drought stress and subsequent recovery. J Plant Physiol 170(11): 1028–38.
18. Martre P, Morillon R, Barrieu F, North GB, Nobel PS, et al. (2002) Plasma membrane aquaporins play a significant role during recovery from water deficit. Plant Physiol 130(4): 2101–10.
19. Laur J, Hacke UG (2014) Exploring *Picea glauca* aquaporins in the context of needle water uptake and xylem refilling. New Phytol 203 (2): 388–400.
20. North GB, Martre P, Nobel PS (2004) Aquaporins account for variations in hydraulic conductance for metabolically active root regions of *Agave deserti* in wet, dry, and rewetted soil. Plant, Cell Environ 27(2): 219–28.
21. Fouquet R, Léon C, Ollat N, Barrieu F (2008) Identification of grapevine aquaporins and expression analysis in developing berries. Plant Cell Rep 27(9): 1541–50.
22. Quigley F, Rosenberg JM, Shachar-Hill Y, Bohnert HJ (2002) From genome to function: the *Arabidopsis* aquaporins. Genome Biol 3(1): research0001.1–research0001.17.
23. Sakurai J, Ishikawa F, Yamaguchi T, Uemura M, Maeshima M (2005) Identification of 33 rice aquaporin genes and analysis of their expression and function. Plant Cell Physiol 46(9): 1568–77.
24. Gupta AB, Sankararamakrishnan R (2009) Genome-wide analysis of major intrinsic proteins in the tree plant *Populus trichocarpa*: characterization of XIP subfamily of aquaporins from evolutionary perspective. BMC Plant Biol 9: 134.
25. Daniels MJ, Mirkov TE, Chrispeels MJ (1994) The plasma membrane of *Arabidopsis thaliana* contains a mercury-insensitive aquaporin that is a homolog of the tonoplast water channel protein TIP. Plant Physiol 106(4): 1325–33.
26. Postaire O, Tournaire-Roux C, Grondin A, Boursiac Y, Morillon R, et al. (2010) A PIP1 aquaporin contributes to hydrostatic pressure-induced water transport in both the root and rosette of *Arabidopsis*. Plant Physiol 152(3): 1418–30.
27. Vandeleur RK, Mayo G, Shelden MC, Gilliham M, Kaiser BN, et al. (2009) The role of plasma membrane intrinsic protein aquaporins in water transport through roots: diurnal and drought stress responses reveal different strategies between isohydric and anisohydric cultivars of grapevine. Plant Physiol 149(1): 445–60.
28. Chaumont F, Tyerman SD (2014) Aquaporins: Highly regulated channels controlling plant water relations. Plant Physiol 164: 1600–18.
29. Sack L, Scoffoni C (2012) Measurement of leaf hydraulic conductance and stomatal conductance and their responses to irradiance and dehydration using the Evaporative Flux Method (EFM). J Vis Exp (70): 1–7.
30. Larson PR, Isebrands JG (1971) The plastochron index as applied to developmental studies of cottonwood. Can J For Res 1(1): 1–11.
31. Laur J, Hacke UG (2013) Transpirational demand affects aquaporin expression in poplar roots. J Exp Bot 64(8): 2283–93.
32. Pavy N, Boyle B, Nelson C, Paule C, Giguère I, et al. (2000) Identification of conserved core xylem gene sets: conifer cDNA microarray development, transcript profiling and computational analyses. New Phytol 180(4): 766–86.
33. Almeida-Rodriguez AM, Cooke JEK, Yeh F, Zwiazek JJ (2010) Functional characterization of drought-responsive aquaporins in *Populus balsamifera* and

Populus simonii × *balsamifera* clones with different drought resistance strategies. Physiol Plant 140(4): 321–33.

34. Wilkins O, Nahal H, Foong J, Provart NJ, Campbell MM (2009) Expansion and diversification of the *Populus* R2R3-MYB family of transcription factors. Plant Physiol 149(2): 981–93.
35. Cohen D, Bogeat-Triboulot M-B, Vialet-Chabrand S, Merret R, Courty P-E, et al. (2013) Developmental and environmental regulation of aquaporin gene expression across *Populus* species: Divergence or Redundancy? PLoS One 8(2): e55506.
36. Rutledge RG, Stewart D (2010) Assessing the performance capabilities of LRE-based assays for absolute quantitative real-time PCR. PLoS One 5(3): e9731.
37. Arvidsson S, Kwasniewski M, Riaño-Pachón DM, Mueller-Roeber B (2008) QuantPrime–a flexible tool for reliable high-throughput primer design for quantitative PCR. BMC Bioinformatics 9: 465.
38. Livak KJ, Schmittgen TD (2001) Analysis of relative gene expression data using real-time quantitative PCR and the $2^{-\Delta\Delta CT}$ method. Methods 25(4): 402–8.
39. Almeida-Rodriguez AM, Hacke UG, Laur J (2011) Influence of evaporative demand on aquaporin expression and root hydraulics of hybrid poplar. Plant Cell Environ 34(8): 1318–31.
40. Gong H, Peng Y, Zou C, Wang D (2006) A simple treatment to significantly increase signal specificity in immunohistochemistry. Plant Mol Biol Report 24: 93–101.
41. Kammerloher W, Fischer U, Piechottka GP, Schäffner AR (1994) Water channels in the plant plasma membrane cloned by immunoselection from a mammalian expression system. Plant J 6: 187–99.
42. Sakurai J, Ahamed A, Murai M, Maeshima M, Uemura M (2008) Tissue and cell-specific localization of rice aquaporins and their water transport activities. Plant Cell Physiol 49(1): 30–9.
43. Sparks JP, Black RA (1999) Regulation of water loss in populations of *Populus trichocarpa*: The role of stomatal control in preventing xylem cavitation. Tree Physiol 19: 453–9.
44. Pezeshki SR, Hinckley TM (1982) The stomatal response of red alder and black cottonwood to changing water status. Can J For Res 12(4): 761–71.
45. Trifilò P, Gascó A, Raimondo F, Nardini A, Salleo S (2003) Kinetics of recovery of leaf hydraulic conductance and vein functionality from cavitation-induced embolism in sunflower. J Exp Bot 54(391): 2323–30.
46. Martorell S, Diaz-Espejo A, Medrano H, Ball MC, Choat B (2014) Rapid hydraulic recovery in *Eucalyptus pauciflora* after drought: linkages between stem hydraulics and leaf gas exchange. Plant Cell Environ 37(3): 617–26.
47. Nardini A, Lo Gullo MA, Salleo S (2011) Refilling embolized xylem conduits: is it a matter of phloem unloading? Plant Sci 180(4): 604–11.
48. Christman MA, Sperry JS, Smith DD (2012) Rare pits, large vessels and extreme vulnerability to cavitation in a ring-porous tree species. New Phytol 193: 713–20.
49. Mayr S, Schmid P, Laur J, Rosner S, Charra-Vaskou K, et al. (2014) Uptake of water via branches helps timberline conifers refill embolized xylem in late winter. Plant Physiol 164(4): 1731–40.
50. Secchi F, Zwieniecki MA (2010) Patterns of PIP gene expression in *Populus trichocarpa* during recovery from xylem embolism suggest a major role for the PIP1 aquaporin subfamily as moderators of refilling process. Plant Cell Environ 33(8): 1285–97.
51. Chitarra W, Balestrini R, Vitali M, Pagliarani C, Perrone I, et al. (2014) Gene expression in vessel-associated cells upon xylem embolism repair in *Vitis vinifera* L. petioles. Planta 239: 887–99.
52. Schulte PJ, Hinckley TM (1987) The relationship between guard cell water potential and the aperture of stomata in *Populus*. Plant Cell Environ 10: 313–8.
53. Guyot G, Scoffoni C, Sack L (2012) Combined impacts of irradiance and dehydration on leaf hydraulic conductance: insights into vulnerability and stomatal control. Plant Cell Environ 35(5): 857–71.
54. Nardini A, Salleo S, Raimondo F (2003) Changes in leaf hydraulic conductance correlate with leaf vein embolism in *Cercis siliquastrum* L. Trees 17(6): 529–34.
55. Johnson DM, Mcculloh KA, Woodruff DR, Meinzer FC (2012) Evidence for xylem embolism as a primary factor in dehydration-induced declines in leaf hydraulic conductance. Plant, Cell Environ 35: 760–9.
56. Alexandersson E, Fraysse L, Sjövall-Larsen S, Gustavsson S, Fellert M, et al. (2005) Whole gene family expression and drought stress regulation of aquaporins. Plant Mol Biol 59(3): 469–84.
57. Secchi F, Lovisolo C, Schubert A (2007) Expression of OePIP2.1 aquaporin gene and water relations of *Olea europaea* twigs during drought stress and recovery. Ann Appl Biol 150(2): 163–7.
58. Pou A, Medrano H, Flexas J, Tyerman SD (2012) A putative role for TIP and PIP aquaporins in dynamics of leaf hydraulic and stomatal conductances in grapevine under water stress and re-watering. Plant Cell Environ 36(4): 828–43.
59. Barrieu F, Chaumont F, Chrispeels MJ (1998) High expression of the tonoplast aquaporin ZmTIP1 in epidermal and conducting tissues of maize. Plant Physiol 117(4): 1153–63.
60. Daniels MJ, Chaumont F, Mirkov TE, Chrispeels MJ (1996) Characterization of a new vacuolar membrane aquaporin sensitive to mercury at a unique Site. Plant Cell 8: 587–99.
61. Almeida-Rodriguez AM, Hacke UG (2012) Cellular localization of aquaporin mRNA in hybrid poplar stems. Am J Bot 99(7): 1249–54.
62. Secchi F, Zwieniecki MA (2013) The physiological response of *Populus tremula* x *alba* leaves to the down-regulation of PIP1 aquaporin gene expression under no water stress. Front Plant Sci 4: 507.
63. Kirch H, Vera-estrella R, Golldack D, Quigley F, Michalowski CB, et al. (2000) Expression of water channel proteins in *Mesembryanthemum crystallinum*. Plant Physiol 123: 111–24.
64. Yamada S, Bohnert HJ (2000) Expression of the PIP aquaporin promoter-MipA from the common ice plant in tobacco. Plant Cell Physiol 41(6): 719–25.

6

Modeling the Impact of Soil and Water Conservation on Surface and Ground Water Based on the SCS and Visual Modflow

Hong Wang[1,3], Jian-en Gao[1,2,4,5]*, Shao-long Zhang[4], Meng-jie Zhang[5], Xing-hua Li[4]

1 Institute of Soil and Water Conservation, Chinese Academy of Sciences and Ministry of Water Resources, Yangling, Shaanxi Province, China, **2** Institute of Soil and Water Conservation, Northwest A&F University, Yangling, Shaanxi Province, China, **3** University of Chinese Academy of Sciences, Beijing, China, **4** College of Water Resources and Architectural Engineering, Northwest A&F University, Yangling, Shaanxi Province, China, **5** College of Natural Resources and Environment, Northwest A&F University, Yangling, Shaanxi Province, China

Abstract

Soil and water conservation measures can impact hydrological cycle, but quantitative analysis of this impact is still difficult in a watershed scale. To assess the effect quantitatively, a three-dimensional finite-difference groundwater flow model (MODFLOW) with a surface runoff model–the Soil Conservation Service (SCS) were calibrated and applied based on the artificial rainfall experiments. Then, three soil and water conservation scenarios were simulated on the sand-box model to assess the effect of bare slope changing to grass land and straw mulching on water volume, hydraulic head, runoff process of groundwater and surface water. Under the 120 mm rainfall, 60 mm/h rainfall intensity, 5 m^2 area, 3° slope conditions, the comparative results indicated that the trend was decrease in surface runoff and increase in subsurface runoff coincided with the land-use converted from bare slope to grass land and straw mulching. The simulated mean surface runoff modulus was 3.64×10^{-2} $m^3/m^2/h$ in the bare slope scenario, while the observed values were 1.54×10^{-2} $m^3/m^2/h$ and 0.12×10^{-2} $m^3/m^2/h$ in the lawn and straw mulching scenarios respectively. Compared to the bare slope, the benefits of surface water reduction were 57.8% and 92.4% correspondingly. At the end of simulation period (T = 396 min), the simulated mean groundwater runoff modulus was 2.82×10^{-2} $m^3/m^2/h$ in the bare slope scenario, while the observed volumes were 3.46×10^{-2} $m^3/m^2/h$ and 4.91×10^{-2} $m^3/m^2/h$ in the lawn and straw mulching scenarios respectively. So the benefits of groundwater increase were 22.7% and 60.4% correspondingly. It was concluded that the soil and water conservation played an important role in weakening the surface runoff and strengthening the underground runoff. Meanwhile the quantitative analysis using a modeling approach could provide a thought for the study in a watershed scale to help decision-makers manage water resources.

Editor: Jose Luis Balcazar, Catalan Institute for Water Research (ICRA), Spain

Funding: This paper was supported by National Technology Support Project (2011BAD31B05), by Natural Science Foundation of China (41371276), by the Subject of National Science and Technology Major Project (2009ZX07212-002-003-02) and by Knowledge Innovation Project of Institute of Soil and Water Conservation, CAS & MWR (Soil and Water Conservation Project) (A315021304). The funders had no role in study design, data collection and analysis, decision to publish, or preparation of the manuscript.

Competing Interests: The authors have declared that no competing interests exist.

* E-mail: gaojianen@126.com

Introduction

In recent years, with more and more officials and professionals paying attention to the problems such as soil erosion and water loss, the contradiction between water supply and demand, whether soil and water conservation measures on upper regions would have effect on the amount of water resources of the lower reaches is a scientific issue to be discussed and probed into urgently [1]. For many years, the work of soil and water conservation as same as the Yellow River management, not only obtains great achievement, but also brings some problems [2]. The biggest achievement was no more than soil erosion control in the watershed, and thereby reducing the amount of sediment flowed into the river. Took the Yellow River basin for example, changes in land use and vegetation in the Loess Plateau had a decisive impact on sediment transport of Yellow River [3], even about 60% sediment yield since 10 ka BP occurred during the last 1040 years of the period of estrepement in Loess Plateau. The sediment monitoring at Sanmenxia hydrological station of Yellow River showed the annual average sediment discharge was 1040 million tons during 1950–2010, 589 million tons during 1987–2010, while it was only 198 and 351 million tons in 2009 and 2010 separately [4].

The reduction in the quantity of the sediment discharge by soil and water conservation is no doubt pleasing, but it also brings a new eco-environmental problem for its surface water reduction function. Some studies found that land use/land cover (LULC) change could alter hydrological cycles by affecting ecosystem evaportranspiration, soil infiltration capacity, surface and subsurface flow regimes [5–8]. Just as the study of the relation between control in Loess Plateau and no-flow in the Yellow River showed that the influence of water reduction on no-flow in the lower Yellow River for the comprehensive control changed the condition underlying surface and influenced the water cycle was nonnegligible [9]. Results of soil and water conservation method showed

Figure 1. The experimental flume.

that the mean flood reduced 0.5456 billion tons in He-Long section of middle Yellow River, Jinghe, Beiluohe and Weihe watershed from 1970 to 1996 [10]. The average runoff reduction benefits of soil and water conservation measures for Chabagou, Dalihe and Wudinghe basins in the 1970s were 14.47%, 20.22% and 20.78% respectively [11]. Jain and Mishra et al [12] evaluated the suitability to particular land use, soil type and combination a quantitative evaluation of the existing Soil Conservation Service Curve Number (SCS-CN) model using a large set of rainfall-runoff data from small to large watersheds of the U.S.A. Shi and Yi et al [13] evaluated the effect of land use/cover change on surface runoff by SCS model in Shenzhen region, China. Guo and Wang et al [14] calculated the volume of surface runoff during 5 rainfalls on 5 different kinds of land use types in sloping runoff plots by SCS model.

Soil and water conservation could reduce surface runoff and it is bound to affect groundwater recharge, which is the entry into the saturated zone of water made available at the table surface, together with the associated flow away from the water table within the saturated zone [15]. But less direct study has previously been undertaken to assess this effect quantitatively. It had been shown that groundwater recharge was closely related to land-use types very much. The impact of land-use on distributed groundwater recharge and discharge in the western Jilin, China, using MODFLOW, WetSpass, the Seepage packages, and ArcGIS showed that forest vegetation had the highest recharge, followed by agricultural farmlands and the recharge generally decreased when vegetated forests deteriorate to be other landforms (bush, grassland or bare-land) [16]. A study by Cho and Barone using MODFLOW indicated that subsurface flow regimes could be negatively affected by urbanization due to increased withdrawal and reduced recharge [17]. A number of studies showed the effects of changes in forest cover on groundwater recharge [18–22]. These studies showed a range in the reduction of groundwater recharge beneath trees, from 15% to 90%, compared to that under grass [18]. Thus the impacts of land-use on the atmospheric components of the hydrologic cycle (regional and global) are increasingly being recognized, though those on the subsurface components of the hydrologic cycle, particularly groundwater recharge are not equally known [23].

Therefore special attention should be given to the effect of soil and water conservation on the hydrologic cycle, especially recharge and discharge. Both monitoring and modeling approaches are used for conjunctive investigation of surface water and groundwater. The monitoring approach is expensive and time demanding yet measuring actual changes in stream and groundwater levels over time may lead to more direct estimates of the impact of land development on both surface and subsurface flows [17]. Through the model coupling approach, interaction between surface water and groundwater has become a trend [24,25]. Linkage between MODFLOW and existing surface models such as the SCS method to consider surface water and groundwater interactions could validate the outputs of the model.

The overall goal of this study is to assess the impact of soil and water conservation on the surface and subsurface flow by mathematical models combined with artificial rainfall simulation experiments and gives a quantitative analysis thought simultaneously.

Materials and Methods

SCS Curve Number Method

The runoff equation of the U.S. Soil Conservation Service, commonly called the curve number method, for estimating runoff from rainfall, came into common use in the mid - 1950s [26]. Although the SCS method for runoff estimation has changed little since the 1960s, its popularity has been maintained over the years [27]. The SCS-CN equation, in the typical form [28], is given as:

Table 1. Mechanical composition of experimental materials (%).

Types	1~0.5	0.5~0.25	0.25~0.05	0.05~0.01	0.01~0.005	0.005~0.001	<0.001
Lou soil	0.07	0.65	5.86	49.04	12.18	13.72	18.48
Mixed soil	1.83	24.34	30.91	25.29	4.82	5.53	7.27

$$\bar{Q}_{surf} = \frac{(P-I_a)^2}{(P-I_a+S)} \text{ for } P>I_a;\ \bar{Q}_{surf}=0 \text{ for } P\leq I_a \quad (1)$$

Where $\bar{Q}_{surf}$ is the surface runoff (mm), in this study it is the average runoff depth per hour; P is the precipitation (mm), in this study it is the precipitation per hour; S is the amount of water storage available in the soil profile or the maximum storage (mm), and I_a is the initial abstraction (mm). To reduce the numbers of variables, the empirical relationship $I_a = 0.2S$ was adopted, which then gives the most familiar form of the runoff equation [29]:

$$\bar{Q}_{surf} = \frac{(P-0.2S)^2}{(P+0.8S)} \text{ for } P>0.2S;\ \bar{Q}_{surf}=0 \text{ for } P\leq 0.2S \quad (2)$$

Table 2. Calibrated parameter value and the relative error (RE) (%).

Period	Calibration period	Verification period			
Rainfall Intensity (mm/h)	75	45	90	105	120
RE (%)	0.015	3.41	0.40	2.20	1.77

Because the range of the S value is too large to obtain a suitable value, the dimensionless parameter, curve number (CN), was introduced into this formula. The parameter is related to CN by the relationship:

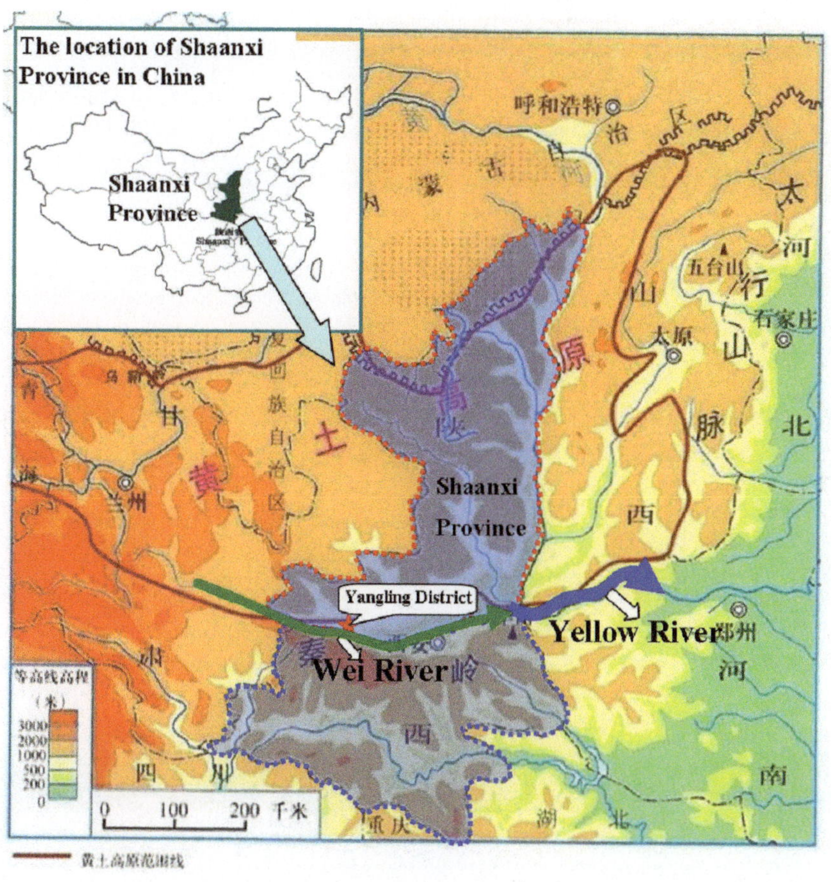

Figure 2. The schematic map of Wei River and Yangling District location.

Table 3. The main calibrated parameters of the MODFLOW model.

Parameters	Kx1(m/s)	Kz1(m/s)	Kx2(m/s)	Kz2(m/s)	Kx3(m/s)	Kz3(m/s)	Sy	Ss
Minimum	1.16×10^{-8}	1.16×10^{-9}	5.79×10^{-7}	5.79×10^{-8}	5.79×10^{-5}	5.79×10^{-6}	0.05	1×10^{-5}
Maximum	5.79×10^{-6}	5.79×10^{-6}	5.79×10^{-5}	5.79×10^{-5}	5.21×10^{-4}	5.21×10^{-4}	0.5	1×10^{-3}
Initial Value	2.90×10^{-6}	2.90×10^{-7}	3.18×10^{-5}	3.18×10^{-6}	2.2×10^{-4}	2.2×10^{-5}	0.2	1×10^{-4}
Calibrated Value	1.12×10^{-7}	2.52×10^{-9}	1.50×10^{-6}	7.47×10^{-7}	3.1×10^{-4}	7.60×10^{-5}	0.36	1.385×10^{-4}

$$CN=\frac{25400}{(S+254)};\ S=25.4\left(\frac{1000}{CN}-10\right) \tag{3}$$

The CN is a comprehensive parameter which related to initial soil moisture, slope, vegetation, soil type and land use status etc [30]. It appeared that the curve numbers were used as a proxy for the retention parameter S in order to scale the curves to a convenient range between zero and 100 [29].

Meanwhile, the runoff amount could be calculated combined the rainfall duration by the relationship:

$$Q_{surf}=\bar{Q}_{surf}\times h \tag{4}$$

Where Q_{surf} is the total runoff of rainfall (120 mm) (mm); h is the rainfall duration (h).

Relative error (RE) was selected as the statistical evaluation index of SCS and it was defined by the equation below:

$$RE=\frac{Q_{cal}-Q_{obs}}{Q_{cal}} \tag{5}$$

Where Q_{cal} was the calculated runoff by SCS model (mm); Q_{obs} was the observed runoff obtained from the rainfall experiment (mm).

Visual Modflow Model

MODFLOW (Modular Three-dimensional Finite-difference Ground-water Flow Model)) is a modular finite-difference ground-water flow model published by the U.S. Geological Survey, which can simulate ground-water flow in a three-dimensional heterogeneous and anisotropic medium. Using the finite-difference method, the domain in which flow is to be simulated is divided into a rectilinear mesh of rows, columns, and layers [31–33]. Visual MODFLOW is also a three-dimensional groundwater flow and contaminant transport modeling application that integrates MODFLOW, MODPATH, MT3DMS, WinPEST, Zone Budget, and so on. Applications include well head capture zone delineation, pumping well optimization, aquifer storage and recovery, groundwater remediation design, simulating natural attenuation, and saltwater intrusion [32].

The three-dimensional movement of ground water of constant density through porous earth material may be described by the partial-differential equation [31]:

$$\frac{\partial}{\partial x}\left(Kxx-\frac{h}{x}\right)+\frac{\partial}{\partial y}\left(Kyy-\frac{h}{y}\right)+\frac{\partial}{\partial z}\left(Kzz-\frac{h}{z}\right)+W =Ss\frac{\partial h}{\partial t} \tag{6}$$

where Kxx, Kyy, and Kzz are values of hydraulic conductivity along the x, y, and z coordinate axes, which are assumed to be parallel to the major axes of hydraulic conductivity (m/s); h is the potentiometric head (m); W is a volumetric flux per unit volume representing sources and/or sinks of water, with $W<0$ for flow out of the ground water system, and $W>0$ for flow into the system; Ss is the specific storage of the porous material (1/m); and t is time (min).

The calibration statistics can be reported in the result plots and the statistical evaluation indexes, including Calibration Residual (CM), Residual Mean (RM), Absolute Residual Mean (ARM), Standard Error of the Estimate (SEE), Root Mean Squared (RMS), Normalized Root Mean Squared (NRMS) and Correlation Coefficient (Cor) [32]. Based on the principle of the modeling and simulated rainfall experiments, a transient three-dimensional groundwater flow model was built using Visual MODFLOW 4.1. The model domain was divided into a 50×265 array of 4×10^{-4} square meter cells uniformly. The model consisted of three individual layers with gradient of 3° and the thicknesses of layers were respectively 0.5, 1 and 98.5 cm from top to bottom. The transient flow simulation was selected as the flow type and the time unit for all parameters was minute. The edge of the model domain was modeled as a no-flow boundary. Actual quantities of groundwater abstraction from drain pipe were treated as flux boundary condition in form of pumping well. Groundwater recharge from rainfall was modeled as recharge package in MODFLOW. The initial ground water level was 0.39 m high and paralleled to the bottom of the model. The initial value and the range of hydraulic conductivity and storage coefficient values at different layers were assigned to each active grid cell by interpolation of discrete property data derived from water releasing test analysis and regional geology data.

Simulated Rainfall Experiments

Experimental conditions and equipment. The simulated rainfall experiments were carried out in the Rainfall Simulation Hall of the State Key Laboratory of Soil Erosion and Dryland Farming on the Loess Plateau during the period of June to October of the year 2012. The simulated rainfall system, with automatic simulation device of under sprinkler, could ensure the kinetic energy of simulated precipitation close to the natural rainfall for the mean fall-height of 18 meters. And the calibration tests showed that rainfall uniformity was greater than 85%. The experiments were conducted in the sand-box model [34,35] (figure 1), 5.3 m×1 m×1 m at the Rainfall Simulation Hall. The

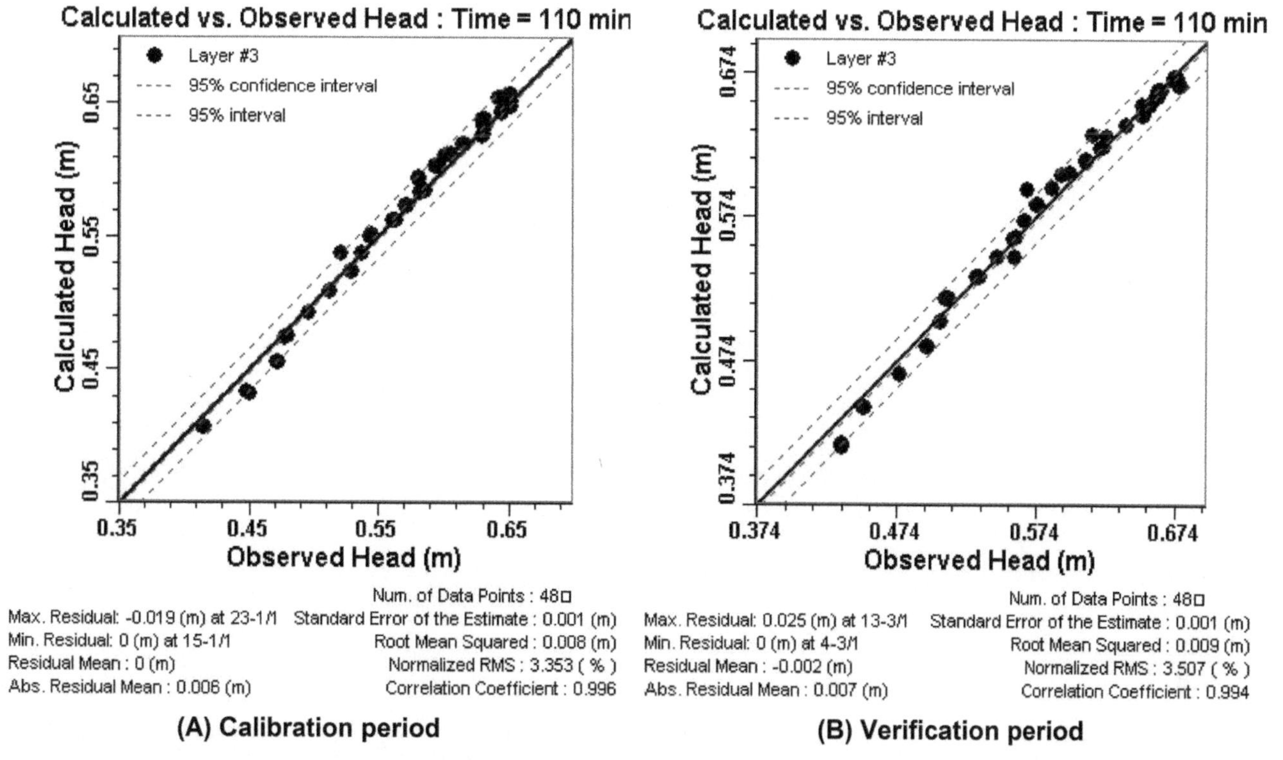

Figure 3. The scatter graph of calculated vs. observed values.

slope of the model could be adjusted manually from 0° to 35°. There were two water tanks, 0.15 m×1 m×1 m in front and back of the model, for the regulation of the groundwater level. Above the front side of the water tank there was one surface water groove and there was one drainage pipe of groundwater in the 0.39 m high at the front side of the water tank.And one hundred and twenty 0.2 m×0.2 m sets of piezometric tubes of level observation were installed in left side of the sand-box model. And two neutron probe access tubes down to a depth of 0.9 m in experimental flume for soil moisture control.

The experimental flume was fixed at an angle of 3° in this study. Three water and soil conservation measures (bare slope, straw mulching and grass land) were considered and the soil surface condition should be roughen to keep the beginning condition consistent every time before experiment. The precipitation for a control to be equal was 120 mm. Six gradient rainfall intensities (45~120 mm/h) with uniform rainfall conditions were simulated for 160 to 60 min correspondingly. The initial ground water level was parallel to the bed of the flume and the distance between two lines was 0.39 m.

Experimental materials and monitoring methods. The test materials included riversand and Lou soil. The riversand samples were dug from the middle and lower reaches of the Wei River bank in Yangling District and the Lou soil was also collected from Yangling District, Shaanxi Province, China. Figure 2 is the schematic map of Wei River and Yangling District location. The mechanical composition of soil particles was shown in table 1. The samples were air-dried for about ten days and sieved through a series of corresponding magnitude sieves. Then the test materials were packed into the flume layer by layer. The experimental flume consists of three individual layers and the thickness was respectively 0.5, 1 and 98.5 cm from top to bottom. The medium sand (0.25 mm ~ 0.5 mm) was paved in the third 98.5 cm deep layer and the soil bulk density was 1.4~1.5 g/cm^3. In the second layer the riversand particle size was less than 0.25 mm and the average soil bulk density was about 1.6 g/cm^3. On the top layer, the soil used was the composite sandy loam, composed of riversand (<0.25 mm) and Lou soil, the weight ratio of this composite soil was about 2:5 and the average soil bulk density was also about 1.6 g/cm^3. The setting of water and soil conservation measures as follows: for the straw mulching, the quantity of arid straw mulching was designed as 0.4 kg/m^2 uniformly each experiment with coverage of 85~90%; for the grassland, the grass was grew on the soil in the flume and its species was ophiopogon japonicus with coverage of 65%~70% and the planting structure was, 10 cm row spacing×5 cm plant spacing×8.2 cm average plant height.

The main monitoring items measured during the experiments were: the surface runoff amount, the surface runoff in the process, underground runoff and groundwater level etc. Surface runoff was collected by collecting buckets at the beginning of runoff yield. The surface runoff in the process was collected for 30 seconds at 5-min intervals during experiments and the runoff amount was collected during the whole process. After groundwater level adjustment, recorded the initial level first and the dynamic changes every 10 minutes with the changes begin for about 4

Table 4. Error indexes of calculated vs. observed values.

Evaluation Indexes	RM/m	ARM/m	SEE/m	RMS/m	NRMS/%	Cor
Calibration	0.0002	0.00727	0.0012	0.0097	4.146	0.996
Verification	0.000375	0.00843	0.00119	0.00981	4.259	0.994

Table 5. The benefits of water reduction in field controlled and uncontrolled small watersheds %.

Water System	Watershed Name	Area (km^2)	Condition	Representative Type Area	Year	Benefits δ(%)
Kuye River	Mengjiagou	2.03	Controlled	Earth-rocky Mountainous Area	1959~1961	42.64
	Yangyagou	1.88	Uncontrolled			
Yuxi River	Qingcaogou	0.373	Controlled	Half sandstorm area	1959~1960	88.19
	Wangjiagou	0.434	Uncontrolled			
	Tiaogou	0.7677	Controlled		1959~1961	16.51
	Lijiagou	0.693	Uncontrolled			
Wuding River	Jiuyuangou	70.1	Controlled	Loess Hilly and Gully Region	1959~1969	16.88
	Peijiamaogou	41.2	Uncontrolled			
	Xiangtagou	0.454	Controlled		1958~1961	23.70
	Tuanyuangou	0.491	Uncontrolled			
	Wangmaozhuanggou	5.967	Controlled		1962~1963	43.70
	Lijiazhaigou	5.45	Uncontrolled			
Yanhe River	Dabiangou	3.7	Controlled		1963~1967	39.63
	Xiaobiangou	3.925	Uncontrolled			
Luohe River	Sigou	4.37	Controlled	gully region of loess plateau	1959~1961	57.98
	Nangou	5.11	Uncontrolled			
Juhe River	Guanzhuanggou	3.39	Controlled		1959~1961	57.98
	Yuanguzhuanggou	2.82	Uncontrolled			
Luohe River	Xingshugou	0.522	Controlled	Loess Hilly and Gully Region	1958~1961	83.62
	Beilougou	0.334	Uncontrolled			
Jinghe River	Fengyugou	1.176	Controlled		1958~1960	14.71
	Wangjiagou	0.874	Uncontrolled			
Chanhe River	Yaojiagou	7.815	Controlled	Terrace region of loess plateau	1960~1961	64.99
	Dicungou	4.4	Uncontrolled			
Average						44.09

hours. Meanwhile, underground runoff was collected by measuring cylinder per 2000 ml uninterruptedly for about 8 hours. Then according to the condition of the underground runoff, the monitoring interval was lengthened gradually until the water level dropped to the initial control water level. Soil moisture (neutron probe method) measurements were taken immediately before and after precipitation to ensure replicability of the initial soil surface conditions. And water temperature was also recorded during each simulation.

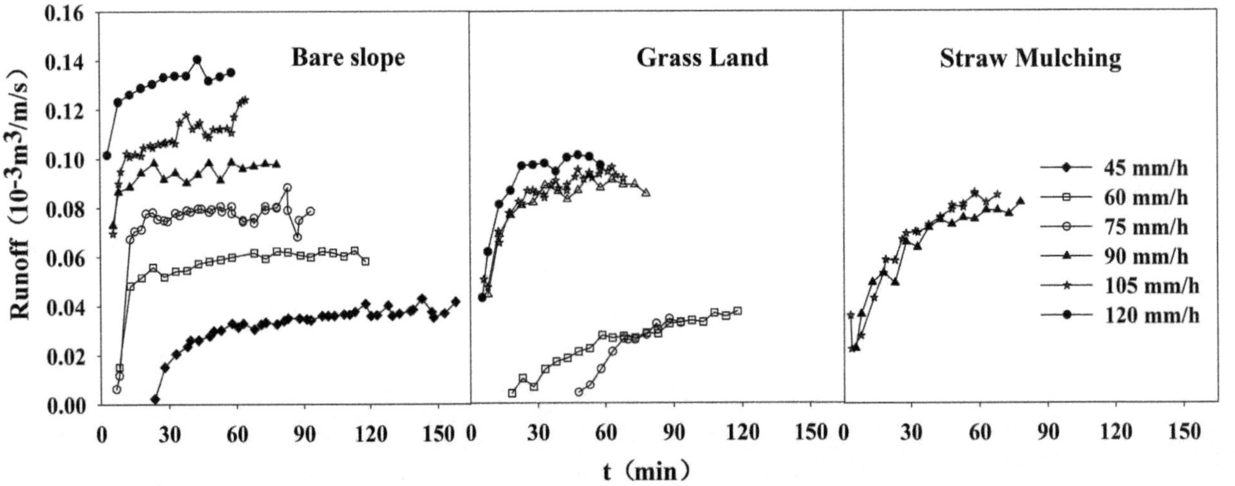

Figure 4. The time-series graphs of the surface runoff under three simulation scenarios.

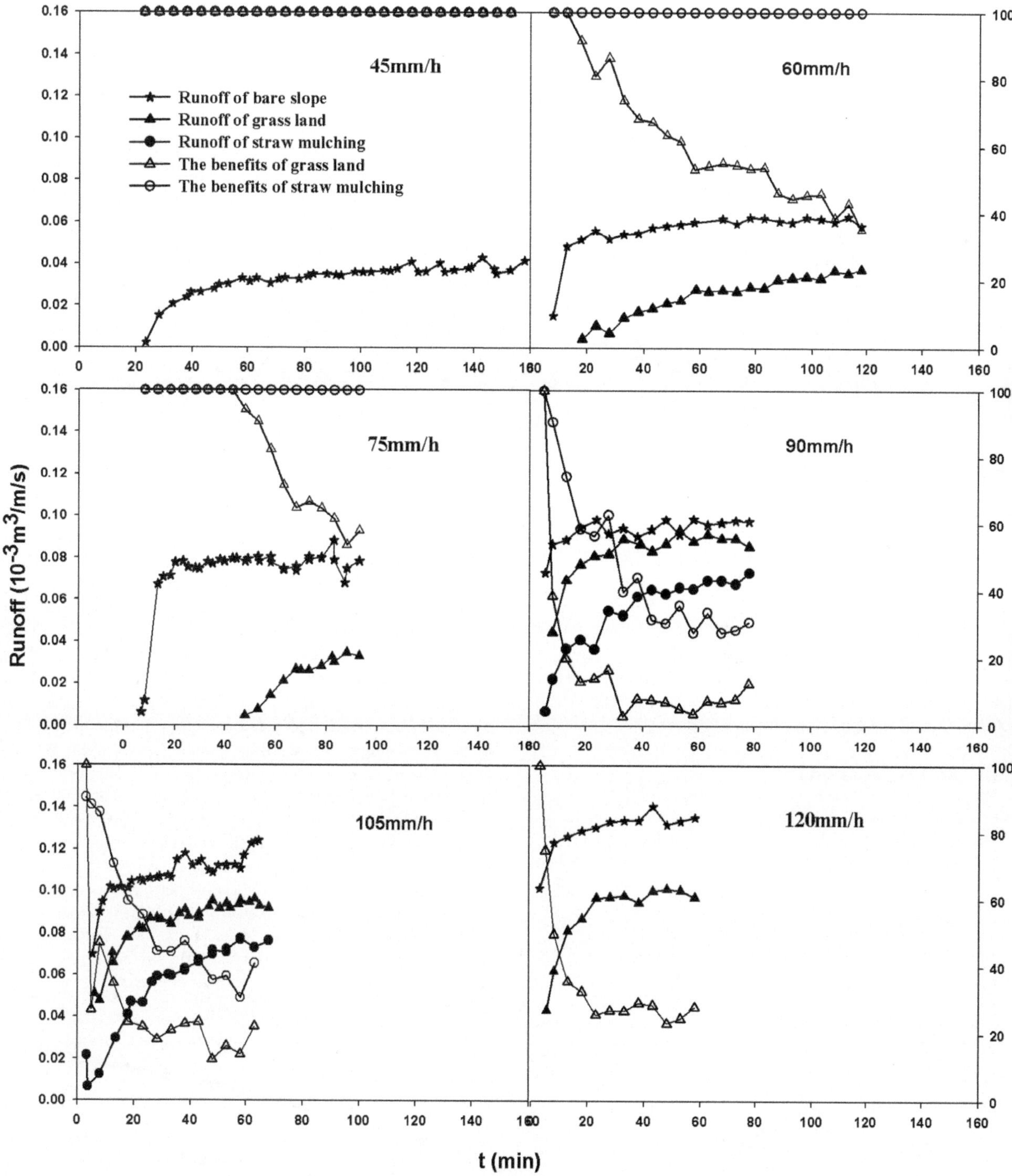

Figure 5. The time-series graphs of the surface runoff and the benefits of surface water reduction.

Ethics Statement

No specific permissions were required for these sampling activities because the location is not privately-owned or protected in any way and the field activities did not involve endangered or protected species.

The Benefits of Water Quantity Change for Soil and Water Conservation Measures

The quantity of surface water and groundwater could be changed by adoption of conservation management practices. Quantitative analysis of this change could be computed by the relationship:

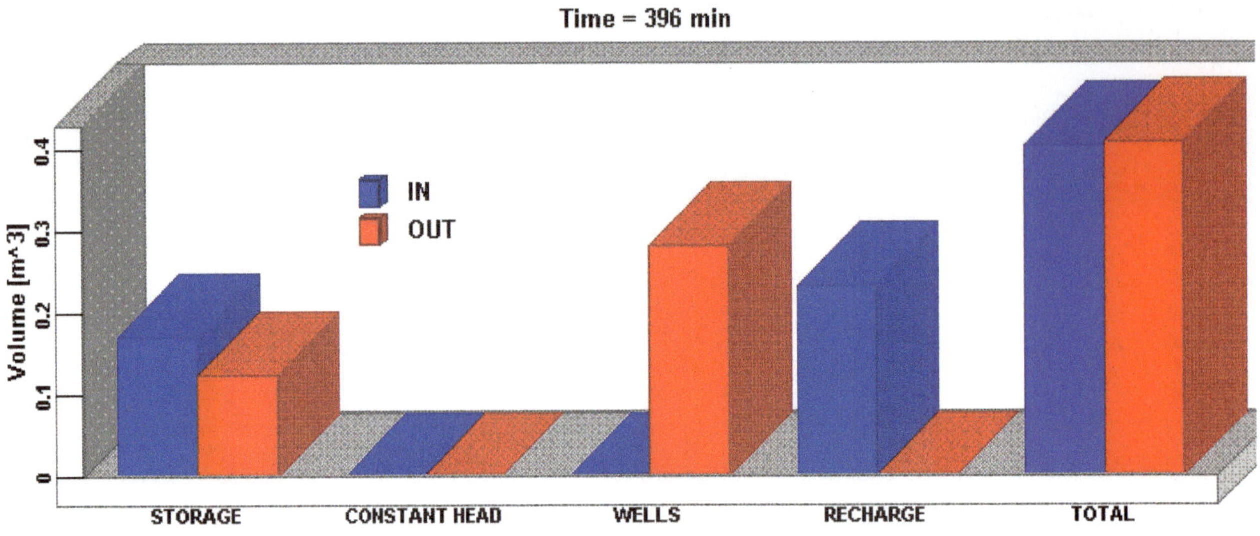

Figure 6. The flow mass balance graph.

$$\delta = \frac{Q_c - Q_{uc}}{Q_{uc}} \times 100\% \tag{7}$$

Where δ is the benefit of water change contributed by soil and water conservation measures, %; Q_c is the runoff modulus in the scenario controlled by soil and water conservation; Q_{uc} is the runoff modulus in the bare slope scenario.

Results and Discussions

Model Calibration and Validation

Model calibration is the process whereby selected model input parameters are adjusted within reasonable limits to produce simulation results that best match the known or measured values [32]. It is the most critical process in building a model, because the quality of the calibration and validation inevitably determines the reliability of any conclusions and recommendations made using the simulation results.

Calibration and Validation of SCS Model

Where concurrent rainfall and runoff data are available, an 'optimal' curve number can be found by calibrating the curve number to the data [29]. Based on the rainfall runoff data obtained from simulated rainfall experiment for the rainfall intensity 75 mm/h of bare slope scenario, the SCS model was calibrated. Observed $\bar{Q}_{surf}$ and P were 50.01 mm and 75 mm respectively. According to eq. (2) and (3), S and CN values were calculated successively. The calibrated CN value was 90.33. Then based on precipitation data from rainfall intensity 45, 90, 105, 120 mm/h of bare slope scenarios respectively, the SCS model were verified by comparing the estimated runoff with in situ measured data.

Based on the statistical analysis of model results, the relative error (RE, eq. 5) in the estimated runoff ranged from 0.40 to 3.41% (table 2) during the verification period respectively. The calibration and verification statistics showed the model did a relatively good job for surface water predicting. Hence, the SCS model was available for the predicting in this study. So surface runoff of bare slope scenario could be simulated by the verified

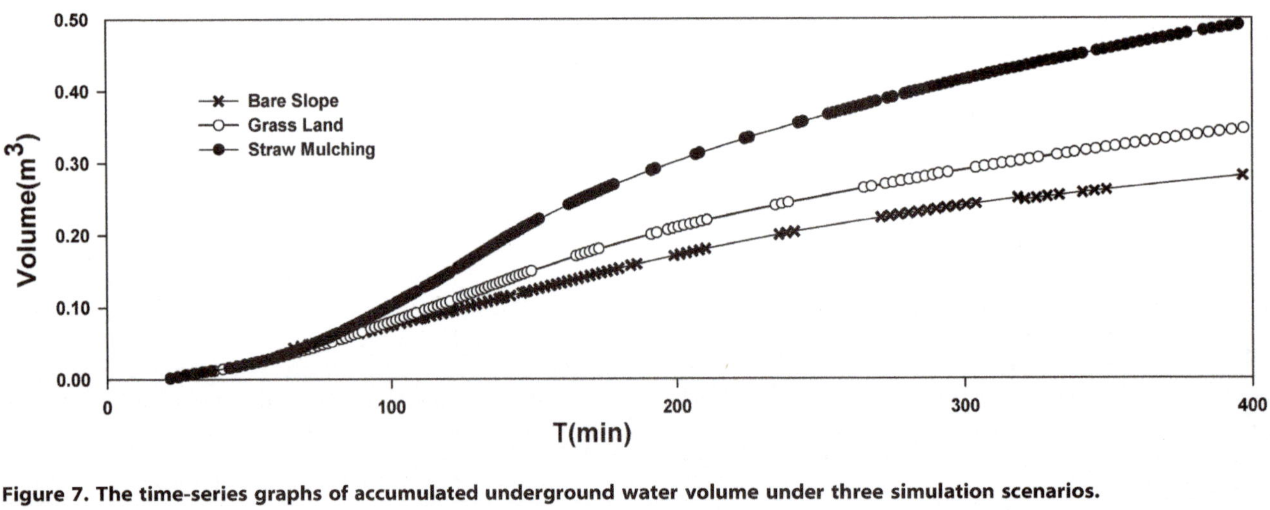

Figure 7. The time-series graphs of accumulated underground water volume under three simulation scenarios.

Table 6. Differences of calculated heads of bare slope vs. observed values of grass land/straw mulching.

Scenario	Average change(cm)	CM_{max}/cm	RM/cm	ARM/cm	RMS/cm	NRMS/%
Grass Land	3.91	−11.8	−3.6	4.1	4.5	14.1
Straw Mulching	7.63	−20.5	−7.3	7.4	7.8	23.4

SCS model and the calculated result was compared with the corresponding observed runoff of lawn and straw mulching experiments to study the impact of soil and water conservation on surface water.

Calibration and Validation of Visual MODFLOW Model

The main parameters of the model were: hydraulic conductivity (Kxx, Kyy, and Kzz), storage (specific storage (Ss), specific yield (Sy)). The initial values and the range of these parameters at different layers were provided by the water releasing test analysis combined with the regional geology data. These data were imported into the constructed MODFLOW model as the initial parameters and adjusted during the process of model calibration. Then based on the simulated rainfall experiment of the bare slope, which intensity was also 75 mm/h and rainfall was 120 mm, a uncertainty analysis was carried out after calibration to quantify the uncertainty in the calibrated model caused by uncertainty in the estimates of aquifer parameters, recharge boundary conditions, initial head conditions and river-aquifer interactions [18]. The uncertainty analysis showed that horizontal conductivity in the third layer (Kx3), specific yield (Sy) were the most sensitive parameters in this model. The main parameters calibrated were shown in table 3. Then to verify the model used another simulated rainfall experiment data of the bare slope, which intensity was 45 mm/h and rainfall was also 120 mm.

The scatter graph of calculated vs. observed values was the default calibration graph. This graph represents a snap-shot in time of the comparison between the values calculated by the model (Y-axis), and the values observed or measured in the field (X-axis) [32]. For example, the scatter graphs at t = 110 min of the calibration period and the verification period (figure 3) showed that most of the data points intersect the 45 degree line on the graph where X = Y. These represented an ideal calibration that simulated hydraulic heads were consistent with the observed heads. The calibration statistics were reported in the footer of the calibration plots window when the calculated vs. observed scatter graph were displayed [32]. Based on these statistical analysis of model results, error indexes including the residual mean (RM), the absolute residual mean (ARM), the standard error of the estimate (SEE), the root mean squared error (RMS), the normalized root mean squared (NRMS) and the correlation coefficient (Cor) for groundwater levels at the location of 48 observation boreholes during the calibration period and the verification period respectively were shown in table 4.

Based upon comparisons the scatter graphs and the calibration statistics between the observed and simulated hydraulic head, we

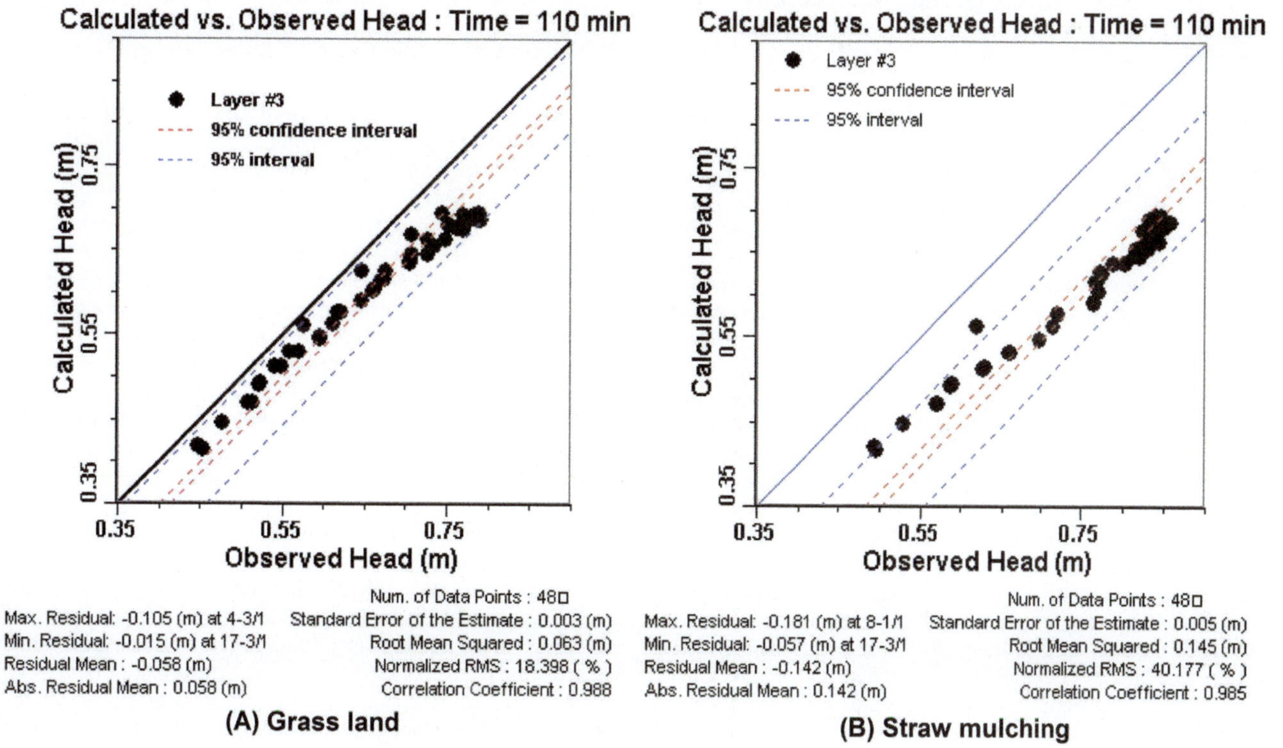

Figure 8. The scatter graph of calculated vs. observed values.

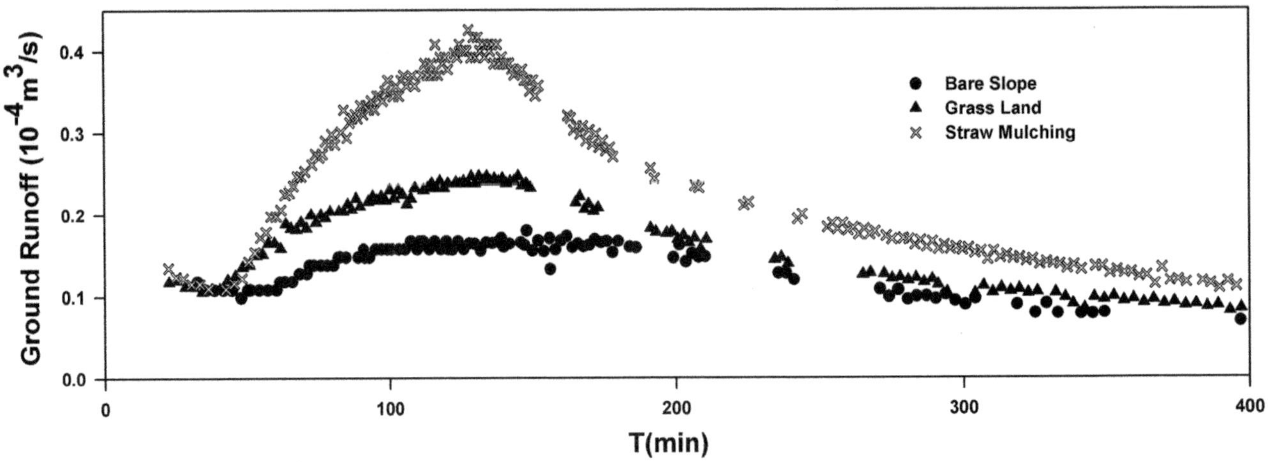

Figure 9. The groundwater runoff of bare slope vs. the discharge of lawn and straw mulching.

could see the groundwater levels simulated by the calibrated groundwater flow model were generally consistent with the physical system represented.

The Impact of Soil and Water Conservation on Surface and Ground Water

Then the calibrated and verified SCS and MODFLOW models were used to simulate rainfall intensity 60 mm/h of bare slope scenario to research the hydrologic impact of soil and water conservation. First, surface runoff amount was simulated by the SCS model and was compared with the corresponding results of lawn and straw mulching experiments to study the impact for the surface water. Second, the calibrated groundwater transient flow model with the input levels obtained from corresponding lawn and straw mulching experiments was used to assess the impact of land change on groundwater.

The Impact of Soil and Water Conservation on Surface Water

The surface runoff modulus amount calculated by SCS model was 0.364 m^3 in bare slope scenario, while the surface runoff observed from the rainfall experiments were 0.154 m^3 and 0.012 m^3 in the lawn and straw mulching scenarios respectively. Compared to the bare slope scenario, the benefits of surface water reduction by these two measures were 57.8% and 92.4% correspondingly. These results indicated that the soil and water conservation played an important role in the benefits of water reduction. These were consistent with the field observation results of the small watersheds. For example, by comparing the 12 groups (24) of controlled and uncontrolled small watersheds (table 5) in Shaanxi Province, China [36], we could see the benefits of surface water reduction by the comprehensive control of soil and water conservation varied from 14.71% to 88.19%, while the corresponding average value was 44.09%.

Not only the amount of surface runoff, but also the runoff process was important for studying the hydrologic impact of soil and water conservation. In this research, the SCS model could reliably estimate runoff amount, but it could not predict runoff process. So the influence for the process of surface runoff was analyzed using simulated rainfall experiments. The time-series graphs (figure 4) were used to evaluate and compare temporal trends in the runoff under three simulation scenarios. The process of runoff would appear to increase first and then stabilize gradually with the passage of rainfall time under three simulation scenarios. This result was consistent with the research that rainfall infiltration rate was decreased first and then stabilize gradually with the passage of rainfall time [37].

Compared with the bare slope scenario, the reduced range of runoff was different in the lawn scenario and in the straw mulching scenario. When rainfall intensity increased from 75 mm/h to 90 mm/h, the latter average runoff was 7.3 times larger than that of the former in the lawn scenario and it started runoff in the straw mulch scenario. During the whole runoff process of the experiments, the benefits of surface water reduction by the grassland varied from 24.3% to 100% and its average value was 58.9%, while the range was 55.3% to 100% and the average value was 84.2% in the straw mulching scenario (figure 5).

The Impact of Soil and Water Conservation on Ground Water

As mentioned above, the calibrated and verified MODFLOW model was then applied to simulating the impact of soil and water conservation scenarios on groundwater recharge and level. The mass balance which is one of the key indicators of a successful simulation [32] was analyzed first. The flow mass balance graph (figure 6) showed the volume of water entering and leaving the system through the flow boundary conditions, and from aquifer storage at the end of simulation period. The total volume flow into the entire system was 0.4059 m^3 which included 0.1719 m^3 storage and 0.2345 m^3 recharge. The total volume flow out-of the model was 0.4070 m^3 which consisted of 0.1252 m^3 storage and 0.2818 m^3 well. The mass balance error for the simulation inflow and outflow was 1.47% $<$2%. The results of the simulation may generally be considered acceptable, provided the model was also calibrated [32].

Figure 7 compared the changes in accumulated underground runoff with scenarios converted from bare slope to grass land and straw mulching. As a result, the cumulative curve of straw mulching was the highest, the grassland took second place, and the bare slope was the lowest. Hence, there was a significant increase for straw mulching, contrasting to the changes for grassland. Then the impact of groundwater cumulant was analyzed quantitatively. At the end of simulation period, the accumulated underground runoff was 0.282 m^3 in the bare slope scenario, 0.346 m^3 in the lawn scenario, and 0.491 m^3 in the straw mulching scenario. Compared to the bare slope scenario, the amount and the benefit of groundwater recharge increased by 0.064 m^3 and 22.7% respectively in the lawn scenario. Accordingly, they were 0.209 m^3 and 60.4% in the straw mulching scenario. It was explicit that groundwater recharge very strongly depended upon land-use type [16].

The groundwater levels responded to the changes in groundwater recharge described as follows. Figure 8 compared the influence of soil and water conservation scenarios on groundwater levels at t = 110 min. Figure 8 (A) was the scatter plots of calculated levels of bare slope scenario vs. observed heads of grass land scenario. And figure 8 (B) was the graph of calculated levels of bare slope scenario vs. observed values of straw mulching scenario. As shown in the figures, the data points were under the X = Y line, so the calculated values were less than the observed values, especially in figure 8 (B). These indicated that groundwater levels increased with land-use converted from bare slope to grassland and even more with land-use converted to straw mulching. To quantify the impact on groundwater heads, the changes between the calculated heads of bare slope scenario and observed levels of grass land and straw mulching scenarios were showed in table 6.

It was explicit from the curve of figure 9 that the soil and water conservations had great influence upon the groundwater runoff. Straw mulching showed the higher runoff, followed by grassland. It was illustrated by further analysis that average flow was 0.14×10^{-4} m^3/s in the bare slope scenario, 0.17×10^{-4} m^3/s in the lawn scenario, and 0.26×10^{-4} m^3/s in the straw mulching scenario, during the simulation period. So the average runoff increased by 1.23 and 1.87 times through measures of lawn and straw mulching.

In summary, both grass and straw mulching scenarios played an important role in reducing the surface runoff and increasing the underground runoff. The effects of measures on redistribution of rainfall runoff was mainly due to the rainfall infiltration rata was changed. The grass and straw mulching dispersed the large raindrops into small raindrops, which reduced the actual rainfall intensity on the ground [38–42] and made more precipitation meet the condition of infiltration. But different soil and water conservation measures had different effects. In this study, straw mulching has more significant hydrological effects than lawn. The main difference between the two scenarios was that the straw mulching scenario had less bare land area compared to lawn scenario.

Conclusions

Based on the principle of water balance, a linked approach for SCS model and Visual MODFLOW was conducted to assess the impact of soil and water conservation on the surface water and groundwater runoff. The calibration results showed that the predicted results matched well with the observed data. Therefore, three land use management scenarios were simulated on the sandbox model to assess the effect of bare slope converted to grass land and straw mulching on water volume, hydraulic head, runoff process of groundwater and surface water.

Soil and water conservation measures could reduce surface runoff effectively. Under the 120 mm rainfall, 60 mm/h rainfall intensity, 5 m^2 area, 3° slope conditions, the comparative results indicated that decrease in surface runoff and increase in subsurface runoff coincided with the land-use converted from bare slope to grass land and straw mulching. Compared to the bare slope, the benefits of surface water reduction by these two measures were 57.8% and 92.4% correspondingly. Not only the individual soil and water conservation measure but also the comprehensive management of the small watershed had a significant benefit of surface water reduction. The comparative results of 12 groups (24) of controlled and uncontrolled small watersheds showed that the benefits of surface water reduction varied from 14.71% to 88.19% while the corresponding average value was 44.09%.

Soil and water conservation measures could promote rainfall recharge groundwater. Under the same condition, compared to the bare slope, the amount and the benefit of groundwater recharge increased by 0.064 m^3 and 22.7% respectively in the lawn scenario. Accordingly, they were 0.209 m^3 and 60.4% in the straw mulching scenario. For the runoff, the average flow of straw mulching was highest, the grassland took second place, the bare slope was the lowest, and the average runoff increased by 1.23 and 1.87 times through measures of lawn and straw mulching.

It was concluded that the soil and water conservation played an important role in weakening the surface runoff and strengthening the underground runoff. Meanwhile the groundwater flow model coupled with the surface water model used in this study, when properly validated, could be used as a tool in the evaluation of soil and water conservation measures on surface water and groundwater resources, and this approach could provide a thought for the study in a watershed scale to help decision-makers manage groundwater resources.

Acknowledgments

We would like to thank Mr. Xiu-quan XU, Mr. Li-zhi JIA, and Mr. Tong ZHANG for their assistant in experiment processing and Miss. Chun-hong ZHAO, Mr. Hong-jie WANG and Mr. Yuan-xing ZHANG for their advices in writing the paper. At the same time, the authors are very grateful for the editors' and reviewer's hard work in reviewing the paper.

Author Contributions

Conceived and designed the experiments: JEG HW. Performed the experiments: HW SLZ MJZ XHL. Analyzed the data: HW JEG. Contributed reagents/materials/analysis tools: JEG HW. Wrote the paper: HW JEG.

References

1. Li ZJ, Zhou PX, Mao LH (2006) General Review on Effects of Soil and Water Conservation Measures on Water Resources in China. Progress in Geography 25: 49–57.
2. Zhu XM, Jiang DS, Zhou PH, Jin ZS (1984) Discussion on the strategic issues of soil and water conservation in the loess region. Bulletin of soil and water conservation 1: 15–18.
3. Ren ME (2006) Sediment Discharge of the Yellow River, China: Past,Present and Future–A Synthesis. Advances in earth science 6: 551–563.
4. Yellow River Conservancy Commission of MWR (2010) Yellow river sediment bulletin, http://www.yellowriver.gov.cn/nishagonggao/2010/index.html.Accessed 2013 March 6.
5. Kim J, Choi J, Choi C, Park S (2013) Impacts of changes in climate and land use/land cover under IPCC RCP scenarios on streamflow in the Hoeya River Basin, Korea. Science of The Total Environment 452: 181–195.
6. Skaggs RW, Amatya DM, Chescheir GM, Blanton CD, Gilliam JW (2006) Effect of drainage and management practices on hydrology of pine plantation. Notes.
7. Sun G, Riedel M, Jackson R, Kolka R, Amatya D, et al. (2004) Influences of management of Southern forests on water quantity and quality. Southern Forest Science, General Technical Report SRS-75 198.
8. Qi S, Sun G, Wang Y, McNulty SG, Moore Myers JA (2009) Streamflow response to climate and landuse changes in a coastal watershed in North Carolina. Trans ASABE 52: 739–749.
9. Li YS (1997) Relation Between Control in Loess Plateau and No-flow in the Yellow River. Bulletin of Soil and Water Conservation 17: 41–45.
10. Ran DC (2006) Water and Sediment Variation and Ecological Protection Measures in the Middle Reach of the Yellow River. Resources Science 28: 93–100.

11. Qi JY Cai QG, Cai L, Sun LY (2011) Scale Effect of Runoff and Sediment Reduction Effects of Soil and Water Conservation Measures in Chabagou, Dalihe and Wudinghe Basins. Progress in Geography 30: 95–102.
12. Jain MK, Mishra SK, Singh VP (2006) Evaluation of AMC-dependent SCS-CN-based models using watershed characteristics. Water Resources Management 20: 531–552.
13. Shi PJ, Yuan Y, Zheng J, Wang JA, GeY, etal. (2007) The effect of land use/cover change on surface runoff in Shenzhen region, China. Catena 69: 31–35.
14. Guo XJ, Wang DJ, Zhuang JQ (2010) Application of SCS model on simulation of the slope -runoff process in dry - hot valley. Science of Soil and Water Conservation 8: 14–18.
15. Freeze RA (1969) The Mechanism of Natural Ground - Water Recharge and Discharge: 1. One - dimensional, Vertical, Unsteady, Unsaturated Flow above a Recharging or Discharging Ground - Water Flow System. Water Resources Research 5: 153–171.
16. Paul MJ (2006) Impact of land-use patterns on distributed groundwater recharge and discharge. Chinese Geographical Science 16: 229–235.
17. Cho J, Barone VA, Mostaghimi S (2009) Simulation of land use impacts on groundwater levels and streamflow in a Virginia watershed. Agricultural water management 96: 1–11.
18. Zhang H, Hiscock KM (2010) Modelling the impact of forest cover on groundwater resources: A case study of the Sherwood Sandstone aquifer in the East Midlands, UK. Journal of Hydrology 392: 136–149.
19. Cooper JD (1980) Measurement of moisture fluxes in unsaturated soil in Thetford Forest. Research Report 66. Institute of Hydrology, Wallingford, UK.
20. Bosch JM, Hewlett JD (1982) A review of catchment experiments to determine the effect of vegetation changes on water yield and evapotranspiration. Journal of Hydrology 55: 3–23.
21. Dams J, Woldeamlak ST, Batelaan O (2008) Predicting land-use change and its impact on the groundwater system of the Kleine Nete catchment, Belgium. Hydrology and Earth System Sciences 12: 1369–1385.
22. Allen A, Chapman D (2001) Impacts of afforestation on groundwater resources and quality. Hydrogeology Journal 9: 390–400.
23. Pouget L, Escaler I, Guiu R, Mc Ennis S, Versini PA (2012) Global Change adaptation in water resources management: The Water Change project. Science of The Total Environment 440: 186–193.
24. Ferrer J, Pérez-Martín MA, Jiménez S, Estrela T, Andreu J (2012) GIS-based models for water quantity and quality assessment in the Júcar River Basin, Spain, including climate change effects. Science of The Total Environment 440: 42–59.
25. Candela L, Tamoh K, Olivares G, Gomez M (2012) Modelling impacts of climate change on water resources in ungauged and data-scarce watersheds. Application to the Siurana catchment (NE Spain). Science of The Total Environment 440: 253–260.
26. Boughton WC, Stone JJ (1985) Variation of runoff with watershed area in a semi-arid location. Journal of Arid Environments 9: 13–25.
27. Yu B (1998) Theoretical justification of SCS method for runoff estimation. Journal of irrigation and drainage engineering 124: 306–310.
28. Rallison RE (1980) Origin and evolution of the SCS runoff equation. Symposium on Watershed Management 1980. ASCE, New York. N, Y. II: 912–924.
29. Boughton WC (1989) A review of the USDA SCS curve number method. Soil Research 27: 511–523.
30. Wang ML, Guo SL, Yi Y (2004) SCS Monthly hydrological model building and its application. Journal of Water Resources Research 2: 002.
31. Harbaugh AW, Banta ER, Hill MC, McDonald MG (2000) MODFLOW-2000, the US Geological Survey modular ground-water model: User guide to modularization concepts and the ground-water flow process. Reston: US Geological Survey.
32. Hydrogeologic Water (2005) Visual MODFLOW v. 4.1 User's Manual. Waterloo, Ontario.
33. Harbaugh AW, McDonald MG (1996) User's documentation for MODFLOW-96, an update to the US Geological Survey modular finite-difference ground-water flow model. United States, Geological Survey.
34. Zhang WZ (1983) The Calculation of the Groundwater Unsteady Flow and Appraisal of Groundwater Resources. Beijing: Science Press.
35. Liang W, Liu L, PAN P (2007) Simulation of nitate-nitrogen transfer in trough scale. Lake Science 19: 710–717.
36. Shaanxi Provincial Bureau of Soil and Water Conservation (1976) Shaanxi Provincial runoff and sediment Data of soil and Water conservation. Xi'an: Shaanxi Provincial Bureau of Soil and Water Conservation.
37. Li YY, Shao MA (2004) Experimental study on characteristics of water transformation on slope land. Journal of hydraulic engineering 4: 48–53.
38. Laflen JM, Colvin TS (1981) Effect of crop residue on soil loss from continuous row cropping. Transactions of the ASAE 24: 1472–1475.
39. West LT, Miller WP, Langdale GW, Bruce RR, Laflen JM, et al. (1991) Cropping system effects on interrill soil loss in the Georgia Piedmont. Soil Science Society of America Journal 55: 460–466.
40. Zhu XM (1960) The impact of vegetation factors on soil and water erosion in Loess area. Acta pedologica sinica 8: 110–121.
41. Tang T, Hao MD, Shan FX (2008) Effects of Straw Mulch Application on Water Loss and Soil Erosion Under Simulated Rainfall. Research of Soil and Water Conservation 15: 9–11.
42. Li M, Yao WY, Li ZB (2005) Progress of the Effect of Grassland and Vegetation for Conserving soil and water on Loess Plateau. Advances in earth science 20: 74–80.

7

Vertical Profiles of Soil Water Content as Influenced by Environmental Factors in a Small Catchment on the Hilly-Gully Loess Plateau

Bing Wang[1], Fenxiang Wen[1], Jiangtao Wu[1], Xiaojun Wang[1]*, Yani Hu[2]

1 College of Environmental Science and Resources, Shanxi University, Taiyuan, China, 2 Library, Hebei University of Science and Technology, Shijiazhuang, China

Abstract

Characterization of soil water content (SWC) profiles at catchment scale has profound implications for understanding hydrological processes of the terrestrial water cycle, thereby contributing to sustainable water management and ecological restoration in arid and semi-arid regions. This study described the vertical profiles of SWC at the small catchment scale on the hilly and gully Loess Plateau in Northeast China, and evaluated the influences of selected environmental factors (land-use type, topography and landform) on average SWC within 300 cm depth. Soils were sampled from 101 points across a small catchment before and after the rainy season. Cluster analysis showed that soil profiles with high-level SWC in a stable trend (from top to bottom) were most commonly present in the catchment, especially in the gully related to terrace. Woodland soil profiles had low-level SWC with vertical variations in a descending or stable trend. Most abandoned farmland and grassland soil profiles had medium-level SWC with vertical variations in varying trends. No soil profiles had low-level SWC with vertical variations in an ascending trend. Multi-regression analysis showed that average SWC was significantly affected by land-use type in different soil layers (0–20, 20–160, and 160–300 cm), generally in descending order of terrace, abandoned farmland, grassland, and woodland. There was a significant negative correlation between average SWC and gradient along the whole profile ($P<0.05$). Landform significantly affected SWC in the surface soil layer (0–20 cm) before the rainy season but throughout the whole profile after the rainy season, with lower levels on the ridge than in the gully. Altitude only strongly affected SWC after the rainy season. The results indicated that land-use type, gradient, landform, and altitude should be considered in spatial SWC estimation and sustainable water management in these small catchments on the Loess Plateau as well as in other complex terrains with similar settings.

Editor: Andrew C. Singer, NERC Centre for Ecology & Hydrology, United Kingdom

Funding: This work was supported by the Natural Science Foundation of China (No. 41201277 and 41101025) and the Natural Science Foundation of Shanxi Province of China (No. 2014011034-2). The funders had no role in study design, data collection and analysis, decision to publish, or preparation of the manuscript.

Competing Interests: The authors have declared that no competing interests exist.

* Email: xjwang@sxu.edu.cn

Introduction

Soil water content (SWC) is a critical factor for plant growth and a determinant of plant distribution in arid and semiarid areas such as China's Loess Plateau [1, 2]. Vertical distribution of SWC can greatly affect soil water movement [3], thereby greatly affecting the biomass production and water use efficiency of plants (e.g., switchgrass) under water stress [4]. Plant-available water stored in the soil profile has a buffering capacity, which, in deep layers, prolongs or alleviates the effects of seasonal or inter-annual drought on plant growth and soil water flux to the atmosphere [5–7]. Research has provided strong evidence that deep soil water depletion plays a key role in sustainable agriculture, ecological restoration, and terrestrial water cycling on the Loess Plateau [8–10]. However, measurement of SWC profiles has been frequently conducted at different spatial scales. The results thus need to be converted before comparison analysis or practical uses. The SWC profile in small catchment is considered to be at a moderate scale for data exchanging. In particular, small catchment is thought to be the basic unit for integrated soil and water loss management in complicated terrain of the Loess Plateau [11, 12]. Characterization of SWC profiles and evaluation of relevant influencing factors at the small catchment scale have implications for hydrological modeling of soil water dynamics and sustainable management of soil water resources in similar areas.

Classical statistics is frequently used to analyze the variability of SWC profiles at the small catchment scale, which involves the estimation of descriptive parameters such as average (mean), variance, standard deviation (STD), and coefficient of variation (CV). Average SWC at individual soil depth intervals or across the whole soil profile is extensively determined. The CV of SWC is also routinely calculated as the temporal variable in a certain period of time or the spatial variable across a specific area. The SWC profile can be divided into distinct intervals by considering its average and CV which exhibit complex spatial-temporal relationships in several plots or watersheds [13–16]. Additionally, ranking method, clustering method, and semivariogram model have been applied for the division of SWC profile [3, 17–20]. However, the above-mentioned methods cannot clearly reflect the variation trend in SWC profiles. Thus, great effort has been made

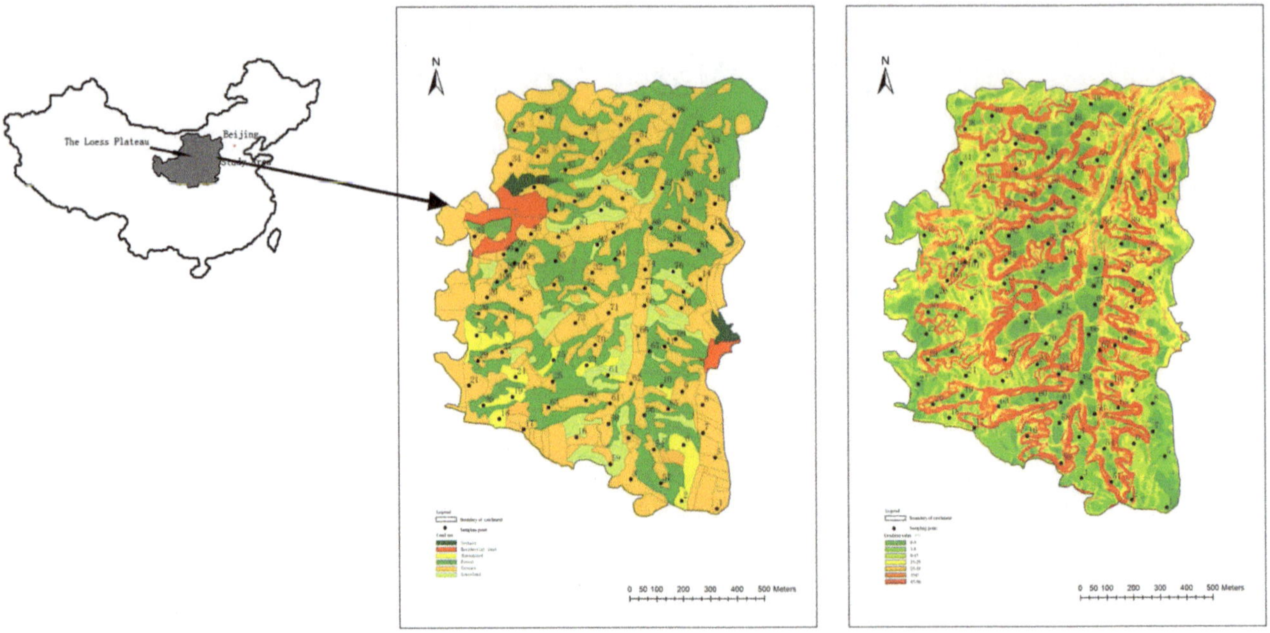

Figure 1. Location of Sanyanjing catchment and distribution of 101 sampling points in the Sanyanjing catchment.

to describe the vertical profiles of SWC through comparing variation curves or variation ranges, in small watersheds related to different land-use types, vegetation species, and/or terrain factors [3, 17, 18, 20–22]. If a massive sample size is involved, however, it becomes difficult to distinguish the vertical profiles and major influencing factors of SWC by direct comparisons.

In recent decades, a great number of studies have been conducted on the spatiotemporal variability of SWC and related influencing factors worldwide. Canton [23] pointed out that wasteland-scale spatial variability of SWC is mainly controlled by surface cover and soil properties in a semi-arid region of Spain, where surface cover counteracts the influence of terrain factors (including gradient, aspect, topographic wetness index, and distance from the river) on SWC distribution. Burnt [24] reported a topographic index which can simulate changes in high-level SWC in a humid climate zone of Devin County, UK. O'loughlin [25] estimated the spatial pattern of SWC distribution in a small catchment using humidity index model based on digital terrain dataset. Hawley [15] discovered that topography is the major factor responsible for the spatial distribution of SWC in an agricultural region, where resultant SWC variation is diminished by vegetation in a moist climate zone in Chickasha, USA. In arid and semi-arid areas, catchment-scale distribution of SWC is strongly affected by land-use/vegetation and topographic indices, e.g., land-use type, soil organic matter content, tillage, soil physical properties, gradient, and aspects [6, 17–19, 21, 26].

Many researchers have focused on the quantification of environmental parameters such as topographic factor, vegetation type, soil texture, and land-use type, in attempt to evaluate their impacts on the variability of SWC. At the small catchment scale, little information is available on the major factors affecting vertical profiles of SWC in cinnamon soil (Haplic Lixisols, FAO) zone on the hilly and gully Loess Plateau [3, 19]. As a regional water reservoir experiencing depletion, the plateau region requires measurements and characterization of deep SWC profiles for the thick soil layer. However, soil sampling for SWC profile analysis at the catchment scale has been commonly conducted at <200 cm depth [18–20, 23]. Wang et al. [21] exceptionally examined SWC along the 0–21 m soil profile on the Loess Plateau, but the reliability of their tests might be affected by a small sample size (11 sites). Deep soil sampling at a larger number of sites and statistical analysis of parameters involving soil depth information will contribute to better understanding of the vertical profiles and influencing factors of SWC.

In the present study, we characterized the vertical profiles of SWC in a small catchment on the Loess Plateau by cluster analysis of two descriptive parameters (mean and regression gradient). Sampling was carried out in the 0–300 cm profile at 101 points throughout the catchment before and after the rainy season, to meet the demand for deep depth, spatial representativeness and temporal comparability. The influences of selected environmental factors (land-use type, topographic factors, and landform) on average SWC were examined by multi-linear regression [6, 26–32]. The results were discussed in order to provide new insights to the vertical profiles and influencing factors of SWC on the Loess Plateau, further providing reference data for sustainable management of water resource in small catchment areas in the semi-arid region with complex terrain.

Material and Methods

1 Site description

This study was conducted in Sanyanjing catchment (112°2′13″, 37°46′23″), which is located on the east margin of the Loess Plateau in Shouyang county, mid-east Shanxi province, China (Figure 1). The catchment has a total area of 1.32 km^2 and the elevation ranges from 1001 to 1160 m. It is a hilly and gully area with mostly deep gully erosion slopes. The landform consists of ridge and gully.

The catchment area has a semi-arid continental climate (Cwa by Koppen Climate Classification) with an average annual precipitation of 474.2 mm (1967–1999). Snow in the winter accounts for ~8% and rainfall in July to September for ~73% of annual precipitation. Monthly average precipitation, potential evapotranspiration and precipitation in 2013 are shown in Figure 2. Annual mean temperature in this area is 8.1°C, with a maximum of 34.7°C and a minimum −20.6°C. The soil type is cinnamon soil (Haplic lixisols, FAO), which consists of 54–62% silt and 10.95–30.15% sand with the bulk density of 1.3–1.4 g/cm^3. Soil texture was measured using a particle size analyzer (SEDIMAT 4–12, UGT, Germany). Soil bulk density was determined through sampling with cutting rings (inner diameter 5.0 cm, volume 100 cm^3) and drying in an oven (105°C, 24 h). The profile of soil texture related to different landforms is listed in Table 1. Maximum soil depth is mostly down to 300 cm on the ridge and bare rock could be rarely seen only on the northern margin of the gully area.

The distribution of land-use types across the catchment is shown in Figure 1. Terrace is the dominant land-use type, accounting for about 60% of the study area. Few terraces had been abandoned for natural restoration of vegetation because of the Grain for Green project since 2000. Grassland is mainly covered with herbs and semi-shrubs, which had never been reclaimed for several decades. About 80% of the woodland is covered with semi-shrubs at steep slopes unsuitable for sampling.

2 Soil sampling

Ethics statement. Sampling activities at the farmland were allowed by the owners. No specific permissions were required at other locations because they were not privately-owned or protected in any way and the field activities did not involve any endangered or protected species.

A total of 101 sampling points were designed in a 150 m×150 m grid throughout the catchment area by considering major land-use types, including terrace (83), abandoned farmland (9), grassland (3), and woodland (6). Soil sampling was carried out during two periods in 2013, from April 29 to May 4 (before the rainy season) and from October 28 to November 1 (after the rainy season). No precipitation occurred during the two sampling periods or a week before sampling. Each sample was taken at 20 cm intervals along the 0–300 cm soil profile using an auger (inner diameter 5.0 cm). The samples were kept in capped aluminum boxes for transportation. Measurement of SWC was conducted using an oven-drying method (105°C, 24 h). At the majority of the sampling points, soils were collected along a vertical profile over 300 cm. A few exceptions were in the north of the gully at the lowest altitude where weathered rock was occasionally encountered. Background information of the 101 sampling points is summarized in Table 2.

3 Data analysis

To identify the variability of SWC profiles at the catchment scale, descriptive parameters were calculated for each profile. Further, we calculated the linear regression coefficient (K value) between SWC and soil depth to represent the variation trend of SWC vertical profiles and the mean value to describe the average level of SWC along the 0–300 cm profile.

The SWC profiles were classified using a combined cluster analysis of the K and mean values. Cluster analysis is the process of grouping a set to data objects into multiple groups (or clusters), so that objects within a cluster share high similarity but are dissimilar to those in other clusters. In this approach, dissimilarities and similarities are assessed based on the attribute values describing the objects and often involve distance measures. Cluster analysis is a statistical classification method for discovering whether the individuals of a population gall into different groups by making quantitative comparisons of multiple characteristics [34]. Here a combined cluster analysis was conducted in three steps: 1) cluster of the mean to three groups, which present the average level of SWC along the vertical soil profile, 2) cluster of K to three groups, which reflect the variation trend of SWC profiles (top to bottom), and 3) combination of the two sets of groups into nine new groups using the between-groups linkage method with squared Euclidean distance criteria [34].

For group comparisons, SWC profiles (0–300 cm) of the same group were averaged and re-plotted. The average curves of SWC were compared between groups to identify the major factors influencing SWC in individual soil layers (0–20, 20–160, and 160–300 cm). On the basis of cluster analysis, the influences of land-use type, topography and landform on average SWC in individual soil layers were examined by multi-regression analysis. The independent variables were land-use type, landform type, Sin(gradient), Sin(aspect), flow accumulation (calculated cell numbers to a grid cell from surrounding cells with the ArcGIS hydrology analysis

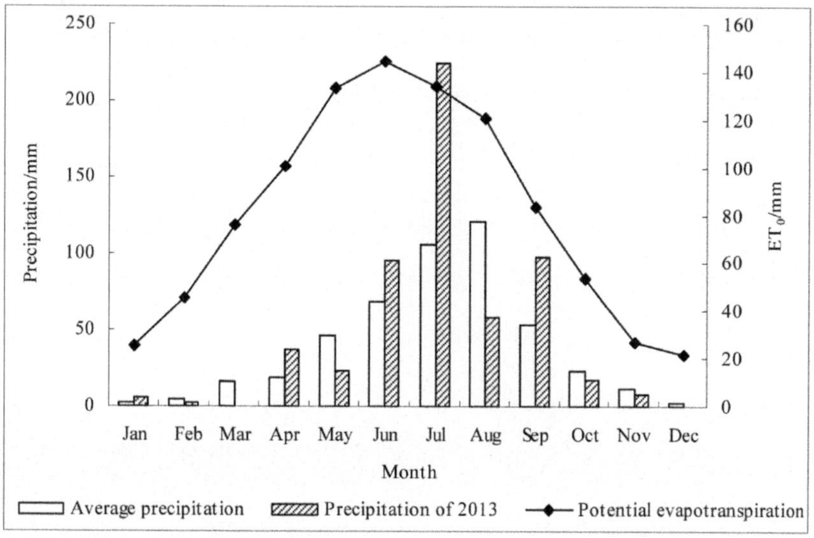

Figure 2. Average annual precipitation and potential evapotranspiration of 1967–1999 and precipitation in 2013 in the Sanyanjing catchment, Shanxi province, China.

Table 1. Soil texture in vertical profiles related to different landforms in the Sanyanjing catchment in Shanxi province, China.

Point description	Soil depth (cm)	Sand (%)		Silt (%)		Clay (%)
		>0.05 mm (%)	0.05–0.02 mm (%)	0.02–0.0063 mm (%)	0.0063–0.002 mm (%)	<0.002 mm (%)
	0–20	26.05	30.40	20.60	7.00	15.95
	20–40	21.35	34.80	20.80	6.90	16.15
	40–60	30.25	25.90	21.60	4.50	17.75
	60–80	28.25	28.80	21.70	2.80	18.45
	80–100	18.45	38.00	20.80	6.70	16.05
	100–120	26.45	30.10	20.40	5.20	17.85
Terrace at	120–140	26.95	27.90	22.00	4.40	18.75
ridge	140–160	16.35	32.60	24.20	8.10	18.75
(Point 33)	160–180	27.75	22.40	22.70	7.90	19.25
	180–200	36.55	17.80	20.60	8.00	17.05
	200–220	18.45	29.00	24.30	9.00	19.25
	220–240	20.45	28.50	23.20	8.60	19.25
	240–260	22.05	27.30	23.90	8.80	17.95
	260–280	31.35	20.30	22.20	9.40	16.75
	280–300	24.65	29.80	20.70	5.90	18.95
	0–20	30.15	20.00	23.20	6.40	20.25
	20–40	27.75	21.80	21.90	6.10	22.45
	40–60	25.95	26.80	20.50	4.50	22.25
	60–80	23.85	27.09	21.20	1.20	26.65
	80–100	0.95	31.00	20.10	20.10	27.85
	100–120	10.95	23.40	18.40	18.40	28.85
Terrace at	120–140	7.45	27.30	19.20	19.20	26.85
ridge	140–160	17.45	19.50	18.90	18.90	25.25
(Point 39)	160–180	10.45	25.80	18.20	18.20	27.35
	180–200	15.75	24.70	17.00	17.00	25.55
	200–220	13.35	21.40	18.60	18.60	28.05
	220–240	20.85	8.80	22.00	22.00	26.35
	240–260	20.25	12.40	20.00	20.00	27.35
	260–280	12.15	23.70	19.10	19.10	25.95
	280–300	24.95	29.00	20.50	0.50	25.05

module), and elevation. The former two factors were categorical variables converted into dummy variables before introduced into the regression analysis; and the latter four factors were continuous variables produced using digital elevation model at 1-m resolution.

SWC data were statistically analyzed in SPSS13.0 (SPSS Inc., Chicago, IL, USA), and topographic features were analyzed in ArcGIS 10.0 (ESRI, Redlands, CA, USA). SWC profiles were drawn in Microsoft Excel 2010 (Microsoft Corp., Redmond, WA, USA) and then clustered in SPSS 13.0 by considering descriptive parameters (maximum, minimum, mean, CV, STD, and K). Multi-regression analysis was performed in SPSS 13.0, with a *P*-value less than 0.05 considered statistically significant.

Results

1 Vertical profiles and descriptive parameters of SWC

The vertical profiles (0–300 cm) of SWC at 101 sampling points before the rainy season were drawn (Figure 3). These SWC profiles showed dynamic variations across the catchment study area, with substantial differences in the soil layers below 100 cm. At a few sampling points, there were obvious soil water depletion (e.g., 67, 75, 85, and 94) and an increasing trend (top to bottom) of SWC (e.g., 12 and 36). High degrees of soil desiccation were rarely detected in the lower soil layers, and low SWC was mainly found in the lower soil layers of woodland.

Descriptive parameters such as maximum, minimum, mean, and CV, STD are commonly used to reveal the spatial-temporal variability of SWC. However, these parameters cannot reflect the variation trend of SWC vertical profiles. To this end, the K value of SWC to profile depth was introduced for quantification of variation trend of SWC vertical profiles (Figure 4). Results showed that before the rainy season, SWC substantially varied between 5.87% and 34.72%, whereas the mean, STD, CV, and K values respectively ranged from 10.57% to 21.76%, 0.47 to 4.53, 3% to 24%, and −0.0405 to 0.0274 along the vertical soil profile (0–

Table 2. Background information of 101 soil sampling points in the Sanyanjing catchment study area in Shanxi province, China.

Land-use type	Vegetation	Landform type	Soil profile/cm	Sampling points
Terrace (n = 83)	Maize	Ridge	300	1, 3–5, 7–9, 12–18, 21–22, 25–26, 28, 32–35, 37, 51, 53, 91, 95, 96, 98
			260	99
		Gully	300	29, 30, 38–46, 48–50, 54–60, 62, 63, 65, 66, 68–74, 77–84, 87–90
			280	47
			260	31, 61
			220	52
			160	86
	Millet	Ridge	300	11
	Maize +five-year-walnut	Gully	300	27
Abandoned farmland	Subshrubs + herbs	Ridge	300	6, 10, 23
(n = 9)			240	2
	Subshrubs + herbs + few ulmus pumila	Gully	220	19
	Robinia peseudoacacia + subshrubs + herbs	Ridge	300	24
	Robinia peseudoacacia+ subshrubs + herbs	Gully	300	64
	Herbs + few almond-apricot	Gully	300	20
	Poplar + subshrubs +herbs	Gully	300	75
Grassland	Subshrubs +herbs	Ridge	300	92, 93
(n = 3)		Gully	300	76
Woodland	Poplar	Gully	300	97
(n = 6)			280	85
	Poplar + herbs	Gully	140	101
	Poplar + subshrubs	Gully	280	100
	Poplar + subshrubs + herbs	Gully	300	67, 94

300 cm). The ranges of the parameters after the rainy season were generally similar with those before the rainy season.

According to the division criteria of Nielson [24], CV in the range of 10–100% indicates moderate variability. Thus, the vertical variability of SWC at all sampling points in Sanyanjing catchment (Figure 4) can be classified to the medium degree. K is the linear regression coefficient between SWC and soil depth. A positive value of K indicates that SWC increases with increasing soil depth. Inversely, a negative value of K indicates that SWC decreases with increasing soil depth. The positive and negative K values of SWC data (Figure 4) are indicative of different variation trends of SWC vertical profiles in the catchment.

2 Clustering of SWC profiles

The vertical profile of SWC across the catchment can be described more clearly using cluster analysis. The 101 SWC profiles before the rainy season were classified into the first three groups by considering the mean value of SWC (Figure 5a), and the second three groups by considering the K value of SWC to soil depth (Figure 5b). The mean value of SWC ranged from 10.57% to 13.13% (low level), 14.15% to 16.86% (medium level), and 17.13% to 21.76% (high level) in the first three groups, whereas the K value of SWC to soil depth ranged from −0.0405 to 0.0106 (decreasing trend), 0.0144 to 0.0163 (stable trend), and 0.0194 to 0.0274 (increasing trend) in the second three groups. By combining the two cluster series, we obtained nine groups of SWC profiles (Table 3).

Before the rainy season, vertical profiles of SWC in groups 1–3 featured low-level SWC (Table 3). In group 1, SWC decreased along the vertical profile (0–300 cm) in woodland (2) and abandoned farmland (1) located in the gully area. In group 2, SWC remained stable along the vertical profile in terrace (2) and grassland (1) located on the ridge as well as woodland (1) located in the gully (Table 4). No sampling points were classified into group 3 with increasing SWC along the vertical profile.

Vertical profiles of SWC in groups 4–6 featured medium-level SWC (Table 3). In group 4, SWC decreased along the vertical profile on the ridge related to terrace (4) as well as in the gully related to abandoned farmland (1) and woodland (1) located in the gully. In group 5, SWC remained stable along the vertical profile in terrace (7) mostly located in the gully, few terrace (3) and abandoned farmland (1) located on the ridge, and grassland (1) and woodland (1) located in the gully. In group 6, SWC increased along the vertical profile in terrace (2) located on the ridge (Table 4).

Vertical profiles of SWC in groups 7–9 featured high-level SWC (Table 3). In group 7, SWC decreased along the vertical profile in terrace (5), grassland (1), and abandoned farmland (1) located on the ridge, and terrace (2), abandoned farmland (1) and woodland (1) located in the gully. In group 8, SWC remained stable along

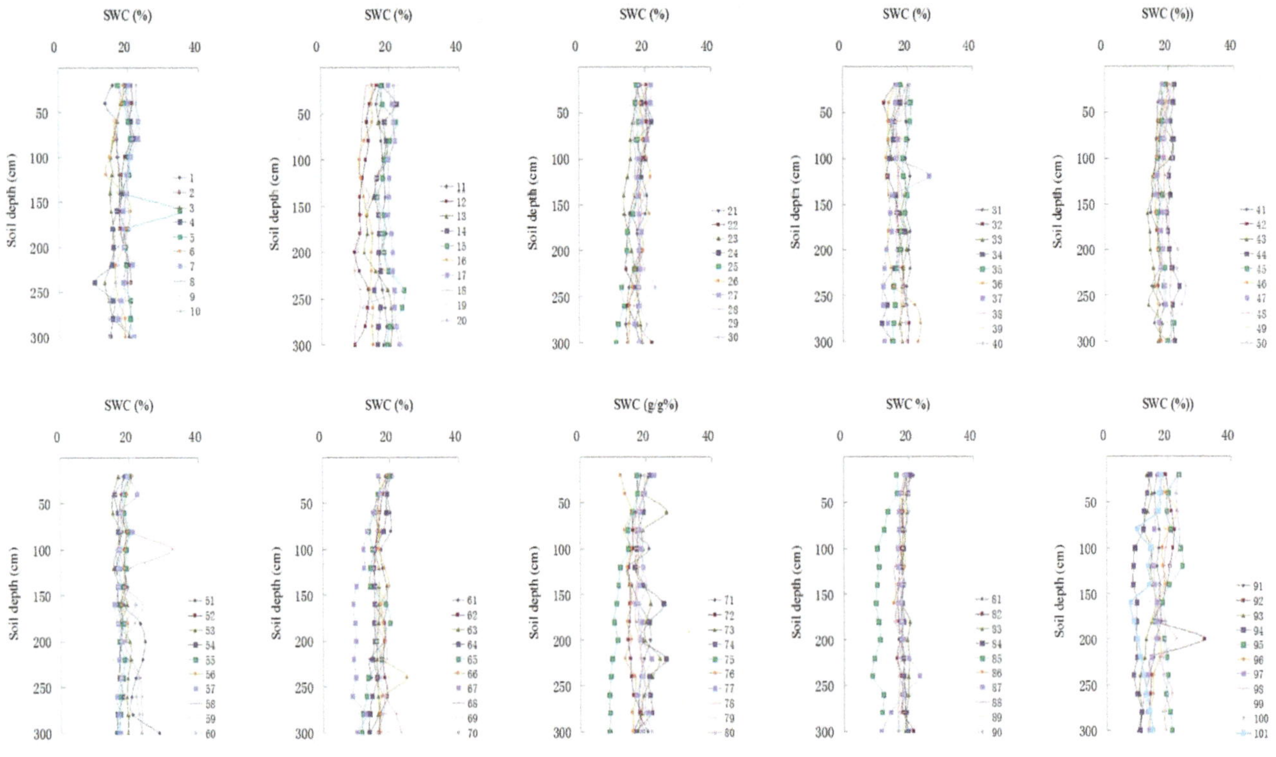

Figure 3. Vertical profiles of soil water at 101 sampling points in the Sanyanjing catchment.

the vertical profile at up to 58 of 101 sampling points, far more than other groups. Most sampling points of group 2 were located in the gully related to terrace (41), and few were on the ridge related to terrace (14) and abandoned farmland (3). In group 9, SWC increased along the vertical profile in terrace located on the ridge (3) and in the gully (1) (Table 4).

Similar grouping of SWC profiles was obtained with data collected after the rainy season. Overall, soil profiles of group 8 with high-level SWC in a stable trend were most commonly present in the catchment, more after the rainy season than before the rainy season. Group 3 of SWC profiles with low level and increasing trend was absent in the study area.

3 The relationships between average SWC and selected environmental factors

According to cluster analysis, there were nine combinations of SWC profiles in terms of average level and variation trend. We averaged SWC profiles of the same group and plotted the average curves (Figure 6), to examine differences of SWC profiles among various types. From Figure 4, we divided the whole soil profile (0–30 cm) into three layers (0–20, 20–160, and 160–300 cm) for multiple linear regression analysis. The results showed that selected environmental factors had significant linear correlations with average SWC at individual layers of 0–20 cm ($P<0.001$, $R^2 = 0.30$; $P<0.001$, $R^2 = 0.37$), 20–160 cm ($P = 0.01$, $R^2 = 0.19$; $P<0.001$, $R^2 = 0.39$), and 160–300 cm ($P<0.001$, $R^2 = 0.32$; $P< 0.001$, $R^2 = 0.43$; Table 5).

Before the rainy season, average SWC in the lower soil layer (10–20 cm) was significantly lower in grassland and woodland than in terrace, with no significant difference between abandoned farmland and terrace ($P_{D51} = 0.109$, $P_{D52} = 0.003$, $P_{D53} = 0.047$, $P_{X1} = 0.013$, and $P_{D61} = 0.005$; Table 5). Additionally, average SWC decreased with increasing gradient, with higher levels on the ridge than in the gully.

In the lower soil layer (20–160 cm), average SWC decreased significantly with increasing gradient ($P = 0.026$; Table 5), and was significantly lower in woodland than in the other three types of land-use types, with no significant differences among the latter three types. Other environmental factors had no significant linear correlation with average SWC ($P>0.05$).

In the deeper soil layer (160–300 cm), there also existed a significantly negative correlation between Sin(gradient) and average SWC ($P = 0.001$; Table 5). Average SWC obviously increased with increasing gradient and was significantly higher in terrace than in abandoned farmland, grassland, and woodland (in descending order).

Similar results can be seen in the data collected after the rainy season. That is, land-use type was the major environmental factor affecting average SWC, whereas landform and altitude strongly affected average SWC only in specific periods and soil layers (Table 5). In the whole vertical profile (0–300 cm), SWC occurred at high levels from upper to deeper layers in terrace, with the lowest level in woodland. Compared with data of terrace, average SWC was relatively low in grassland and woodland in the top (0–20 cm) and deeper soil layers (160–300 cm), with significantly low levels in abandoned farmland soils in the deeper layer only. In the deeper soil layer (160–300 cm), average SWC varied with different land-use types in descending order of terrace > abandoned farmland > grassland > woodland.

Discussion

1 Vertical profiles of SWC at the catchment scale

According to Wang [21], the variability of SWC (as indicated by the CV) varies notably across the whole Loess Plateau, i.e., 15% in

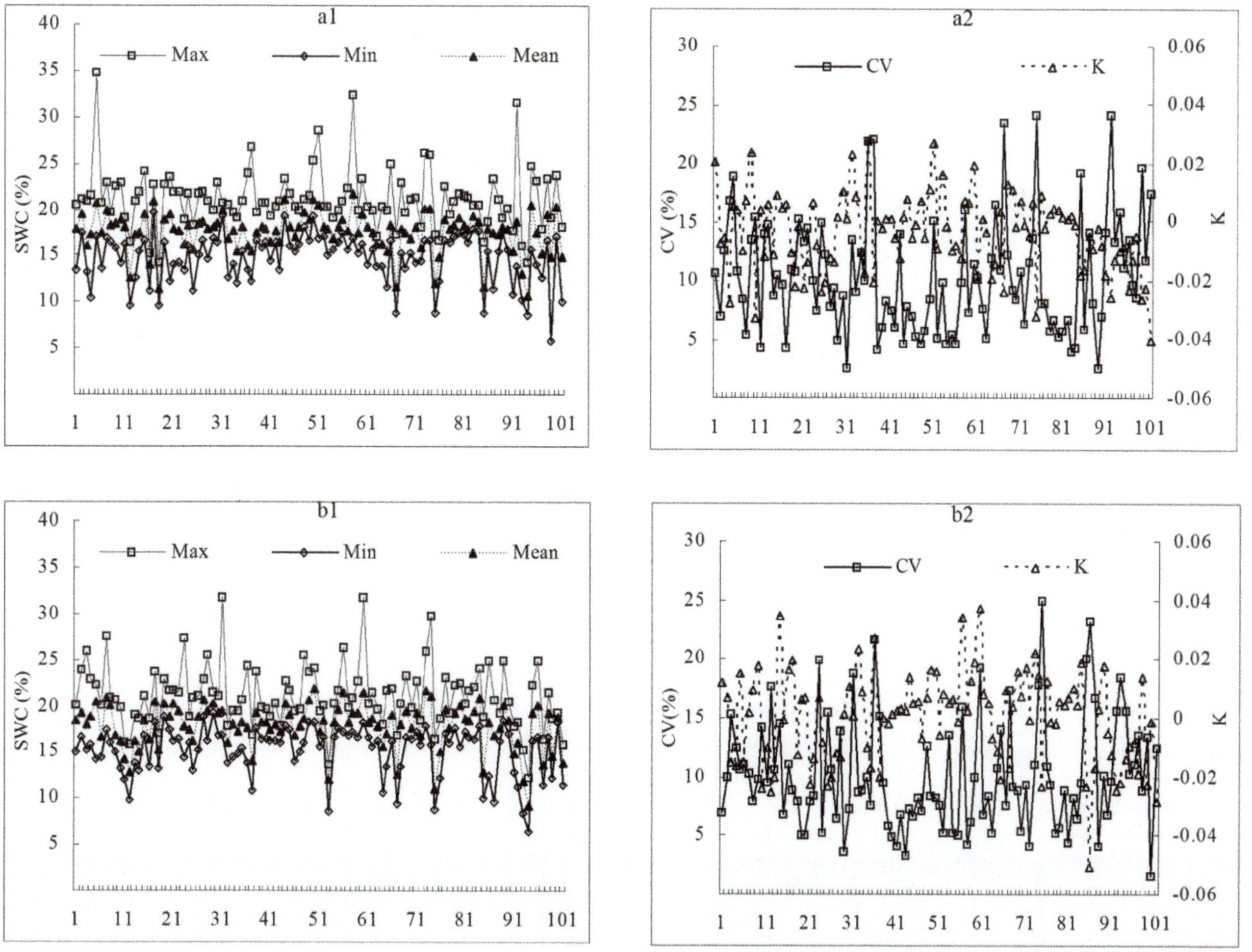

Figure 4. Statistical parameters of soil water content at 101 sampling points across the Sanyanjing catchment. (a. before the rainy season; and b. after the rainy season.)

Changwu and 55% in Shenmu. In the small catchment of Sanyanjing, SWC profiles exhibited weak and medium degrees of variability at 0–300 cm depth [33], with CV in the range of 3–24% (Figure 2). The lower variability of SWC profiles in our study area may be related to the higher SWC levels across the catchment (Pearson correlation coefficient between average SWC and CV,

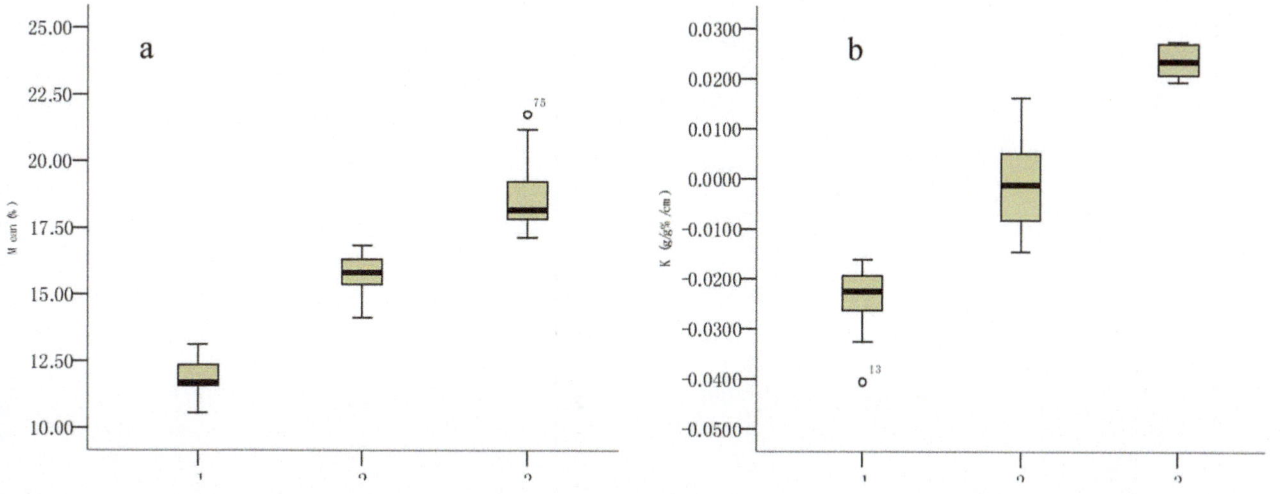

Figure 5. Grouping of 101 vertical soil water profiles in the Sanyanjing catchment before the rainy season by cluster analysis of the mean value (a) and regression gradient (K, b).

Table 3. Combined grouping of 101 vertical profiles of soil water content (0–300 cm) in Sanyanjing catchment by cluster analysis of the mean value and regression gradient.

Cluster by mean	Cluster by K	Combined grouping	Quantity of points		Point Nos.	
			Before the rainy season	After the rainy season	Before the rainy season	After the rainy season
1	1	1	3	0	65,67, 85	-
	2	2	4	1	12,18,93,94	94
	3	3	0	0	-	-
2	1	4	6	15	25,37,64,91,99,101	11,12,18,37,53,64,67,75,76,85,91,93,97,99,101
	2	5	13	0	3,16,23,34,41,43,56,63,65,70,72,76,97	3,16,23,34,41, 43,56,63,65,70,72,76,97
	3	6	2	1	32,36	14
3	1	7	11	1	4,10,19,21,26,35,51,86,92,96,100	86
	2	8	58	80	2,5–8,11,13–15,17, 20,22,24,27–31,33, 38–40,42,44–50,52, 53,55,56–59,62,66, 68,69,71,73,74, 77–84,87–90,95,98	2–10,13,15,16, 17,19–36,38–52,54,55,57–59, 61–63,65,66,68,69–74, 77–84, 87–90,92,95,96,98,100
	3	9	4	3	1,9,51,60	1, 51,60

Table 4. Grouping richness of 101 vertical profiles of SWC (0–300 cm) in relation to different land-use types in Sanyanjing catchment before the rainy season.

Land-use type	Grouping								
	1	2	3	4	5	6	7	8	9
Terrace	0	2	0	4	10	2	7	54	4
Abandoned	1	0	0	1	1	0	2	0	0
Grassland	0	1	0	0	1	0	1	0	0
Woodland	2	1	0	1	1	0	1	0	0

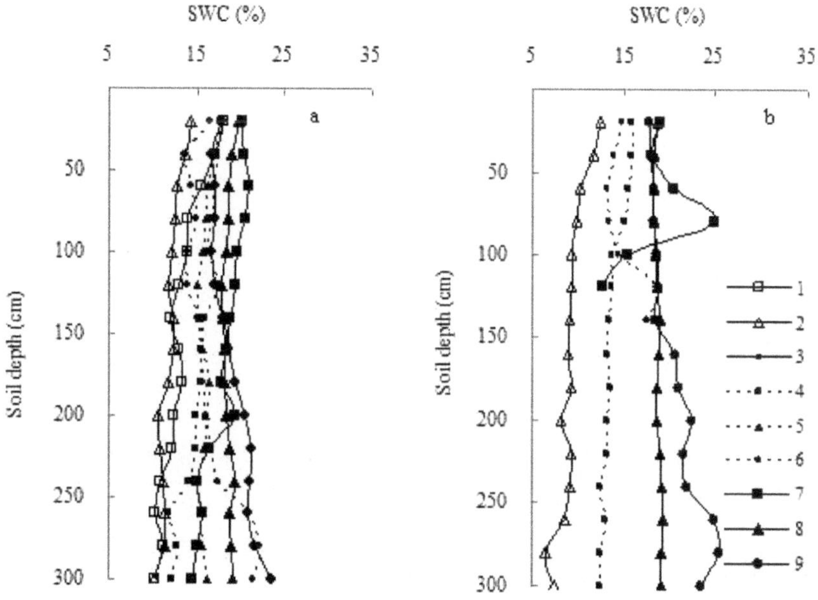

Figure 6. Vertical soil water profiles in relation to different groups in the Sanyanjing catchment study area before and after the rainy season (a. before the rainy season; and b. after the rainy season).

−0.40; $P<0.01$). Qiu [17] found that wetter soil with greater vertical variations in an increasing trend along the SWC profile (mean 13.03%; and STD, 2.3%) is representative in a dry year in Danangou catchment on the Loess Plateau, where the land-use pattern (including slope farmland, terrace, and orchard) differs from that in our study area.

Cluster analysis of the mean and K values provides a clear description for the overall variability of SWC in the vertical profile. Based on combined grouping, the 101 vertical SWC profiles were classified into nine groups with high, medium, and low levels associated with increasing, stable, and decreasing trends (Table 3). More than half of the SWC profiles were obtained from terrace soils in the gully and classified into group 8 (58/101 before the rainy season and 80/101 after the rainy season) with high-level SWC in a stable trend (Tables 3, 4). Despite that all sampling points of woodland were also located in the gully, their average SWC remained the lowest among different land-use types and mostly descended along the vertical profile (Table 4). The above differences can be attributed to the lower soil water consumption by maize crop in the terrace, which generally has shallower root distribution and less above-ground biomass than trees in the woodland. Our observations coincide with previous findings on the Loess Plateau that soil water conditions of terrace, gully farmland, and dam land are better than that of artificial woodland. The latter land-use type is associated with soil desiccation, especially in deep soil layers [2, 22, 35, 36].

Although the cluster analysis divided vertical SWC profiles into nine groups, only eight types were present in the Sanyanjing catchment and no sampling points were classified into group 3 (i.e., low-level SWC with an increasing trend from top to bottom). According to previous research in semi-arid regions, if SWC occurs at low level in the upper soil layer, deep-root crops, shrubs, and trees will consume more soil water in the deeper soil layers through root extraction [37–40]. Additionally, it is hard to achieve soil water recharge in the deeper soil layers by precipitation infiltration because the depth of soil water infiltration is shallow. Therefore, soil desiccation exists in the lower soil layer in case of no groundwater recharge [8–10]. These mechanisms explain the absence of high-level SWC with an increasing trend along the vertical profile in the small catchment of Sanyanjing (Table 3).

2 Effects of environmental factors on average SWC at the catchment scale

Consistent with cluster analysis (Table 3), multiple regression analysis showed that land-use type had a significant effect on soil water status in the small catchment of Sanyanjing (Table 4, 5). This result coincides with the data previously reported in small catchments on the Loess Plateau [17, 18, 22, 23, 29]. For example, Zhang [23] concluded that average SWC (20–200 cm) descends with different land-use types (farmland > grassland > shrub land > and woodland, n = 80) in the small catchment of Zhifanggou. Bai [42] found that average SWC (0–500 cm) ranges from of 9% to 16% in orchard, gradient farmland, terrace, and grassland, but remains less than 10% in shrub land and most woodland (n = 91) in Nangou catchment in the central area of Loess Plateau, Ansai, Shaanxi. The consistency of the data demonstrates that cluster analysis is a reliable method for characterization of SWC profiles.

The effect of land-use type on SWC can be related to the differences existing in anthropogenic activity and vegetation type [22]. Average SWC was found significantly higher in terrace and abandoned farmland than in grassland and woodland along the 0–300 cm profile (Table 5). Abandoned farmland and terrace are associated with artificial tillage in the surface soil layer, which improves soil porosity and loosens soil structure, further enhancing soil water infiltration [22, 43]. Additionally, soil water consumption by crops is less than that in grassland and woodland due to lower leaf area index [21], contributing to the accumulation of SWC. The above mechanisms account for the greater average of SWC profiles with a stable trend to soil depth in terrace and abandoned farmland.

Difference in root distribution is another factor contributing the effect of land-use type on SWC [40]. In the Sanyanjing catchment, average SWC of woodland was higher in the 0–20 cm soil layer but lower in the 20–160 and 160–300 cm soil

Table 5. Multi-linear regression analysis of soil water content and selected environmental factors in three layers (0–20, 20–160, and 160–300 cm) of the vertical soil profile in the Sanyanjing catchment study area.

				Before the	**rainy season**				
Model		**Y_1**			**Y_2**			**Y_3**	
	Unstandardized coefficients B	Standardized coefficients (Beta)	Sig.	Unstandardized coefficients B	Standardized coefficients (Beta)	Sig.	Unstandardized coefficients B	Standardized coefficients (Beta)	Sig.
Constant	14.714		0.066	6.031		0.498	18.514		0.069
X_1	−2.743	−0.23	0.013	−2.751	−0.22	0.026	−4.471	−0.296	0.001
X_2	−0.112	−0.037	0.68	0.042	0.013	0.889	−0.261	−0.068	0.449
X_3	4.86E−06	0.025	0.783	−7.80E−06	−0.038	0.694	1.20E−05	−0.05	0.578
X_4	0.005	0.081	0.507	0.011	0.177	0.175	0	0.005	0.97
D_{51}	1.09	0.146	0.109	0.019	0.002	0.98	−0.788	−0.084	0.354
D_{52}	−3.472	−0.278	0.003	−0.845	−0.064	0.503	−2.553	−0.162	0.073
D_{53}	−1.715	−0.191	0.047	−3.216	−0.342	0.001	−4.745	−0.384	0
D_{61}	−1.512	−0.348	0.005	−0.607	−0.134	0.307	−0.833	−0.151	0.212
	$R^2=0.30$	($P<0.001$)		$R^2=0.19$	($P=0.001$)		$R^2=0.32$	($P<0.001$)	
				After the	**rainy season**				
Model		Y_1			Y_2			Y_3	
	Unstandardized coefficients B	Standardized coefficients (Beta)	Sig.	Unstandardized coefficients B	Standardized coefficients (Beta)	Sig.	Unstandardized coefficients B	Standardized coefficients (Beta)	Sig.
Constant	11.426		0.111	−5.811		0.452	−9.045		0.41
X_1	−2.01	−0.178	0.041	−2.763	−0.221	0.01	−3.688	−0.206	0.015
X_2	0.313	0.11	0.201	−0.132	−0.042	0.618	0.294	0.063	0.44
X_3	6.30E−06	−0.034	0.692	1.80E−05	−0.089	0.285	1.60E−05	−0.054	0.508
X_4	0.007	0.125	0.278	0.023	0.359	0.002	0.027	0.291	0.009
D_{51}	−0.622	−0.088	0.307	−1.161	−0.149	0.08	−1.81	0.162	0.053
D_{52}	−2.374	−0.2	0.021	−2.372	−0.181	0.032	−4.166	−0.222	0.008
D_{53}	−2.599	−0.305	0.001	−4.807	−0.51	0	−6.862	−0.467	0
D_{61}	−2.082	−0.506	0	−1.512	−0.332	0.004	−3.14	−0.476	0
	$R^2=0.37$	($P<0.001$)		$R^2=0.39$	($P=0.001$)		$R^2=0.43$	($P<0.001$)	

Dependent Variable: Y_1 (soil water content of 0–20 cm layer).
Y_2 (average soil water content of 20–160 cm layer).
Y_3 (average soil water content of 160–300 cm layer).
Independent Variables: X_1 = Sin(gradient), X_2 = Sin(aspect), X_3 = flowaccu, X_4 = elevation.
Dummy Variables: X_5 = terrace, (D_{51}, D_{52}, D_{53}) = (0,0,0); X_5 = abandoned farmland, (D_{51}, D_{52}, D_{53}) = (1,0,0).
X_5 = grassland, (D_{51}, D_{52}, D_{53}) = (0,1,0); X_5 = woodland, (D_{51}, D_{52}, D_{53}) = (0,0,1).
X_6 = ridge, (D_{61}) = 1; X_6 = gully, (D_{61}) = 0.
D represents sub-variable; binary variables 0 and 1 for the absence and presence of some land-use type or landform, respectively.

layers than data of grassland (Table 5). The varying trends of SWC profiles between grassland and woodland can be related to different distribution of root system in individual soil layers and stratified root extraction of soil water. The rooting depth of maize crop is reported to be approximately 100 cm and most maize roots are distributed in the soil layer of 0–20 cm, shallower than average rooting depths in grassland (20–60 cm) and woodland (20–100 cm) [40]. Diverse root distribution patterns can lead to different levels of soil water consumption by plants, contributing to great variability of SWC level.

In addition to land-use type, topographic factors strongly affected SWC in the study area (Table 4). This is because the distribution of wind and solar radiation varies with different topographic conditions, leading to different levels of soil evaporation, runoff on gradient, and soil water infiltration [41]. Gradient negatively affected SWC in the soil layers of 0–20, 20–160, and 160–300 cm (Table 5), possibly due to the increased runoff with increasing gradient and resultant reduction of precipitation infiltration [17, 27, 29, 41]. Other topographic factors including aspect and flow accumulation had no significant effects on average SWC in the three soil layers (Table 5). Similarly, Gómez [29] referred that aspect has no obvious influence on SWC in burned and unburned areas. Shi [19] suggested that aspect and catchment area significantly affect SWC during the wet period only, whereas elevation has a significant effect on SWC in arid and humid periods but not in semi-arid and semi-humid periods. In the present study, we found the effect of elevation on SWC of the three layers varying with the period of time and being significant after the rainy season only.

As for the landform type, location of sampling points significantly affected SWC only in the surface layer (0–20 cm) before the rainy season and throughout all the three layers (0–20, 20–160, and 160–300 cm) after the rainy season, with greater values in the gully than on the ridge (Table 5). The effect of landform type on SWC can be related to different levels of soil evaporation as affected by wind strength and solar radiation and soil physical properties. Similarly, Zhang [22] suggested that average SWC descends with different landforms as gully > terrace > slop land > hill top.

Overall, land-use type is the most significant factor affecting SWC while topographic factors and landform type are interacting jointly at the catchment-scale. Because the impact of environmental factors on SWC varies in different periods, it is necessary to increase the observation frequency, in order to better understand the spatiotemporal distribution and influencing factors of SWC in the small catchment. Such work will provide reference data for selecting reasonable environmental parameters in catchment scale SWC simulation over different periods of time.

Conclusions

In this study, cluster analysis enables catchment-scale characterization of soil water profiles in terms of average level and variation trend along the vertical profile, allowing for simple and clear interpretation of the results. A total of nine groups of soil water profiles are recognized but those with low-level soil water content and a decreasing trend are not present in the Sanyanjing catchment. Land-use type, gradient, landform type, and altitude are the major environmental factors significantly influencing average soil water content in the hilly and gully catchment with complex terrain. The former two factors strongly affect soil water content along the 0–300 cm soil profile, whereas effects exerted by the latter two factors vary by soil layer and season.

Understanding the vertical profile of soil water content and evaluation of related major influencing factors in individual soil layers can help with sustainable land use and water management in catchment areas on the hilly and gully Loess Plateau as well as in arid and semi-arid areas with complex terrain. For better estimation of soil water profiles in small catchments, other factors such as fertilization, coverage, and soil physical properties may be considered with respect to specific soil layers.

Acknowledgments

We give our thanks to Prof./Dr. Wenzhao Liu (Institute of Soil and Water Conservation, Northwest A & F University, Yangling) for suggestions on experimental arrangement.

Author Contributions

Conceived and designed the experiments: BW FW JW XW YH. Performed the experiments: BW FW JW XW YH. Analyzed the data: BW FW JW XW YH. Contributed reagents/materials/analysis tools: BW FW JW XW YH. Wrote the paper: BW FW JW XW YH.

References

1. Engelbrecht BMJ, Comita LS, Condit R, Kursar TA, Tyree MT, et al. (2007) Drought sensitivity shapes species distribution patterns in tropical forests. Nature 447: 80–82.
2. Yang L, Wei W, Mo BR, Chen LD (2011) Soil water under different artificial vegetation restoration in the semi-hilly region of the Loess Plateau. Acta Ecologica Sinica 31: 3060–3068 (in Chinese with English abstract).
3. Xing G, Zhang XM, Fei XL, Wu YX (2012) Study on soil moisture content under different land use types in Sunjiacha basin. Agricultural Research in the Arid Areas 30: 225–229 (in Chinese with English abstract).
4. Li JW, Zuo HT, Li QF, Fan XF, Hou XC (2011) Effect of soil water spatial distribution pattern on switchgrass during first growing season. Acta Agrista Sinica 19: 43–50 (in Chinese with English abstract).
5. Jipp PH, Nepstad DC, Cassel DK, Carvalho C (1998) Deep soil moisture storage and transpiration in forests and pastures of seasonally-dry Amazonia. Climatic Change 39: 395–412.
6. Grassini P, You JS, Hubbard KG, Cassman KG (2010) Soil water recharge in a semi-arid temperate climate in the central US Great Plains. Agricultural Water Management 97: 1063–1069.
7. Markewitz D, Devine S, Davidson EA, Brando P, Nepstad DC (2010) Soil moisture depletion under simulated drought in the Amazon: impacts on deep root uptake. New Phytologist 187: 592–607.
8. Li YS (2001) Fluctuation of yield on high-yield field and desiccation of the soil on dryland. Acta Pedologica Sinica 38: 353–356 (in Chinese with English abstract).
9. Huang MB, Dang TH, Gallichand J, Goulet M (2003) Effect of increased fertilizer applications to wheat crop on soil-water depletion in the Loess Plateau, China. Agricultural Water Management 58: 267–278.
10. Liu WZ, Zhang XC, Dang TH, Zhu OY, Li Z, et al. (2010) Soil water dynamics and deep soil recharge in a record wet year in the southern Loess Plateau of China. Agricultural Water Management 97: 1133–1138.
11. Beldring S, Gottschalk L, Seibert J, Tallaksen LM (1999) Distribution of soil moisture and groundwater levels at patch and catchment scales. Agricultural and Forest Meteorology 98–99: 305–324.
12. Li B, Rodell M (2013) Spatial variability and its scale dependency of observed and modeled soil moisture over different climate regions. Hydrology and Earth System Sciences 17: 1177–1188.
13. Henninger DL, Petersen GW, Engman ET (1976) Surface soil moisture within a watershed: Variations, factors influencing, and relationship to surface runoff. Soil Science Society of American Journal 40: 773–776.
14. Jones EB, Owe M, Schmugge TJ (1982) Soil moisture variation patterns observed in Hand county, South Dakota. Water Recources Bulletin 18: 949–954.
15. Hawley ME, Jackson TJ, Mccuen RH (1983) Surface soil moisture variation on small agricultural watersheds. Journal of Hydrology 62: 179–200.
16. Robinson M, Dean TJ (1993) Measurement of near surface soil water content using a capacitance probe. Hydrological Processes 7: 77–86.
17. Qiu Y, Fu BJ, Wang J, Chen LD (2000) Quantitative analysis of relationships between spatial and temporal variation of soil moisture content and

environmental factors at a gully catchment. Acta Ecologica Sinica 20: 741–747 (in Chinese with English abstract).
18. Zeng C, Shao MA, Wang QJ, Zhang J (2011) Effects of land use on temporal-spatial variability of soil water and soil-water conservation. Acta Agriculturae Scandinavica Section B-Soil and Plant Science 61: 1–13.
19. Shi ZH, Zhu HD, Chen J, Fang NF, Ai L (2012) Spatial heterogeneity of soil moisture and its relationships with environmental factors at small catchment level. Chinese Journal of Applied Ecology 23: 889–895 (in Chinese with English abstract).
20. Chen LD, Huang ZL, Gong J, Fu BJ, Huang YL (2007) The effect of land cover/vegetation on soil water dynamic in the hilly area of the loess plateau, China. Catena 70: 200–208.
21. Wang YQ, Shao MA, Liu ZP, Orton R (2013) Regional-scale variation and distribution patterns of soil saturated hydraulic conductivities in surface and subsurface layers in the loessial soils of China. Journal of Hydrology 487: 13–23.
22. Zhang R, Cao H, Wang YQ, Huang CQ, Tan WF (2012) spatial variability of soil moisture and its influence factors in watershed of gully region on the loess plateau. Research of Soil and Water Conservation 19: 52–58 (in Chinese with English abstract).
23. Canton Y, Sole-benet A, Domingo F (2004) Temporal and spatial patterns of soil moisture in semiarid badlands of SE Spain. Journal of Hydrology 285: 199–214.
24. Burnt TP, Butcher DP (1985) Topographic controls of soil moisture distributions. Journal of Soil Science 36: 469–486.
25. O'loughlin EM (1981) Saturation regions in catchments and their relations to soil and topographic properties. Journal of Hydrology 53: 229–246.
26. Huang J, WU P, Zhao XN (2012) Effects of rainfall intensity, underlying surface and slope gradient on soil infiltration under simulated rainfall experiments. Catena 104: 93–102
27. Qiu Y, FU BJ, Wang J, Chen LD (2003) Spatiotemporal prediction of soil moisture content using multiple-linear regression in a small catchment of the Loess Plateau, China. Catena 54: 173–195.
28. Qiu Y, Fu B, Wang J, Chen L, Meng Q, Zhang Y (2010) Spatial prediction of soil moisture content using multiple-linear regressions in a gully catchment of the Loess Plateau, China. Journal of Arid Environments 74: 208–220.
29. Gómez-Plaza A, Martínez-Mena M, Albaladejo J, Castillo VM (2001) Factors regulating spatial distribution of soil water content in small semiarid catchments. Journal of Hydrology 253: 211–226.
30. Dripps WR, Bradbury KR (2007) A simple daily soil-water balance model for estimating the spatial and temporal distribution of groundwater recharge in temperate humid areas. Hydrogeology Journal 15: 433–444.
31. Yao XL, Fu BJ, Lu YH, Sun FX, Wang S, et al. (2013) Comparison of four spatial interpolation methods for estimating soil moisture in a complex terrain catchment. PLoS One 8(1): e54660.
32. Wang MB, Li HJ (1995) Quantitative study on the soil water dynamics of various forest plantations in the loess plateau region in northwestern Shanxi. Acta Ecologica Sinica 15: 172–184 (in Chinese with English abstract).
33. Wang YQ, Zhang XC, Han FP (2008) Profile variability of soil properties in check dam on the Loess Plateau and its functions. Environmental Science 29: 1020–1026 (in Chinese with English abstract).
34. Jain AK (2010) Data clustering: 50 years beyond K-means. Pattern Recognition Letters 31: 651–666.
35. Huang YL, Chen LD, Fu BJ, Wang YL (2005) Spatial pattern of soil water and its influencing factors in gully catchment of the Loess Plateau. Journal of Natural Resources 20: 483–492 (in Chinese with English abstract).
36. Zou JL, Shao MA, Gong SH (2011) Effects of different vegetation and soil types on profile variability of soil moisture. Research of Soil and Water Conservation 18: 12–17 (in Chinese with English abstract).
37. Kizito F, Dragila M,Se'ne M, Lufafa A, Diedhiou I, et al. (2006) Seasonal soil water variation and root patterns between two semi-arid shrubs co-existing with Pearl millet in Senegal, West Africa. Journal of Arid Environments 67: 436–455.
38. Li J, Chen B, Li XF, Zhao YJ, Ciren YJ, et al. (2008) Effects of deep soil desiccation on artificial forestlands in different vegetation zones on the Loess Plateau, China. Acta Ecologica Sinica 28: 1429–1445 (in Chinese with English abstract).
39. Cheng LP, Liu WZ (2013) Long term effects of farming system on soil water content and dry soil layer in deep loess profile of Loess Tableland in China. Journal of Integrative Agriculture. 13(6): 1382–1392.
40. Wang XZ, Jiao F (2011) Partition of soil moisture profiles based on sequential clustering method. Journal of Northwest A &F University 39: 191–201,196 (in Chinese with English abstract).
41. Fu XL, Shao MA, Wei XR, Wang HM, Zeng C (2013) Effects of monovegetation restoration types on soil water distribution and balance on a hillslope in northern Loess Plateau of China. Journal of Hydrologic Engineering 18: 413–421.
42. Bai TL, Yang QK, Shen J (2009) Soil variability of soil moisture vertical distribution and related affecting factors in hilly and gully watershed region of Loess Plateau. Chinese Journal of Ecology 28: 2508–2514 (in Chinese with English abstract).
43. Lian G, Guo XD, Fu BJ, Hu CX (2006) Spatial variability of bulk density and soil water in a small catchment of the Loess Plateau. Acta Ecologica Sinica 26: 647–654 (in Chinese with English abstract).

8

Sensitivity and Tolerance of Riparian Arthropod Communities to Altered Water Resources along a Drying River

Kevin E. McCluney[¤]*, John L. Sabo

School of Life Sciences, Arizona State University, Tempe, AZ, United States of America

Abstract

Background: Rivers around the world are drying with increasing frequency, but little is known about effects on terrestrial animal communities. Previous research along the San Pedro River in southeastern AZ, USA, suggests that changes in the availability of water resources associated with river drying lead to changes in predator abundance, community composition, diversity, and abundance of particular taxa of arthropods, but these observations have not yet been tested manipulatively.

Methods and Results: In this study, we constructed artificial pools in the stream bed adjacent to a drying section of the San Pedro River and maintained them as the river dried. We compared pitfall trapped arthropods near artificial pools to adjacent control sites where surface waters temporarily dried. Assemblage composition changed differentially at multiple taxonomic levels, resulting in different assemblages at pools than at control sites, with multiple taxa and richness of carabid beetle genera increasing at pools but not at controls that dried. On the other hand, predator biomass, particularly wolf spiders, and diversity of orders and families were consistently higher at control sites that dried. These results suggest an important role for colonization dynamics of pools, as well as the ability of certain taxa, particularly burrowing wolf spiders, to withstand periods of temporary drying.

Conclusions: Overall, we found some agreement between this manipulative study of water resources and a previous analysis of river drying that showed shifts in composition, changes in diversity, and declines in abundance of certain taxa (e.g. carabid beetles). However, colonization dynamics of pools, as well as compensatory strategies of predatory wolf spiders seem to have led to patterns that do not match previous research, with control sites maintaining high diversity, despite drying. Tolerance of river drying by some species may allow persistence of substantial diversity in the face of short-term drying. The long-term effects of drying remain to be investigated.

Editor: Bernd Schierwater, University of Veterinary Medicine Hanover, Germany

Funding: This study was funded by EPA STAR grant FP-91649201 to K. McCluney and the Brian Daniel Corrigan Foundation. The funders had no role in study design, data collection and analysis, decision to publish, or preparation of the manuscript.

Competing Interests: The authors have declared that no competing interests exist.

* Email: kmcclun@bgsu.edu

¤ Current address: Department of Biological Sciences, Bowling Green State University, Bowling Green, OH, United States of America

Introduction

Human activities are dramatically altering the distribution of freshwater across the Earth's surface and these changes may have important effects on both aquatic and terrestrial ecosystems worldwide [1–4]. Along the unregulated San Pedro River, in AZ, USA, groundwater and river waters have declined in recent decades, converting some once perennial reaches to reaches with only intermittent flows [5,6]. Whereas the impacts of these water declines on aquatic animals [7–12] and on riparian (streamside) vegetation have been relatively well investigated [5,13,14], our understanding of the impacts on terrestrial animal communities remains limited.

Evidence of the effects of river drying on aquatic ecosystems could provide us with several hypotheses about how riparian animals might respond to river drying. For instance, stream and river drying events have been shown to have strong, long-term effects on aquatic community structure and diversity [7,15,16]. Additionally, even short-term drying events seem to reduce aquatic food chain length in rivers across the US [8]. However, refugia in the hyporheic zone and migration may modulate these effects, providing some degree of resilience, with community dynamics related to life-history strategies, e.g. [7,11,12,17].

Rivers provide many important resources to terrestrial consumers. Many consumers rely on subsidies of emergent aquatic insects for energy and nutrients [18–20]. Similarly, some riparian herbivores rely on algal subsidies [21]. Additionally, the river water itself may be an important resource that could limit survival and performance [22–24]. This is especially true in arid lands and may also hold during droughts in mesic biomes when rivers are the sole source of free water [25,26].

Here we focus on the effects of river drying on terrestrial riparian arthropod communities. Terrestrial arthropods play key functional roles in ecosystems, influencing rates of decomposition [27] and altering emergence of certain groups of aquatic insects in ways that depend on the identity of the consumer involved [28]. Arthropods are also near the base of the food web and thus are important for the support of higher consumers, including bird species found along the San Pedro River that attract tourism to the region [29].

Studies on other rivers in arid [30–32] and mesic [33–35] regions have examined the influence of flood alteration or flow regulation on terrestrial arthropods. Additionally, some research has investigated how desiccation tolerance influences species distributions along gradients of soil moisture and other habitat conditions [25,26]. These studies found that carabid beetles are often particularly sensitive to alterations of flood regime, showing increases in abundance and diversity with flood events [31,32] or shifts in community composition between rivers that differ in flood regime [35] or habitats that differ in flood disturbance [33]. Because many riparian carabids derive most of their nutrients from emergent aquatic insects, river drying may impact this group most severely by altering both food and water resources [19]. Some studies also suggest that spiders are often less influenced by changes to flow regime than many other groups of riparian arthropods [31,36], despite the opposite reported in other studies [33] and the finding that riparian spider distributions often match differences in desiccation tolerance [25,26]. The influence of river drying events specifically on riparian arthropods remains poorly studied, but see [37,38].

Previous observational analyses along the San Pedro River, which experiences seasonal drying, found differences in predator abundance, familial community composition, familial diversity, and abundance of some groups of arthropods between dry and flowing river reaches [39]. Similar, but stronger patterns were observed for genera within the family Carabidae. Analyses of association of arthropods with environmental parameters in that study suggested that water resources were of prime importance in structuring the community. Although this previous work shed light on how river drying may influence arthropod communities, only a manipulative approach can provide direct causal evidence of the effects of water resources on arthropods. Thus, in this paper, we ask if previously observed differences between dry and flowing reaches are attributable to changes in water resources.

We analyze results from a comparison of pitfall trapped arthropods between artificial pools, constructed within the active river channel, and nearby controls, during a drying event, manipulatively testing if previously observed differences in riparian arthropod communities between dry and flowing river sections were caused by differences in water resources. Specifically, we compare dry stream-bed habitats which were near flowing river initially, but which dried by the end of the study, to habitats where we constructed and maintained artificial pools of water as the river dried, supplementing water resources. These pools are unlikely to be a perfect replication of flowing river conditions, but should function to manipulate water resource availability. As a measure of change in community composition, we assessed differences in the Bray-Curtis dissimilarity index applied to changes in biomass or abundance over time. We also examined changes in α-diversity and the abundance and biomass of key groups of arthropods suggested by multivariate statistical analyses. Our analyses investigated responses at multiple taxonomic/functional levels including trophic groups, orders, families, and genera within the beetle family Carabidae.

We predicted differential changes in community composition as the river dries, with seasonal increases in diversity at pools only. We also predicted increases in the abundance or biomass of particular carabid beetle genera near pools, as well as increases in other key groups of arthropods, like wolf spiders (Lycosidae), field crickets (Gryllidae), or aerial arthropods. These predictions are based on 1) results from the previous analyses of observational data in this system [39], 2) evidence of partial reliance on surface water of crickets and spiders in this system, based on stable water isotope analysis [39], 3) the effects of changes in riverine resources on riparian arthropods [18,23,36], and 4) inference from previous examinations of the influence of flow regime alterations on terrestrial arthropod communities [30–32].

Methods

Ethics Statement

No specific organization regulates research on invertebrates, but care was taken to minimize unnecessary harm. No species used in this research were considered to be endangered or protected. All necessary permits were obtained for the described field studies. In particular, we received permission from the US Bureau of Land Management and a scientific collecting permit from the State of Arizona, Game and Fish Department (SP736471).

Study Site

Our study occurred in the active channel along a drying reach of the upper San Pedro River approximately 1.5 km in length near Boquillas Ranch House (31°41′50.95″ N, 110°10′57.15″ W) in the San Pedro Riparian National Conservation Area, managed by the US Bureau of Land Management (BLM). This site is located approximately 15 km downstream from population centers that derive municipal water from the Sierra-Vista sub-watershed groundwater aquifer [6]. Hydrologic studies have linked the aquifer in the region of groundwater pumping to river base flows near our study site [6]. The combination of groundwater pumping with changes in local precipitation regime has led to decreased rates of recharge and may have contributed to recently observed river drying [6,40]. In the year prior to conducting our research (2005), this reach of river was flowing at the beginning of the spring/summer dry season (April), but dried before the arrival of the summer rainy season (July).

The San Pedro River originates in Mexico and flows north through AZ, USA for 160 km, eventually joining the Gila River, part of the Colorado River drainage [5,41]. The river floodplain can be quite wide, in places extending up to several hundred meters from the river [42]. Where perennial, this floodplain is often dominated by cottonwood (*Populus fremontii*) and willow (*Salix gooddingii*) trees, but becomes increasingly dominated by tamarisk (*Tamarix* spp.) as groundwater declines and river flows become more intermittent [43]. The uplands near our study site were dominated by plants of Chihuahuan desert. The active river channel was characterized by sand, gravel, and cobble bars, which became exposed at base flow or as the river dried. The river is also temporally variable, with large flood pulses in the summer rainy season (July-September) which can cause substantial disturbance to the floodplain, but which did not occur during our study period.

The San Pedro River has extremely high richness of birds, mammals, and reptiles, and supports endangered species such as the southwest willow flycatcher [5]. Additionally, it is one of the last free-flowing rivers in the western US [41] and may provide an important stopover for migratory birds [44]. Thus, this area is of considerable conservation concern.

Artificial Pools

In the last two weeks of April 2006, we constructed 10 artificial pools in the active channel of the San Pedro River, within several meters of the flowing river (wetted channel). Longitudinally, pools were spaced ~150 m apart, with a control site for comparison in between every two pools (~75 m from either pool), with one extra on the end (10 control sites). Control sites were also initially located in the active channel within several meters of the flowing river (wetted channel). Pools were lined with 114 L (30-gal) Beckett preformed pond liners (Model PP1035) approximately 1 m in diameter and 36 cm deep. Liners were dug into the streambed and refilled with the removed substrate, leaving a slight depression in the middle. Reference control sites were also disturbed by digging, which may or may not have adequately mimicked the disturbance and structure added by the pools. Pools were filled automatically by gravity from nearby tanks (see Text S1, Figure 1). Pools were maintained for approximately 2.5 months as the river dried, until the final sampling on 25 June 2006. Flows continuously declined during this period and rainfall was minimal.

Flowing Reference Sites

Although this experiment only directly examined differences between pools and dry areas, we also sampled along still flowing sections of the river and report limited information from these sites for reference. At time 0, all sites were flowing but we also marked flowing reference sites approximately 3 km upstream. Since many of these sites dried more quickly than expected we added additional flowing sites upstream as drying progressed. Graphs show comparisons across all sites and dates, but statistical analyses focus on differences on the final sampling date to deal with inconsistent sampling.

Pitfall Sampling

Sticky pitfall traps were used for all trapping to avoid biases associated with liquid traps across gradients of water availability (K. McCluney, unpublished data, see Text S1 for details). The traps were constructed using 16-oz (473 mL) cups lined with Tangle-trap (The Tanglefoot Company, Grand Rapids, MI) on the bottom 4 cm of the cup and open on the top. Traps were prepared in advance and stored in quart-sized (946 mL) Ziploc bags (see Text S1 for details).

We sampled four times in the late spring and early summer of 2006 (14 May 2006 to 25 June 2006). Every site received two traps, one within 0.5 m to the east of the object of interest (pool, flowing river, dry area) and one within 0.5 m to the west. Cups were buried so that the ground was level with the top of the cup. Traps were immediately open upon placement and left for approximately 24 hours. Traps were removed between sampling events.

Traps were processed by freezing, then soaking with baby oil to dissolve the Tangle-trap followed by filtering (0.5 mm) and collection and identification of anything identifiable as an arthropod with the naked eye. Most arthropods were identified to the family level and all lengths were measured to the nearest 0.5 mm from the tip of the head to the tip of the abdomen. Biomasses of adults were estimated from these measurements using published values for riparian arthropods in California, USA [45]. Direct gravimetric methods were not possible due to residue associated with the pitfall trapping and processing techniques. We also identified all carabids to genus. Identification was aided by Borrer et al. [46], Ubick et al. [47] and Arnett and Thomas [48].

Other Measurements and Sampling

Aquatic insect samples were collected from pools (still waters) and flowing sites by jab and sweep methods, sweeping a standard aquarium net (~13×15 cm) three times, gently scraping the bottom. These aquatic samples were frozen until identification. Samples were defrosted and arthropods were picked out of the samples and identified to order or family.

Data Processing and Statistics

We excluded very small arthropods (less than 1.5 mm) and all collembolans from our pitfall trap data set prior to analyses, due to potential biases in our sample processing procedures. We also excluded crayfish (virile, *Orconectes virilis*, and red swamp, *Procambarus clarkii*) caught in some traps from all analyses and excluded unidentifiable arthropods from community level analyses. We averaged the two traps per site, to produce an estimate of abundance or biomass per trap. Finally, we removed three pool sites and three dry sites, where the river never completely dried, from our primary analyses (see Text S1 for more details).

We employed several statistical methods, all in the statistical program R (v2.15.1). When possible we used likelihood ratio tests of longitudinal linear mixed effects models (LME) of each response metric, with site as a random effect (intercept), and an explicit

Figure 1. Experimental setup. A. An artificial pool near the beginning of the experiment. B. Pitfall trapping near an artificial pool at the end of the experiment.

consideration of multiple plausible temporal variance-covariance structures (compound symmetry, autoregressive, or unstructured). This is an analogous approach to standard RmANOVA, but allows specification of alternative variance-covariance structures. Response metrics analyzed in this manner included diversity and biomass, whenever variance and normality assumptions were met. For some response metrics, this form of analysis was not possible, due to current limitations of longitudinal mixed effects models in R. In particular, different temporal variance-covariance structures cannot currently be specified for tests of multivariate community responses, or for non-gaussian response distributions (e.g., Poisson count data). Therefore, for multivariate community responses, for abundance, and for biomass responses that severely violated assumptions of normality and equal variance (transformations ineffective), we tested for treatment effects on the differences between initial (14 May 2006) and final (25 June 2006) sampling dates. When those temporal difference tests were significant, we tested values from the final date to detect whether responses to treatments diverged or converged. These tests examined whether the response metric changed differentially between treatments across time and whether they converged or diverged, avoiding effects of temporal autocorrelation. However, this approach is less powerful than linear mixed effects modeling. An overview of our analysis approach can be found in Figure 2.

Our first analysis was for changes (final – initial) in community composition over time with permutational multivariate analysis of variance using distance matrices (adonis/PERMANOVA) in the VEGAN package of R v. 2.15.1. These tests took place on transformed data, where we eliminated negative values or those less than 1 by adding a fixed integer and then applied the natural log. Upon finding significant changes in community composition, we followed with a permutational multivariate analalysis of variance on the final date. Whenever we found a significant difference on the final date, we examined non-metric multi-dimensional scaling (nMDS) plots of the community using the VEGAN package of R. Similarity percentages (Simper) analysis was used to identify the most influential taxa, added to NMDS plots.

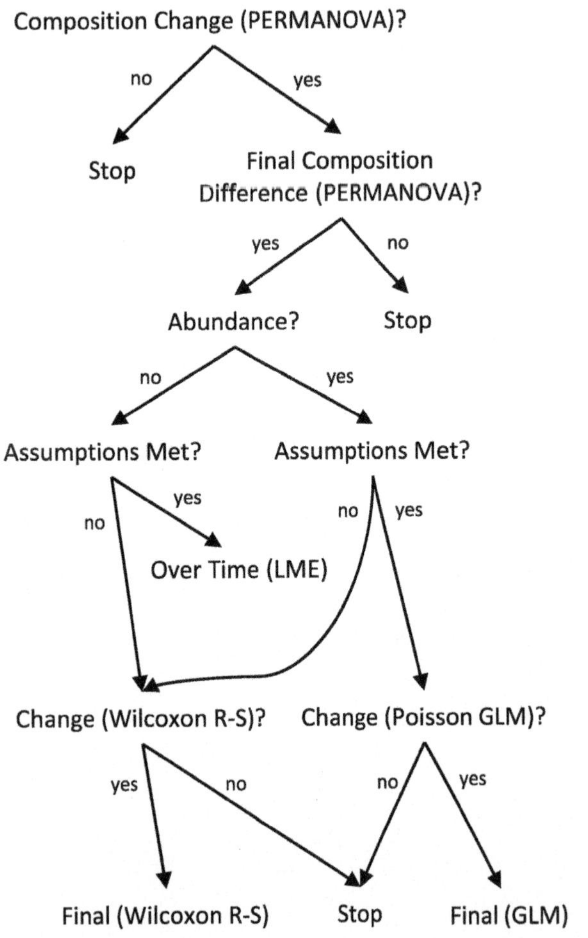

Figure 2. Statistical analysis decision tree.

If community tests suggested both significant differences in how pools and dry sites changed over time and a significant difference in the final communities, we tested for changes in abundance or biomass of individual taxonomic groups. We analyzed differences in changes in abundance and final abundance using generalized linear models (GLM) with either a Poisson or quasi-Poisson distribution, the latter of which helps with modeling overdispersed count data [49]. Since trap counts were averaged at each site, we rounded up to the nearest integer prior to fitting Poisson or quasi-Poisson glm models. For analyses of biomass that met assumptions of normality and equal variance, we conducted likelihood ratio tests for longitudinal mixed models, with time and planned treatment as fixed effects and site as a random effect (intercept), using the lme function in the nlme package in R. Likelihood ratio tests were performed on the change in likelihood when dropping each factor from a model, one at a time, following [50]. When biomass responses were non-normal or had unequal variance, which could not be remedied with transformation, we analyzed only the difference between initial and final and if significant, then the final values using non-parametric Wilcoxon rank-sum tests in R. We followed similar techniques to test for changes in Shannon diversity (H), richness (S), and Pielou's evenness (J), using the VEGAN package and either longitudinal mixed models or glm models, as appropriate.

We compared our artificial pools to flowing sites using 1) aquatic insect samples from a single date, 1 June 2006, 2) multivariate examinations of changes in assemblage composition of aquatic and riparian arthropods, 3) the total abundance and biomass of all pitfall trapped arthropods, and 4) the abundance and biomass of bombardier beetles (Carabidae:*Brachinus*). Due to a lack of dry sites at the beginning of the experiment and the necessity of changing the location of flowing sites throughout the experiment, we analyzed only differences on the final date. To see if bombardier beetles (*Brachinus*), which have ectoparasitic larvae of aquatic dytiscid beetles, were attracted to pools with more dytiscids, we tested for correlations between the two in our pools using Spearman correlations in R v. 2.9.0.

We conducted all relevant analyses at four ecological or taxonomic levels: trophic group, order, family, and genera (only of beetles in the family Carabidae). All of these analyses followed the statistical approaches described above.

Initial nMDS plots revealed that the ground-dwelling arthropod community at one pool site (Pool 3) was very different from all the others for both abundance and biomass and that this site had a large influence on results. Compared to the other pool sites, this site had extremely low abundances of large carabid beetles in the *Chlaenius, Agonum,* and *Brachinus* genera. Additionally, this site seemed to have particularly high abundances of ants on both of the last two dates. Reznikova and Dorosheva [51] found that many carabids tend to avoid high concentrations of ants. Thus, in the body of the paper we report results with this site removed, but report full results in the appendix (Table S1). We also found that one dry site (Dry 9) was very different from other dry sites and had a strong influence on results. Compared to other dry sites, this site had particularly high numbers of carabid beetles in the genus *Brachinus*. Thus we also remove this site from our main analysis, but report results with both sites included in the appendix, for comparison (Table S1). Thus our main analyses were conducted on 6 pools and 6 dry sites and the analyses in the appendix were conducted on 7 pools and 7 dry sites. Comparisons of pools to flowing sites on the final date were conducted on 4 flowing sites and 10 pools, and comparisons of aquatic insect abundance on 1 Jun 2006 were on 3 flowing sites and 7 pools.

Results

Total abundance

We did not observe a significant change in total abundance of all pitfall-trapped arthropods between artificial pools and control sites that dried ($\chi^2 = 0.05$, df = 1, p = 0.826, Table S1).

Trophic groups

We found no significant difference in the change in the assemblage of trophic groups between artificial pools and control sites that dried (PERMANOVA Abundance: $F_{1,11} = 0.64$, p = 0.571; PERMANOVA Biomass: $F_{1,11} = 1.67$, p = 0.070, Table S1). Despite the lack of a detected effect on trophic group composition, we examined changes in abundance and biomass of predators and diversity of trophic groups in order to make comparisons with a previous observational study [22] that found more predators along flowing than dry sections of this river. In this study, we found no treatment effect on abundance (Poisson GLM: $\chi^2 = 0.02$, df = 1, p = 0.889, Table S1), but we found slightly higher predator biomass at control sites than at artificial pools (LME: $\chi^2 = 10.07$, df = 1, p = 0.002, Table 1; Figure 3A). We also found higher Shannon's diversity, richness, and Pielou's evenness of trophic groups at control sites than at artificial pools (Shannon's diversity LME: $\chi^2 = 9.33$, df = 1, p = 0.002, Richness LME: $\chi^2 = 7.77$, df = 1, p = 0.005, Pielou's evenness LME: $\chi^2 = 6.72$, df = 1, p = 0.010, Table 1; Figures 3B, S1, & S2). All of these differences were consistent across time (no time x treatment interaction, Table 1), suggesting that control sites started out with higher predator biomass and trophic diversity and maintained it across time.

Orders

The assemblage of orders of arthropods changed between the initial and final dates differentially for artificial pools and controls that dried (PERMANOVA Biomass: $F_{1,11} = 2.52$, p = 0.001). The final date showed differences between these sites, showing evidence of divergence (PERMANOVA: $F_{1,11} = 4.66$, p = 0.025, Figure 4A). We only found a significant contribution of beetles (Coleoptera) to changes at these sites (LME: $\chi^2 = 8.25$, df = 1, p = 0.004, Table 1, Figure 5A), with an increase at artificial pools and a decrease at control sites that dried. We also found a significant time by treatment interactive effect on Shannon's diversity of orders, with a complicated pattern over time, but overall a greater decline in diversity at pool sites than at control sites (LME: $\chi^2 = 5.00$, df = 1, p = 0.025, Figure 6A). We note that in each analysis we report, here and below, a lack of a detectable univariate response for any particular taxon does not mean that this taxon was not influential. Differences in multiple rare taxa could have resulted in multivariate differences between sites, but changes in abundance or biomass individually could be undetectable due to low sampling. Non-metric multi-dimensional scaling graphs and Simper analysis provide clues to other taxa that may have influenced community level differences. Lepidoptera, Orthoptera, and Araneae may also have played important roles in the community level differences between pools and dry sites (Figure 4A).

Families

The assemblage of families of arthropods changed between the initial and final dates differentially for pools and controls that dried (PERMANOVA Biomass: $F_{1,11} = 2.64$, p = 0.002). The final date showed differences between these sites, showing evidence of divergence (PERMANOVA Biomass: $F_{1,11} = 3.40$, p = 0.030, Figure 4B). Carabid ground beetles (Carabidae; LME: $\chi^2 = 7.68$, df = 1, p = 0.006, Table 1, Figure S3), wolf spiders (Lycosidae; LME: $\chi^2 = 9.52$, df = 1, p = 0.002, Table 1, Figure 6C), and owlet moths (Noctuidae; Wilcoxon R–S: W = 31.5, p = 0.026, Figure 5B, Table S1) contributed significantly to these differences. Carabids and owlet moths increased in biomass at artificial pools, while wolf spiders had consistently higher biomass at control sites, from the beginning of the experiment. Simper analysis suggested that field crickets (Gryllidae) and ants (Formicidae) may also have been important in community differences, but univariate analyses failed to show a response. Shannon's diversity and Pielou's evenness were consistently higher at control sites than at artificial pools, from the beginning of the experiment to the end (LME Shannon's diversity: $\chi^2 = 6.00$, df = 1, p = 0.014, Figure 6B, Pielou's evenness:, Table 1, Figure S4).

Carabid genera

The assemblage of carabid ground beetles changed differentially between artificial pools and controls sites that dried (PERMANOVA Abundance: $F_{1,11} = 4.23$, p = 0.012, PERMANOVA Biomass: $F_{1,11} = 3.01$, p = 0.003). Differences between treatments on the final date suggested divergence (PERMANOVA Abundance: $F_{1,11} = 5.69$, p = 0.012, Figure 4C, PERMANOVA Biomass: $F_{1,11} = 7.81$, p = 0.005, Figure 4D). The genera *Brachinus* (bombardier beetles) and *Chlaenius* contributed significantly to the change in relative abundance and biomass, with increases at pools, but decreases at control sites that dried (LME *Brachinus* Biomass: $\chi^2 = 6.33$, df = 1, p = 0.012, Figure 5C, Table 1, Wilcoxon R–S Change in *Chlaenius* Biomass: W = 31.5 p = 0.026, Wilcoxon R–S Final *Chlaenius* Biomass: W = 25.5, p = 0.182, Figure 5E, Table S1, GLM Change in *Brachinus* Abundance: $\chi^2 = 13.41$, df = 1, p<0.001, GLM Final *Brachinus* Abundance: $\chi^2 = 23.66$, df = 1, p<0.001, Figure 5D, GLM Change in *Chlaenius* Abundance: $\chi^2 = 1.97$, df = 1, p = 0.010, GLM Final *Chlaenius* Abundance: $\chi^2 = 1.05$, df = 1, p = 0.253, Figure 5F, Table S1). Simper analysis suggested the genus *Syntomus* may have also contributed to differences in abundance. The richness of Carabidae genera increased significantly more at artificial pool sites than at controls that dried (LME: $\chi^2 = 2.34$, df = 1, p = 0.037, Figure 5G, Table 1). There was no significant correlation between bombardier beetles (*Brachinus*) and diving beetles (Dytiscidae) in our pools (Spearman: r = 0.72, S = 9.86, n = 6, p = 0.11).

Comparison to flowing reference sites

The assemblage of orders of aquatic insects differed between artificial pools and flowing sites on 1 Jun 2006 (PERMANOVA Abundance: $F_{1,9} = 7.57$, p = 0.009, Table S2, Figure 7A), with Odonata, Hemiptera, and Coleoptera most influential according to Simper analysis. Total abundance of all aquatic insects was higher at flowing sites (GLM: $\chi^2 = 4.26$, df = 1, p = 0.039, Figure 7B). The composition and relative abundance of orders of pitfall trapped arthropods also differed significantly between artificial pools and flowing sites on the final sampling date

Table 1. Results of likelihood ratio tests for the effect of removing each fixed effect term from a full longitudinal linear mixed effects model (following Bolker et al [50]).

Model component removed (Fixed Effects)	df	ΔAIC	LRT (χ^2)	p-value
Predator Biomass				
-Time * treatment	1	−1.95	0.05	0.816
-Time	1	−1.89	0.11	0.738
-Treatment	*1*	*8.08*	*10.07*	***0.002***
Trophic Group Shannon's Diversity				
-Time * treatment	1	−1.75	0.25	0.616
-Time	1	1.56	3.56	0.059
-Treatment	*1*	*7.33*	*9.33*	***0.002***
Trophic Group Pielou's Evenness				
-Time * treatment	1	−1.87	0.12	0.720
-Time	*1*	*6.32*	*8.31*	***0.004***
-Treatment	*1*	*4.72*	*6.72*	***0.010***
Trophic Group Richness				
-Time * treatment	1	−1.92	0.08	0.784
-Time	*1*	*2.80*	*4.80*	***0.028***
-Treatment	*1*	*5.77*	*7.77*	***0.005***
Coleoptera Biomass				
*-Time * treatment*	*1*	*6.26*	*8.25*	***0.004***
Order Shannon's Diversity				
*-Time * treatment*	*1*	*3.00*	*5.11*	***0.025***
Carabidae Biomass				
*-Time * treatment*	*1*	*5.68*	*7.68*	***0.006***
Lycosidae Biomas				
-Time * treatment	1	−1.92	0.08	0.776
-Time	1	−1.87	0.13	0.723
-Treatment	*1*	*7.53*	*9.52*	***0.002***
Family Shannon's Diversity				
-Time * treatment	1	−1.70	0.30	0.582
-Time	1	−1.53	0.47	0.492
-Treatment	*1*	*4.00*	*6.00*	***0.014***
Family Pielou's Evenness				
-Time * treatment	1	−1.96	0.04	0.848
-Time	1	0.60	2.60	0.107
-Treatment	*1*	*5.56*	*7.56*	***0.006***
Brachinus Biomas				
*-Time * treatment*	*1*	*4.33*	*6.33*	***0.012***
Carabid Genera Richness				
*-Time * treatment*	*1*	*2.34*	*4.34*	***0.037***

*df = degrees freedom, ΔAIC = change in AIC associated with removal of each model term, LRT (χ^2) = the χ^2 test statistic associated with the change in likelihood with removal of each model term. All models share a single random effect of trap location (example specification: *lme(log(Pred.bio+1) ~ samp.day*Treatment, random = ~1 | Location, data = PoolCompAll.d.r, correlation = corCompSymm(form = ~1 | Location), method = "ML")*). Only responses with a treatment effect are shown here.

(PERMANOVA Abundance: $F_{1,13} = 2.38$, $p = 0.035$, Figure 7C), as did genera of carabid beetles (PERMANOVA Abundance: $F_{1,13} = 20.41$, $p = 0.001$, Figure 7E). However we did not find a difference in the composition and relative abundance of pitfall-trapped families between these sites (PERMANOVA Abundance: $F_{1,13} = 1.37$, $p = 0.216$). On the final sampling date, the total abundance (GLM: $\chi^2 = 14.13$, df = 2, $p = 0.001$) and biomass (GLM: $F_{2,18} = 6.17$, $p = 0.009$) of all pitfall trapped arthropods was higher at flowing sites than at either dry (Tukey's: Abundance $p < 0.001$, Biomass $p = 0.001$) or artificial pool sites (Tukey's: Abundance $p = 0.007$, Biomass $p = 0.038$), which were equivalent (Tukey's: Abundance $p = 0.525$, Biomass $p = 0.272$, Figure 7D). On the final sampling date, bombardier beetle abundance (GLM: $\chi^2 = 63.30$, df = 2, $p < 0.001$) differed significantly between dry, artificial pool, and flowing sites, with flowing sites having highest abundance, artificial pools intermediate abundance, and dry sites

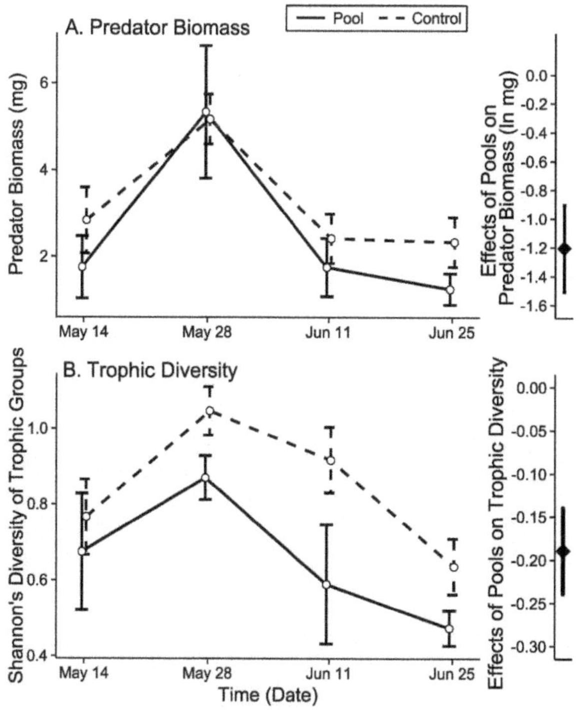

Figure 3. Trophic group responses to experimental treatments. Predator biomass (A) and trophic group diversity (B) of pitfall-trapped arthropods were significantly and consistently higher at control sites that dried than at artificial pools (no time x treatment, Table 1). Fig. S1 and S2 show similar patterns for trophic group richness and evenness. Error bars are standard error. "Effect" of pools on each response is the difference in parameter estimates from mixed effects modeling.

the lowest abundance (Tukey's: Flowing-Dry $p<0.001$, Flowing-Pool $p<0.001$, Dry-Pool $p<0.001$, Figure 7F). Biomass (GLM: $F_{2,18}=9.55$, $p=0.001$) was similar, except that we did not detect differences between artificial pools and flowing sites (Tukey's: Flowing-Dry $p<0.001$, Flowing-Pool $p=0.230$, Dry-Pool $p=0.003$).

Discussion

Groundwater pumping, climate change, and regional droughts can alter surface flows in rivers, but the effects of river drying on riparian animal communities have not been well studied. We document several effects of altered water resources on riparian communities: 1) shifts in community composition, 2) changes in richness and diversity at certain taxonomic levels and 3) decreased abundance and/or biomass of some groups of arthropods, including carabid beetles and owlet moths. Our results suggest that water resource availability, movement and colonization dynamics, and tolerance mechanisms interact to influence patterns of arthropod community structure and diversity along drying rivers.

The current study supports a mechanistic role for water resources in some of the patterns observed in a previous observational study along this section of river [22]. In particular, the abundance of carabid beetles in the genera Brachinus (bombardier beetles) and Chlaenius, as well as the overall richness of carabid beetle genera increased at artificial pools, but declined at control sites that dried. These two genera drove similar patterns found in the biomass of ground beetles (Carabidae) and beetles more generally (Coleoptera). Thus, water resources seem to have important effects on beetles, particularly two genera of carabids.

We found several surprising differences between our current water resource manipulation using pools and the previous comparison of flowing and dry sites. Most notably, artificial pools had lower diversity of trophic groups, orders, and families, as well as lower biomasses of predators, particularly wolf spiders. This contrasts with the finding from the previous study that flowing sites had higher diversity and higher predator and wolf spider abundance. We hypothesize that these differences could be due to several important interacting factors. First, artificial pools were newly constructed water sources, as opposed to flowing river sites. Thus, colonization dynamics likely played a role. More time may have been needed for the communities at these artificial pools to reach an equilibrium. Second, while many arthropods probably migrate in response to river drying, some may be less prone, *sensu* [52]. Predators, particularly wolf spiders, maintained high biomass at control sites, even as they dried. Our samples were dominated by the beach wolf spider (*Arctosa littoralis*), a long-lived, large-bodied species capable of digging burrows. We suggest that due to the likely ability of this species to reduce water demands with burrows [53] and to meet water demands by consuming moist prey [23] (Figure S1), *A. littoralis* may have reduced rates of migration compared with other species. Both through positive effects of predation on diversity [54,55] and the potential for other rare species to exhibit similar river drying tolerant behavior, overall diversity could remain high at dry sites for a short time as a legacy of previous flows. Pools, on the other hand, were constructed at least a meter or two from the river and may not have started out with quite the same community as the control sites. We suggest that the low diversity at pools was likely due to the combined factors of incomplete colonization near pools and of philopatry to previously flowing areas by some species that had behavioral adaptations to drying. However, long-term equilibrium may differ and even over the short term, diversity was less at dry sites than flowing reference sites [22]. In general, the patterns of diversity and community dynamics found in this study are consistent with metacommunity theory, where diversity patterns are driven by the interplay of dispersal and species interactions [56]. Our findings agree with evidence of differential sensitivity and resilience of aquatic species to river drying events [7,12,17], but these terrestrial communities may be less sensitive, with some species able to persist on terrestrial resources alone.

Both owlet moths and some genera within the family Carabidae (e.g., *Brachinus*) responded particularly strongly and positively to pools. These taxa are mobile and known to be tied to water resources. The positive response of *Brachinus* to pools may be due to 1) attraction to aquatic beetles (e.g. Dytiscidae) that serve as hosts for ecto-parasitic larvae of *Brachinus* [57,58], 2) attraction to emergent aquatic insects as a source of food [58], 3) attraction to terrestrial prey such as aerial insects that had greater biomass near pools [58], or 4) attraction to increased water availability [59]. On 1 June 2006, we found significantly higher total abundance of aquatic insects at flowing reference sites than at our pools, as well as differences in community composition. We also tested for correlations between *Brachinus* and Dytiscidae abundance in artificial pools on this date, but found no significant correlation. At the final sampling, the mean abundance of *Brachinus* at flowing reference sites was greater than that at our artificial pools (Figure 7f). This suggests that our pools had some positive effects on *Brachinus*, but did not completely replicate flowing river resources.

Owlet moths have been found in other southwestern riparian areas [60] and many species of Lepidoptera have been observed

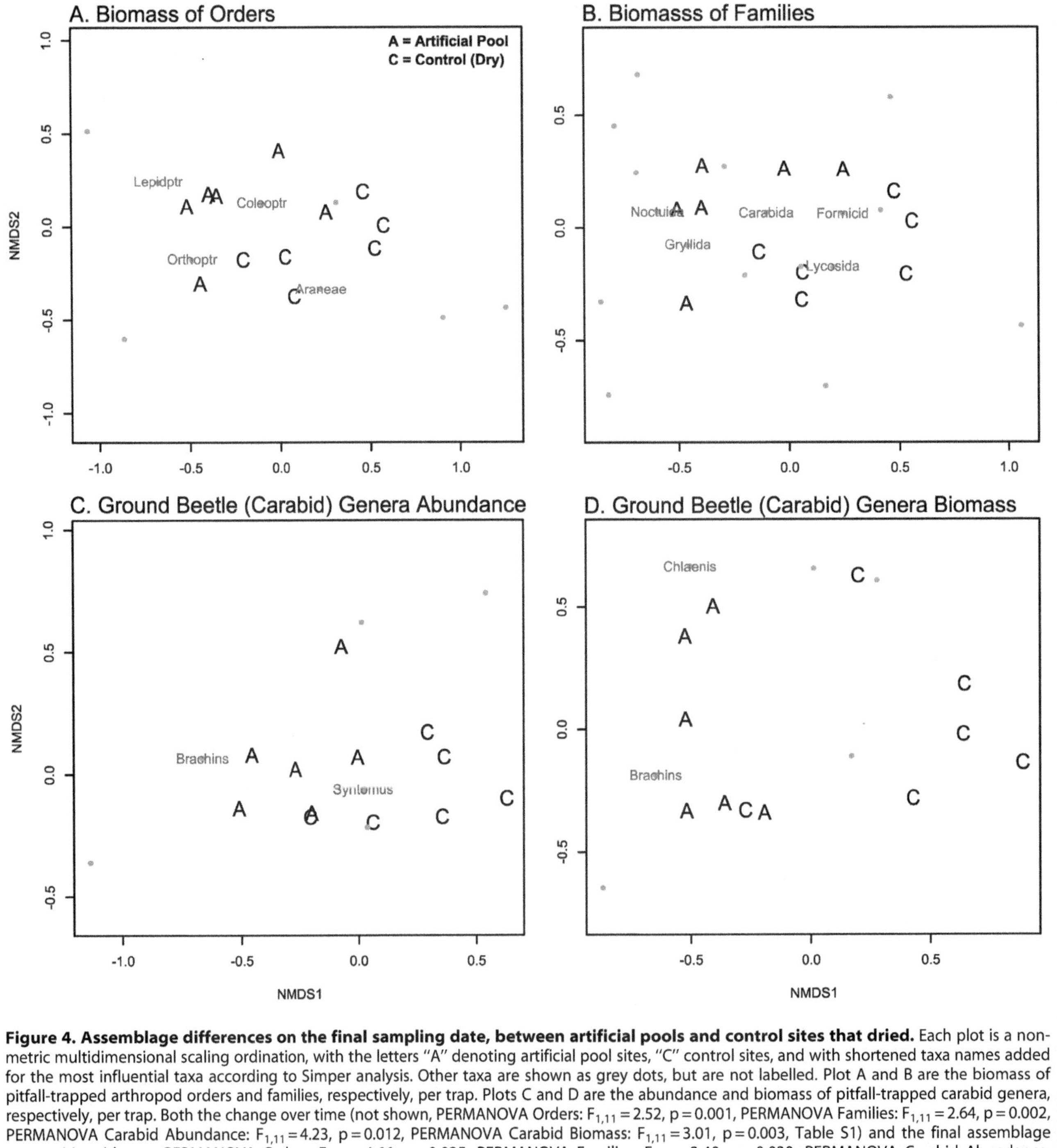

Figure 4. Assemblage differences on the final sampling date, between artificial pools and control sites that dried. Each plot is a non-metric multidimensional scaling ordination, with the letters "A" denoting artificial pool sites, "C" control sites, and with shortened taxa names added for the most influential taxa according to Simper analysis. Other taxa are shown as grey dots, but are not labelled. Plot A and B are the biomass of pitfall-trapped arthropod orders and families, respectively, per trap. Plots C and D are the abundance and biomass of pitfall-trapped carabid genera, respectively, per trap. Both the change over time (not shown, PERMANOVA Orders: $F_{1,11} = 2.52$, $p = 0.001$, PERMANOVA Families: $F_{1,11} = 2.64$, $p = 0.002$, PERMANOVA Carabid Abundance: $F_{1,11} = 4.23$, $p = 0.012$, PERMANOVA Carabid Biomass: $F_{1,11} = 3.01$, $p = 0.003$, Table S1) and the final assemblage composition (shown, PERMANOVA Orders: $F_{1,11} = 4.66$, $p = 0.025$, PERMANOVA Families: $F_{1,11} = 3.40$, $p = 0.030$, PERMANOVA Carabid Abundance: $F_{1,11} = 5.69$, $p = 0.012$, PERMANOVA Carabid Biomass: $F_{1,11} = 7.81$, $p = 0.005$, Table S1) were significantly different between artificial pools and control sites that dried.

puddling (aggregating near puddles) in a variety of other studies [61] and in this study (K. McCluney, personal observations; Figure S2). In addition to attraction to water resources, it is also thought that Lepidoptera that exhibit puddling behavior and drinking may be seeking sodium ions [62]. Thus, pools may provide multiple resources to these animals. While Noctuidae and other aerial insects may not reside near pools, their mobility may allow them to regularly visit nearby water sources, behaviorally responding to changes in water availability more quickly than ground-dwelling arthropods.

In addition to owlet moths, other Lepidoptera and winged Hymenoptera were observed to regularly visit artificial pools to drink (K. McCluney, personal observation; Figure S3). Of the Hymenoptera, bees and parasitic wasps seemed particularly frequent visitors to pools (e.g., photograph of a tarantula hawk, family Pompilidae, drinking, Figure S7). These aerial taxa were rarely collected in pitfall traps, and thus we may have missed one

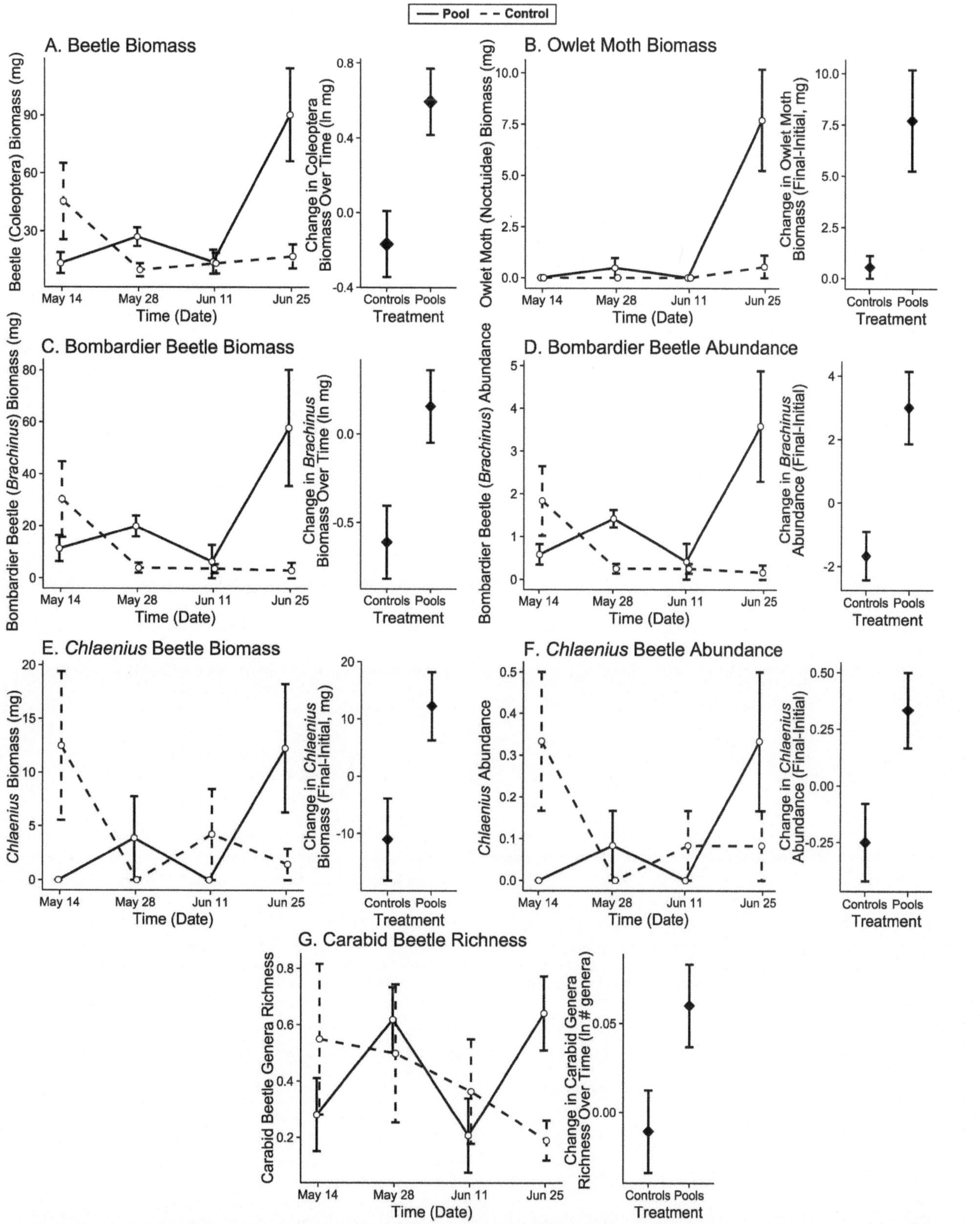

Figure 5. Pitfall trapped arthropods with positive responses to artificial pools. Beetle (A, Table 1) and owlet moth (B, Table S1) biomass increased at artificial pools, but either declined or did not increase at control sites that dried. The biomass (C, E, Table 1) and abundance (D, F, Table S1) of two genera of ground beetles (*Brachinus*, C&D, *Chlaenius*, E&F) had a similar response to overall beetle biomass. The richness of genera of carabid beetles also increased at pools, but declined at control sites as they dried (G). Error bars are standard error. "Change" of each response over time is derived from parameter estimates from mixed effects modeling.

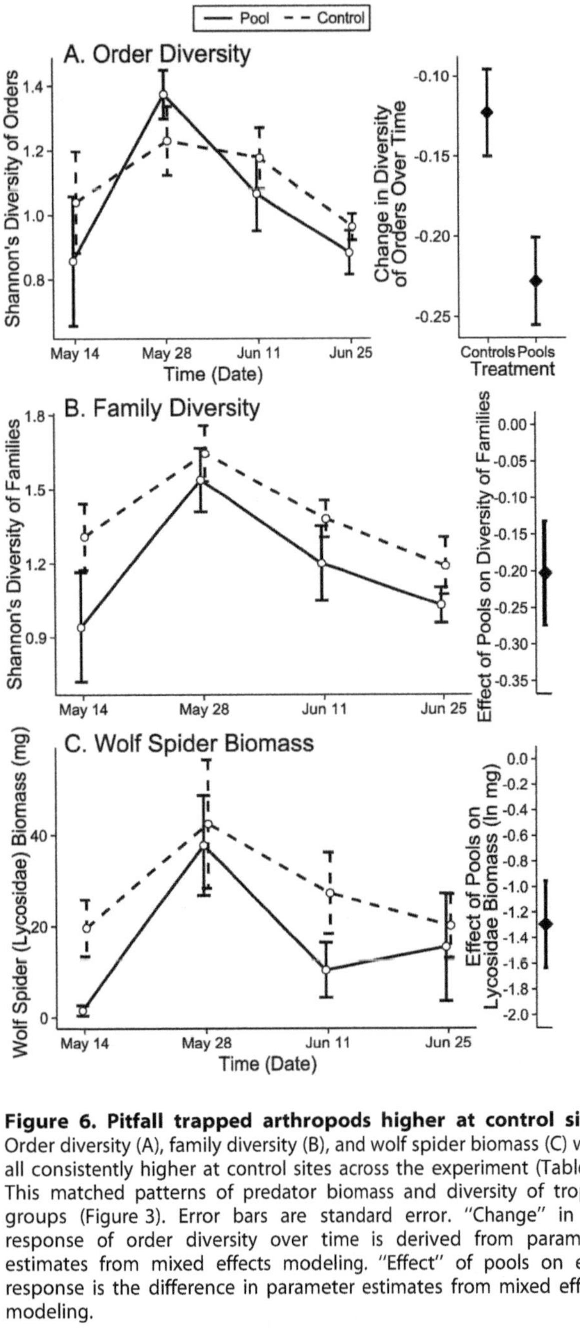

Figure 6. Pitfall trapped arthropods higher at control sites. Order diversity (A), family diversity (B), and wolf spider biomass (C) were all consistently higher at control sites across the experiment (Table 1). This matched patterns of predator biomass and diversity of trophic groups (Figure 3). Error bars are standard error. "Change" in the response of order diversity over time is derived from parameter estimates from mixed effects modeling. "Effect" of pools on each response is the difference in parameter estimates from mixed effects modeling.

factor influencing ground arthropod communities. This may have contributed to the lower biomass of wolf spiders (Lycosidae) at pools than at dry sites. Increased visits to pools by parasitic spider wasps (Pompilidae) could have caused decreases in large spiders through parasitism or through behavioral avoidance of pools by spiders. Unfortunately, we poorly sampled strong fliers like spider wasps, so we cannot test this hypothesis here.

Other aerial consumers also used our pools. Motion-activated cameras recorded several omnivorous, insect-eating bird species drinking from these pools (bird species: house finch, lesser goldfinch, lazuli bunting, song sparrow, Figure S8). Toads were also found in and near pools (Figure S9). Thus, pools attracted a variety of insectivorous predators and could have reduced the potential positive effects of pools on arthropods. Temporary pools of water may represent a source of limited resources, but may also present great risk of predation and thus these pools may be involved in foraging games surrounding water resources [63].

Despite previous work in our study system that pinpoints water as a resource responsible for structuring communities [23,42], the role of energy and nutrients cannot be dismissed. The importance of riverine subsidies of energy and nutrients has recently received great attention [18–21,28,36,64,65]. In our study, aquatic insect abundances were lower at pools than at flowing reference sites, suggesting possible energetic components to the differences in terrestrial arthropods between pools and flowing sites. However, the response of some taxa (e.g. noctuid moths) was probably not influenced by aquatic insects.

As rivers dry, many riparian consumers may be forced to meet water demands solely by consuming moist food [23,42]. Aquatic insects are not only a source of energy and nutrients, but also a source of water (trophic and metabolic) to terrestrial predators. Thus, while decreases in energy or nutrient subsidies associated with river drying may have contributed to our observations, meeting water demands may be a more immediate limitation [23,66], playing a role in the response to drying.

Caveats

We note that we did not detect significant differences for many of the responses previously reported to differ between flowing and dry sites along this river [22]. This lack of detection could be due to real differences between flowing sites and pools, as reported here, but low sample sizes in this intensive manipulative experiment could have also contributed. Additionally, many taxa had low collection frequencies and zeros, which limited our ability to detect changes. With greater capture levels or replication, we may have had greater ability to discern patterns statistically.

Another factor potentially influencing our results is the location of our experiments. Pools and traps were located along the active channel of the river, in areas with high concentrations of gravel, sand, and cobble bars and little vegetation. Floodplains in this system tend to be leaf litter dominated with an overstory of cottonwood and willow trees and may harbor different species in different abundances. For instance, field crickets (Gryllidae) appear to be more abundant in floodplains (K. McCluney, personal observations). Also, while the large, beach wolf spider, *Arctosa littoralis*, is often found in river channels, it is rarely found in floodplains, instead being replaced by the large wolf spider *Hogna antelucana* (K. McCluney, personal observations). Thus, by sampling the active channel, we may have missed sampling key areas of abundance for certain taxa, such as Gryllidae, and thus had insufficient capture rates of these taxa to observe significant differences in univariate analyses.

Broader implications of river drying for riparian communities

It is important to note that this particular reach of the San Pedro River currently dries for only 1–2 months per year and the cottonwood-willow forest of the floodplain is abundant. Thus, we examined only short-term effects of river drying. Other studies have shown substantial changes in vegetation composition, structure, and abundance in other sections of this river that dry more frequently and in years when groundwater levels drop [13,14,67]. Further Sabo et al [42] and McCluney and Sabo [23] suggest that the cricket *Gryllus alogus* (the most abundant species of Gryllidae in our samples) may depend on greenfall from these forests to meet water demands. Thus, long-term changes in the plant community associated with river drying may result in more dramatic changes to the ground-dwelling arthropod community.

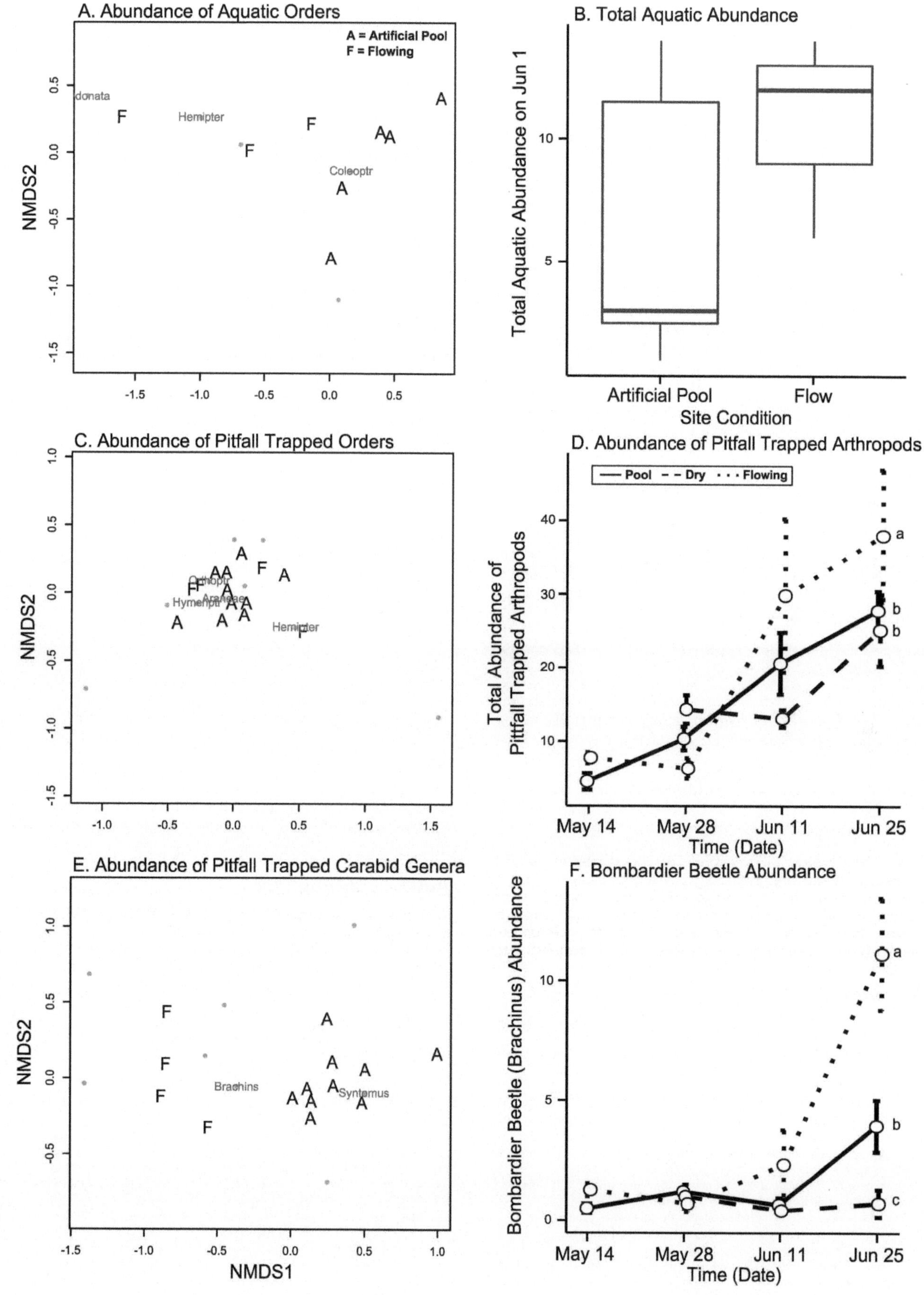
A. Abundance of Aquatic Orders
A = Artificial Pool
F = Flowing
donata
Hemipter
Coleoptr
NMDS2
B. Total Aquatic Abundance
Total Aquatic Abundance on Jun 1
Artificial Pool
Flow
Site Condition
C. Abundance of Pitfall Trapped Orders
Orthoptr
Hymenptr
Araneae
Hemipter
NMDS2
D. Abundance of Pitfall Trapped Arthropods
Pool
Dry
Flowing
Total Abundance of Pittfall Trapped Arthropods
May 14
May 28
Jun 11
Jun 25
Time (Date)
E. Abundance of Pitfall Trapped Carabid Genera
Brachins
Syntomus
NMDS2
NMDS1
F. Bombardier Beetle Abundance
Bombardier Beetle (Brachinus) Abundance
May 14
May 28
Jun 11
Jun 25
Time (Date)

Figure 7. Comparisons of artificial pools to flowing reference sites. Plots A, C, and E show significant differences (Table S2) between assemblage composition at artificial pool ("A") and flowing ("F") sites using nonmetric multidimensional scaling plots of the abundance of orders of aquatic insects (plot A), pitfall trapped arthropod orders (plot C), and pitfall trapped carabid genera (plot E). Plots B, D, and F show significant differences in the total abundance of aquatic insects (plot B), all pitfall trapped terrestrial arthropods (plot D), and bombardier beetles (*Brachinus*, plot F, Table S2). Plot B is a standard box plot and error bars in plots D and F are standard error. In plots D and F, small letters denote significant pairwise Tukey's differences between site conditions on the final date.

Overall, our research suggests that in the short-term, river drying and temporary pools change the riparian arthropod community via negative effects of declining water resources on noctuid moths and carabid beetles. However, given the dynamic nature of these systems, there may be some resilience to short-term drying events. Effects of more severe dewatering and increased frequency or duration of river drying may differ.

Due to their dynamic nature and high concentrations of resources, riparian areas are often essential for animal communities, especially in dryland climates like southeastern AZ, USA [44,68,69]. Further, riparian animals provide important ecosystem services (e.g. recreational bird watching) with both clear direct and indirect monetary values [29]. Thus, streamside areas are of critical conservation concern. We know from extensive previous research that drastic changes to flow regimes can greatly alter riparian vegetation [5,14]. However, here we show that even short drying events may have direct effects on terrestrial animal communities through alteration of water resources, but some degree of resistance and resilience is clear. Managing rivers for the benefit of multiple users will require incorporating an understanding of the effects of river drying on terrestrial animal communities.

Supporting Information

Figure S1 Richness of trophic groups of pitfall-trapped arthropods. Error bars are SE. See Table 1.

Figure S2 Evenness of trophic groups of pitfall-trapped arthropods. Error bars are SE. See Table 1.

Figure S3 Biomass of carabid beetles. Error bars are SE. See Table 1.

Figure S4 Pielou's evenness of families of pitfall-trapped arthropods. Error bars are SE. See Table 1.

Figure S5 The beach wolf spider (*Arctosa littoralis*) consuming an adult female damp-loving field cricket (*Gryllus alogus*) along a dry section of the San Pedro River, near the study site.

Figure S6 Lepidoptera puddling at an artificial pool.

Figure S7 Bees and a tarantula hawk (family Pompillidae) drinking from an artificial pool.

Figure S8 Omnivorous insect-eating birds drinking from pools. Song sparrow, house finch, lesser goldfinch, lazuli bunting.

Figure S9 A toad using one of the artificial pools.

Table S1 Comparison of results with and without including 2 sites (one pool and one dry) with uncharacteristic distributions of carabid beetles and/or ants.

Table S2 Results of tests comparing artificial pools to flowing reference sites.

Acknowledgments

We wish to thank S. Anderson of Gray Hawk Nature Center for accommodations, support, and bird identification and the US Bureau of Land Management for permission to work on lands under their management. We thank Y. Marusenko, L. Thompson, and N. McCluney for assistance in the field, M. Tseng for assistance with arthropod identification, and C. Soykan and C. Bang for helpful reviews.

Author Contributions

Conceived and designed the experiments: KM JS. Performed the experiments: KM. Analyzed the data: KM. Contributed reagents/materials/analysis tools: KM JS. Wrote the paper: KM JS.

References

1. IPCC (2007) The AR4 Synthesis Report. Valencia: Intergovernmental Panel on Climate Change.
2. Gleick PH (2002) The World's Water 2002–2003: The Biennial Report On Freshwater Resources: Island Press. 310 p.
3. Pearce F (2006) When the Rivers Run Dry: Water - The Defining Crisis of the Twenty-First Century. Boston: Beacon Press.
4. McCluney KE, Belnap J, Collins SL, Gonzalez AL, Hagen EM, et al. (2012) Shifting species interactions in terrestrial dryland ecosystems under altered water availability and climate change. Biological Reviews of the Cambridge Philosophical Society 87: 563–582.
5. Stromberg JC, Tellmann B (2009) Ecology and Conservation of the San Pedro River. Tucson: University of Arizona Press. 656 p.
6. Pool D, Coes A (1999) Hydrogeologic Investigations of the Sierra Vista Subwatershed of the Upper San Pedro Basin, Southeast Arizona. In: U.S. Departments of the Interior UGS, editor.
7. Sponseller RA, Grimm NB, Boulton AJ, Sabo JL (2010) Responses of macroinvertebrate communities to long-term flow variability in a Sonoran Desert stream. Global Change Biology 16: 2891–2900.
8. Sabo JL, Finlay JC, Kennedy T, Post DM (2010) The Role of Discharge Variation in Scaling of Drainage Area and Food Chain Length in Rivers. Science 330: 965–967.
9. Clarke A, Mac Nally R, Bond N, Lake PS (2010) Flow permanence affects aquatic macroinvertebrate diversity and community structure in three headwater streams in a forested catchment. Canadian Journal of Fisheries and Aquatic Sciences 67: 1649–1657.
10. Arscott DB, Larned S, Scarsbrook MR, Lambert P (2010) Aquatic invertebrate community structure along an intermittence gradient: Selwyn River, New Zealand. Journal of the North American Benthological Society 29: 530–545.
11. Larned ST, Datry T, Arscott DB, Tockner K (2010) Emerging concepts in temporary-river ecology. Freshwater Biology 55: 717–738.
12. Stanley EH, Buschman DL, Boulton AJ, Grimm NB, Fisher SG (1994) Invertebrate Resistance and Resilience to Intermittency in a Desert Stream. American Midland Naturalist 131: 288–300.
13. Stromberg JC, Tiller R, Richter B (1996) Effects of groundwater decline on riparian vegetation of semiarid regions: The San Pedro, Arizona. Ecological Applications 6: 113–131.

14. Lite SJ, Stromberg JC (2005) Surface water and ground-water thresholds for maintaining Populus-Salix forests, San Pedro River, Arizona. Biological Conservation 125: 153–167.
15. Bogan MT, Boersma KS, Lytle DA (2013) Flow intermittency alters longitudinal patterns of invertebrate diversity and assemblage composition in an arid-land stream network. Freshwater Biology 58: 1016–1028.
16. Bogan MT, Lytle DA (2011) Severe drought drives novel community trajectories in desert stream pools. Freshwater Biology 56: 2070–2081.
17. Boersma KS, Bogan MT, Henrichs BA, Lytle DA (2014) Invertebrate assemblages of pools in arid-land streams have high functional redundancy and are resistant to severe drying. Freshwater Biology 59: 491–501.
18. Sabo JL, Power ME (2002) River-watershed exchange: Effects of riverine subsidies on riparian lizards and their terrestrial prey. Ecology 83: 1860–1869.
19. Paetzold A, Schubert CJ, Tockner K (2005) Aquatic terrestrial linkages along a braided-river: Riparian arthropods feeding on aquatic insects. Ecosystems 8: 748–759.
20. Sanzone DM, Meyer JL, Marti E, Gardiner EP, Tank JL, et al. (2003) Carbon and nitrogen transfer from a desert stream to riparian predators. Oecologia 134: 238–250.
21. Bastow JL, Sabo JL, Finlay JC, Power ME (2002) A basal aquatic-terrestrial trophic link in rivers: algal subsidies via shore-dwelling grasshoppers. Oecologia 131: 261–268.
22. McCluney KE, Sabo JL (2012) River drying lowers the diversity and alters the composition of an assemblage of desert riparian arthropods. Freshwater Biology 57: 91–103.
23. McCluney KE, Sabo JL (2009) Water availability directly determines per capita consumption at two trophic levels. Ecology 90: 1463–1469.
24. McCluney KE, Date RC (2008) The effects of hydration on growth of the house cricket, *Acheta domesticus*. Journal Of Insect Science 8: 1–9.
25. DeVito J, Meik JM, Gerson MM, Formanowicz DR Jr (2004) Physiological tolerances of three sympatric riparian wolf spiders (Araneae: Lycosidae) correspond with microhabitat distributions. Canadian journal of zoology 82: 1119–1125.
26. Graham AK, Buddle CM, Spence JR (2003) Habitat affinities of spiders living near a freshwater pond. Journal of Arachnology 31: 78–89.
27. Andersen DC, Nelson SM (2006) Flood pattern and weather determine Populus leaf litter breakdown and nitrogen dynamics on a cold desert floodplain. Journal of Arid Environments 64: 626–650.
28. Paetzold A, Tockner K (2005) Effectsd of riparian arthropod predation on the biomass and abundance of aquatic insect emergence. Journal of the North American Benthological Society 24: 395–402.
29. Colby B, Orr P (2005) Economic tradeoffs in preserving riparian habitat. Natural Resources Journal 45: 15–32.
30. Ellis LM, Molles MC, Crawford CS, Heinzelmann F (2000) Surface-active arthropod communities in native and exotic riparian vegetation in the middle Rio Grande Valley, New Mexico. Southwestern Naturalist 45: 456–471.
31. Ellis LM, Crawford CS, Molles MC (2001) Influence of annual flooding on terrestrial arthropod assemblages of a Rio Grande riparian forest. Regulated Rivers-Research & Management 17: 1–20.
32. Cartron JLE, Molles MC, Schuetz JF, Crawford CS, Dahm CN (2003) Ground arthropods as potential indicators of flooding regime in the riparian forest of the middle Rio Grande, New Mexico. Environmental Entomology 32: 1075–1084.
33. Lambeets K, Hendrickx F, Vanacker S, Van Looy K, Maelfait JP, et al. (2008) Assemblage structure and conservation value of spiders and carabid beetles from restored lowland river banks. Biodiversity and Conservation 17: 3133–3148.
34. Paetzold A, Yoshimura C, Tockner K (2008) Riparian arthropod responses to flow regulation and river channelization. Journal of Applied Ecology 45: 894–903.
35. Bonn A, Hagen K, Wohlgemuth-Von Reiche D (2002) The significance of flood regimes for carabid beetle and spider communities in riparian habitats - A comparison of three major rivers in Germany. River Research and Applications 18: 43–64.
36. Paetzold A, Bernet JF, Tockner K (2006) Consumer-specific responses to riverine subsidy pulses in a riparian arthropod assemblage. Freshwater Biology 51: 1103–1115.
37. Greenwood MJ, McIntosh AR (2010) Low river flow alters the biomass and population structure of a riparian predatory invertebrate. Freshwater Biology 55: 2062–2076.
38. Corti R, Datry T (2014) Drying of a temperate, intermittent river has little effect on adjacent riparian arthropod communities. Freshwater Biology 59: 666–678.
39. McCluney KE (2010) Impacts of altered water resources on terrestrial animal communities using stable water isotopes and a water web approach. Tempe: Arizona State University.
40. McKinnon S (2007) Worries increase of San Pedro River's health as it runs dry. The Arizona Republic. Phoenix.
41. Glennon R (2002) Water Follies: Groundwater Pumping and the Fate of America's Fresh Waters. Washington, DC: Island Press.
42. Sabo JL, McCluney KE, Keller AC, Marusenko YY, Soykan CU (2008) Greenfall links groundwater to aboveground food webs in desert river floodplains. Ecological Monographs 78: 615–631.
43. Stromberg J (1998) Dynamics of Fremont cottonwood (Populus fremontii) and saltcedar (Tamarix chinensis) populations along the San Pedro River, Arizona. Journal of Arid Environments 40: 133–155.
44. Skagen SK, Melcher CP, Howe WH, Knopf FL (1998) Comparative use of riparian corridors and oases by migrating birds in southeast Arizona. Conservation Biology 12: 896–909.
45. Sabo JL, Bastow JL, Power ME (2002) Length-mass relationships for adult aquatic and terrestrial invertebrates in a California watershed. Journal of the North American Benthological Society 21: 336–343.
46. Borror DJ, Triplehorn CA, Johnson NF (1992) An Introduction to the Study of Insects. Orlando: Sanders College Publishing. 875 p.
47. Ubick D, Paquin P, Cushing PE, Roth V, editors (2005) Spiders of North America: an identification manual: American Arachnological Society. 377 p.
48. Arnett RH Jr, Thomas MC, editors (2000) American Beetles, Volume I: Archostemata, Myxophaga, Adephaga, Polyphaga: Staphyliniformia. Boca Raton, FL: CRC Press LLC. 443 p.
49. Potts JM, Elith J (2006) Comparing species abundance models. Ecological Modelling 199: 153–163.
50. Bolker BM, Brooks ME, Clark CJ, Geange SW, Poulsen JR, et al. (2009) Generalized linear mixed models: a practical guide for ecology and evolution. Trends in Ecology & Evolution 24: 127–135.
51. Reznikova Z, Dorosheva H (2004) Impacts of red wood ants Formica polyctena on the spatial distribution and behavioural patterns of ground beetles (Carabidae). Pedobiologia 48: 15–21.
52. Rothenbücher J, Schaefer M (2006) Submersion tolerance in floodplain arthropod communities. Basic and Applied Ecology 7: 398–408.
53. Humphreys W (1975) The influence of burrowing and thermoregulatory behaviour on the water relations of Geolycosa godeffroyi (Araneae: Lycosidae), an Australian wolf spider. Oecologia 21: 291–311.
54. Holt RD (1977) Predation, Apparent Competition, and Structure of Prey Communities. Theoretical Population Biology 12: 197–229.
55. Chesson P (2000) Mechanisms of maintenance of species diversity. Annual Review of Ecology and Systematics 31: 343–+.
56. Loreau M, Mouquet N (1999) Immigration and the maintenance of local species diversity. American Naturalist 154: 427–440.
57. Juliano SA (1984) Multiple Feeding and Aggression among Larvae of Brachinus-Lateralis Coleoptera Carabidae. Coleopterists Bulletin 38: 358–360.
58. Juliano SA (1985) Habitat Associations, Resources, and Predators of an Assemblage of Brachinus (Coleoptera, Carabidae) from Southeastern Arizona. Canadian Journal of Zoology-Revue Canadienne De Zoologie 63: 1683–1691.
59. Juliano SA (1986) Resistance to Desiccation and Starvation of 2 Species of Brachinus (Coleoptera, Carabidae) from Southeastern Arizona. Canadian Journal of Zoology-Revue Canadienne De Zoologie 64: 73–80.
60. Kearsley MJ, Cobb N, Yard H, Lightfoot D, Brantley S, et al. (2003) Inventory and Monitoring of Terrestrial Riparian Resources in the Colorado River corridor of Grand Canyon: An Integrative Approach. 01 WRAG 0044 (NAU) and 01 WRAG 0034 (HYC) 01 WRAG 0044 (NAU) and 01 WRAG 0034 (HYC). 22–51 p.
61. Scoble MJ (1995) The Lepidoptera: Form, Function and Diversity: Oxford University Press. 404 p.
62. Arms K, Feeny P, Lederhouse RC (1974) Sodium: Stimulus for Puddling Behavior by Tiger Swallowtail Butterflies, Papilio glaucus. Science 185: 372–374.
63. Valeix M, Fritz H, Matsika R, Matsvimbo F, Madzikanda H (2008) The role of water abundance, thermoregulation, perceived predation risk and interference competition in water access by African herbivores. African Journal of Ecology 46: 402–410.
64. Sabo JL, Power ME (2002) Numerical response of lizards to aquatic insects and short-term consequences for terrestrial prey. Ecology 83: 3023–3036.
65. Nakano S, Murakami M (2001) Reciprocal subsidies: Dynamic interdependence between terrestrial and aquatic food webs. Proceedings of the National Academy of Sciences of the United States of America 98: 166–170.
66. Hadley NF (1994) Water Relations of Terrestrial Arthropods. San Diego: Academic Press.
67. Stromberg JC, Bagstad KJ, Leenhouts JM, Lite SJ, Makings E (2005) Effects of stream flow intermittency on riparian vegetation of a semiarid region river (San Pedro River, Arizona). River Research and Applications 21: 925–938.
68. Sabo JL, Sponseller R, Dixon M, Gade K, Harms T, et al. (2005) Riparian zones increase regional species richness by harboring different, not more, species. Ecology 86: 56–62.
69. Stromberg J, McCluney K, Dixon M, Meixner T (2013) Dryland Riparian Ecosystems in the American Southwest: Sensitivity and Resilience to Climatic Extremes. Ecosystems: 1–5.

9

Iodine Deficiency and Excess Coexist in China and Induce Thyroid Dysfunction and Disease

Yang Du[9], Yanhui Gao[9], Fangang Meng, Shoujun Liu, Zhipeng Fan, Junhua Wu, Dianjun Sun*

Center for Endemic Disease Control, Chinese Center for Disease Control and Prevention, Harbin Medical University; Key Lab of Etiology and Epidemiology, Education Bureau of Heilongjiang Province & Ministry of Health (23618504), Harbin, China

Abstract

Background: In spite of the salt iodization, iodine deficiency disorders (IDD) have not been sustainably eliminated in China. There are coastal areas with low iodized salt coverage rates (iodine nutrition is inadequate) and other areas with excessive amounts of iodine in the drinking water.

Objective: This study aimed to clarify the association of iodine deficiencies resulting from a low coverage rate of iodized salt, excess iodine intake from drinking water with thyroid function and disease in adults.

Design: A cross-sectional study was conducted in adults in different iodine nutrition areas in three provinces in China.

Results: The prevalence of thyroid nodules was 15.52%, 8.66% and 22.17% in the iodine excess, sufficient and deficient groups, respectively. The prevalence of subclinical hypothyroidism was 20.09%, 10.41%, and 2.25% in the excess, sufficient and deficient iodine groups, respectively. The prevalence of subclinical hyperthyroidism and overt hyperthyroidism in the iodine deficient group was higher than that in the iodine excess group ($\chi^2=9.302$, $p=0.002$) and iodine sufficient group ($\chi^2=7.553$, $p=0.006$). Thyroid-stimulating hormone (TSH) was significantly correlated with excess iodine intake ($\beta=1.764, P=0.001$) and deficient iodine intake ($\beta=-1.219$, $P=0.028$).

Conclusions: Thyroid nodules are more likely to be present in the iodine excess and deficient areas than in the iodine sufficient areas. Subclinical hyperthyroidism and overt hyperthyroidism are more likely to be prevalent in the iodine deficient areas than in the iodine excess or sufficient areas. Subclinical hypothyroidism is more likely to be prevalent in the high iodine intake areas than in the iodine deficient or sufficient areas. Median TSH may be deemed as an alternative indicator for monitoring the iodine nutrition status of the adult population in iodine excess and deficient areas.

Editor: Eliseo A. Eugenin, Rutgers University, United States of America

Funding: This study was supported by the Special Fund for National Health Scientific Research in the Public Interest by the Ministry of Health of China (No. 201202012). The funders had no role in study design, data collection and analysis, decision to publish, or preparation of the manuscript.

Competing Interests: The authors have declared that no competing interests exist.

* Email: hrbmusdj@163.com

[9] These authors contributed equally to this work.

Introduction

Iodine is an essential trace element required for the synthesis of thyroid hormones (including thyroxine and triiodothyronine), and it is indispensable for normal growth and fetal brain development in human beings [1,2]. Both iodine deficiency and excess intake may lead to thyroid dysfunction or disease. Iodine deficiency has widespread implications, which include cretinism, endemic goiter, intellectual impairments, increased pregnancy loss and infant mortality [3,4,5,6,7]. Excess iodine intake may lead to endemic goiter and thyroid dysfunction [8,9]. It is reported that thirty countries remain iodine deficient and ten countries have excessive iodine intake across the world [10]. The best strategy to prevent iodine deficiency disorders (IDD) is the addition of iodine to dietary salt in most countries [11]. To date, IDD has been primarily controlled in China. A common reason for iodine excess intake in many countries is the over iodization of salt and/or poor monitoring of salt iodization [12]. However, unlike other iodine excess countries, China has excess iodine areas resulting from high iodine concentrations in water [13]. Moreover, there are iodine deficient areas due to the low coverage rate of iodized table salt along the coast. Previous studies of iodine excess or deficiency have primarily focused on the vulnerable populations such as children, pregnant women and lactating women [14,15,16,17,18]. The last study in China was performed in Henan, Hebei, Shanxi and Shandong province (four main excess iodine areas by historical

Yellow River flushing, and are representative of Chinese iodine excess area) and published in 2014, this study focused on the iodine excess in the vulnerable population, and suggested that excessive iodine intake was associated with impaired thyroid function of children, pregnant women and lactating women [19]. Studies on the impact of iodine deficiency or excess on the adults are scarce. Therefore, the main objective of this study is to clarify the relationship between adult iodine malnutrition (including iodine deficiency due to low coverage rate of iodized salt and excess iodine intake from drinking water) and thyroid abnormalities (dysfunction and disease).

Subjects and Methods

Subjects

According to the recent national IDD surveillance data and water iodine surveillance data, three populations with different iodine nutrition levels were selected in three provinces. Fengrun and Yangfang village in Qingxu County, Dongwenzhuang village in Xiaodian District of Shanxi Province, and Caizhuang village in Cao County of Shandong Province were selected as areas where residents consume high iodine water. Shuangzhai, Houhuangtai, Lizhuang, Anrong and Hutong villages in Shanyin County and Xicuan village in Xiaodian District of Shanxi Province were selected as areas where residents have sufficient iodine nutrition. Tieshangang District in Beihai City, Guangxi Autonomous Region, was selected as a coastal area with low coverage rates of iodized salt and regarded as an area with iodine deficiency. The income of most participants in these areas is similar (4000–5000 RMB yuan/person/year, equivalent to USD 640–800/person/year). A cluster random sampling method was adopted in these areas. All participants who had lived in the region for more than 10 years and were older than 18 years were eligible for recruitment. Subjects taking anti-thyroid drugs during the time of this study were excluded.

Methods

Questionnaire. A standard questionnaire was designed to obtain demographic information, including name, gender, age, personal or family history of thyroid disease, smoking and drinking habits, economic income and source of drinking water and drinking duration. For the smoking habit, participants were divided into four grades: non smoker, mild smokers (1–10 cigarettes/day), moderate smokers (11–20 cigarettes/day) and heavy smokers (>20 cigarettes/day). For the alcohol consumption habit, participants were divided into three categories: non-drinking, occasional light consumption (<1 time/week) and heavy consumption (≥1 times/week). The investigation was performed face to face in dialect by well-trained staff.

Drinking Water Collection and Analysis. In the preselected high iodine water and iodine sufficient villages, at least four water samples were collected from the east, west, south, north and central directions of each village with scattered water supplies; one water sample was collected from each village with centralized water supplies. In the preselected iodine deficient area (Tieshangang District), each participant provided a water sample. Each water sample was more than 15 ml and kept at 4°C until analysis. The iodine concentration of the drinking-water was determined by arsenic–cerium catalytic spectrophotometry [20].

Urine Sample Collection and Analysis. Morning spot urine samples were collected in clean plastic tubes for each participant. Urinary iodine concentration was determined by the acid digestion method ($As^{3+}-Ce^{4+}$catalytic spectrophotometry [21]). The internal quality control samples for urine iodine and water iodine were provided by the China National Iodine Deficiency Disorders Reference Laboratory.

Coverage Rate of Iodized Salt. In the preselected iodine excess or sufficient areas, the iodized salt coverage rates were calculated by the Information Management System of National Iodized Salt Surveillance, which is managed by our organization. In the preselected iodine deficient area (Tieshangang District), each participant provided a table salt sample of at least 100 g. The salt iodine levels were determined using the General Test Method of the salt industry for the determination of iodine [22]. The standard for salt iodine level in Shandong, Shanxi province and Guangxi Autonomous Region is 25 mg/kg. The salt type was defined as follows: non-iodized salt, salt iodine less than 5 mg/kg, qualified iodized salt, salt iodine more than 18 mg/kg and less than 50 mg/kg, others, unqualified iodized salt.

Blood Sample and Analysis. Blood samples were collected from the cubital vein and stored in a glass tube. The serum samples were prepared by centrifugation (3000 r/m, 10 mins) after standing for an hour and were frozen at −70°C until determination. Serum thyroid-stimulating hormone (TSH), free triiodothyronine (FT3), free tetraiodothyronine (FT4), thyroid peroxidase antibody (TPOAb) and thyroglobulin antibody (TGAb) were tested. These indexes were tested by chemiluminescent immunoassay (Fluorometric magnetic beads enzyme immunoassay, TOSOH Corporation, Japan). The thyrotropin receptor antibody (TRAb) was tested by chemiluminescent immunoassay (Snibe Co., Ltd, China).

Thyroid Ultrasound. Thyroid ultrasonography was performed using a 7.5-MHz transducer by one experienced examiner (mainly measure the thyroid volume, nodule diameter and echo). The thyroid lobe volume was calculated by measuring the depth (d), the width (w) and the length (l) of each lobe by the following formula: $V\ (ml) = 0.479 \times d \times w \times l\ (mm)/1000$. The thyroid volume was recorded as the sum of both lobes. The normal volume for adult males and females is less than 25 ml or 18 ml, respectively [23].

Diagnostic Criteria for Thyroid Diseases. The diagnostic criteria for thyroid diseases were listed in Table 1 [24,25].

Statistical Analysis. All data were recorded with Epi-Data3.1, the data processing and statistical analyses were performed using IBM SPSS statistics version 19 and SigmaPlot 12.5 software. Comparison of the proportions among the three groups was performed by the chi-square test ($\alpha = 0.05$); if the null hypothesis was rejected, pairwise comparisons were performed ($\alpha = 0.0125$). Normally distributed data were expressed as the mean ±SD; abnormally distributed data were described as the median with the 25th and 75th percentiles. Binary logistic regression analysis was used to calculate the odds ratio (OR) and 95% Confidence Intervals (CIs) for thyroid diseases by iodine nutrition status and other risk factors. The dependent variables entered into the logistic regression models were goiter, autoimmune thyroiditis, subclinical hyperthyroidism and overt hyperthyroidism, nodules and subclinical hypothyroidism. Gender, age, smoking status, alcohol consumption, goiter, nodules, iodine status, TPOAb positive and TGAb positive were entered as explanatory variables. First, univariate logistic regression was performed to select the potentially significant variables for the multivariate model. Then, based on the results of univariate logistic analysis, the significant variables, gender and age were put into the multivariate logistic model. The pairwise comparisons of the TSH values among the three groups were compared by the Kruskal-Wallis test. The relativity between TSH and urinary iodine concentration was tested using the Spearman rank correlation.

Table 1. Diagnostic criteria for the thyroid diseases included in this study.

Thyroid disease	Diagnostic criteria*
Goiter	Thyroid volume >25 ml (male) or >18 ml (female)
Diffuse	Goiter without nodules
Nodular	Goiter with nodules >10 mm in diameter
Single nodule	Normal thyroid volume with a single nodule >3 mm in diameter
Multiple nodules	Normal thyroid volume with >2 nodules >3 mm in diameter
Overt hypothyroidism	TSH>5.57 mIU/L, FT4<10.06 pmol/L
Subclinical hypothyroidism	TSH>5.57 mIU/L, FT4 within the normal range
Over hyperthyroidism	TSH<0.38 mIU/L; FT4>23.99 pmol/L, FT3>5.85 pmol/L, or both
Subclinical hyperthyroidism	TSH<0.38 mIU/L, FT3 and FT4 within the normal ranges
Graves' disease	Hyperthyroidism. Diffuse goiter on ultrasound. TPOAb>2.6 IU/ml or TRAb>30 IU/ml
Autoimmune thyroiditis	Hypothyroidism or subclinical hypothyroidism. TPOAb>2.6 IU/ml or TgAb>14.58 IU/ml
Hashimoto's thyroiditis	With goiter
Atrophic thyroiditis	Without goiter

*Reference ranges: FT3, 2.77–5.85 pmol/L; FT4, 10.06–23.99 pmol/L; TSH, 0.38–5.57 mIU/L; TPOAb, 0–2.6 IU/ml; TgAb, 0–14.58 IU/ml; TRAb, 0.11–30 IU/ml. Reference ranges were provided by the second affiliated hospital of Harbin Medical University.

Ethical Statement. The study was approved by the Ethical Review Board of Harbin Medical University (HMUIRB 20120021). Written informed consent was obtained from all participants ahead of the survey. No specific permits were required for the locations or activities associated with the drinking water sample collection in this field study. The locations were not privately owned or protected in any way and this field study did not involve endangered or protected species.

Results

A total of 2147 adults were recruited for this survey, the sample sizes were presented in Table 2. The demographic characteristics of the participants were presented in Table 3.

Iodine Concentration in Urine Sample

Each village was a unit, the villages were divided into excess, sufficient and deficient groups according to the median urinary iodine (MUI) of each village. The grouping criteria was as follows: iodine excess group, MUI>400 μg/L; iodine sufficient group, 100 μg/L<MUI<300 μg/L; and iodine deficient group, MUI< 100 μg/L. The MUI of the participants was 750.18 μg/L, 228.70 μg/L and 62.03 μg/L for iodine excess, sufficient and deficient groups, respectively, as shown in Table 2.

Iodine Concentration in Drinking Water

The detailed iodine concentrations in the drinking water of each survey site were presented in Table 2. The median iodine concentration of the drinking water ranged from 272.98 μg/L to 629.90 μg/L in the iodine excess group and 7.69 μg/L to 77.77 μg/L in the iodine sufficient group. The median in the iodine deficient group was 2.84 μg/L.

Coverage Rate of Iodized Table Salt

In Shanyin County, the coverage rates of iodized salt and qualified iodized salt were 99.67% and 98%, respectively. In high iodine water areas of Xiaodian District and Cao County, the coverage rates of iodized table salt were 0% and 1.67%, respectively. In the iodine deficient areas of Tieshangang, the coverage rates of iodized salt and qualified iodized salt were 24.29% and 17.14%, respectively.

The Prevalence of Thyroid Diseases

Nodule. The prevalence of thyroid nodules in the different iodine nutrition groups was presented in Table 4. For single and multiple nodules, the prevalence of thyroid nodules differed significantly among the three groups ($\chi^2 = 39.779$, $p = 0$), between the iodine deficient group and iodine sufficient group ($\chi^2 = 39.234$, $p = 0$), between the iodine excess group and iodine sufficient group ($\chi^2 = 13.964$, $p = 0$), and between the iodine excess group and iodine deficient group ($\chi^2 = 11.125$, $p = 0.001$). The prevalence of thyroid nodules for males and females was shown in Figure 1. In males, a significant difference in the prevalence of thyroid nodules was found among the three groups ($\chi^2 = 18.401$, $p = 0$), between the iodine deficient group and iodine sufficient group ($\chi^2 = 13.67$, $p = 0$), and between the iodine deficient group and iodine excess group ($\chi^2 = 11.916$, $p = 0.001$), but not between the iodine excess group and iodine sufficient group ($\chi^2 = 0.389$, $p = 0.533$). In females, the prevalence of thyroid nodules differed significantly among the three groups ($\chi^2 = 21.463$, $p = 0$), between the iodine excess group and iodine sufficient group ($\chi^2 = 11.775$, $p = 0.001$), and between the iodine deficient group and iodine sufficient group ($\chi^2 = 21.629$, $p = 0$), but not between iodine excess group and iodine deficient group ($\chi^2 = 2.998$, $p = 0.083$). The prevalence of single nodule differed significantly among the three groups ($\chi^2 = 20.624$, $p = 0$), between the iodine excess group and iodine sufficient group ($\chi^2 = 10.999$, $p = 0.001$), and between the iodine deficient group and iodine sufficient group ($\chi^2 = 20.809$, $p = 0$), but not between the iodine excess group and iodine deficient group ($\chi^2 = 2.96$, $p = 0.085$). For multiple nodules, the prevalence of nodules in the iodine deficient group was higher than that in other groups (Iodine sufficient group vs. Iodine deficient group, $\chi^2 = 17.918$, $p = 0$; Iodine excess group vs. Iodine deficient group, $\chi^2 = 12.933$, $p = 0$), as shown in Table 4. Logistic-regression analysis indicated that female or elder age, excess iodine intake or iodine deficiency, smoking 11–20 cigarettes per day and goiter were risk factors for developing thyroid nodules (Table 5).

Table 2. A summary of the samples collected and the iodine nutrition status of the survey sites.

Survey sites		Participant numbers	Water samples		Urine samples		Blood samples	Ultrasound samples	Table salt samples
			n	Median (Interquartile range) (μg/L)	n	Median (Interquartile range) (μg/L)			
Iodine excess group	Caizhuang	285	246	272.98 (171.95,393.33)	281	473.38 (270.51,758.51)	282	285	—
	Dongwenzhuang	156	1	629.90	145	1183.96 (793.26,1557.11)	156	156	—
	Fengrun	145	1	571.20	108	896.55 (525.03,1264.95)	137	139	—
	Yangfang	344	1	460.20	254	823.90 (573.05, 1219.88)	286	335	—
	Sum	930	249	—	788	750.18 (451.84, 1188.55)	861	915	—
Iodine sufficient group	Xicuan	122	2	77.77	116	264.72 (195.85, 380.04)	122	121	—
	Shuangzhai	68	20	23.15 (9.94, 32.96)	56	272.41 (175.95, 328.72)	56	66	—
	Houhuangtai	163	20	10.90 (1.62, 19.52)	111	188.45 (112.48, 258.75)	146	158	—
	Lizhuang	65	20	13.13 (1.0, 30.92)	53	229.50 (108.39, 288.78)	61	63	—
	Anrong	86	25	7.69 (1.46, 11.37)	71	239.50 (124.10, 324.50)	80	78	—
	Hutong	46	20	15.29 (10.51, 28.02)	43	209.99 (126.94, 293.20)	44	45	—
	Sum	550	107	—	450	228.70 (134.13, 322.73)	509	531	—
Iodine deficient group	Tieshangang	667	336	2.84 (1.93,5.87)	642	62.03 (36.24, 99.10)	667	636	280

Table 3. Demographic characteristics of the participants in the three different iodine nutrition groups.

Characteristics		Iodine excess group	Iodine sufficient group	Iodine deficient group
Participants number		930	550	667
Gender				
	Male	281	212	157
	Female	649	338	510
Age				
	Mean ±SD	55.46±12.44	57.71±12.11	51.89±15.10
	≤40	181	91	247
	41–60	432	254	214
	61-	317	205	206
Education level※				
	Illiteracy	238	91	61
	Primary school	396	282	197
	Junior high school	227	140	217
	Senior high school	55	16	120
	College	14	5	72
Marital status▲				
	Single	4	2	27
	Married	914	533	614
	Divorced/widowed	12	2	26
Smoking★				
	None	789	432	628
	1–10	29	33	21
	11–20	29	3	12
	>20	66	77	6
Alcohol consumption△				
	None	809	490	592
	Light	65	28	51
	Heavy	37	22	24

※16 missing.
▲13 missing.
★22 missing.
△29 missing.

Goiter. No significant difference was found in the prevalence of goiter among the three groups ($\chi^2 = 1.871$, $p = 0.392$). Logistic-regression analysis indicated that female, TPOAb positive, TGAb positive and having nodules were risk factors for developing goiter (Table 5).

Hypothyroidism. The prevalence of overt hypothyroidism was 2.56%, 1.18% and 1.05% for the iodine excess, sufficient and deficient groups, respectively. A significant difference was found among the three groups ($\chi^2 = 6.215$, $p = 0.045$). However, there were no significant differences in the pairwise comparisons (Iodine excess group vs. Iodine sufficient group, $\chi^2 = 3.027$, $p = 0.082$; Iodine excess group vs. Iodine deficient group, $\chi^2 = 4.576$, $p = 0.032 > 0.0125$; Iodine sufficient group vs. Iodine deficient group, $\chi^2 = 0.044$, $p = 0.834$). No significant difference was found among the three groups in males or females (males, $p = 0.253$; females, $p = 0.183$) as shown in Table 6.

The prevalence of subclinical hypothyroidism was 20.09%, 10.41%, and 2.25% for the iodine excess, sufficient and deficient iodine groups, respectively. There were significant differences in the prevalence of subclinical hypothyroidism among the comparisons in males or females. In males, the prevalence of subclinical hypothyroidism differed significantly among all three groups ($\chi^2 = 24.159$, $p = 0$), between the iodine excess group and iodine sufficient group ($\chi^2 = 9.315$, $p = 0.002$), and between the iodine excess group and iodine deficient group ($\chi^2 = 18.911$, $p = 0$), but not between the iodine sufficient group and iodine deficient group ($\chi^2 = 3.763$, $p = 0.052$). In females, there were significant differences among the three groups ($\chi^2 = 95.232$, $p = 0$), between the iodine excess group and iodine sufficient group ($\chi^2 = 10.816$, $p = 0.001$), and between the iodine sufficient group and iodine deficient group ($\chi^2 = 37.349$, $p = 0$), and between the iodine excess group and iodine deficient group ($\chi^2 = 95.295$, $p = 0$). Logistic-regression results indicated that females were more susceptible to subclinical hypothyroidism than males, and people living in the excess and sufficient areas were more susceptible to subclinical hypothyroidism than those in the deficient areas. Thyroid nodules and TPOAb positive were risk factors for subclinical hypothyroidism (Table 5).

Table 4. Prevalence of thyroid nodules and goiter within the different iodine nutrition groups.

Thyroid diseases	Iodine excess group number of cases (percent)	Iodine sufficient group number of cases (percent)	Iodine deficient group number of cases (percent)
Nodules	142 (15.52)	46 (8.66)	141 (22.17)
Single nodule	121 (13.22)	40 (7.53)	104 (16.35)
Multiple nodules	21 (2.30)	6 (1.13)	37 (5.82)
Goiter	23 (2.51)	12 (2.26)	22 (3.46)
Diffuse	21 (2.30)	9 (1.69)	15 (2.36)
Nodular	2 (0.22)	3 (0.56)	7 (1.10)

Hyperthyroidism. There were no significant differences in the prevalence of overt hyperthyroidism ($\chi^2 = 3.590$, $p = 0.166$) and Graves' disease ($p = 0.593$, Fisher's exact test) among the three groups. The prevalence of subclinical hyperthyroidism differed significantly among the three groups ($\chi^2 = 11.596$, $p = 0.003$); significant differences were found between the iodine sufficient group and deficient group ($\chi^2 = 7.538$, $p = 0.006$), but not between the iodine excess group and deficient group ($\chi^2 = 6.045$, $p = 0.014 > 0.0125$).

The prevalence of subclinical hyperthyroidism and overt hyperthyroidism was 1.16%, 0.98% and 3.44% for the iodine excess, sufficient and deficient groups, respectively. The prevalence of subclinical hyperthyroidism and overt hyperthyroidism differed significantly among the three groups ($\chi^2 = 13.628$, $p = 0.001$), between the iodine deficient group and iodine excess group ($\chi^2 = 9.302$, $p = 0.002$), and between the iodine deficient group and iodine sufficient group ($\chi^2 = 7.553$, $p = 0.006$). Logistic-regression results indicated that female, living in an iodine deficient area, and young or middle aged were risk factors for developing overt hyperthyroidism or subclinical hyperthyroidism.

Autoimmune Thyroiditis. The prevalence of autoimmune thyroiditis differed significantly among the three groups ($\chi^2 = 21.564$, $p = 0$), between the iodine excess group and iodine deficient group ($\chi^2 = 21.805$, $p = 0$), and between the iodine sufficient group and iodine deficient group ($\chi^2 = 9.249$, $p = 0.002$). The prevalence of atrophic thyroiditis differed among the three groups ($\chi^2 = 15.808$, $p = 0$), between the iodine sufficient group and iodine deficient group ($\chi^2 = 7.615$, $p = 0.006$), and between the iodine excess group and iodine deficient group ($\chi^2 = 15.964$, $p = 0$), but not between the iodine excess group and iodine sufficient group ($\chi^2 = 1.003$, $p = 0.317$) (Table 6).

TSH Level

The results of the Kruskal-Wallis test indicated that the TSH levels differed significantly among the three groups (among all the three groups, $p = 0$; Iodine excess group vs. Iodine sufficient group, $p = 0$; Iodine excess group vs. Iodine deficient group, $p = 0$; Iodine sufficient group vs. Iodine deficient group, $p = 0$). Figure 2 showed the correlation between the MUI and median TSH, with the increasing MUI correlating with the increasing median TSH. By spearman correlation analysis, the urinary iodine concentration was positively related with the TSH value (r = 0.414, $p = 0$). By multiple linear regression analysis, after adjusting for gender and age, the TSH value was significantly correlated with excess iodine

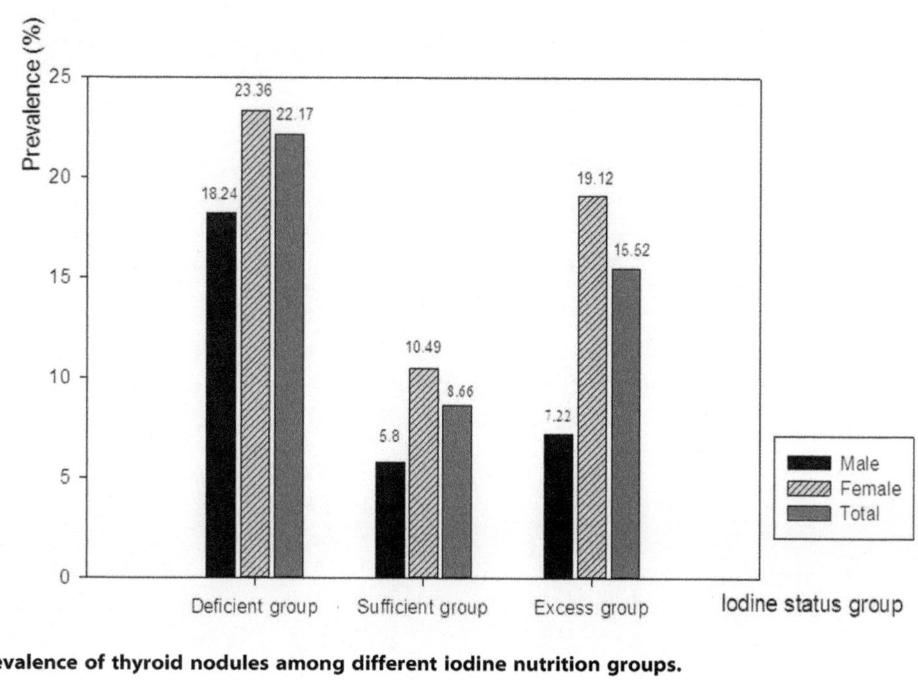

Figure 1. The prevalence of thyroid nodules among different iodine nutrition groups.

Table 5. Associations between thyroid diseases and different iodine nutrition status and other potential risk factors.

Factors		Sub-group Number	Goiter	Autoimmune thyroiditis	Subclinical hyperthyroidism and overt hyperthyroidism	Nodules	Subclinical hypothyroidism
			OR (95% CI)	OR (95% CI)	OR (95% CI)	OR (95% CI)	OR (95% CI)
Gender	Female	1497	1	1	1	1	1
	Male	650	0.251 (0.097, 0.647)**	0.849 (0.558, 1.292)	0.214 (0.050, 0.909)*	0.478 (0.332, 0.689)**	0.666 (0.473, 0.940)*
Age	≤40	519	1	1	1	1	1
	41–60	900	0.687 (0.319, 1.478)	1.472 (0.856, 2.531)	0.575 (0.280, 1.180)	1.602 (1.146, 2.240)**	1.363 (0.907, 2.046)
	61–	728	1.583 (0.775, 3.233)	1.465 (0.830, 2.588)	0.216 (0.071, 0.653)**	1.913 (1.354, 2.702)**	1.186 (0.768, 1.832)
Iodine status	Sufficient	550				1	
	Excess	930				1.865 (1.299, 2.678)**	
	Deficient	667				2.973 (2.046, 4.321)**	
Iodine status	Deficient	667	1	1	1		1
	Sufficient	550	0.979 (0.448, 2.143)	2.162 (1.168, 4.001)*	0.401 (0.146, 1.102)		5.791 (3.161, 10.611)**
	Excess	930	0.981 (0.524, 1.836)	3.012 (1.742, 5.209)**	0.404 (0.186, 0.880)*		11.717 (6.742, 20.362)**
Smoking	Never	1849				1	
	1–10	83				1.502 (0.708, 3.186)	
	11–20	44				2.530 (1.054, 6.071)*	
	>20	149				0.940 (0.455, 1.940)	
Alcohol consumption	None	1891				1	
	Light	144				0.709 (0.364, 1.383)	
	Heavy	83				0.717 (0.298, 1.725)	
TPOAb	Negative	1686	1		1		1
	Positive	351	2.582 (1.274, 5.233)**		2.124 (0.879, 5.130)		2.20 (1.454, 3.327)**
TGAb	Negative	1640	1		1		1
	Positive	397	2.373 (1.171, 4.812)*		2.289 (0.952, 5.501)		1.398 (0.935, 2.091)
Thyroid nodules	No	1753	1	1			1
	Yes	329	4.383 (2.455, 7.826)**	0.492 (0.252, 0.960)*			1.769 (1.222, 2.561)**
Goiter	No	2025				1	1
	Yes	57				3.477 (1.997, 6.055)**	1.918 (0.963, 3.818)

*$p<0.05$.
**$p<0.01$.

intake ($\beta = 1.764, P = 0.001$), iodine deficient intake ($\beta = -1.219$, $P = 0.028$) and TGAb positivity ($\beta = 3.038$, $P = 0$) (Table 7).

Discussion

Interestingly, our findings indicated that thyroid nodules were likely to be prevalent when the population has excess or deficient iodine intake. The correlation between iodine deficiency and thyroid nodules has been reported in many countries [26,27,28,29]. Takahashi et al. found that moderate iodine deficiency might be responsible for the increase of thyroid nodules in the Marshallese population [27]. The study performed by Gutekunst et al. concluded that moderate dietary iodine deficiency might cause both goiters and thyroid nodules [28]. However, research on the relationship between excess iodine intake and thyroid nodules is scarce. Recently, a cross-sectional study on the relationship between non-iodized salt and thyroid nodules conducted in Hangzhou (a foreland city in China) found a significant correlation between low urine iodine concentration and high thyroid nodules prevalence. However, in contrast to our findings, the Hangzhou study did not find a significant relationship between high and excess iodine intake and thyroid nodules [30]. In that study, the excessive iodine intake population selection didn't depend on the MUI of the population, but depended on the individual urine iodine concentration [30]. In fact, Hangzhou is another environmental iodine deficient area [31]. Future study is needed to investigate the relationship between excess iodine intake and the prevalence of thyroid nodules. Our study demonstrated that smoking 11–20 cigarettes per day was significantly related with the presence of thyroid nodules. The main reason for this might be the sample size of the 11–20 cigarettes per day group, which was much smaller than the other groups.

The results indicated that subclinical hypothyroidism was more likely to be prevalent in high iodine intake areas than in deficient or sufficient areas, while overt hyperthyroidism and subclinical hyperthyroidism were more likely to be prevalent in iodine deficient areas than in excess or sufficient areas. Likewise, our study results are similar to Andersen's study [32], in which they found more hypothyroidism and less hyperthyroidism in elderly people (aged 75–80) with sufficient iodine nutrition compared to those with mild iodine deficiency. A study conducted in China found that children had a higher concentration of TSH and a higher prevalence of thyroid diseases in the excess iodine area than in the iodine adequate area, especially for subclinical hypothyroidism [16].

It's well documented in previous studies that the prevalence of goiter in school children is higher in iodine excess or deficient areas than in iodine sufficient areas [4,33,34,35,36]; the prevalence of goiter in school-age children is usually used as a major indicator of IDD surveillance [37]. The results of this study, which included both diffuse and nodular goiter, indicated that goiter prevalence was low (<5%) in all three groups, regardless of iodine excess or deficiency. In addition, females were more susceptible to goiter. In Teng's study, the prevalence of goiter in adults in the more than adequate iodine area was higher than that in the excess iodine areas [25]. The goiter prevalence was different between adults and school children. Based on the above evidence, it could be concluded that thyroid goiter might not be an accurate indicator for assessing the iodine status in adults.

The results of this study indicated that iodine nutrition status play an important role in the TSH level, and iodine excess or deficiency influenced the TSH level. It is reported that neonatal blood TSH concentrations are an indicator for monitoring iodine deficiency in many countries [38,39,40]. Based on our survey,

Table 6. Prevalence of thyroid disease within the different iodine nutrition groups.

Thyroid disease	Iodine excess group number of cases (percent)			Iodine sufficient group number of cases (percent)			Iodine deficient group number of cases (percent)		
	Total (n = 861)	Male (n = 256)	Female (n = 605)	Total (n = 509)	Male (n = 197)	Female (n = 312)	Total (n = 667)	Male (n = 157)	Female (n = 510)
Subclinical hypothyroidism	173 (20.09)	39 (15.23)	134 (22.15)	53 (10.41)	12 (6.09)	41 (13.14)	15 (2.25)	3 (1.91)	12 (2.35)
Overt hypothyroidism	22 (2.56)	6 (2.34)	16 (2.64)	6 (1.18)	1 (0.51)	5 (1.60)	7 (1.05)	1 (0.64)	6 (1.18)
Subclinical hyperthyroidism and overt hyperthyroidism	10 (1.16)	1 (0.39)	9 (1.49)	5 (0.98)	0	5 (1.60)	23 (3.44)	1 (0.64)	22 (4.31)
Subclinical hyperthyroidism	5 (0.58)	1 (0.39)	4 (0.66)	1 (0.20)	0	1 (0.32)	13 (1.95)	0	13 (2.55)
Overt hyperthyroidism	5 (0.58)	0	5 (0.83)	4 (0.79)	0	4 (1.28)	10 (1.50)	1 (0.64)	9 (1.76)
Graves' disease	3 (0.35)	0	3 (0.50)	3 (0.59)	0	3 (0.96)	5 (0.75)	1 (0.64)	4 (0.78)
Autoimmune thyroiditis	70 (8.13)	23 (8.98)	47 (7.77)	31 (6.09)	8 (4.06)	23 (7.37)	17 (2.55)	4 (2.55)	13 (2.55)
Hashimoto's thyroiditis	9 (1.05)	2 (0.78)	7 (1.16)	2 (0.39)	0	2 (0.64)	0	0	0
Atrophic thyroiditis	61 (7.08)	21 (8.20)	40 (6.61)	29 (5.70)	8 (4.06)	21 (6.73)	17 (2.55)	4 (2.55)	13 (2.55)

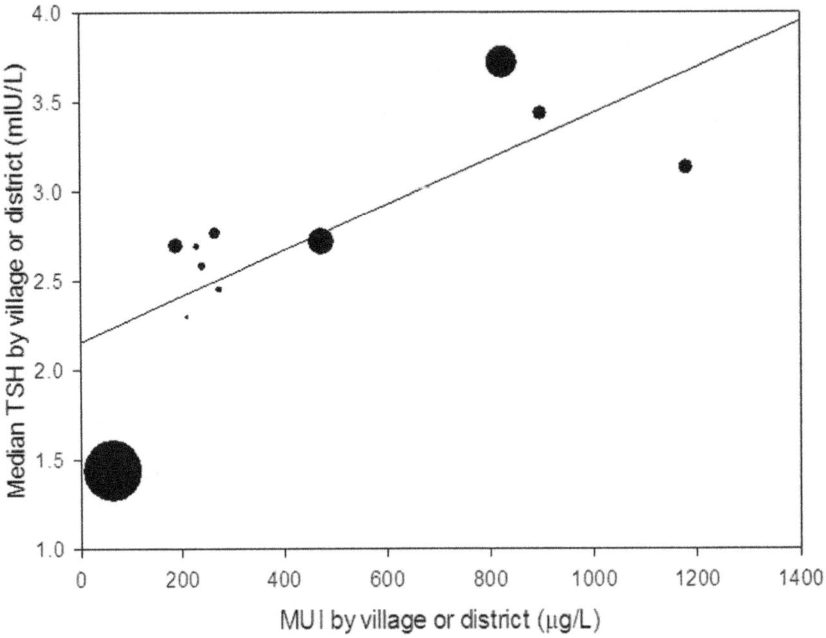

Figure 2. Bubble plot (clustered by village or district) of median TSH vs. MUI with a linear trend line. The size of the bubbles reflects the sample size for each village or district.

except for MUI, median TSH could be deemed as an alternative indicator for monitoring the iodine nutrition status of the adult population in iodine excess or deficient areas. Our results suggested that autoimmune thyroiditis might be prevalent in iodine excess or sufficient areas, which was in line with the previous study [25].

IDD was widespread in China until 1994 when universal salt iodization (USI) was carried out as a mandatory effective strategy to prevent IDD. Afterward, according to the IDD Surveillance Report of China, the government adjusted the iodine concentration in iodized salt three times to maintain the iodine nutrition at an appropriate level. However, at present, there are extra environmental iodine deficient areas with low iodized salt coverage in China and iodine deficiency remains a public health problem. The results of the present study indicate that Tieshangang District of Beihai City is a mild iodine deficient area with a high prevalence of thyroid nodules, overt hyperthyroidism and subclinical hyperthyroidism. The coverage rates of iodized salt and qualified iodized salt are low in Tieshangang District, and the extra environmental iodine nutrition is deficient, such as drinking

Table 7. Simple and multivariate linear regression analyses of TSH on determinants.

Variables	*β*	95% CI	*P*
Simple linear regression			
Age	0.038	0.007–0.068	0.016
Gender	0.371	−0.526–1.268	0.471
Smoking	0.353	−0.027–0.733	0.069
Drinking	−0.394	−1.308–0.52	0.398
Iodine status	1.574	1.101–2.047	0
TPOAb positive	3.041	1.96–4.121	0
TGAb positive	3.815	2.791–4.839	0
Multivariate linear regression			
Age	0.023	−0.008–0.054	0.14
Gender	0.484	−0.415–1.384	0.291
TGAb positive	3.038	1.787–4.288	0
TPOAb positive	1.307	−0.001–2.616	0.050
Excess group*	1.764	0.746–2.781	0.001
Deficient group*	−1.219	−2.307– −0.131	0.028

*Dump variable: the reference is the sufficient group.

water. In the coastal areas, there are illegal private salt plants that produce coarse salt (non-iodized salt), the price of course salt is lower than that of iodized salt, and thus it is easier to obtain for the households living in the coastal areas. This situation in the coastal areas highlights the necessity of consolidating the USI and ensuring a high coverage rate of qualified iodized salt, which are indispensable for preventing IDD, especially in coastal areas in China. Adequate iodine nutrition protects the health of the thyroid and prevents a high prevalence of thyroid disease. There are 30 million people living in the high iodine water areas distributed in 734 towns in China [13]. One of the effective measures to prevent excess iodine intake is to sell non-iodized salt in the market in high iodine water areas; the water authorities seldom pay attention to the iodine concentration in the water improvement program. This study demonstrated that the current intervention of supplying non-iodized salt was not enough to prevent excess iodine intake in high iodine water areas; therefore, changing to a safe water source may be the most effective measure for the prevention of the harm resulted from excess iodine intake.

Currently, the subjects of regular iodine nutrition surveillance are children, pregnant women and lactating women. However, our study suggested that both excess and deficient iodine intake could cause thyroid dysfunction or disease in adults; therefore, the adult population should also be under surveillance.

The present study has some limitations. First, the proportion of gender among the three different iodine nutrition groups was different, which might result in a bias for the prevalence of the thyroid disease; however, we applied the multivariable regression analysis to exclude the potential influence of confounders. Second, our survey sites were selected from three provinces in China, the dietary habits may be different in these sites, and we did not conduct a dietary survey, which may affect the results of this study.

In conclusion, iodine deficiency and iodine excess coexist in China. We have demonstrated that both iodine excess and deficiency induce thyroid dysfunction or diseases. The high iodine water areas should decrease iodine intake by changing to a safe water source without providing non-iodized salt, whereas in coastal areas, the non-iodized salt should be forbidden and we should encourage the household to buy iodized salt. These measures should be adopted to make sure that the MUI is maintained at an optimal level (100–199 μg/L) as suggested by the WHO [37]. Our study has provided important baseline data for monitoring the iodine nutrition and thyroid function in iodine excess or deficient areas. In view of our findings, it is recommended that, in the future, except for children, pregnant women and lactating women, an increased emphasis should be placed on monitoring the iodine status of the adult population, especially in the endemic regions.

Acknowledgments

We thank all the organizations that participated in this survey: the provincial (municipality) Centers for Disease Control (CDC) and the Centers for Endemic Disease Control in Shandong, Shanxi and Guangxi, and the residents of Cao County, Shanyin County, Qingxu County, Xiaodian District and Tieshangang District who participated in this survey.

Author Contributions

Conceived and designed the experiments: DS. Performed the experiments: FM SL ZF JW. Analyzed the data: YG. Wrote the paper: YD.

References

1. Zimmermann MB (2009) Iodine deficiency. Endocrine reviews 30: 376–408.
2. Zimmermann MB, Jooste PL, Pandav CS (2008) Iodine-deficiency disorders. Lancet 372: 1251–1262.
3. Mitka M (2013) Even mild iodine deficiency during gestation may impair brain function in children. JAMA 309: 2428–2428.
4. District C (2013) Prevalence of Goitre in School Children of Chamarajanagar District, Karnataka, India. Journal of Clinical and Diagnostic Research 7: 2807–2809.
5. Matthys B, Davlatmamadova M, Karimova G, Jean-Richard V, Zimmermann MB, et al. (2013) Iodine nutritional status and risk factors for goitre among schoolchildren in South Tajikistan. BMC endocrine disorders 13: 50.
6. Zimmermann MB (2012) The effects of iodine deficiency in pregnancy and infancy. Paediatric and perinatal epidemiology 26: 108–117.
7. Pearce EN (2012) Effects of iodine deficiency in pregnancy. Journal of Trace Elements in Medicine and Biology 26: 131–133.
8. Roti E, Uberti ED (2001) Iodine excess and hyperthyroidism. Thyroid 11: 493–500.
9. Li M, Liu DR, Qu CY, Zhang PY, Qian QD, et al. (1987) Endemic goitre in central China caused by excessive iodine intake. Lancet 2: 257–259.
10. Pearce EN, Andersson M, Zimmermann MB (2013) Global iodine nutrition: where do we stand in 2013? Thyroid 23: 523–528.
11. Andersson M, de Benoist B, Rogers L (2010) Epidemiology of iodine deficiency: salt iodisation and iodine status. Best Practice & Research Clinical Endocrinology & Metabolism 24: 1–11.
12. Zimmermann MB (2013) Iodine deficiency and excess in children: worldwide status in 2013. Endocrine Practice 19: 839–846.
13. Shen H, Liu S, Sun D, Zhang S, Su X, et al. (2011) Geographical distribution of drinking-water with high iodine level and association between high iodine level in drinking-water and goitre: a Chinese national investigation. British Journal of Nutrition 106: 243–247.
14. Lv S, Xu D, Wang Y, Chong Z, Du Y, et al. (2014) Goitre prevalence and epidemiological features in children living in areas with mildly excessive iodine in drinking-water. British Journal of Nutrition 111: 86–92.
15. Kapil U, Prakash S, Sareen N, Bhadoria A, Khenduja P, et al. (2014) Status of iodine deficiency among pregnant mothers in Himachal Pradesh, India. Public health nutrition 17: 1971–1974.
16. Sang Z, Chen W, Shen J, Tan L, Zhao N, et al. (2013) Long-Term Exposure to Excessive Iodine from Water Is Associated with Thyroid Dysfunction in Children. The Journal of nutrition 143: 2038–2043.
17. Sang Z, Wei W, Zhao N, Zhang G, Chen W, et al. (2012) Thyroid dysfunction during late gestation is associated with excessive iodine intake in pregnant women. The Journal of Clinical Endocrinology & Metabolism 97: E1363–E1369.
18. Lv S, Zhao J, Xu D, Chong Z, Jia L, et al. (2012) An epidemiological survey of children's iodine nutrition and goitre status in regions with mildly excessive iodine in drinking water in Hebei Province, China. Public health nutrition 15: 1168–1173.
19. Liu P, Liu L, Shen H, Jia Q, Wang J, et al. (2014) The Standard, Intervention Measures and Health Risk for High Water Iodine Areas. PLoS One 9: e89608.
20. Jin Y, E X, Chen Y, Zhang L, Chen C, et al. (2006) Standard exanination methods for drinking water-Nonmetal parameters(GB/T5750.5–2006). Beijing, China: China Criteria Publishing House.
21. Yan Y, Zhang Y, Liu L, Liu J, Li W, et al. (2006) Method for determination of iodine in urine by As $^{3+}$-Ce^{4+} catalytic spectrophotometry (WS/T 107–2006). Beijing, China: People's Medical Publishing House.
22. Tong Y, Huo J (2012) General test method in salt industry: Determination of iodine (GB/T 13025.7–2012). Beijing, China: China Criteria Publishing House.
23. Liu S, Chen Z, Jia Q, Su X, Zhang S (2007) Diagnostic criterion of endemic goiter(WS 276–2007). Beijing, China: People's Medical Publishing House.
24. Zou S, Wu F, Guo C, Song J, Huang C, et al. (2012) Iodine nutrition and the prevalence of thyroid disease after salt iodization: a cross-sectional survey in Shanghai, a coastal area in China. PLoS One 7: e40718.
25. Teng W, Shan Z, Teng X, Guan H, Li Y, et al. (2006) Effect of iodine intake on thyroid diseases in China. N Engl J Med 354: 2783–2793.
26. Yu X, Fan C, Shan Z, Teng X, Guan H, et al. (2008) A five-year follow-up study of goiter and thyroid nodules in three regions with different iodine intakes in China. J Endocrinol Invest 31: 243–250.
27. Takahashi T, Fujimori K, Simon SL, Bechtner G, Edwards R, et al. (1999) Thyroid nodules, thyroid function and dietary iodine in the Marshall islands. International journal of epidemiology 28: 742–749.
28. Gutekunst R, Smolarek H, Hasenpusch U, Stubbe P, Friedrich H-J, et al. (1986) Goitre epidemiology: thyroid volume, iodine excretion, thyroglobulin and thyrotropin in Germany and Sweden. Acta Endocrinologica 112: 494–501.
29. Szabolcs I, Podoba J, Feldkamp J, Dohan O, Farkas I, et al. (1997) Comparative screening for thyroid disorders in old age in areas of iodine deficiency, long-term iodine prophylaxis and abundant iodine intake. Clin Endocrinol (Oxf) 47: 87–92.
30. Chen Z, Xu W, Huang Y, Jin X, Deng J, et al. (2013) Associations of noniodized salt and thyroid nodule among the Chinese population: a large cross-sectional study. The American journal of clinical nutrition 98: 684–692.

31. Lu Y-l, Wang N-j, Zhu L, Wang G-x, Wu H, et al. (2005) Investigation of iodine concentration in salt, water and soil along the coast of Zhejiang, China. Journal of Zhejiang University SCIENCE B 6: 1200–1205.
32. Andersen S, Iversen F, Terpling S, Pedersen KM, Gustenhoff P, et al. (2009) More hypothyroidism and less hyperthyroidism with sufficient iodine nutrition compared to mild iodine deficiency—A comparative population-based study of older people. Maturitas 64: 126–131.
33. Mesele M, Degu G, Gebrehiwot H (2014) Prevalence and associated factors of goiter among rural children aged 6-12 years old in Northwest Ethiopia, cross-sectional study. BMC public health 14: 130.
34. Ahmed M, Zama SY, Nagarajarao V, Khan MA (2014) Iodine deficiency in children: A comparative study in two districts of south-interior Karnataka, India. Journal of family & community medicine 21: 48–52.
35. Medani AMM, Elnour AA, Saeed AM (2013) Excessive iodine intake, water chemicals and endemic goitre in a Sudanese coastal area. Public health nutrition 16: 1586–1592.
36. Henjum S, Barikmo I, Gjerlaug AK, Mohamed-Lehabib A, Oshaug A, et al. (2010) Endemic goitre and excessive iodine in urine and drinking water among Saharawi refugee children. Public health nutrition 13: 1472–1477.
37. WHO UNICEF, ICCIDD (2007) Assessment of iodine deficiency disorders and monitoring their elimination: a guide for programme managers.
38. Evans C, Nix AB, Hillier S, Moat SJ (2014) Neonatal blood TSH concentration in Wales (UK): an indicator of iodine sufficiency. Clinical Endocrinology 81: 606–609.
39. Eltom A, Eltom M, Idris M, Gebre-Medhin M (2001) Thyroid function in the newborn in relation to maternal thyroid status during labour in a mild iodine deficiency endemic area in Sudan. Clinical endocrinology 55: 485–490.
40. Sukkhojaiwaratkul D, Mahachoklertwattana P, Poomthavorn P, Panburana P, Chailurkit L-o, et al. (2014) Effects of maternal iodine supplementation during pregnancy and lactation on iodine status and neonatal thyroid-stimulating hormone. Journal of Perinatology 34: 594–598.

10

Effects of Climate Variability and Accelerated Forest Thinning on Watershed-Scale Runoff in Southwestern USA Ponderosa Pine Forests

Marcos D. Robles[1]*, Robert M. Marshall[1], Frances O'Donnell[2], Edward B. Smith[3], Jeanmarie A. Haney[1], David F. Gori[4]

1 The Nature Conservancy Center for Science and Public Policy, Tucson, Arizona, United States of America, 2 Northern Arizona University, Flagstaff, Arizona, United States of America, 3 The Nature Conservancy in California, Sacramento, California, United States of America, 4 The Nature Conservancy in New Mexico, Santa Fe, New Mexico, United States of America

Abstract

The recent mortality of up to 20% of forests and woodlands in the southwestern United States, along with declining stream flows and projected future water shortages, heightens the need to understand how management practices can enhance forest resilience and functioning under unprecedented scales of drought and wildfire. To address this challenge, a combination of mechanical thinning and fire treatments are planned for 238,000 hectares (588,000 acres) of ponderosa pine (*Pinus ponderosa*) forests across central Arizona, USA. Mechanical thinning can increase runoff at fine scales, as well as reduce fire risk and tree water stress during drought, but the effects of this practice have not been studied at scales commensurate with recent forest disturbances or under a highly variable climate. Modifying a historical runoff model, we constructed scenarios to estimate increases in runoff from thinning ponderosa pine at the landscape and watershed scales based on driving variables: pace, extent and intensity of forest treatments and variability in winter precipitation. We found that runoff on thinned forests was about 20% greater than unthinned forests, regardless of whether treatments occurred in a drought or pluvial period. The magnitude of this increase is similar to observed declines in snowpack for the region, suggesting that accelerated thinning may lessen runoff losses due to warming effects. Gains in runoff were temporary (six years after treatment) and modest when compared to mean annual runoff from the study watersheds (0–3%). Nonetheless gains observed during drought periods could play a role in augmenting river flows on a seasonal basis, improving conditions for water-dependent natural resources, as well as benefit water supplies for downstream communities. Results of this study and others suggest that accelerated forest thinning at large scales could improve the water balance and resilience of forests and sustain the ecosystem services they provide.

Editor: Julia A. Jones, Oregon State University, United States of America

Funding: The authors have no support or funding to report.

Competing Interests: The authors have declared that no competing interests exist.

* Email: mrobles@tnc.org

Introduction

As we transition into a warmer century, managing forests in the southwestern United States for resilience is more urgent and necessary than ever. Forest conditions across the region have declined significantly. In ponderosa pine (*Pinus ponderosa*) forests of central Arizona, stand densities range from 2 to 44 times greater than during pre-settlement conditions and total basal areas range from 2 to 4 times greater [1–3]. Severe wildfires and drought have caused high tree mortality rates across 14–18% of forests and woodlands in Arizona and New Mexico [4] and these results do not include mortality from two high-severity wildfires in 2011, the Wallow and the Las Conchas wildfires, which burned 217,000 and 63,500 ha (535,000 and 157,000 acres) in Arizona and New Mexico, respectively. Persistent drought conditions have also adversely impacted water-dependent habitats and species within and downstream of forest uplands, as evidenced by heightened mortality of a riparian tree species in central Arizona [5] and declines in number of species and densities of native fish in the Gila River in southwestern New Mexico [6], [7]. Researchers suggest that warming is amplifying the size and severity of wildfire- and drought-mortality events and decreasing snow pack levels [8–11]. Suggested mechanisms involve changes in the water cycle: increased evapotranspiration losses, extended water-stress periods, earlier snowmelt, and lengthened fire seasons. These changes and trends suggest a widespread drying of forests and an increasing risk of uncharacteristic fire and competition-induced water stress and mortality.

At the same time, many communities in the western U.S. are facing critical water shortages and river flows in some basins have diminished or been lost altogether due to unsustainable water management practices [12–15]. Recent studies at the scale of both the Colorado River basin and the state of Arizona concluded that water supplies for many communities will be inadequate to meet future projected demands [16],[17]. Moreover, elevated temperatures are predicted to directly impact river flows. A recent study estimated that for every one-degree Celsius rise in temperature, Colorado river flows will decline by 3–10% [18].

In this study, we explored the ways in which a new era of forest management, including accelerating the pace and scale of forest thinning, could be used to address declining conditions in forests in central and northern Arizona. At about 6,000 hectares per year (15,000 acres/year), the current pace of mechanical thinning in these forests does not begin to match the magnitude of recent forest disturbances. However, a new congressionally funded program called the Four Forest Restoration Initiative (hereafter "4FRI") will accelerate the use of mechanical thinning and prescribed burns across four national forests, treating 238,000 ha (588,000 acres) in the first analysis area over the next 10 years. The objective of the 4FRI project is "...to re-establish forest structure, pattern, and composition, which would lead to increases in forest resiliency and function," including reductions in fire risk and improved watershed function [19]. Mechanical thinning alone or in combination with prescribed fires reduced fire risk, increased runoff, and improved tree water-stress at the plot scale [20–23]. The 4FRI project provided the opportunity to evaluate the effectiveness of forest thinning to improve forest resilience and ecosystem functioning at larger scales.

The objectives of this study were to: (1) investigate how variation in climate and the pace and extent of thinning of ponderosa pine forests affects runoff in the Salt and Verde watersheds of central Arizona; and (2) explore the management implications of using accelerated thinning to improve forest resilience in a time of declining forest and water conditions. We chose the Salt-Verde watersheds because forests from these watersheds provide water for 1.5 million people in Phoenix, which is the 6th largest city in the United States. We focused on ponderosa pine because this forest type produces 50% of the runoff in these watersheds even though it accounts for only 20% of the area [24]. We modified a forest runoff model developed from historical paired watershed experiments in the Beaver Creek sub-watershed within our study area [20] and ran the model in multiple scenarios to estimate additional annual runoff from mechanical thinning. The scenarios simulated forest treatment at two scales (a) landscape-scale thinning planned within the first analysis area of the 4FRI project and (b) watershed-scale thinning in forests across the Salt-Verde watersheds (Figure 1). Scenarios were designed to account for variability in winter precipitation, as well as the pace and extent of forest treatments. We estimated runoff in periods with below-average precipitation, herein referred to as droughts, and in periods with above-average precipitation, herein referred to as pluvials, using a model of 20th century precipitation in ponderosa pine forests in the region [25]. We believe this is the first attempt to estimate the influence of mechanical thinning on runoff over multi-year broad-scale restoration projects that accounts for the effects of climate variability.

Materials and Methods

The subsequent sections provide detailed information on the study area, the development of the modified runoff model, the method we used to simulate variability in winter precipitation, the 4FRI runoff scenarios, and the Salt-Verde runoff scenarios. All statistical analyses and graphics were completed using the statistical software SigmaPlot 11 (Systat software Inc, San Jose, CA), except for regression model fitting which was performed using Matlab 2013 (Mathworks, Natick, MA). Methods and results are reported in SI units, but for accessibility to forest and water managers, we reported English units in parentheses within the text and included all study figures and tables in English units as File S1. Area and water volume amounts were rounded off to three significant figures.

Study Area

Ponderosa pine forests occupy approximately 0.68 million ha (1.68 million acres) of the 3.46 million ha (8.56 million acres) within the Salt-Verde watersheds in central Arizona (Figure 1). A portion of these forests, all within the Verde watershed, will be mechanically thinned in the 4FRI project in the next 10 years. Within the Salt-Verde watersheds, ponderosa pine forests grow at mid-elevations, from 1,800 to 2,600 m (5,900–8,600 feet) [26], [27], on soils derived from basalt (59%) and sedimentary rock (41%) [28]. Mean annual precipitation for ponderosa forests in the southwestern United States ranges from 510 to 760 mm (20 to 30 inches) and is highly variable [29]. Sixty percent of annual precipitation falls primarily as snow in the winter and spring months from October to April, and the melting of the resulting snowpack accounts for 80 to 90% of annual stream flow [29].

Forest Treatment Runoff Model

For the purposes of this study, we were interested in estimating the additional runoff that becomes available due to forest thinning on an annual basis, not absolute values of watershed runoff. We define "additional runoff due to thinning" as the portion of precipitation that appears as surface water at the sub-watershed outlet and that is directly attributable to mechanical thinning treatments. This additional runoff can be considered as "in-place" or "in-situ" as we did not attempt to model runoff accumulation, routing, groundwater outflows or inflows, or channel losses in downstream watersheds.

We modified the Baker-Kovner logistical regression equation from the historical Beaver Creek sub-watershed experiments to calculate annual runoff in our study [20]. We selected this model because the Beaver Creek sub-watershed is actually within our study area (Figure 1), and because the model was based upon measured empirical data. These experiments and others like it demonstrated that reductions in forest basal area of 30–100% in ponderosa forests on basalt-derived soils increased runoff between 15–40% for up to six years after the initial treatment [29]. Runoff increases were negligible for basal area reductions below 30%, six years after treatments, and for years with winter precipitation below 230 mm (9 inches) [20], [29]. The original model predicted annual watershed runoff based on three independent variables: winter precipitation (Oct-Apr), forest basal area, and a solar insolation index, and explained a fair amount of variability in total watershed runoff [20].

To derive increases in runoff directly associated with thinning using the paired watershed method, researchers in the Beaver Creek experiments first established a relationship between stream flows in the control and treated watersheds before treatments [30]. This relationship was used to predict what would have been baseline flows in the treated watershed. The difference between measured flows in the treated watershed and this predicted value were attributed to thinning effects. We compared data calculated in this fashion in the Beaver Creek experiments [30], [31] and one additional nearby site, Castle Creek [32], to output generated by the Baker-Kovner model. The output from the model was calculated as the difference in post-treatment and pre-treatment watershed runoff using post- and pre-treatment basal areas respectively and holding all other variables constant. In this comparison, we found the fit of the model output to the portion of runoff due to thinning to be much poorer ($r^2 = 0.43$, Figure 2a)

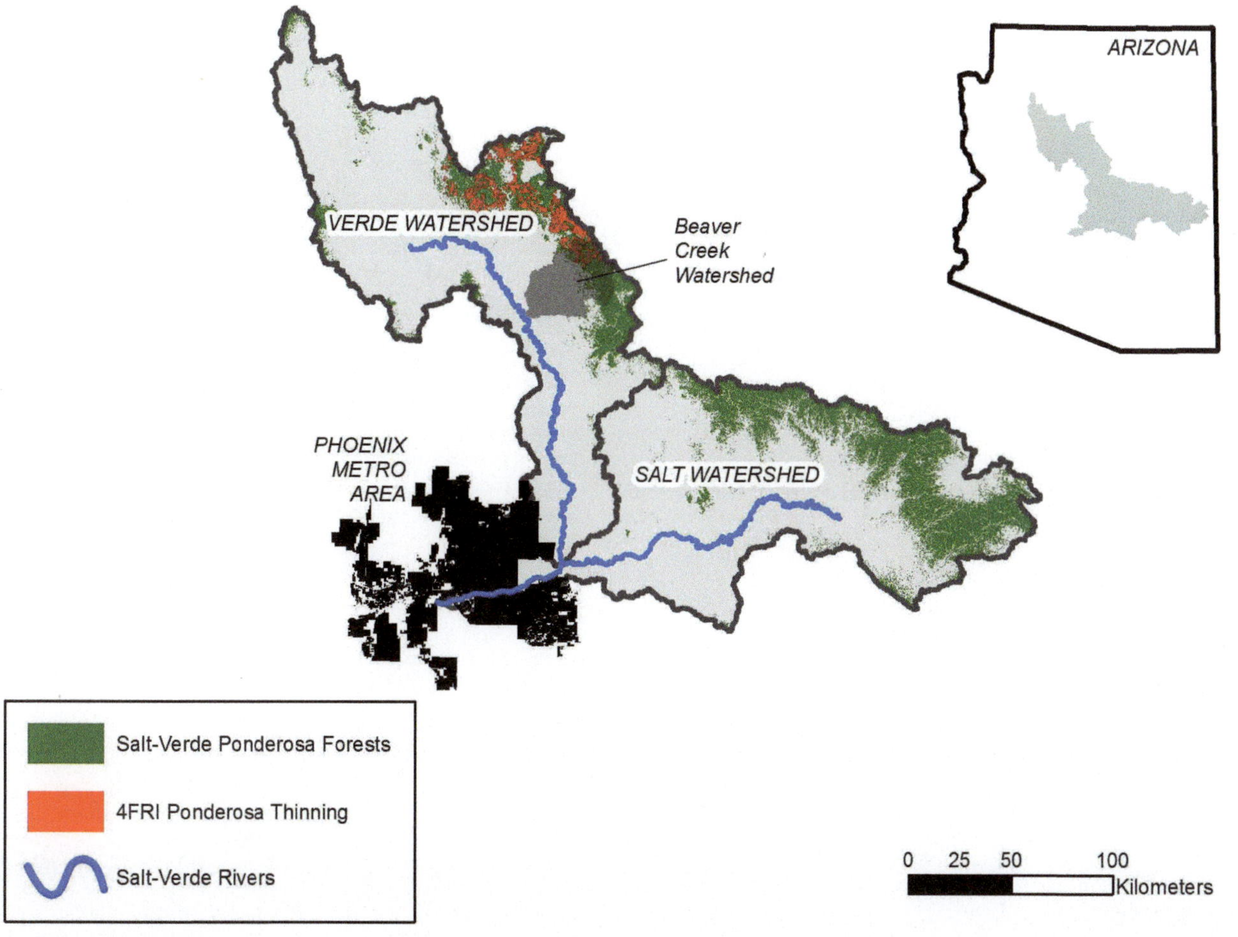

Figure 1. Map of Study Area. Map showing ponderosa pine forests in Salt-Verde watersheds in central Arizona, including those forests that are slated for mechanical thinning within the 4FRI project. Runoff from snowmelt in these forests is primary source of flow to Salt-Verde rivers which in turn are major sources of water for communities in the Phoenix Metro Area. Study used runoff model developed from experimental studies conducted in Beaver Creek watershed [20]. *Inset*: Location of study area in Arizona.

than model fit to total watershed runoff ($r^2 = 0.69$). We concluded that the original regression model was relatively insensitive to the direct effects of forest treatments, including time since treatment, on runoff.

We used a stepwise regression procedure to develop a revised model, beginning with an initial model that included the independent variables (winter precipitation and basal area) that were found to be significant in historical experiments [20]. Input data for all independent variables used to develop the model were derived from empirical data collected during the Beaver Creek and Castle Creek historical experiments and archived in databases or published reports [30–32]. We added independent variables and used an F-test to test the null hypothesis that the coefficient of each term was zero at the 95% significance level. To ensure that the model was not over-fitted, we repeated the stepwise regression procedure with 70% of the original data, selected randomly. We determined that the structure of the model did not change and provided a good fit to the remaining 30% of the data. We examined the residuals from the model and found no discernible patterns of dispersion or bias. Including terms for slope, percent of watershed area with a south-facing aspect, and mean, minimum, and maximum winter temperatures did not significantly improve the model.

The final model included three terms: winter precipitation, an interaction term between winter precipitation and years since treatment, and an interaction term between winter precipitation and basal area. It can be expressed as:

$$R_{treatment} = -28.464 + 0.148 * P - 0.015 * P * Y - 0.092 * P * [\exp(-BA_1/10.33) - \exp(-BA_2/10.33)]$$

$r^2 = 0.67$

where

$R_{treatment}$ = Increase in Annual Runoff attributed to Forest Thinning in mm

P = Total Winter Precipitation (Oct-Apr) in mm

Y = Years since Treatment

BA_1 = Basal Area before Treatment in m^2/ha

BA_2 = Basal Area after Treatment in m^2/ha

Figure 2b shows that the revised model explained more of the variability in predicting additional runoff associated with forest treatments ($r^2 = 0.67$) than did the original model. The model predicted runoff increases in a linear fashion with winter precipitation and basal area reductions, and decreases linearly

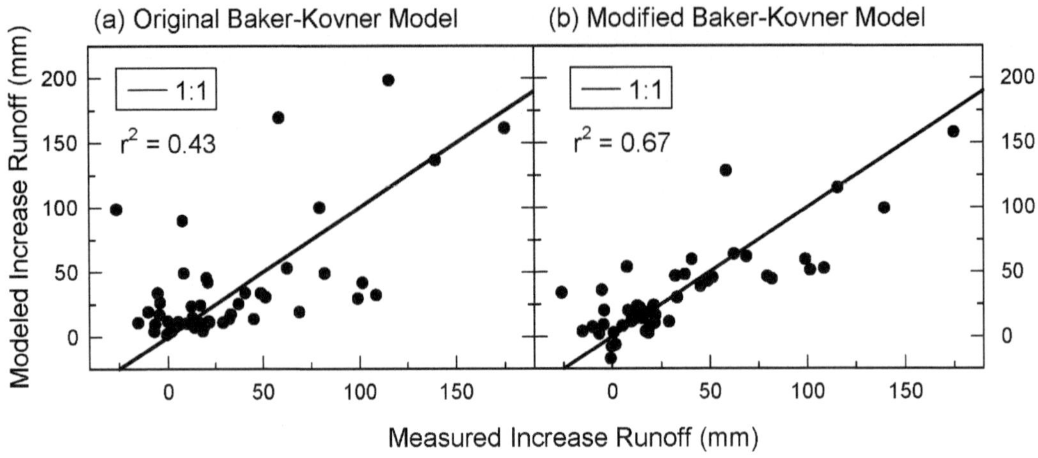

Figure 2. Comparison of models to observed runoff. Fit of (a) original [20] and (b) modified Baker-Kovner regression model output to increases in runoff associated with forests treatments in central Arizona, from Beaver Creek [30],[31] and Castle Creek [32] watersheds.

with years since treatment (Figure 3). To compare the original and modified Baker-Kovner models in English Units, see File S2.

We did use the original Baker-Kovner regression equation in this study to calculate total watershed runoff for forests in their current condition so that we could estimate the percentage increase in runoff associated with treatments. For these calculations, we used the average value for the solar insolation index, $r = 0.7$, reported from the Beaver Creek experiments [20]. In a sensitivity analysis, we found that calculating absolute runoff based on the low ($r = 0.66$) and high values ($r = 0.74$) of this index changed the estimated increase from treatments by only 1–2 percentage points.

Winter Precipitation

In order to account for inter-annual and decadal variability of winter precipitation, we used the PRISM model [25] to extract pluvials and droughts from the 20th century and inserted levels of winter precipitation from these periods into our runoff scenarios. We first extracted total winter precipitation, summed across October to April, for every year from 1900–2012 from the PRISM modeled dataset [25] across 2 scales: (a) ponderosa pine forests within the Verde watershed and (b) these forests within the Salt-Verde watersheds. We selected these geographies to be most representative of conditions for the 4FRI and Salt-Verde runoff scenarios, respectively. From 1900–2012, mean winter precipitation in Verde ponderosa pine forests was 394 mm (15.5 inches) with a range of 99–815 mm (3.9–32.1 inches) (Figure 4). For Salt-Verde forests, the mean was 368 mm (14.5 inches) and the range was 89–747 mm (3.5–29.4 inches) for this same period of time.

To test the accuracy of these modeled data, we compared values from the Verde ponderosa pine PRISM modeled data to measurements of total winter precipitation recorded during the original Beaver Creek experiments [31]. Given that we found the distribution of the data to be non-normal, we ran the non-parametric Spearman's rank order correlation. We found a strong association between modeled and observed data ($Rho = 0.9577$, $p<0.001$) although the model underestimated measured winter precipitation over the time period by approximately 25 mm (1 inch) and consistently under-estimated precipitation in wet winters (Figure 4 inset).

We calculated mean winter precipitation across 15-, 25-, and 35-year periods from this modeled data and used this information to select droughts and pluvials. For each scenario, we moved the start and end years that thinning was simulated to occur. For example, time periods for our 15-year scenarios included 1944–1958, 1945–1959, and 1946–1960 (examples illustrated as shaded areas in Figure 4). Using this method, we found two instances of pluvials – early 20th century and late 1970s to late-1990s – and two droughts – during the 1950s and the current drought.

Estimating runoff from 4FRI project

We used the modified regression model to calculate additional annual runoff from individual forests stands that will be thinned in the first analysis area of the 4FRI project, established an annual thinning schedule, and then summed increases in runoff from individual stands across treatment years. We ran a total of 26 scenarios where in each scenario we inserted a different 15-year winter precipitation sequence.

We obtained a Geographic Information System (GIS) dataset of alternative C in the draft environmental impact statement for the first analysis area of the 4FRI project [19]. For each stand to be mechanically thinned, this dataset contained estimates of current basal areas and desired post-treatment basal areas. Forest prescriptions for "group-selection" stands called for two post-treatment basal areas, one for "open" areas that would be evenly thinned and another for "group" areas where thinning would be minimal to enhance wildlife habitats by retaining greater tree densities. For these stands, we calculated a post-treatment basal area that was an average of these two basal areas, weighted by the proportional area of the open and group areas. We selected 4,064 ponderosa pine stands within the Verde watershed slated for mechanical thinning, where the prescription will reduce basal areas by at least 30%. They ranged in size from 0.4 to 217 ha (1 to 536 acres), for a total of 61,900 ha (153,000 acres). The prescriptions within these stands called for reductions of basal area between 30% and 70% with a mean reduction of 48% (Figure 5).

Based on consultation with US Forest Service staff, we found that it was reasonable to assume that an equal number of hectares will be mechanically thinned every year across the 10-year treatment period. So for the purposes of our 4FRI scenarios, we assigned each of these stands to one of ten cohorts until the area of each cohort equaled 1/10th of total area or approximately 6,190 ha (15,300 acres). Figure 6a illustrates the treatment schedule for

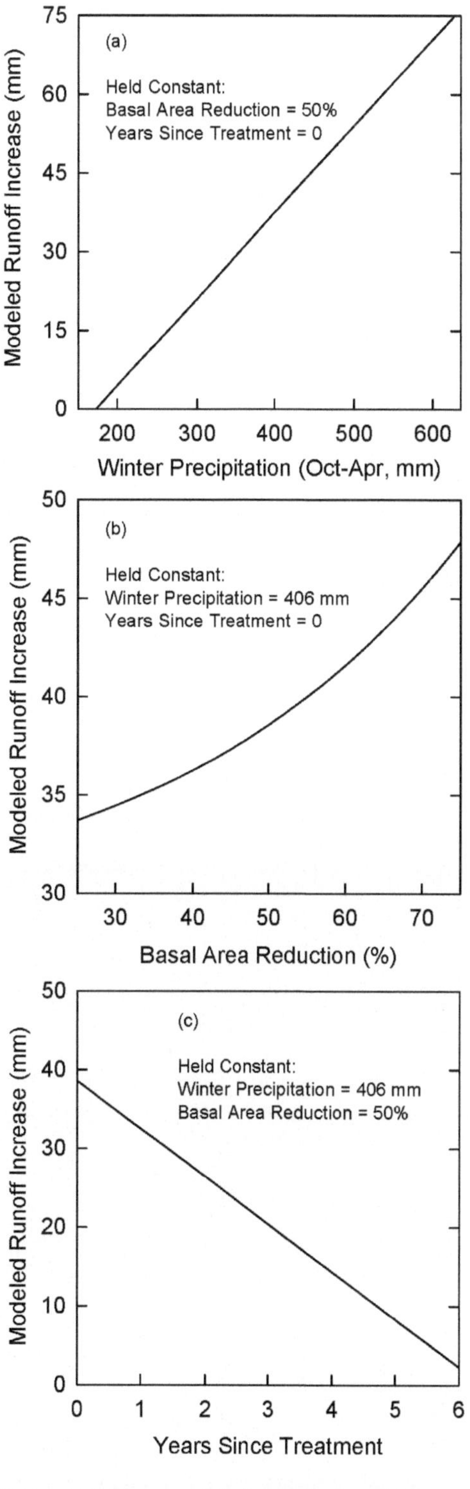

Figure 3. Influence of independent variables on runoff model. Relationships between model output to values for independent variables, including (a) winter precipitation, (b) percent basal area reduction, and (c) years since treatment. In all cases, other independent variables are held constant in order to view relationship of independent variable plotted on X-axis to model output.

the 15-year 4FRI scenarios reported in this study. Cohorts were treated consecutively in the first 10 years and it was assumed that cohorts influenced runoff for 6 years. So, for example, stands treated in cohort 1, contributed to additional runoff in scenario years 1-6 and stands treated in cohort 10 contributed from years 10–15.

We calculated additional runoff associated with thinning and total watershed runoff by inserting values of the independent variables into the revised and original Baker-Kovner regression models, respectively. Pre-treatment and post-treatment basal area values were derived from the 4FRI stand data. Years since treatment ranged from y = 0 to 5. As described previously, winter precipitation values were droughts and pluvials drawn from the PRISM model for ponderosa forests in the Verde watershed. We set runoff increase to zero in years when winter precipitation was less than 230 mm (9 inches). We transformed the unit of runoff from mm to volume (million m^3) using the area of each stand. We estimated annual increases in runoff at the landscape scale by summing stand level amounts for each scenario year, and calculated summary statistics (mean, median, max, cumulative) that allowed for a comparison of scenarios.

Estimating runoff from Salt-Verde watersheds

Unlike the first analysis area of the 4FRI project, there was not a planned forest restoration project across the larger geography of the Salt-Verde watersheds. We developed a range of estimates for the extent, pace, and intensity of forest thinning that could be conducted over this larger geography, grouped these estimates into runoff scenarios, and ran the scenarios using the revised and original regression models to estimate additional runoff from treatments and total watershed runoff. To estimate the potential areal extent of thinning, we subtracted from the total forested areas those land uses that are typically considered unsuitable for mechanical thinning. We bracketed this initial area estimate with lower values to account for the lack of comprehensive spatial information. Assuming that the pace of treatments at this scale could take longer, we constructed 15-, 25-, and 35-year runoff scenarios. Finally, we assumed that the intensity of forest thinning at this scale would be similar to the range of basal area reductions planned for the first analysis are of the 4FRI project. The remaining parts of this section describe these steps in greater detail.

Based on a methodology developed in a previous study that estimated wood supply on four National Forests in northern Arizona [33], we subtracted from the total area of ponderosa pine in the watersheds – 0.68 million ha (1.68 million acres) – those areas that are typically excluded from mechanical thinning projects due to steepness, restrictions in land management, previous treatments, or considerations of soil, habitat, or wildlife conditions (Table 1). We adopted 6 of the 7 "exclusion" criteria from that study and added one additional factor. We removed one criterion – Northern goshawk nest areas –because the mean basal area reduction for these areas in the 4FRI project was greater than 30% [19] suggesting that thinning in these areas could result in additional runoff based on the results of the Beaver Creek experiments [20]. We added one criterion – high severity burn patches – because a nearby study on the recovery of ponderosa pine forests after wildfire demonstrated that mature forests suitable for thinning would not likely develop in areas that had burned with high severity within the timeframe of this study [34]. While two of the three 4FRI prescriptions for Mexican Spotted Owl Protected Activity Centers (MSO PAC) would result in negligible changes in basal area, one would allow for basal area reductions greater than 30% [19]. However, we were not able to differentiate among these prescriptions from the data we obtained, so we opted for a conservative estimate for this factor by excluding MSO PAC areas from our analysis. Compiling the publicly available geospatial data

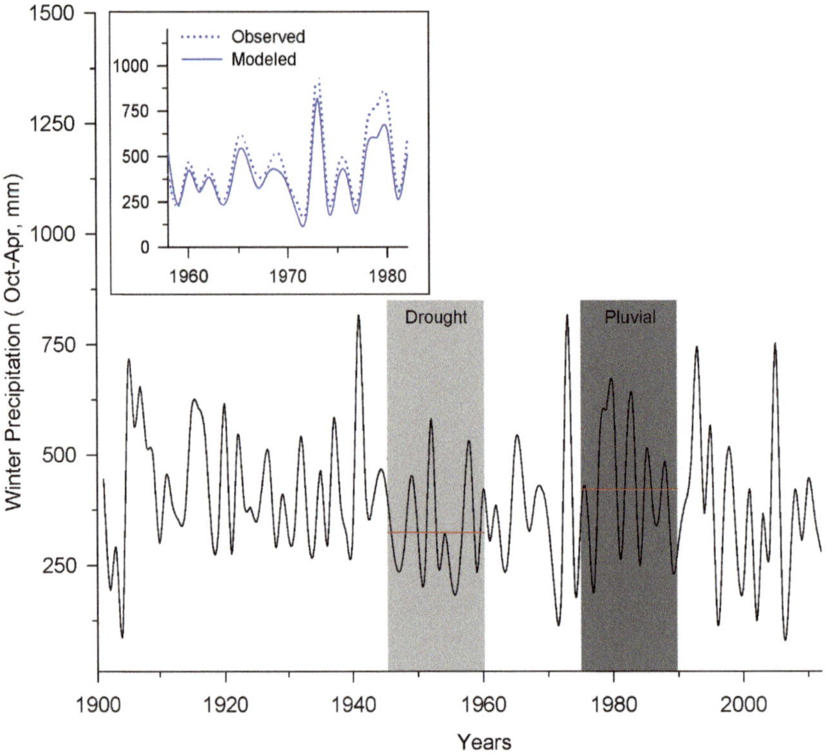

Figure 4. Variability in winter precipitation in ponderosa pine forests. Estimates of historical winter precipitation from 1900–2012 in ponderosa pine forests within Verde watershed from PRISM model [25]. Shaded areas are examples of 15-year droughts and pluvials that were used in study scenarios; horizontal red lines represent mean winter precipitation within these shaded areas. *Inset*: Comparison of measured winter precipitation observed during the historical Beaver Creek watershed experiments [20] from 1958–1982 versus modeled winter precipitation data shown in main figure.

for these factors [26], [27], [33], [35–38], we found that 401,000 ha (992,000 acres), or 59% of ponderosa pine forests, were available for mechanical thinning after accounting for these exclusion areas. The percentage of forests excluded in this estimate – 41% – compared favorably with the average of 37% of forests that were excluded in dozens of restoration projects evaluated in the northern Arizona [33]. We bracketed this high estimate with a moderate estimate – 45% or 301,000 ha (743,000 acres) – and a low estimate – 30% or 204,000 ha (505,000 acres) – because quantitative geo-spatial data on some exclusion areas was missing or incomplete. As illustrated graphically for the moderate estimate in Figure 6b-d, we divided these areas into 10-, 20-, and 30-year treatment cohorts assuming that equal areas would be thinned every year.

We scaled up the forest condition information contained in the 4FRI stand data using a proportion metric, assuming that these stands were representative of stand-level forest conditions across the Salt-Verde watersheds. For example, 4FRI stands that comprised 1,500-ha in a 15,000-ha 4FRI treatment cohort were estimated to be 3,000-ha in a 30,000-ha Salt-Verde cohort. If the thinning prescription for these stands called for a reduction in basal area from 33 to 17 m^2/ha, then the Salt-Verde scenarios calculated additional runoff from 3,000-ha that are thinned from 33 to 17 m^2/ha. After scaling up the area of each stand to the watershed scale, the procedure for estimating runoff in the Salt-Verde scenarios was the same as described for the 4FRI scenarios. We ran 147 scenarios in total, 67 fifteen-year, 52 twenty-five year, and 28 thirty-five year scenarios.

Results

Four Forest Restoration Initiative Runoff Scenarios

Mechanical thinning of ponderosa pine forests in the first analysis area of the 4FRI project – 6,190 ha/year for ten years totaling 61,900 ha (15,300 ac/year; 153,000 acres total) – increased mean annual runoff from 3.13 million m^3 (2,540 acre-feet) in a simulated drought to 7.27 million m^3 (5,890 acre-feet) in a pluvial. Differences in winter precipitation over the 15-year simulation periods significantly influenced runoff gains. A difference in mean winter precipitation of only 130 mm (5 inches), from 330 mm (13 inches) in drought scenario to 460 mm (18 inches) in a pluvial scenario, resulted in a doubling of the annual increase in runoff from treatments (Figure 7).

Inter-annual variability in winter precipitation was also important. Figure 8 illustrates year to year increases in runoff for those scenarios that resulted in the lowest and highest gains in runoff. Years with high winter precipitation played a disproportionate role in additional runoff in both droughts and pluvials. In the majority of scenarios ran (15 out of 26), winter precipitation from only 5 of the 15 years accounted for at least 75% of the increased runoff. Cumulative increases in runoff across 15-year periods resulted in a total increase from 54.3 to 111 million m^3 (44,000–89,800 acre-feet) (Figure 9). Runoff from thinned forests was approximately 20% greater than unthinned forests (as estimated using original Baker-Kovner regression model) in both droughts and pluvials (data not shown).

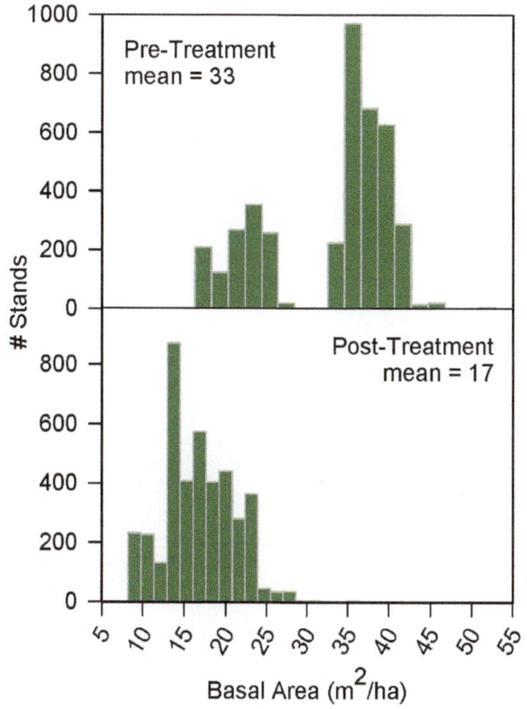

Figure 5. Ponderosa pine basal area reductions in 4FRI project. Histograms showing (top) pre-thinning and (bottom) desired post-thinning basal areas (in m^2/ha) of ponderosa pine stands in the first analysis area of the 4FRI project (excluding stands where basal area reduction <=30%).

Salt-Verde Runoff Scenarios

Depending on winter precipitation and the forest treatment schedule, mean annual increases in runoff from thinning of ponderosa forests across the Salt-Verde watersheds ranged from 4.76 to 15.0 million m^3 (3,860–12,200 acre-feet) over a 35-year treatment period, 6.18 to 23.4 million m^3 (5,010 to 19,000 acre-feet) over 25 years, and 9.23 to 42.8 million m^3 (7,480 to 34,700 acre-feet) over 15 years (Table 2). Similar to the 4FRI scenarios, additional runoff in the Salt-Verde watersheds was 1.6–2.3 times greater in pluvials than in droughts. Regardless of whether the scenarios occurred in a drought or pluvial, cumulative runoff gains in thinned forests were 20-26% greater than unthinned forests. Cumulative gains ranged from 167 to 525 million m^3 (135,000–426,000 acre-feet) in the 35-year scenarios, 154 to 585 million m^3 (125,000–474,000 acre-feet) in 25-year scenarios, and 138 to 643 million m^3 (112,000–521,000 acre-feet) in 15-year scenarios. In both droughts and pluvials, runoff increased in a positive linear fashion with increases in the pace and the extent of forest thinning (Figure 10). See File S1 to see study figures and tables in English units.

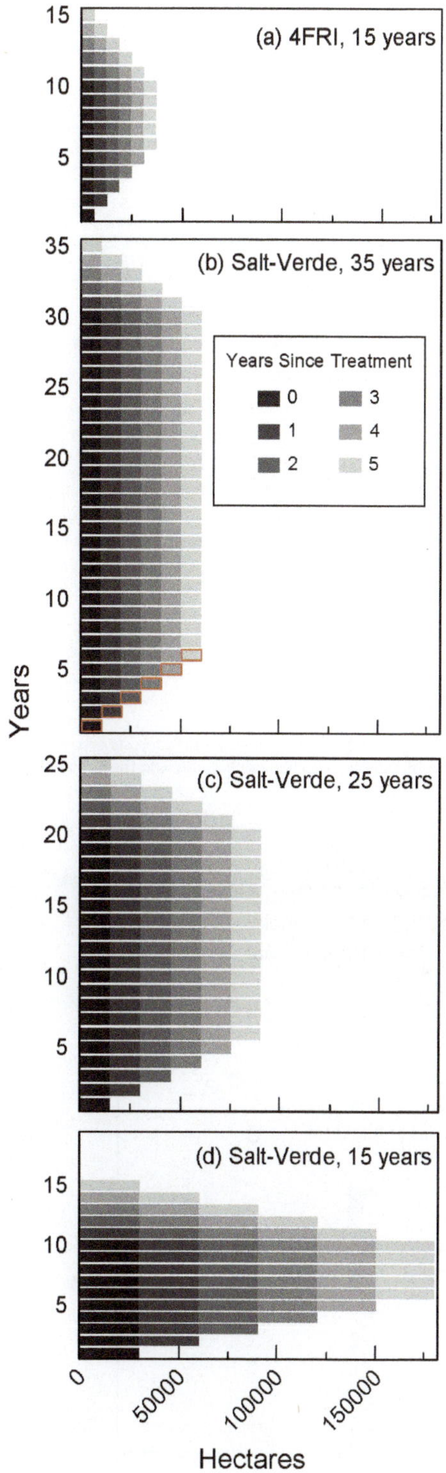

Figure 6. Forest treatment schedules for study scenarios. Graphical depiction of mechanical thinning treatment schedules for (a) 15-year 4FRI scenarios and (b-d) 35-year, 25-year, and 15-year Salt-Verde moderate thinning scenarios (total thinned area was 301,000 ha or 743,000 acres). Scenarios assumed consecutive treatments for 10-, 20-, and 30-year treatment periods shown as black bars in the bottom left portion of each of the figures. Bars outlined in red in (b) show the contribution of one cohort of stands through six years in the scenario.

Discussion

This study demonstrated that the pace, extent, and intensity of forest thinning that is planned under the first analysis area of 4FRI and at larger scales could measurably increase runoff in ponderosa pine forests in central Arizona. Modeled runoff from thinned forests was approximately 20% greater than runoff from unthinned forests which is within the range of 10–40% increases in runoff demonstrated in the Beaver Creek experiments [20], [29]. The Salt-Verde runoff scenarios showed that levels of

Table 1. Estimate of hectares of ponderosa pine forests available to mechanical thinning in Salt-Verde watersheds.

Category	Hectares	%	*Sources*
Salt-Verde Watersheds Ponderosa Pine Forests	*681,000*	-	[27]
Exclusions			
Steep Slopes greater than or equal to 40%	68,000	10%	[26]
Hi-Severity Burn Patches within Wallow and Rodeo-Chedeski Fires	43,700	6%	[35]
Specially Designated Areas including Wilderness Areas, Research Natural Areas, State Wildlife Management Areas[a]	21,800	3%	[36],[37]
Streamside Management Zones within 100 feet perennial reaches in national forests and 200 feet in tribal areas	8,050	1%	[38]
Mexican Spotted Owl Protected Activity Centers where owls have been found to be nesting[b]	57,500	8%	[33]
Erodible Soils where thinning is unsuitable due to erosion risk or rocky conditions[c]	93,100	14%	[33],[37]
Completed Treatments that have been thinned in last 10 years[d]	35,200	5%	[37]
Sub-Total Exclusions (accounting for overlap between layers)	*280,000*	*41%*	
Sub-Total Forest Available Mechanical Thinning	*401,000*	*59%*	

[a] Estimate derived from USDA Forest Service Southwestern Region Wilderness Status layer and Forest Level Special Interest Management Areas layers [37].
[b] Data available for Coconino, Apache-Sitgreaves, Kaibab, and Tonto National Forests only. Added qualitative estimate for this exclusion factor for tribal lands and Prescott National Forest assuming land excluded in these land units would be proportional to lands excluded for this factor on the four national forests above.
[c] List of excluded soils for Coconino, Apache-Sitgreaves, Kaibab, and Tonto National Forests based on Hampton et al. [33]. This list of excluded soils was augmented based on consultation with USDA Forest Service southwestern region soil scientists. Geospatial data for these soils derived from USDA Forest Service Forest Level Terrestrial Ecological Units layers [37]. Added qualitative estimate for this exclusion factor for tribal lands and Prescott National Forest where geospatial data was unavailable, assuming land excluded in these land units would be proportional to lands excluded for this factor on the four national forests above.
[d] Geospatial data for this factor derived from USDA Forest Service Forest Level Activities Layer joined to FACTs table for Coconino, Apache-Sitgreaves, Kaibab, Tonto, and Prescott National Forests [37]. Added qualitative estimate for this exclusion factor for tribal lands based on consultation with Bureau of Indian Affairs forestry personnel.
Summary of hectares and percentages of ponderosa pine forests within Salt-Verde watersheds in relation to potential for thinning treatments, including total hectares, those areas typically excluded from mechanical thinning, and the remaining hectares available for mechanical thinning. Estimates rounded off to three significant digits.

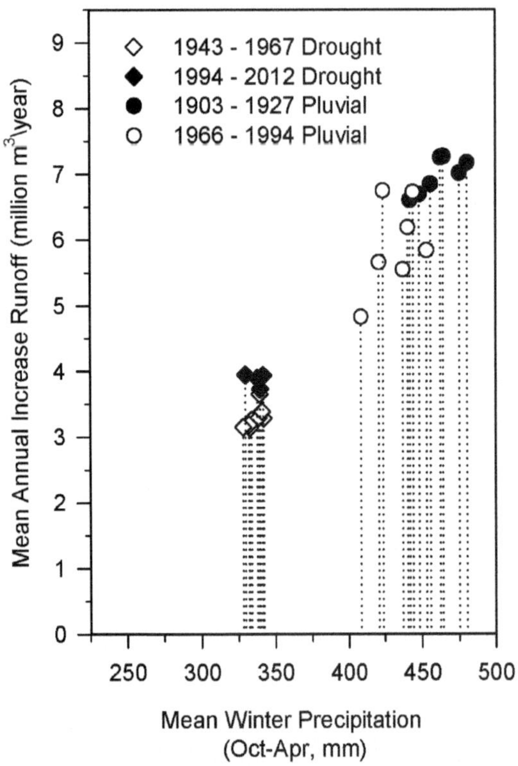

Figure 7. Increases in mean annual runoff from thinning in 4FRI project. Results from 26 scenarios with varying levels of winter precipitation showing increases in mean annual runoff associated with mechanical thinning of ponderosa pine forests in the first analysis area of the 4FRI project. In order to compare scenarios, only increases in *mean* annual runoff are shown. Annual variability in runoff for two of these scenarios is shown in Figure 8.

additional runoff increased proportionally with increases in the pace and extent of thinning. Runoff gains occurred in droughts and pluvials, a surprising outcome that underscored the importance of inter-annual and decadal variability in precipitation. These increases in runoff would likely improve conditions for water-dependent natural resources, such as cienegas, riparian areas, and aquatic habitats, which are vulnerable to low flows that are experienced seasonally, especially in summer months, and also during droughts [5–7]. Runoff gains also could provide incidental benefits to the water supply of downstream users, but the increases were more modest when compared to total runoff from the Salt-Verde watersheds. They would comprise a 0–3% increase from mean annual surface flows from the Salt-Verde rivers of 1.39 billion m^3 (1.13 million acre-feet) and approximately a 1-9% increase of Salt-Verde flows supplied to municipal users in the Phoenix Metro Area on an annual basis [39].

Consistent with the historical experiments, thinning effects on runoff were temporary in this study. Additional runoff would cease six years after the multi-year thinning schedules in our scenarios. Beaver Creek investigators speculated that thinning effects were short-lived because of subsequent regrowth of understory species [29]. We note that prescribed fires are planned as subsequent treatments for all stands that are initially thinned within the first analysis area of 4FRI [19]. These maintenance treatments should provide additional information as to how effective subsequent actions are in terms of removing understory vegetation and sustaining runoff gains.

Our study adds to previous research on the potential for increases in runoff associated with thinning of ponderosa pine forests in the Southwest. Using a simulation model, Brown and Fogel [40] found that thinning approximately 9,300 ha (23,000 acres) of ponderosa pine per year over a 10-year period produced increases in runoff that ranged from 0.28 to 11.8 million m^3/year (230–9,600 acre-feet/year). This range overlapped with but was

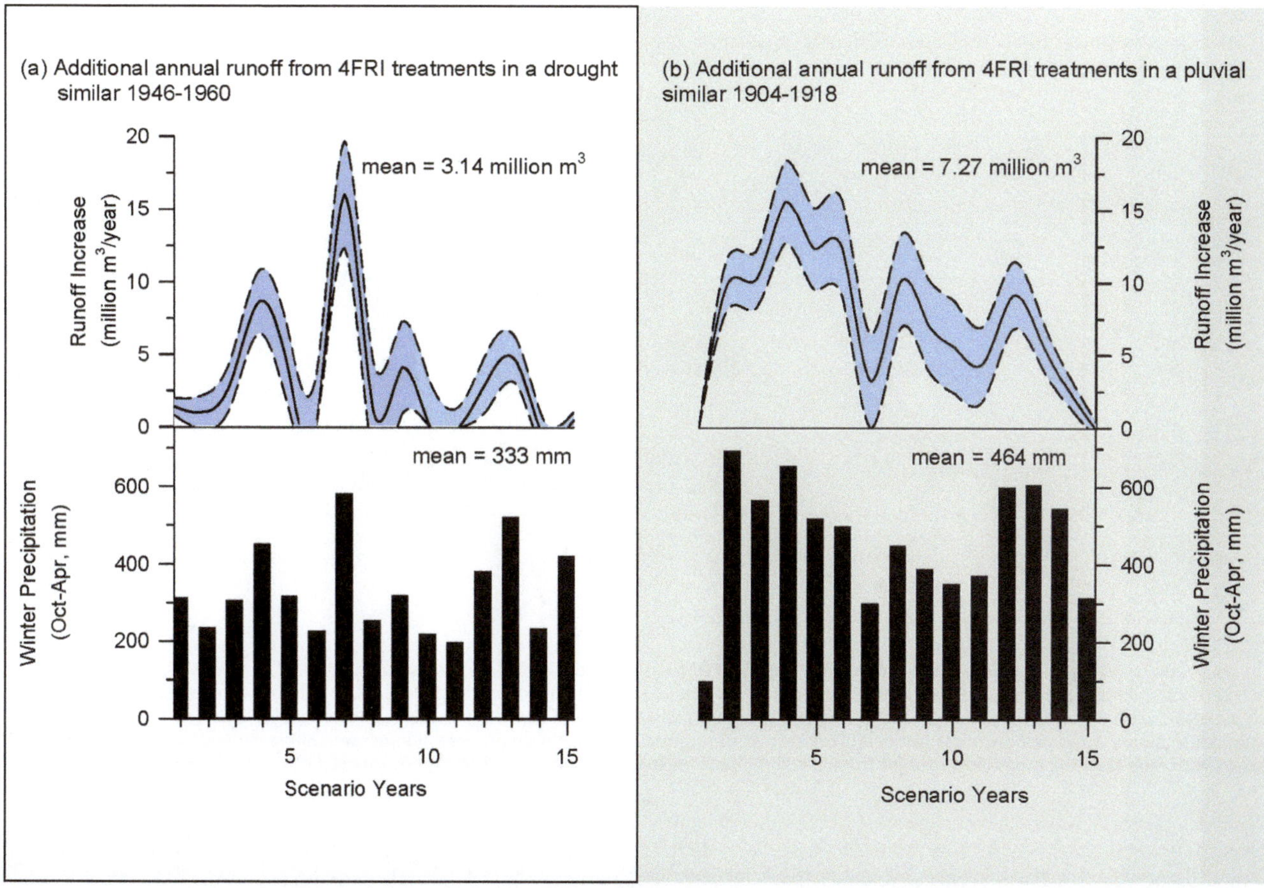

Figure 8. Year to year variability in runoff increases from thinning in 4FRI project. Modeled increases in annual runoff associated with mechanical thinning of ponderosa pine forests in the first analysis area of the 4FRI project during the (a) drought and (b) pluvial that produced lowest and highest gains in runoff respectively. Top panes show increases in annual runoff in million m^3/year. Solid black lines are output values from regression model; dotted lines and blue areas represent 90% confidence intervals. Bottom panes show corresponding winter precipitation values (Oct-Apr, mm) used as one of the independent variables to calculate runoff.

lower than the 6.18 to 11.9 million m^3/year (5,010–9,660 acre-feet/year) increases in runoff for the two Salt-Verde runoff scenarios with annual thinning of approximately 10,000 ha (25,000 acres) (Table 2). One potential reason for this difference was that Brown and Fogel [40] modeled the treatment-runoff relationship to be constant over time, 18 mm (0.71 inches), and did not evaluate variability in precipitation but instead assumed average winter precipitation of 396 mm (15.6 inches). Using similar values for winter precipitation and basal area reduction, our model simulated the runoff response to vary depending on years since treatment, from a high of 38 mm (1.5 inches) in the year immediately following treatments to a low approaching zero mm after six years (Figure 3c).

Study Limitations

In this study, we revised a statistical runoff model from the historical Beaver Creek watershed experiments to explore how variations in climate and forest thinning would affect runoff. To the extent possible, we limited the application of the runoff model to the range of climatic and environmental conditions under which the original data were collected and model developed. However, our modeling approach was constrained by several uncertainties that will require further investigation. The model was limited by the scope of forest practices conducted in the Beaver Creek watershed experiments [20], [30–32]. These experiments used thinning techniques such as strip-thinning and patch clearing that had a similar range of basal area reductions but a different spatial configuration from contemporary thinning prescriptions. They did not measure the effects of maintenance treatments on runoff. They measured "in-place" runoff at sub-watershed outlets and did not measure other aspects of the water budget, such as evapotranspiration, soil moisture, surface routing, and groundwater recharge. The revised model represented statistical relationships between annual runoff and winter precipitation, forest basal area reductions and time since treatment, but these relationships were not necessarily related to physical processes or functions that control water balance [20]. We summarized how important factors that are not directly modeled may increase or decrease the range or runoff due to thinning that is documented in this study in File S3.

Of all the factors where uncertainties in forest hydrology and runoff remain, perhaps the most significant were the effects of soil types, long-term climate variability, and climate change. In terms of soil types, the revised regression model in this study was derived from experimental thinning on basalt-derived soils that support 59% of ponderosa pine forests in the Salt-Verde watersheds. The runoff effects of thinning on the remaining 41% of forests on sedimentary-derived soils are unknown. Two studies demonstrated that runoff from untreated ponderosa forests on sedimentary-derived soils can be

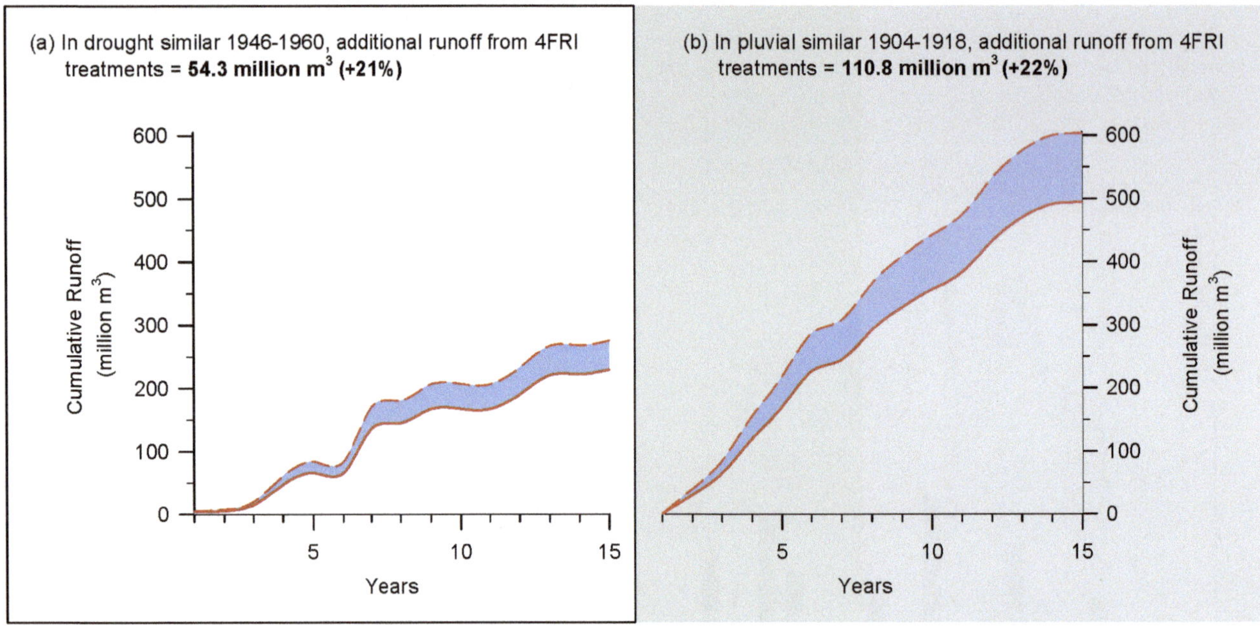

Figure 9. Cumulative runoff increases from thinning in 4FRI project. Estimates of cumulative increases in runoff (million m^3) from planned mechanical thinning of ponderosa pine forests in the first analysis area of the 4FRI project under (a) drought and (b) pluvial conditions. Solid red lines are estimates of cumulative runoff under current forest conditions using original Baker-Kovner regression model [20]. Dotted red lines represent increases in cumulative runoff associated with 4FRI treatments using modified Baker-Kovner regression model. Difference between these two values, shown with blue shading, is additional runoff from forest thinning treatments. Estimated increases in runoff ceased after 15 years.

substantially less than runoff from volcanic soils [41], [42]. Several properties of sedimentary-derived soils relative to basalt-derived soils may lead to greater water permeability and infiltration, including lower clay content, extensive fracturing, and deeper soil horizons [30], [43], [44]. Depending on a number of factors, increased infiltration may lead to increased recharge, which we would expect to result in increased discharge to surface water at some location downstream.

A nearby paired-watershed study in mixed conifer forests on sedimentary-derived soils found that overstory removal of 30-80% led to a slightly higher range of increased runoff, 30–110%, that persisted for more years than the Beaver Creek experiments [45],[46]. Direct comparisons with ponderosa pine forests are not possible because this study was conducted within a higher-elevation mixed-conifer forest that had higher initial basal areas and higher mean annual precipitation.

Relative to climate, this study's scenarios captured the high end of variability associated with pluvials but probably not the low end of variability associated with droughts. Specifically, the early 20th century pluvial may have been the wettest pluvial that has occurred in the last 12 centuries [47], whereas the two occurrences of 20th century drought were eclipsed in terms of duration and severity by several multi-decade "mega-droughts" [48–51]. Output from global circulation models indicates that climate variability will continue to be an important characteristic of the region in the future [52], but that climate change may increase the risk of extreme climatic events such as multi-decade droughts and extreme winter precipitation [53], [54]. Some global circulation models also project that mean winter precipitation in the Southwest will decline by up to 10% [52], but it may take many years to detect effects on stream flows because of precipitation variability [55]. The net effect of changes in precipitation on forest condition and hydrology still needs to be resolved and will require more investigation.

Unlike precipitation, the effects of anthropogenic warming on forest hydrology are clearer. Studies have found that warmer temperatures in recent decades help explain a downward trend in snowpack in the western United States, even after patterns of natural climate variability have been considered [9], [56]. In our revision of the historical runoff model, we attempted to add several parameters of temperature – seasonal means, maximums, and minimums – as explanatory variables to the revised regression model but none were significant. This suggests that the temporal and spatial range of winter temperatures that occurred during the years when the watershed experiments were conducted was not sufficient to add explanatory value to predicting runoff. In a separate trend analysis of winter temperatures (Oct-Apr), we found that temperatures in the months of March and April in the last 25 years, 1988–2012, were significantly warmer by 2 and 1 degrees Celsius (3.5 and 2 degrees Fahrenheit), respectively, than temperatures for the same months in the 25 years when the experiments were conducted, 1958–1982 (see File S4 for more detailed information). The magnitude of observed declines in snowpack in the Southwest, in the range of 20%, is similar to the increases in runoff associated with thinning from this study, suggesting that accelerated thinning may at least offset or ameliorate runoff losses due to climate change.

Management Implications: Accelerated Forest Thinning to Improve Resilience

On balance, we believe there is enough evidence for managers to explore the ways in which accelerated forest thinning can restore forest resilience and functioning to ameliorate the drying of ponderosa pine forests associated with warming and past management activities. Below we summarize the management implications of this and related studies.

Table 2. Annual runoff increases from thinning in Salt-Verde watersheds.

Forest Management: Thinning[a]				**Winter Precipitation Regime**			**Runoff: Annual Increase**			**Runoff: Cumulative Increase**	
Scenario Name	*Total Hectares*	*Hectares per Year*	*Effective Hectares per Year*	*Type*	*Similar to 20th Century Period*	*Mean (mm/yr)*	*Mean (million m^3)*	*Median (million m^3)*	*Maximum (million m^3)*	*Total (million m^3)*	*%*
35-low	204,000	6,800	35,000	Drought	1942–1976	340	4.76	2.99	17.5	167	23%
				Pluvial	1965–1999	396	7.65	5.85	25.7	268	26%
35-mid	301,000	10,000	51,400	Drought	1942–1976	340	7.01	4.39	25.8	245	23%
				Pluvial	1965–1999	396	11.3	8.61	37.7	393	26%
35-high	401,000	13,400	68,800	Drought	1942–1976	340	9.35	5.86	34.4	327	23%
				Pluvial	1965–1999	396	15.0	11.5	50.4	525	26%
25-low	204,000	10,200	49,000	Drought	1942–1966	338	6.18	4.06	26.3	154	20%
				Pluvial	1975–1999	411	11.9	9.65	33.6	299	22%
25-mid	301,000	15,100	72,000	Drought	1942–1966	338	9.09	5.97	38.7	227	20%
				Pluvial	1975–1999	411	17.5	14.2	49.3	438	22%
25-high	401,000	20,100	96,300	Drought	1942–1966	338	12.1	7.98	51.7	303	20%
				Pluvial	1975–1999	411	23.4	19.0	66.0	585	22%
15-low	204,000	20,400	81,700	Drought	1945–1959	315	9.23	5.95	52.3	138	20%
				Pluvial	1975–1989	422	21.8	8.31	60.1	327	22%
15-mid	301,000	30,100	120,000	Drought	1945–1959	315	13.6	8.75	77.0	204	20%
				Pluvial	1975–1989	422	32.1	12.2	88.4	481	22%
15-high	401,000	40,100	161,000	Drought	1945–1959	315	18.1	11.7	103	271	20%
				Pluvial	1975–1989	422	42.8	16.3	118	643	22%

[a] 'Scenario Name' indicates number of years simulated in each scenario (15, 25, 35 years) and extent of forest thinned (low = 30% of forest thinned; mid = 45%; high = 60%). 'Total Hectares' are total number of hectares thinned in each scenario. 'Hectares per Year' are number of hectares thinned every year for 10 consecutive years (15-year scenarios), 20 consecutive years (25-year scenarios), and 30 consecutive years (35-year scenarios). 'Effective Hectares per Year' are the average number of thinned hectares across all years of the scenario that contribute to additional runoff (see figure 6).

Estimates of additional runoff associated with thinning of ponderosa pine forests across Salt-Verde watersheds accounting for variation in treatment periods, total forest extent treated, and drought and pluvial periods of winter precipitation (Oct-Apr). Estimates rounded off to three significant digits.

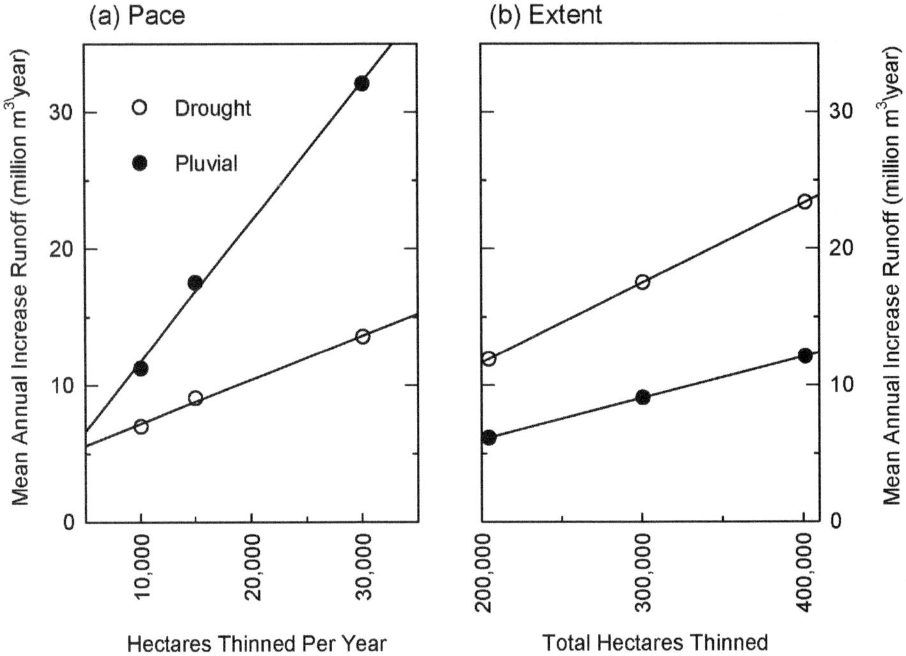

Figure 10. Scale effects of thinning on runoff in Salt-Verde watersheds. Effects of increasing (a) pace and (b) extent of thinning treatments in ponderosa pine forests in Salt-Verde watersheds on increases in mean annual runoff (million m^3/year). In (a) total area thinned is held constant at 301,000 ha (743,000 acres) (scenarios: 35mid, 25mid, 15mid) to show influence of increasing the area thinned per year. In (b) duration of thinning treatments is held constant at 25 years (scenarios: 25low, 25mid, 25high) to show influence of increasing the total area thinned across the scenario. In order to illustrate scale effects, only increases in *mean* annual runoff are shown. Statistics describing annual variability in runoff gains are shown in Table 2 and illustrated graphically for 4FRI scenario in Figure 8.

1. *Maintain & improve ecosystem function* – The current paradigm of forest management has moved away from a debate about managing forests to address one societal need, such as downstream water supply, to recognition that management strategies need to address multiple challenges [29], [57]. The 4FRI restoration project demonstrates this approach with planned benefits that include "...improved vegetation biodiversity, wildlife habitat, soil productivity, and watershed function" [19]. The impaired functioning and heightened vulnerability of southwestern forests to large-scale disturbances increases the urgency to improve ecosystem functioning, and to measure the effectiveness and trade-offs associated with these management activities using robust adaptive management and monitoring protocols. Such a focus may result in more options and flexibility in subsequent years, whereas failure to maintain functioning ecosystems may eliminate future options if soil and hydrologic processes are irreversibly altered.

Our study demonstrated that accelerated thinning can improve surface water runoff, a key ecosystem function, and provided indirect evidence of benefits for soil moisture and productivity. Another study found that average soil water content on low density ponderosa forests (250 trees/ha) was substantially higher than on high density forests (2,710 trees/ha), although these differences in forest density were not a result of mechanical thinning [58]. Researchers in the historical Beaver Creek experiments hypothesized that lower tree densities associated with thinning would reduce evapotranspiration losses, and thereby allow more water to be available for soil moisture, groundwater recharge, and surface water runoff [29]. Although we were unable to evaluate thinning effects on evapotranspiration, Dore et al. [59] found that light thinning (40% basal area reduction) in a ponderosa pine forest reduced stand-level evapotranspiration by 12% over four years, but the effect diminished over time and was not detectable in the fourth post-harvest year. This same study found that removal of all trees in a wildfire-burned site reduced annual evapotranspiration by 20%.

Other studies demonstrated that mechanical thinning can reduce the vulnerability of forests to uncharacteristic crown fires. Using a fire simulation model, Cochrane et al. [60] found that landscape-scale mechanical thinning that preceded actual wildfires could have reduced the average size of six wildfires in ponderosa pine and mixed conifer forests in California and the southwestern United States by an average of 18% (range 0.3 to 65%). A meta-analysis of 54 experimental studies showed that mechanical thinning significantly reduced the susceptibility of western ponderosa and Jeffrey pine forests to crown fires [23]. These studies describe the primary mechanism by which thinning reduces fire risk as a redistribution of fuels, thereby reducing fire spread rates and crowning behavior.

2. *Apply management tools at scale* – Managing forests at a scale that is comparable to recent forest disturbances will be another key factor for improving resilience. At 238,000 ha (588,000 acres) of ponderosa pine to be treated with mechanical thinning and prescribed fire, the 4FRI project is approaching the scale of these amplified disturbances. Our study demonstrated that potential increases in runoff associated with thinning were scale dependent: runoff increased in a linear fashion as the pace and extent of thinning increased in the Salt-Verde runoff scenarios (Figure 10). Recent efforts to reduce fire risk with landscape-scale treatments of ponderosa pine forests have been successful. For example, treatments completed before the largest wildfire in Arizona history, the Wallow Fire in 2011, effectively protected communities and towns by reducing fire severity before it reached key residential areas [61].

Further, accelerated thinning may open up opportunities to use other management tools at an accelerated pace. For example, wildland fire use has been an effective and inexpensive tool for maintaining forest resilience across large areas in remote forests and woodlands – the Gila Wilderness in New Mexico and Grand Canyon National Park in northern Arizona [62], [63] – but has been used less frequently in forests near population centers and towns. Completion of treatments in the first analysis area of the 4FRI project could have a multiplying effect on maintaining forest resilience if the thinning itself allows for increased use of wildland fire or prescribed fire over many more hectares.

3. *Leverage knowledge of climate variability* – The Beaver Creek watershed experiments and others like it demonstrated that winter precipitation was the best predictor of additional runoff from forest thinning [20], [30]. We add to this knowledge by examining how variability in winter precipitation could influence management effectiveness. Our scenarios demonstrated that gains in runoff from thinning could occur even under drought conditions. In simulated droughts, 1 out of 3 years had above-average precipitation, and these years accounted for 75% of the runoff gains in these scenarios. Even though these gains were substantially less than runoff gains during pluvials, additional runoff under drought conditions could help lessen the impact of predicted climate-driven losses in hydrologic connectivity important to native fish in the Salt and Verde watersheds [64].

This study also demonstrated that the same thinning schedule could result in a doubling in additional runoff if conducted in a pluvial typical of 20th century variability as compared to a drought. Scientists have developed decision support tools to help managers understand the likelihood that precipitation regimes will shift from wet to dry conditions or vice-versa [65], [66]. Knowledge of whether management activities are taking place in a pluvial or a drought could improve the ability of managers to predict the likelihood of success for various objectives. For example, thinning in a pluvial might meet substantive objectives to improve runoff, whereas thinning in a drought might be more significant for reducing risks of catastrophic crown fires or drought mortality.

With the notable exception of a mega-drought, the droughts and pluvials examined in this study were characteristic of climate variability in the region. An examination of hydro-climatic trends in the Salt-Verde watersheds revealed that temperatures are non-stationary, increasing significantly in recent decades, but the same is not true for winter precipitation or resulting stream flows [55]. Rather, winter precipitation has fluctuated between droughts and pluvials in these watersheds in the last century with no evidence of directional change, a pattern that has reoccurred in the southwestern United States for the last 1000 years [50]. These precipitation cycles have been linked to persistent anomalies in sea surface temperatures that vary across inter-annual and decadal time steps [67], [68]. If past patterns of precipitation variability remain stable in the near term, then it is probable that precipitation and flows in the Salt-Verde watersheds will shift into wetter conditions within the timeframe examined in this study [66].

4. *Building Knowledge to Reduce Uncertainties* – Landscape-scale restoration projects like 4FRI present the opportunity to learn about the influence of accelerated thinning on forest water budgets and resilience using modern forestry techniques and under a changing and variable climate. Stakeholders in the 4FRI project are developing a robust adaptive management and monitoring program so that progress towards objectives can be measured and timely adjustments to management made. Additionally, a paired-watershed experiment to evaluate the effects of forest treatments, including prescribed fire maintenance treatments, on watershed properties is planned within the first analysis area of 4FRI. It will monitor runoff, precipitation, evapotranspiration, snow water storage, soil moisture, groundwater recharge, and sediment yield, thereby reducing our knowledge gaps regarding the effects of forest management on the water cycle.

Conclusions

The widespread pattern of forest drying, evidenced by ecosystem level moisture deficits and larger and more severe wildfires, indicates we are entering a new era of forest management. The 4FRI project is an example of a restoration effort that is addressing this challenge in two ways: establishing objectives to improve resilience and increasing the scale at which management tools are applied. Our study demonstrated that this type of project can increase runoff by approximately 20% compared to unthinned forests, even under simulated drought conditions. If treatments at this scale are completed and repeated over the next several decades, increases in runoff could help offset the current and projected declines in snowpack and stream flow due to warming while improving the resilience of forest stands. As an incidental benefit in an era of dwindling water supplies and projected water shortages, forest thinning could play a role in augmenting river flows on a seasonal basis, improving conditions for water-dependent ecosystems, and benefitting the water supplies of downstream communities. Accelerated forest thinning to reduce water stress and wildfire risk is one of the only management options under our control, and it is probably the most critical to apply over the short term. Because the effects are temporary, it is important to frame accelerated thinning as a reset of ecosystem resilience, not a permanent cure. Rather, accelerated forest thinning creates opportunities to apply other management tools that restore forest resilience at broader scales, whereas management actions at the current pace will likely not be sufficient to address recent and projected changes in forest conditions.

Supporting Information

File S1 Methods and results figures and tables in English units.

File S2 Original and revised Baker-Kovner regression models in English units.

File S3 Additional factors that influence runoff.

File S4 Temperature trend analysis.

Acknowledgments

We thank Daniel Majka for his early contributions and input into this study. We thank Dan Huebner and Karen Koestner for providing data from the historical Beaver Creek experiments and Shaula Hedwall and Mark Nigrelli for providing data and expertise from the Four Forest Restoration Initiative and the Northern Arizona Wood Supply Study. We thank Kit MacDonald, Chris Nelson, Estella Smith, Rory Steinke, Joe Ringelberg, and Andrea Thode for providing data and expertise of forest thinning exclusions. We thank Julio Bentancourt, Michael Crimmins, and Gregg Garfin for input on climate variability and climate change, Charlie Ester for input on water management in the Salt-Verde River watersheds, and Robert Steidl for advice on statistical analyses. We thank Pat Graham, Molly Hunter, Dale Turner and Sue Sitko for critical review of this paper.

Author Contributions

Conceived and designed the experiments: MR RM FO JH. Performed the experiments: MR FO. Analyzed the data: MR FO. Contributed reagents/materials/analysis tools: MR FO JH. Wrote the paper: MR RM FO ES JH DG.

References

1. Covington WW, Moore MM (1994) Southwestern ponderosa pine forest structure: Changes since Euro-American settlement. Journal of Forestry 92(1): 39–47.
2. Fule PZ, Heinlein TA, Covington WW, Moore MM (2003) Assessing fire regimes on Grand Canyon landscapes with fire-scar and fire-record data. International Journal of Wildland Fire 12: 129–145.
3. Cocke AE, Fule PZ, Crouse JE (2005) Forest change on a steep mountain gradient after extended fire exclusion: San Francisco Peaks, Arizona, USA. Journal of Applied Ecology 42(5): 814–823.
4. Williams AP, Allen CD, Millar CI, Swetnam TW, Michaelsen J, et al. (2011) Forest responses to increasing aridity and warmth in the southwestern United States. Proceedings of the National Academy of Sciences 107(50): 21289–21294.
5. Gitlin AR, Sthultz CM, Bowker MA, Stumpf S, Paxton KL, et al. (2006) Mortality gradients within and among dominant plant populations as barometers of ecosystem change during extreme drought. Conservation Biology 20(5): 1477–1486.
6. Propst DL, Gido KB, Stefferud JA (2008) Natural flow regimes, nonnative fishes, and native fish persistence in arid-land river systems. Ecological Applications 18(5): 1236–1252.
7. Stefferud JA, Gido KB, Propst DL (2011) Spatially variable response of native fish assemblages to discharge, predators and habitat characteristics in an arid-land river. Freshwater Biology 56: 1403–1416.
8. Breshears DD, Cobb NS, Rich PM, Price KP, Allen CD, et al. (2005) Regional vegetation die-off in response to global-change-type drought. Proceedings of the National Academy of Sciences 102(42): 15144–15148.
9. Mote PW (2006) Climate-driven variability and trends in mountain snowpack in western North America. Journal of Climate 19: 6209–6220.
10. Westerling AL, Hidalgo HG, Cayan DR, Swetnam TW (2006) Warming and earlier spring increase Western US forest wildfire activity. Science 313: 940–943.
11. van Mantgem PJ, Stephenson NL, Byrne JC, Daniels LD, Franklin JF, et al. (2009) Widespread increase of tree mortality rates in the Western United States. Science 323: 521–524.
12. Thomas BE, Pool DR (2006) Trends in Streamflow of the San Pedro River, Southeastern Arizona, and Regional Trends in Precipitation and Streamflow in Southeastern Arizona and Southwestern New Mexico. U.S. Geological Survey Professional Paper 1712. Reston, VA: U.S. Geological Survey. 79 p.
13. Timmons D, Springer A (2006) Prescott AMA groundwater flow model update report. Available: http://www.azwater.gov/AzDWR/Hydrology/Modeling/documents/Prescott_Model_Update_Report.pdf. Accessed 6 May 2014.
14. Webb RH, Turner RM, Leake SA. (2007) The Ribbon of Green: Change in Riparian Condition in the Southwestern United States. Tucson, AZ: The University of Arizona Press. 480 p.
15. Marshall RM, Robles MD, Majka DR, Haney JA (2010) Sustainable Water Management in the Southwestern United States: Reality or Rhetoric? PLoS ONE 5(7): e11687. doi:10.1371/journal.pone.0011687.
16. Water Resources Development Commission (2011) Water Resources Development Commission Final Report: Volume II Committee Reports. Phoenix, AZ: Water Resources Development Commission. 685 p.
17. United States Department of Interior Bureau of Reclamation (2012) Colorado River Basin Water Supply and Demand Study: Study Report. Washington DC: U.S. Department of Interior, Bureau of Reclamation. 95p.
18. Vano JA, Udall B, Cayan DR, Overpeck JT, Brekke LD, et al. (2014) Understanding uncertainties in future Colorado River streamflow. Bulletin of the American Meteorological Society 95(1): 59–78.
19. United States Department of Agriculture Forest Service (2013) Draft environmental impact statement for the Four-Forest Restoration Initiative. Flagstaff, AZ: USDA Forest Service, Southwestern Region, Coconino and Kaibab National Forests. 744 p.
20. Brown HE, Baker MB, Rogers JJ, Clary WP, Kovner JL, et al. (1974) Opportunities for increasing water yields and other multiple uses on ponderosa pine forest lands. USDA Forest Service Research Paper RM-129. Fort Collins, CO: U.S. Department of Agriculture, Forest Service, Rocky Mountain Forest and Range Experiment Station. 36 p.
21. Zausen GL, Kolb TE, Bailey JD, Wagner MR (2005) Long-term impacts of stand management on ponderosa pine physiology and bark beetle abundance in northern Arizona: A replicated landscape study. Forest Ecology and Management 218: 291–305.
22. Stephens SL, Moghaddas JJ, Edminster C, Fiedler CE, Haase S, et al. (2009) Fire treatment effects on vegetation structure, fuels, and potential fire severity in western US forests. Ecological Applications 19(2): 305–320.
23. Fule PZ, Crouse JE, Roccaforte JP, Kalies EL (2012) Do thinning and/or burning treatments in western USA ponderosa pine or Jeffrey pine-dominated forests help restore natural fire behavior? Forest Ecology and Management 269: 68–81.
24. Barr GW (1956) Recovering rainfall. Tucson, AZ: Department of Agricultural Economics, University of Arizona. 33 p.
25. Daly C, Halbleib M, Smith JI, Gibson WP, Doggett MK, et al. (2008) Physiographically-sensitive mapping of climatological temperature and precipitation across the conterminous United States. International Journal of Climatology 28(15): 2031–2064.
26. Gesch D, Oimoen M, Greenlee S, Nelson C, Steuck M, et al. (2002) The National Elevation Dataset. Photogrammetric Engineering and Remote Sensing 68(1): 5–11.
27. US Geological Survey, Gap Analysis Program (2010) National Land Cover, Version 1. US Geological Survey. Available: http://gapanalysis.usgs.gov/gaplandcover/viewer/. Accessed 2012 Dec 23.
28. Arizona Geological Survey (2011) Geologic Map of Arizona. Available: http://www.azgs.az.gov/services_azgeomap.shtml. Accessed 2013 Apr 9.
29. Baker MB (2003) Hydrology. In: P . Friederici, editor. Ecological Restoration of Southwestern Ponderosa Pine Forests. Washington DC: Ecological Restoration Institute, Northern Arizona University, Island Press. pp. 161–174.
30. Baker MB (1986) Effects of ponderosa pine treatments on water yield in Arizona. Water Resources Research 22(1): 67–73.
31. Neary D (2011) Managing Arid and Semi-Arid Watersheds: Beaver Creek Watershed. U.S. Department of Agriculture, Forest Service, Rocky Mountain Research Station. Available: http://www.rmrs.nau.edu/watersheds/highlands/watersheds/beavercreek/. Accessed 2012 Oct 29.
32. Rich LR (1972) Managing a ponderosa pine forest to increase water yield. Water Resources Research 8(2): 422–428.
33. Hampton HM, Sesnie SE, Bailey JD, Snider GB (2011) Estimating regional wood supply based on stakeholder consensus for forest restoration in Northern Arizona. Journal of Forestry: 15–26.
34. Haire SL, McGarigal K (2010) Effects of landscape patterns of fire severity on regenerating ponderosa pine forests (*Pinus ponderosa*) in New Mexico and Arizona, USA. Landscape Ecology 25: 1055–1069.
35. Eidenshink J, Schwind B, Brewer K, Zhu Z, Quayle B, et al. (2007) A project for monitoring trends in burn severity. Journal of Association of Fire Ecology 3(1): 3–21.
36. Conservation Biology Institute (2010) Protected Areas Database of the United States, Version 2. Available: http://consbio.org/products/projects/pad-us-cbi-edition. Accessed 2013 Apr 9.
37. United States Department of Agriculture Forest Service (2013) GIS datasets for the Southwestern Region, United States Department of Agriculture, Forest Service. Available: http://www.fs.usda.gov/detail/r3/landmanagement/gis/?cid=stelprdb5202474. Accessed 2013 Apr 9.
38. Turner DS, List MD. 2007. Habitat mapping and conservation analysis to identify critical streams for Arizona's native fish. Aquatic Conservation: Marine and Freshwater Systems 17: 737–748.
39. Arizona Department of Water Resources (2010) Arizona Water Atlas Volume 8: Active Management Areas Water Atlas. Available: http://www.azwater.gov/azdwr/StatewidePlanning/WaterAtlas/ActiveManagementAreas/default.htm. Accessed 2014 May 5.
40. Brown TC, Fogel MM (1987) Use of streamflow increases from vegetation management in the Verde River Basin. Water Resources Bulletin 23(6): 1149–1160.
41. Campbell RE, Baker MB, Ffolliott PF, Larson FR, Avery CC (1977) Wildfire effects on a ponderosa pine ecosystem: An Arizona case study. USDA Forest Service Research Paper RM-191. Fort Collins, CO: U.S. Department of Agriculture, Forest Service, Rocky Mountain Forest and Range Experiment Station. 12 p.
42. Ffolliott PF, Baker MB (1977) Characteristics of Arizona ponderosa pine stands on sandstone soils. USDA Forest Service General Technical Report RM-44. Fort Collins, CO: U.S. Department of Agriculture, Forest Service, Rocky Mountain Forest and Range Experiment Station. 7 p.
43. Pearson GA (1931) Forest types in the Southwest as determined by climate and soil. Technical Bulletin No. 247. Washington DC: U.S. Department of Agriculture, Forest Service, Southwestern Forest and Range Experiment Station. 144 p.
44. Bosch JM, Hewlett JD (1982) A review of catchment experiments to determine the effect of vegetation changes on water yield and evapotranspiration. Journal of Hydrology 55: 3–23.
45. Rich LR, Gottfried GJ (1976) Water yields resulting from treatments on the Workman Creek experimental watersheds in Central Arizona. Water Resources Research 12(5): 1053–1060.
46. Hibbert AR, Gottfried GJ (1987) Streamflow responses to forest treatments on two Arizona mixed conifer watersheds. In: Troendle CA, Kaufmann MR, Hamre RH, Winokur RP editors. Management of subalpine forest: Building on 50 years of research. USDA Forest Service General Technical Report RM-149. Silver Creek, CO: U.S. Department of Agriculture, Forest Service, Rocky Mountain Forest and Range Experiment Station. pp. 189–194.

47. Woodhouse CA, Kunkel KE, Easterling DR, Cook ER (2005) The twentieth-century pluvial in the western United States. Geophysical Research Letters 32: L07701, doi:10.1029/2005GL022413.
48. Woodhouse CA, Overpeck JT (1998) 2000 years of drought variability in the central United States. Bulletin of the American Metereological Society 79(12): 2693–2714.
49. Stahle DW, Cook ER, Cleaveland MK, Therrell D, Meko DM, et al. (2000) Tree-ring data document 16th century megadrought over North America. Eos 81(12): 121–125.
50. Ni F, Cavazos T, Hughes MK, Comrie AC, Funkhouser G (2002) Cool-season precipitation in the southwestern USA since AD 1000: Comparison of linear and nonlinear techniques for reconstruction. International Journal of Climatology 22: 1645–1662.
51. Cook ER, Seager R, Cane MA, Stahle DW (2007) North American drought: Reconstructions, causes, and consequences. Earth-Science Reviews 81: 93–134.
52. Garfin G, Jardine A, Merideth R, Black M, LeRoy S, editors (2013) Assessment of Climate Change in the Southwest United States: A Report Prepared for the National Climate Assessment. A report by the Southwest Climate Alliance. Washington DC: Island Press. 531 p.
53. Ault TR, Cole JE, Overpeck JT, Pederson GT, Meko DM (2014) Assessing the risk of persistent drought using climate model simulations and paleoclimate data. Journal of Climate e-View doi:http://dx.doi.org/10.1175/JCLI-D-12-00282.1.
54. Dominguez F, Rivera E, Lettenmaier DP, Castro CL (2012). Changes in winter precipitation extremes for the western United States under a warmer climate as simulated by regional climate models. Geophysical Research Letters (39) L05803, doi:10.1029/2011GL050762.
55. Murphy KW, Ellis AW (2014) An assessment of the stationarity of climate and stream flow in watersheds of the Colorado River Basin. Journal of Hydrology 509: 454–473.
56. Pierce DW, Barnett TP, Hidalgo HG, Das T, Bonfils C, et al. (2008) Attribution of declining Western U.S. snowpack to human effects. Journal of Climate 21: 6425–6444.
57. Grant GE, Tague CL, Allen CD (2013) Watering the forest for the trees: An emerging priority for managing water in forest landscapes. Frontiers in Ecology and the Environment 11: 314–321.
58. Zou CB, Breshears DD, Newman BD, Wilcox BP, Gard MO, et al. (2008) Soil water dynamics under low-versus high-ponderosa pine tree density: Ecohydrological functioning and restoration implications. Ecohydrology 1: 309–315.
59. Dore S, Montes-Helu M, Hart SC, Hungate BA, Koch GW, et al. (2012) Recovery of ponderosa pine ecosystem carbon and water fluxes from thinning and stand-replacing fire. Global Change Biology 18: 3171–3185.
60. Cochrane MA, Moran CJ, Wimberly MC, Baer AD, Finney MA, et al. (2012) Estimation of wildfire size and risk changes due to fuel treatments. International Journal of Wildland Fire 21(4): 357–267.
61. Kennedy MC, Johnson MC (2014) Fuel treatment prescriptions alter spatial patterns of fire severity around the wildland-urban interface during the Wallow Fire, Arizona, USA. Forest Ecology and Management 318: 122–132.
62. Fule PZ, Laughlin DC (2007) Wildland fire effects on forest structure over an altitudinal gradient, Grand Canyon National Park, USA. Journal of Applied Ecology 44(1): 136–146.
63. Hunter ME, Iniguez JM, Lentile LB (2011) Short- and long-term effects on fuels, forest structure, and wildfire potential from prescribed fire and resource benefit fire in southwestern forests, USA. Fire Ecology 7(3): 108–121.
64. Jaeger KL, Olden JD, Pelland NA (2014) Climate change poised to threaten hydrologic connectivity and endemic fishes in dryland streams. Proceedings of the National Academy of Sciences. doi: 10.1073/pnas.1320890111.
65. Enfield DB, Cid-Serrano L (2006) Projecting the risk of future climate shifts. International Journal of Climatology 26: 885–895.
66. Gangopadhyay S, McCabe GJ (2010) Predicting regime shifts in flow of the Colorado River. Geophysical Research Letters 37: L20706, doi:10.1029/2010GL044513.
67. Sheppard PR, Comrie AC, Packin GD, Angersbach K, Hughes MK (2002) The climate of the Southwest. Climate Research 21: 219–238.
68. Mantua NJ, Hare SR (2002) The Pacific Decadal Oscillation. Journal of Oceanography 58: 35–44.

11

Photosynthetic Diffusional Constraints Affect Yield in Drought Stressed Rice Cultivars during Flowering

Marco Lauteri[1], Matthew Haworth[2]*, Rachid Serraj[3,4], Maria Cristina Monteverdi[5], Mauro Centritto[2]

1 Institute of Agro-Environmental and Forest Biology, National Research Council, Porano, Italy, **2** Trees and Timber Institute (IVALSA - Istituto per la valorizzazione del legno e delle specie arboree), National Research Council, Florence, Italy, **3** Consultative Group on International Agricultural Research, Independent Science and Partnership Council Secretariat, Food and Agriculture Organisation of the United Nations, Rome, Italy, **4** International Rice Research Institute, Los Banos, Philippines, **5** Forestry Research Centre, National Research Council, Arezzo, Italy

Abstract

Global production of rice (*Oryza sativa*) grain is limited by water availability and the low 'leaf-level' photosynthetic capacity of many cultivars. *Oryza sativa* is extremely susceptible to water-deficits; therefore, predicted increases in the frequency and duration of drought events, combined with future rises in global temperatures and food demand, necessitate the development of more productive and drought tolerant cultivars. We investigated the underlying physiological, isotopic and morphological responses to water-deficit in seven common varieties of *O. sativa*, subjected to prolonged drought of varying intensities, for phenotyping purposes in open field conditions. Significant variation was observed in leaf-level photosynthesis rates (*A*) under both water treatments. Yield and *A* were influenced by the conductance of the mesophyll layer to CO_2 (g_m) and not by stomatal conductance (g_s). Mesophyll conductance declined during drought to differing extents among the cultivars; those varieties that maintained g_m during water-deficit sustained *A* and yield to a greater extent. However, the variety with the highest g_m and yield under well-watered conditions (IR55419-04) was distinct from the most effective cultivar under drought (Vandana). Mesophyll conductance most effectively characterises the photosynthetic capacity and yield of *O. sativa* cultivars under both well-watered and water-deficit conditions; however, the desired attributes of high g_m during optimal growth conditions and the capacity for g_m to remain constant during water-deficit may be mutually exclusive. Nonetheless, future genetic and physiological studies aimed at enhancing *O. sativa* yield and drought stress tolerance should investigate the biochemistry and morphology of the interface between the sub-stomatal pore and mesophyll layer.

Editor: Wagner L. Araujo, Universidade Federal de Vicosa, Brazil

Funding: This work was funded by: Ministero dell'Istruzione, dell'Università e della Ricerca of Italy (PRIN) grant: "Going to the root of plant productivity: how the rhizosphere interact with the aboveground armament for indirect and direct defense against abiotic and biotic stressors (PRO-ROOT)" URL: http://hubmiur.pubblica.istruzione.it/web/universita/home and Marie Curie Intra European Fellowship (2010-275626) - HAWORTH - Evoutionary adaptation to atmospheric carbon dioxide, URL: http://ec.europa.eu/research/mariecurieactions/about-mca/actions/ief/index_en.htm. The funders had no role in study design, data collection and analysis, decision to publish, or preparation of the manuscript.

Competing Interests: The authors have declared that no competing interests exist.

* Email: matthew.haworth@hotmail.com

Introduction

Global production of rice (*Oryza sativa*) is limited by the availability of freshwater [1] and the tolerance of rice cultivars to drought stress [2]. Nonetheless, *O. sativa* is cultivated in a diverse range of climates and habitats, with lowland varieties generally considered to be more productive, but less tolerant of drought, than their upland counterparts [3]. Selective breeding of *O. sativa* varieties has increased the grain production of individual plants by 70% since 1965 [4]. However, these increases in productivity have been restricted by the comparatively low 'leaf-level' photosynthetic capacity of *O. sativa* relative to other C3 crops [5,6]. Phenotypic screening of the photosynthetic and gas-exchange characteristics of *O. sativa* varieties under water deficit conditions may permit the identification of attributes that confer both drought tolerance and high productivity [7,8].

Yield in varieties of *O. sativa* are closely related to photosynthesis under both well-watered and drought conditions [7,9]. The uptake of CO_2 for photosynthesis involves two major resistance steps at the stomata and at the surface of the mesophyll layer [10,11]. Stomatal (g_s) and mesophyll conductance (g_m) to CO_2 often decrease in unison in response to environmental stresses [7,12–14]. Upland and lowland *O. sativa* cultivars exposed to a soil moisture gradient ranging from high water-availability to severe water-deficit exhibited identical reductions in both g_s and g_m that correlated with the rate of photosynthesis (*A*) and yield. Most notably, while there was no observable difference in *A*, g_s or g_m response to drought between the upland and the lowland varieties, those cultivars that exhibited the highest values of total conductance to CO_2 (g_t: where $g_t = [g_s * g_m]/[g_s + g_m]$) supported higher *A* and yield under all levels of water availability [7]. Those *O. sativa* varieties with higher capacities for CO_2 uptake are therefore the most likely to offer the greatest potential improvement of leaf-level photosynthesis towards achieving enhanced yields [9,15,16]. However, an increase in the capacity for leaf gas exchange may be accompanied by reduced drought tolerance [17,18].

These differences in the photosynthetic and leaf gas-exchange capacities of *O. sativa* varieties have been ascribed to genetic variation [9]. The quantitative trait loci responsible for leaf gas-exchange and photosynthetic parameters in *O. sativa* tend to be found in adjacent regions of the plant genome; indicative of the exertion of selective pressures favouring coordination and high heritability of photosynthetic behaviour [19,20]. Therefore, effective screening and analysis of the photosynthetic and gas exchange characteristics of *O. sativa* varieties under stress conditions has the potential to identify functional traits associated with optimisation of water use efficiency (*WUE*) and the maintenance of yield and *A* under drought [21]. Rates of photosynthesis are not only determined by diffusive CO_2-uptake, but also by the activity of ribulose-1,5-bisphosphate (RubisCO) [22]. However, exposure of thirteen hydroponically grown *O. sativa* varieties to physiological drought did not suggest that the capacity of RubisCO carboxylation (Vc_{max}) or the maximum rate of electron transport required for ribulose-1,5-bisphosphate regeneration (J_{max}) were associated with *A* or grain yield. Rather, *A* and yield were largely determined by g_s and g_m setting the limits for diffusional CO_2 uptake [9], consistent with field based studies [7]. Furthermore, the intrinsic water-use efficiency (A/g_s) of the *O. sativa* varieties was closely related to the ratio of g_m to g_s [9], and this attribute showed the greatest variability between cultivars, and thus potential in terms of possible improvements in yield and growth through maximising *WUE* [5,9,21]. In theory, those varieties with high g_m to g_s ratios should be most suited to maintaining photosynthesis and yield under drought stress conditions, while high values of both g_m and g_s should be favourable for varieties grown under constant high water availability [21].

The observation that *A* and yield of *O. sativa* closely correlate to g_m [7,9,16] has led to a focus on the interface between the leaf mesophyll cells and internal sub-stomatal air space in the improvement of rice productivity [23]. Adachi et al. [16] crossed two varieties of *O. sativa* to produce a new cultivar with higher values of g_m, which were associated with modifications of the mesophyll structure that increased the surface area available for CO_2 uptake; thus demonstrating the potential for enhanced productivity through increased g_m. In addition to morphological adaptations [15], differences in g_m between *O. sativa* varieties are also likely affected by biochemical factors [23] such as the abundance of cooporin proteins responsible for active transport of CO_2 [24,25] (Kaldenhoff, 2012). Furthermore, the abundance and activity of these proteins has been observed to decline under water-stress [26]. Exposure to drought is associated with declines in values of *A*, g_s and g_m [11,27]. Sustained impairment of *A* following the alleviation of drought stress is commonly related with continued biochemical constraints to the uptake and assimilation of CO_2 [28]. However, experimental evidence suggests that *O. sativa* does not display prolonged reductions in *A* following the cessation of drought, and that the observed constraints to *A* under water-stress are largely physical impediments to CO_2 diffusion [7,9,29,30]. This suggests that artificial selection of those *O. sativa* varieties with high g_m would enhance *A* while not negatively affecting drought tolerance or recovery.

Rice production is sensitive to water deficit at different developmental stages of plant growth. For example, early drought stress can affect leaf expansion [3], and later drought stress during flowering or grain-filling stages can disrupt grain formation and development [31,32]. The photosynthetic and yield responses of seven *O. sativa* varieties (five upland and two lowland varieties chosen to represent genetic variability) were analysed in response to sustained drought. As different *O. sativa* varieties have different phenologies (for example Van forms flower inflorescences relatively earlier than Moro), through the use of a drip irrigation system it was possible to apply the stress at the same developmental stage for each variety. This has the advantage over a previous study into the drought tolerance of *O. sativa* varieties conducted at the IRRI in that all varieties experienced drought at the same phenological stage, therefore eliminating a crucial source of error whereby early flowering varieties outperformed late developing cultivars [cf. 7]. Mesophyll conductance to CO_2 is of key interest to the productivity and drought tolerance of *O. sativa* [7,16], however it is not possible to directly measure g_m, and due to the uncertainties associated with its calculation [33] (particularly in the use of the variable J method under stress conditions) it is preferable to simultaneously use two independent methodologies in the estimation of g_m [34]. We therefore employed two independent methodologies in the estimation of g_m: firstly, based upon the synchronous measurement of leaf gas exchange and chlorophyll fluorescence parameters (variable J), and secondly, the analysis of carbon isotope discrimination ($\Delta^{13}C$) in recently formed metabolites to assess their effectiveness in an analysis of *O. sativa* drought tolerance in the field. Further details of the variable J and '$\Delta^{13}C$ of recently synthesised sugars' methods of g_m analysis are presented in the Material and Methods section. This study aimed to: 1) assess the effect of g_s and g_m on *A* and yield in well-watered and drought conditions; 2) analyse differences in g_m calculation between the 'variable J' [35] and '$\Delta^{13}C$ of recently synthesised sugars' [15] methods, and; 3) identify *O. sativa* varieties that exhibit drought tolerance and/or high productivity, and may be suitable for further genomic/biochemical development to further enhance grain production.

Materials and Methods

Plant material and growth conditions

The experiment was conducted at the International Rice Research Institute Experimental Station at Los Banos, Philippines (14° 11′N 121° 15′E, 21 m above sea level) during the dry season (January to May). Rainfall was minimal, however, temperature and leaf to air vapour pressure difference (*VPD*) increased during the experimental period. The soil at the experimental site is a neutral Andaqueptic Haplaquol (pH = 6.5). Seven *O. sativa* cultivars were studied: Apo, IR55419-04 (IR55), IR64, IR71525-19-1-1 (IR71), Moroberekan (Moro), PSBRc80 (PS80) and Vandana (Van). Moroberekan failed to set or produce grains and seemed to suffer from disease and heat sensitivity, particularly under drought conditions and it was not possible to record physiological measurements for this variety under water deficit stress conditions. Seeds were sown in dry soil at a rate of 80 kg ha^{-1} in rows spaced ~25 cm apart. Fertilisation with nitrogen was conducted on two occasions at application rates of 90 and 120 kg ha^{-1}, alongside fertilisation of 30 kg ha^{-1} of phosphate and potassium. Weeds were controlled by the application of standard agricultural herbicides and manual picking. Pesticides were used to control insect pests; in particular stem borers.

Experimental plots for each *O. sativa* variety measured 3 m×2 m, with adjacent irrigated and drought-stressed treatments. The plots were arranged in a randomized block design. A sprinkler system was initially used to provide equal irrigation levels to all plants until flower (panicle) development, when the water deficit treatment was instigated. During the drought stress period, the well-watered treatments were irrigated by use of a drip irrigation system, with soil water potential maintained above -0.04 MPa (monitored using tensiometers placed at soil depths of 15 and 30 cm) – the level at which the more sensitive rice cultivars begin to exhibit a reduction in transpiration [36].

Leaf gas exchange measurements

Leaf gas-exchange and fluorescence of plants grown under both well-watered treatment or stress conditions were simultaneously measured with a LI-6400-40 leaf chamber fluorometer (Li-Cor, Inc., Nebraska, USA) on flag leaves enclosed in a temperature, light and humidity controlled cuvette. The measurements were made *in situ* between 11.00 and 15.00, in saturating PPFD (1400 μmol $m^{-2}s^{-1}$), with relative humidity ranging between 45–55%, and a leaf temperature of 30°C. Instantaneous measurements of steady-state A, g_s, internal [CO_2] (C_i) and the quantum yield of PSII in the light (ΔF/Fm') were made on six to 14 plants per treatment. Measurements of dark respiration (R_d) were also made at ambient CO_2 concentration in the dark on the same leaves. Instantaneous transpiration efficiency was calculated as the ratio of A to g_s.

Carbon isotope discrimination

Three of the same flag leaves used for *in situ* gas exchange measurements were collected in the evening, oven dried at 70°C until their weight remained constant and were then stored until soluble sugar extraction, following the protocol of Richter et al. [37]. Leaves were finely ground in liquid nitrogen and shaken for 60 min in water at room temperature. After centrifugation (15 min at 5000 g), the supernatant was sequentially mixed with cationic (Dowex-50) and anionic (Dowex-1) exchange resins. The residual solution of purified soluble sugars was freeze-dried and carbon isotope composition ($\delta^{13}C$) determined using a mass-spectrometer. Additional isotope ratio mass spectrometry analyses were performed on the pellet samples that remained after the extraction of soluble sugars. Samples of approximately 1 mg were used for $\delta^{13}C$ analysis. Stable isotope ratios were determined using a continuous-flow triple-collector isotope ratio mass spectrometer (ISOPRIME, GV, Manchester, UK). Solid samples were quantitatively combusted in an elemental analyser (Model NA 1500, Carlo Erba, Milan, Italy) and CO_2 was transferred in helium flow to the mass spectrometer. Calculations of carbon isotope discrimination ($\Delta^{13}C$) of solid samples were undertaken following the protocol of Farquhar et al. [38], assuming the carbon isotopic composition of CO_2 in air (δ_{air}) to be −8.0‰.

Estimation of mesophyll conductance

Mesophyll conductance to CO_2 diffusion is the inverse of the total resistance encountered in the transport of CO_2 across the leaf mesophyll. Estimation of g_m is subject to a number of assumptions and sources of error, and it is preferable to employ two independent approaches in its calculation [34,39]. In this study g_m was calculated using the 'variable J' and '$\Delta^{13}C$ of recently synthesised sugars' methods. The calculation of g_m using the variable J method is based on simultaneous measurements of gas-exchange and chlorophyll fluorescence parameters as described by Harley et al. [35] and Loreto et al. [40] (equations 1 and 2):

$$g_m = \frac{A}{Ci - \frac{\Gamma * [J + 8 * (A + R_d)]}{J - 4 * (A + R_d)}} \quad (1)$$

where the electron transport rate (J_F) is calculated from fluorescence [41]:

$$J_F = PPFD * \frac{\Delta F}{F_m} * \alpha * \beta \quad (2)$$

.

Where F_m is the fluorescence maximum and the partitioning factor (β) between photosystems I and II was considered to be 0.5 and leaf absorbance (α) was measured directly [42]. The variable J method is sensitive to the estimation of the CO_2 compensation point to photorespiration (Γ^*) and R_d. For the results of this study to be directly comparable with those of Centritto et al. [7], the value of Γ^* used in the gas exchange algorithm was calculated using the Rubisco specific factor for an annual herb [43], consistent with previously published *O. sativa* Γ^* values [44]. As measurements were conducted under field conditions it was not feasible to produce robust measurements of respiration in the light (R_l); respiration in the dark was therefore used in the calculation of gm, as R_l and R_d have been observed to be broadly consistent when recycling of CO_2 is taken into consideration [45]. Measurements of R_d were performed at ambient [CO_2] concentration in the dark on the same leaves by switching off the light in the leaf cuvette; when CO_2 release from the leaf had become stable for approximately five to ten minutes this was recorded and considered to represent R_d. To calibrate the values of J calculated by fluorescence (J_F) and gas-exchange (J_A), measurements were also conducted under non-photorespiratory conditions using 1% [O_2] [35,40]. Electron transport rate and R_d values are presented in Figures S1 and S2. Total conductance to CO_2 (g_t) was calculated as: $g_t = g_s * g_m/(g_s + g_m)$. The concentration of CO_2 at the chloroplast site (C_c) was then calculated from mesophyll conductance values [46].

The method described by Lauteri et al. [47] for 'on-line' determinations was adapted to calculate g_m by using the $\Delta^{13}C$ of recently synthesized sugars [48]. Discrepancies between $\Delta^{13}C$ of leaf soluble carbohydrates (Δ_{obs}) and $\Delta^{13}C$ expected on the basis of gas-exchange measurements (Δ_{exp}), allow the estimation of g_m utilising the approach of Evans et al. [49] that was subsequently modified by Lloyd et al. [50]:

$$g_m = \frac{(b - b_s - \alpha_1) * {}^{A}/_{C_a}}{(\Delta_{exp} - \Delta_{obs}) - (f\,\Gamma*/C_a)} \quad (3)$$

Where: b is the discrimination associated with carboxylation reactions, taken to be 27.5‰; b_s is the fractionation occurring when CO_2 enters solutions (1.1‰ at 25°C); a_1 is the fractionation during diffusion in water (0.7‰); f is the fractionation associated with photorespiration, and; p_a is the partial pressure of CO_2 in air. Two g_m calculations were performed and compared, since two values were taken into account for the fractionation factor f: 7‰ [50] and 0‰ [48,51]. The closest relationship between g_m estimates ($R^2 = 0.911$) using the variable J and $\Delta^{13}C$ of recently synthesized sugars methods in all varieties was found when using a f value of 0‰.

Harvesting

Plants were harvested at physiological maturity by sampling 1 m of the 2 central rows of each plot. Data collected included biomass, plant height, anthesis date, tiller number, straw production, grain yield, and harvest index (*HI*).

Statistical analysis

Analyses of variance were conducted using SAS to generate least squares means for each entry. Leaf gas exchange data were tested using a simple factorial ANOVA (three-way maximum interactions) and, where appropriate, the treatment means of leaf properties and gas exchange parameters were compared using a Tukey *post-hoc* test.

Results

Drought stress induced reductions in yield, A and rates of gas-exchange across all the *O. sativa* varieties analysed in this study. The duration of time required for 50% of plants to develop flowers (flower days) did not correlate with yield in well-watered plants, but those varieties with a shorter flowering day period exhibited enhanced yield under drought stress (Fig. 1a). Varieties with shorter flowering days allocated a higher proportion of their total biomass to yield (harvest index: *HI*) under control and drought treatments (Fig. 1b). Positive allometric correlations between plant height and both yield (Fig. 1c) and *HI* (Fig. 1d) were observed, both under watered or drought conditions.

The carbon isotopic composition of the leaves of the *O. sativa* varieties became increasingly enriched in the heavier ^{13}C isotope following water-stress, with the exception of IR64. Changes in the carbon isotopic composition of soluble sugars induced by the drought treatment were less consistent, with three varieties exhibiting increases and four decreases in the ratio of stable carbon isotopes (Table 1). Despite the lack of *genotype x water treatment* interaction, the varieties yielded differences in the 'long-term' isotopic signal of the bulk leaf material; some genotypes showed wide variation (up to $-1.5‰$ in pellets of Van) while others were more negligible (e.g. 0.2‰ in pellets of IR64). Drought caused also different variation in $\Delta^{13}C$ of soluble sugars: up to $-3.5‰$ in PS80 but 0.2‰ in IR71. Under both well-watered and rainfed conditions, soluble sugars showed generally lower values of carbon isotopic discrimination when compared with the average $\Delta^{13}C$ values of the flag leaf pellets. Significant differences were observed in the carbon isotopic discrimination values of of leaf pellets between varieties, however this was not replicated in the analysis of soluble sugars (Table 1).

A strong positive correlation was observed between the two estimates of g_m in the *O. sativa* varieties grown under control and drought conditions ($R^2 = 0.930$; $P<0.01$) (Fig. 2), excluding IR64 and PS80. Indeed, the lowland varieties IR64 and PS80 grown under well-watered conditions exhibited high values of g_m derived from the application of the isotopic methodology. It was therefore decided to utilise g_m measurements resulting from the variable J method in the subsequent analyses [an extended discussion of the contrasting merits of the approaches is given in 7]. Rates of A, g_s and g_m all showed consistent declines under water-stress conditions (Fig. 3). IR55 showed the highest yield and g_m under well-watered conditions. Variety IR71 exhibited the highest general

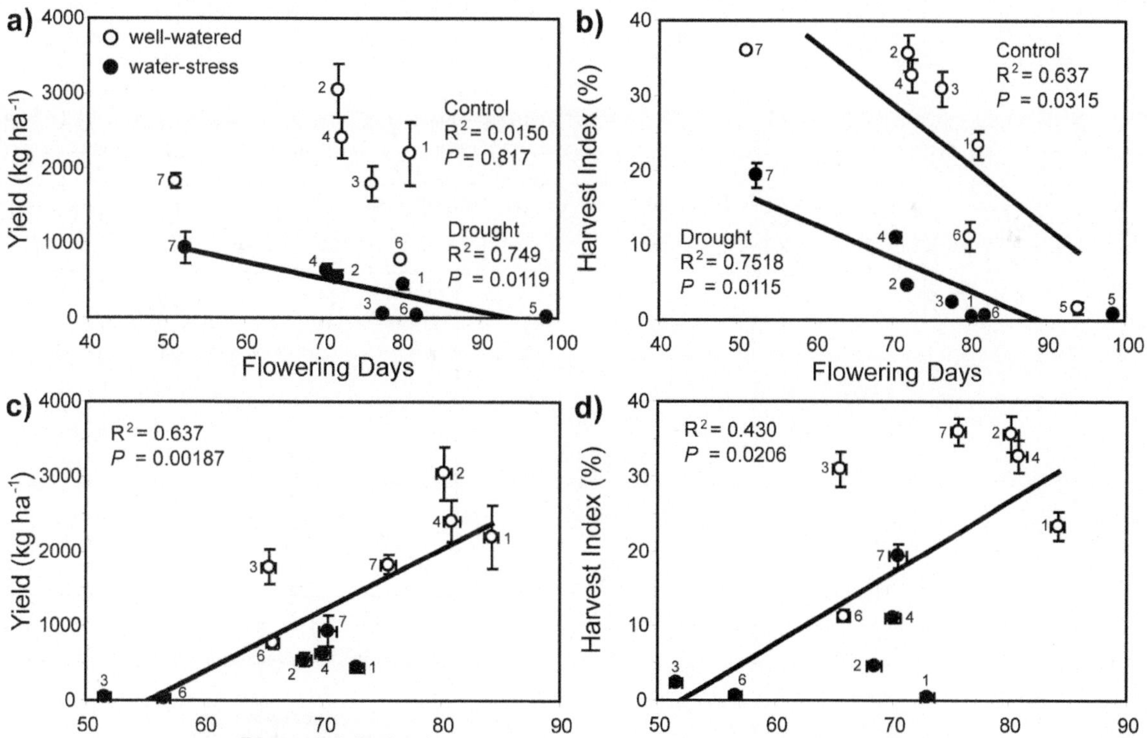

Figure 1. The performance of seven rice varieties under well-watered (open symbols) and drought conditions (closed symbols): a) relationship between yield and flowering days (period of time for 50% of plants to develop flowers) under full-water (linear regression: $R^2=0.0150$; $F_{1,4}=0.0609$; $P=0.817$) and drought (linear regression: $R^2=0.749$; $F_{1,5}=14.905$; $P=0.0119$); b) allometric relationship between yield and plant height (linear regression: $R^2=0.637$; $F_{1,10}=17.536$; $P=0.00187$); c) relationship between harvest index (*HI*: dry weight of grain relative to dry total plant biomass) and flowering days under full-water (linear regression: $R^2=0.637$; $F_{1,5}=8.760$; $P=0.0315$) and drought (linear regression: $R^2=0.752$; $F_{1,5}=15.144$; $P=0.0115$), and; d) relationship between *HI* and plant height (linear regression: $R^2=0.430$; $F_{1,10}=7.548$; $P=0.0206$). Error bars indicate one standard error either side of the mean. Numbers next to data points indicate *Oryza sativa* variety: 1 = Apo; 2 = IR55; 3 = IR64; 4 = IR71; 5 = Moro; 6 = PSBRc80; 7 = Van.

Table 1. Carbon isotope discrimination (‰) of the flag leaf pellet (bulk) and soluble sugars of seven rice cultivars grown under drought and well-watered conditions in ‰.

Genotype	Habitat	Bulk Leaf $\Delta^{13}C$ (‰)			Sugars $\Delta^{13}C$ (‰)		
		Well-watered	Drought	Mean	Well-watered	Drought	Mean
Apo	U	21.6±0.9	20.6±0.3	21.0 ab	18.3±1.3	18.7±0.6	18.5 a
IR55	U	21.4±0.5	20.8±0.0	21.1 a	18.2±0.5	16.2±0.9	17.2 a
IR64	L	20.9±0.2	21.1±0.5	21.0 ab	18.5±0.6	16.6±0.8	17.5 a
IR71	U	21.1±0.2	20.7±0.5	20.9 ab	17.3±1.3	17.5±0.7	17.4 a
Moro	U	19.9±0.4	19.7±0.2	19.8 b	18.3±0.6	19.5±0.2	18.9 a
PS80	L	21.4±0.4	21.2±0.4	21.2 a	19.3±0.3	15.8±0.2	17.2 a
Van	U	21.5±0.3	20.0±0.5	20.7 b	18.1±1.5	15.8±0.4	17.0 a
Mean		21.1 a	20.6 b		18.2 a	17.2 b	

Values indicate the mean of six plants; ± indicates the standard error of mean; different letters indicate significant differences (P≤0.05) among means according to an ANOVA.

values of A, g_s and g_m under both well-watered and drought conditions, while Apo and PS80 showed the lowest values of these photosynthesis and gas-exchange parameters under control and water-stress. The A/g_s ratio of the *O. sativa* varieties either increased (IR71; Van; IR55) or remained constant under drought stress in all varieties with the exception of PS80, where a 42.6% reduction was observed (Fig. 3d). The ratio of $[CO_2]$ within the substomatal pore (C_i) to the external atmospheric $[CO_2]$ (C_a) was largely unaffected by water-stress (Fig. 4a). In contrast, the ratio of the $[CO_2]$ within the chloroplast envelope (C_c) to C_a was diminished in all *O. sativa* varieties during drought treatment (Fig. 4b), indicative of transport limitations of CO_2 in the mesophyll for photosynthesis. *Oryza sativa* variety IR71 showed the smallest proportional reduction of 16.8% in the C_c/C_a ratio, while Apo exhibited the largest reduction of 33.8%.

Yield and A exhibited significant positive correlations with g_s (Fig. 5a and 5b). However, the positive correlations between yield and A with g_m (Fig. 5c and 5d) were stronger than those observed with g_s or total conductance (g_{tot}); suggesting that g_m is the major determinant of yield and A under both well-watered and water-stressed conditions. The yield of the *O. sativa* varieties was strongly related to A under both control and drought treatments (Fig. 5g). In order to gauge the response of these diffusive limitations to CO_2 transport and their effect on yield and A, the proportional change (Δ) in response to drought was calculated. The proportional reduction in yield (Δyield) was not related to Δg_s, Δg_m or Δg_{tot} (Fig. 6a; 6b; 6c). However, ΔA was significantly correlated to Δg_m and Δg_{tot}. IR71 exhibited the highest levels of A and g_m (Fig. 3a and 3c), but also the lowest proportional reduction in g_m and A (Fig. 6d). The highest yields observed under control conditions were found in IR55 (3039.6±359.1 kg ha^{-1}) and IR71 (2407.4±284.4 kg ha^{-1}); under water-stress the yield of IR55 fell to 552.7±80.3 kg ha^{-1} and IR71 to 640.3±93.0 kg ha^{-1}. The lowest Δyield reduction under drought conditions was observed in variety Van (Fig. 6), which also produced the highest yield under

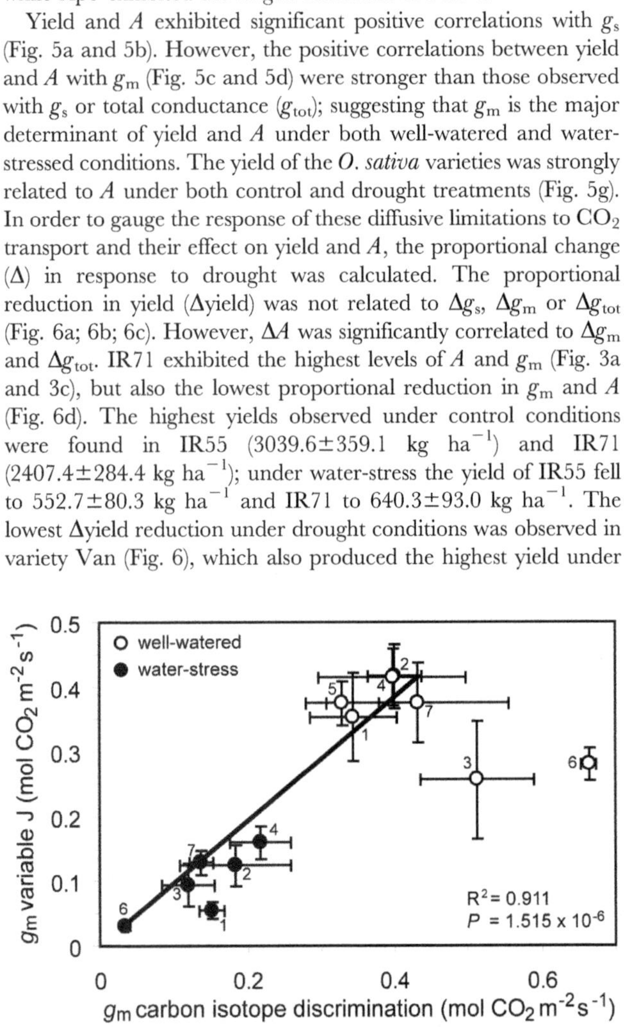

Figure 2. Comparison between the estimates of mesophyll conductance of CO_2 (g_m) obtained by applying two independent methods: the variable J method and the 'δ13C of recently synthesised sugars' method (linear regression: $R^2=0.932$; $F_{1,9}=122.750$; $P=1.515\times10^{-6}$). Each data point represents the average value of three observations based upon the Δ13C of recently synthesised sugars and six to fourteen gas-exchange measurements utilising the variable J method. Error bars as in Figure 1. The regression line excludes the two data points on the right of the graph (IR64 and PS80) with anomalously high g_m derived from the δ13C of recently synthesised sugars. Numbers next to data points indicate *Oryza sativa* variety as in Figure 1.

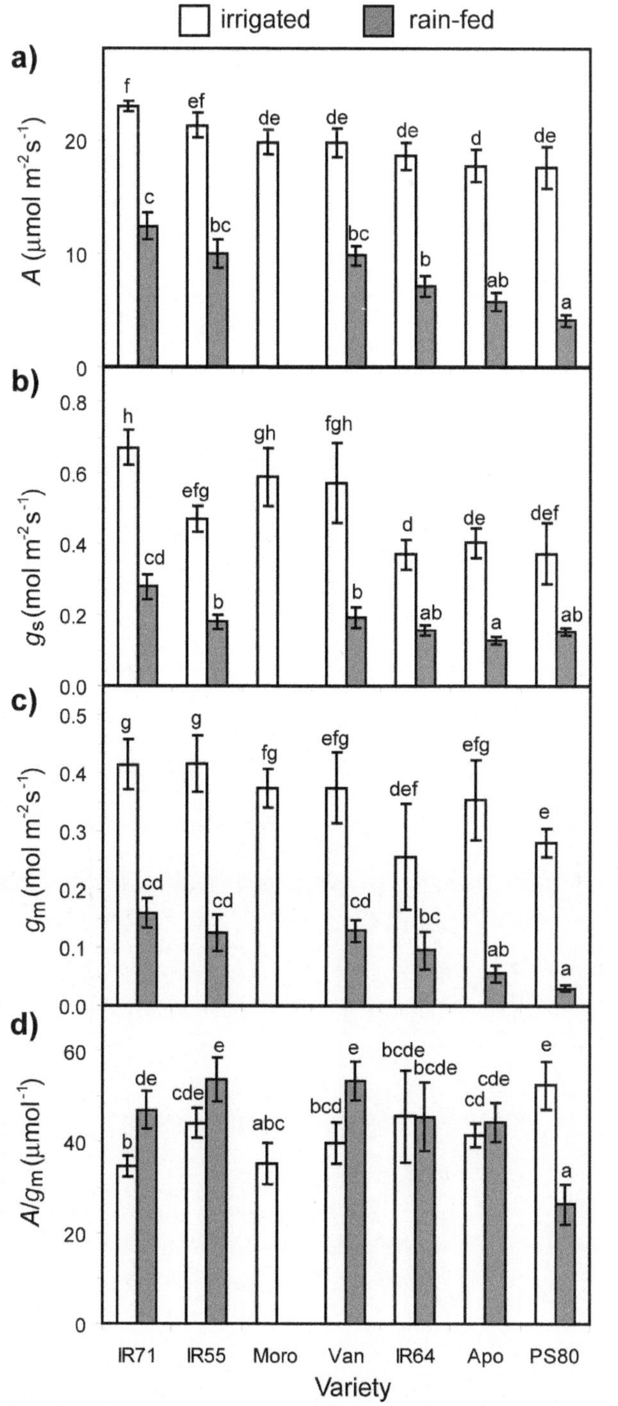

Figure 3. Measurements of (a) photosynthesis rate (*A*), (b) stomatal conductance (g_s), (c) mesophyll conductance (g_m), and (d) intrinsic transpiration efficiency (A/g_s) in control and water-stressed leaves of the seven *Oryza sativa* genotypes. The measurements were made on the flag leaf in saturating PPFD (1400 µmol $m^{-2}s^{-1}$), with relative humidity ranging between 45–55%, and a leaf temperature of 30°C. Data are means of 4 to 7 plants per treatment. Error bars as in Figure 1. Different letters denote significant differences among means derived using a factorial ANOVA and Tukey *post-hoc* test.

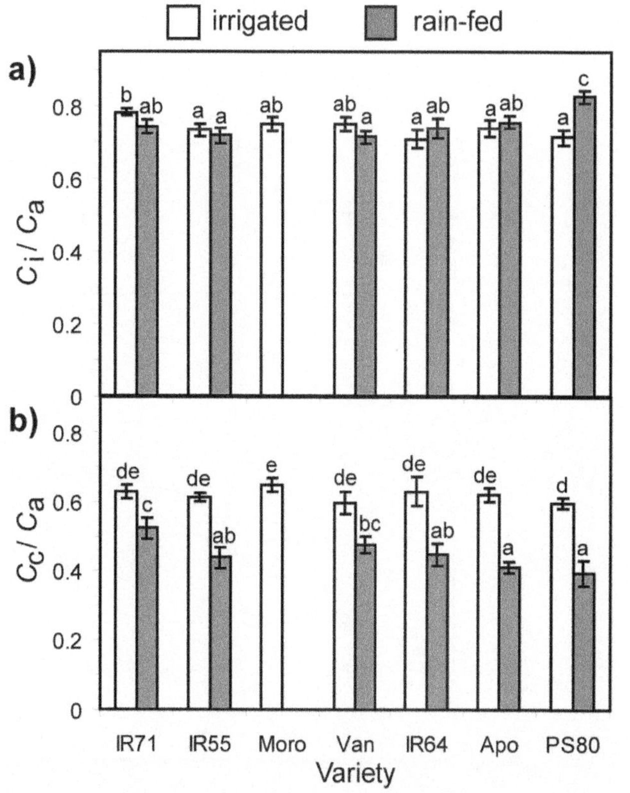

Figure 4. Measurements of (a) the intercellular [CO_2] (C_i) to the ambient [CO_2] (C_a) ratio (C_i/C_a), and (b) the chloroplastic [CO_2] (C_c) to the ambient [CO_2] ratio (C_c/C_a) in control and water-stressed leaves of the seven *Oryza sativa* genotypes. The measurements were made on the flag leaf in saturating PPFD (1400 µmol $m^{-2}s^{-1}$), with relative humidity ranging between 45–55%, and a leaf temperature of 30°C. Data are means of 4 to 7 plants per treatment. Error bars as in Figure 1. Different letters denote significant differences among means derived using a factorial ANOVA and Tukey *post-hoc* test.

water-stress (935.3±206.0 kg ha^{-1}). IR71 exhibited the second lowest Δyield and produced second largest yield during water-stress of the varieties analysed (Fig. 6g). The A/g_s ratio was not significantly related to either yield or *HI* under both well-watered or drought conditions (Fig. 7a and 7c). However, the g_m:g_s ratio was significantly correlated with both yield and *HI* (Fig. 7b and 7d), suggesting that those *O. sativa* varieties with greater g_m/g_s ratio will exhibit enhanced productivity under both well-watered and water-stressed conditions.

Discussion

The seven *O. sativa* varieties exhibited a diverse range of morphological, isotopic, photosynthetic and leaf gas-exchange responses to drought stress. A number of factors determined the yield of the *O. sativa* cultivars analysed in this study. Despite water deficit being instigated at the same developmental stage for all varieties, phenology had an effect on yield under drought as rapidly developing plants maintained yield to a greater extent than slower developers (Fig. 1a) [52]. Despite the type and phenolgical timing of the drought stress being different to that utilised in the investigation of Centritto et al. [7], the effect on *A* and yield observed in the two studies are broadly consistent. Those plants that were able to accumulate biomass rapidly produced higher

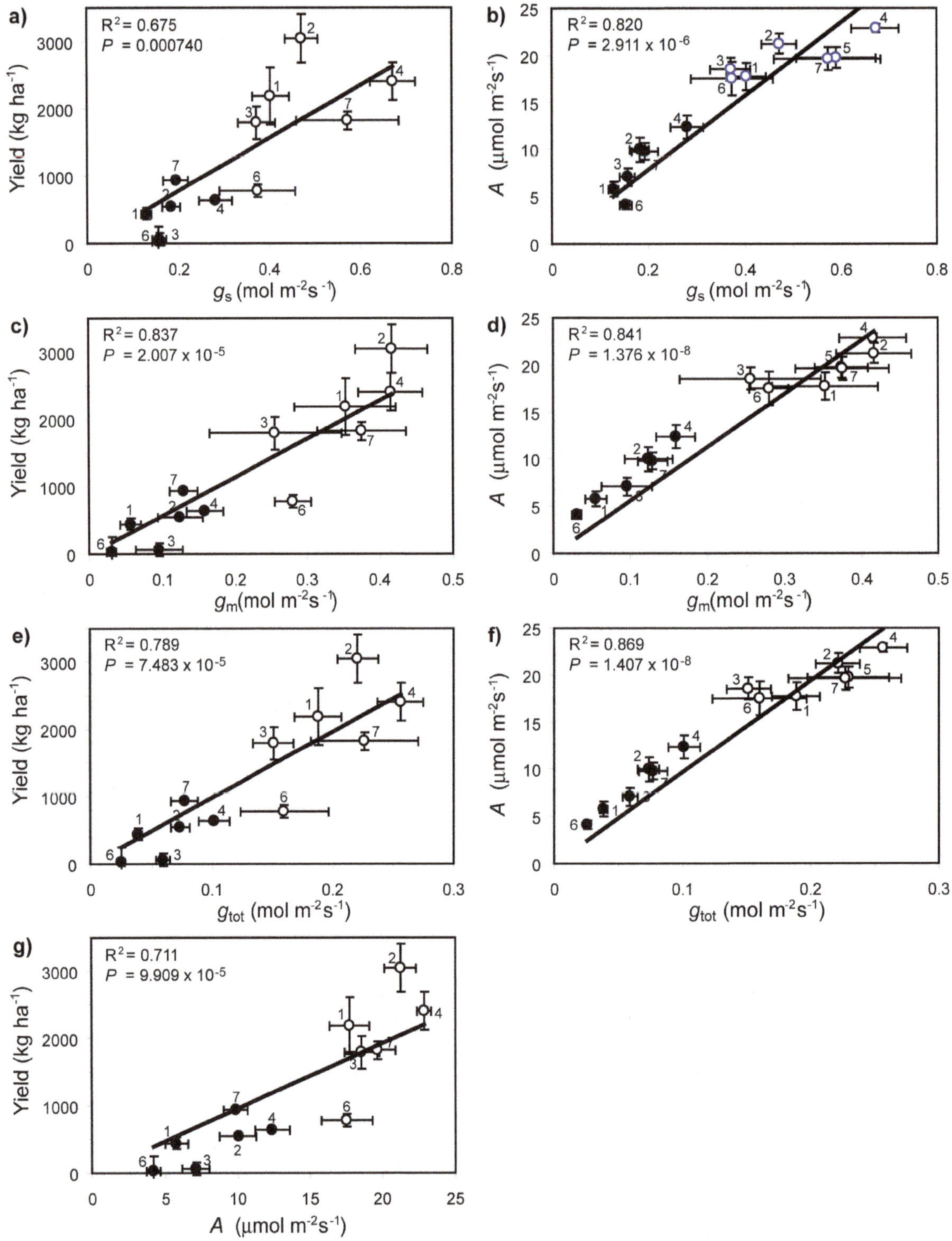

Figure 5. Interaction of diffusive conductance parameters to CO_2 uptake with yield and photosynthesis (*A*) under well-watered (open symbols) and drought conditions (closed symbols): a) relationship between yield and stomatal conductance (g_s) (linear regression: $R^2=0.696$; $F_{1,10}=22.900$; $P=0.000740$); b) relationship between *A* and g_s (linear regression: $R^2=0.873$; $F_{1,11}=75.721$; $P=2.911\times10^{-6}$); c) relationship between yield and mesophyll conductance (g_m) (linear regression: $R^2=0.850$; $F_{1,10}=56.611$; $P=2.911\times10^{-5}$); d) relationship between *A* and g_m (linear regression: $R^2=0.952$; $F_{1,11}=217.071$; $P=1.376\times10^{-8}$); e) relationship between yield and total conductance (g_{tot}) (linear regression: $R^2=0.806$; $F_{1,10}=41.418$; $P=7.483\times10^{-5}$); f) relationship between *A* and g_{tot} (linear regression: $R^2=0.952$; $F_{1,11}=216.173$; $P=1.407\times10^{-8}$), and; g) relationship between yield and *A* (linear regression: $R^2=0.795$; $F_{1,10}=38.665$; $P=9.909\times10^{-5}$). Error bars as in Figure 1. Numbers next to data points indicate *Oryza sativa* variety as in Figure 1.

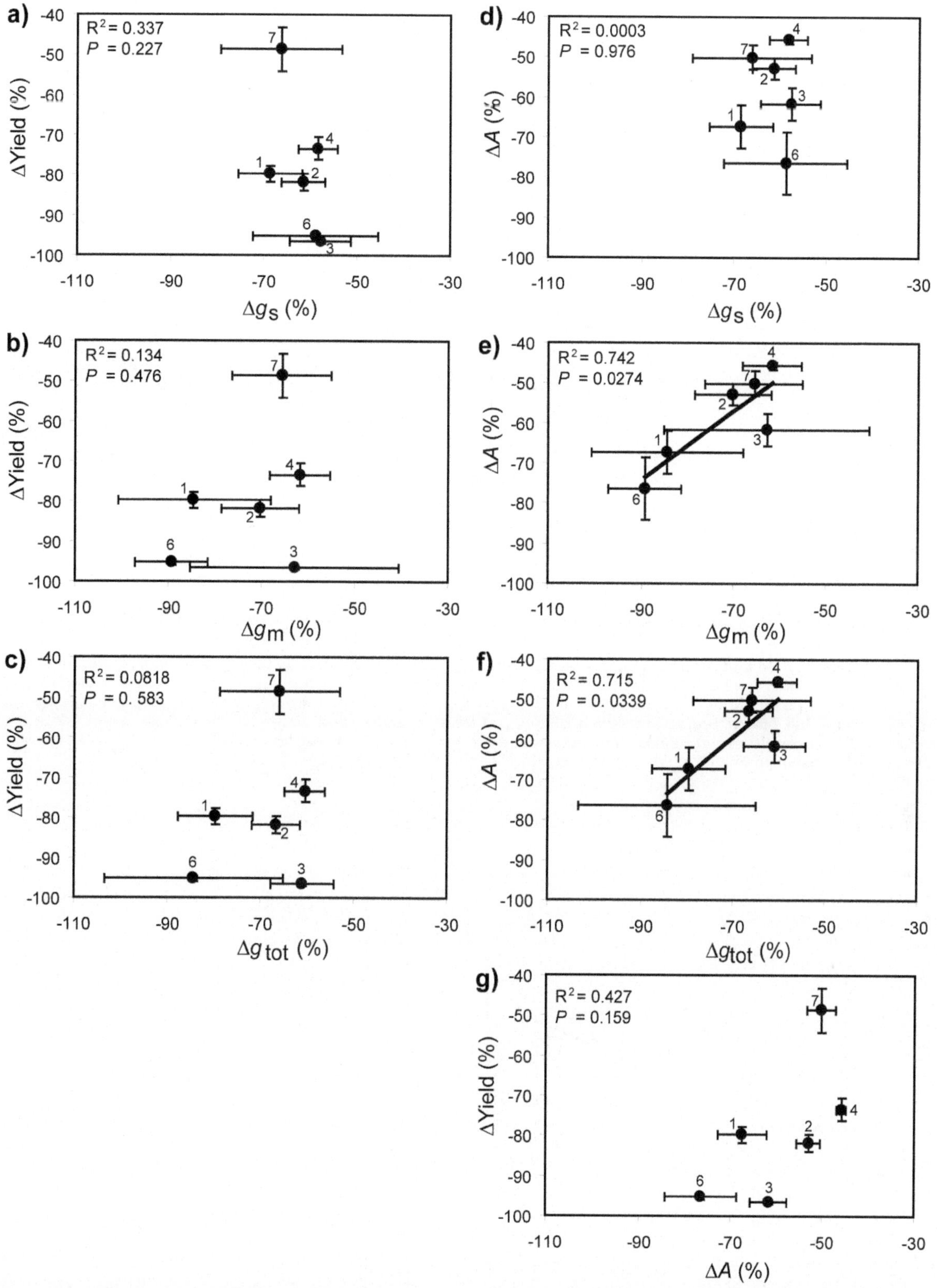

Figure 6. Changes in yield and photosynthesis in relation to modification of diffusive resistances to CO_2 uptake following water-stress. Those varieties that experienced smaller reductions in parameters were more tolerant of drought. a) relationship between Δyield and Δg_s (linear regression: $R^2=0.337$; $F_{1,4}=2.032$; $P=0.227$); b) relationship between Δyield and Δg_m (linear regression: $R^2=0.134$; $F_{1,4}=0.618$; $P=0.476$); c) relationship between Δyield and Δg_{tot} (linear regression: $R^2=0.0818$; $F_{1,4}=0.356$; $P=0.583$); d) relationship between ΔA and Δg_s (linear regression: $R^2=0.0003$; $F_{1,4}=0.00106$; $P=0.976$); e) relationship between ΔA and Δg_m (linear regression: $R^2=0.742$; $F_{1,4}=11.527$; $P=0.0274$); f) relationship between ΔA and Δg_{tot} (linear regression: $R^2=0.715$; $F_{1,4}=10.042$; $P=0.0339$), and; g) relationship between Δyield and ΔA (linear regression: $R^2=0.427$; $F_{1,4}=2.979$; $P=0.159$). Error bars as in Figure 1. Numbers next to data points indicate *Oryza sativa* variety as in Figure 1.

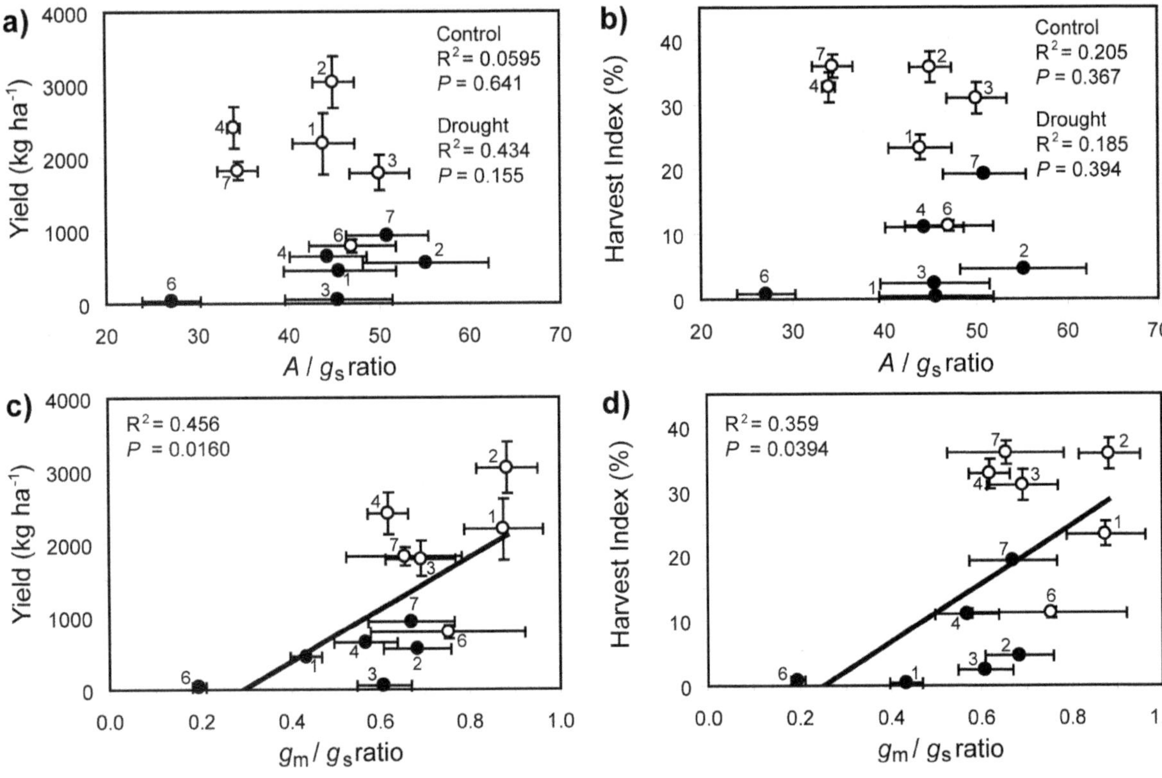

Figure 7. Interaction of yield and *A* with transpiration efficiency (*A*/g_s) and the ratio of g_m to g_s in well-watered (open symbols) and drought conditions (closed symbols): a) relationship between yield and *A*/g_s under full (linear regression: $R^2=0.0595$; $F_{1,4}=0.253$; $P=0.641$) and water-stressed (linear regression: $R^2=0.434$; $F_{1,4}=3.072$; $P=0.155$) conditions; b) relationship between harvest index (*HI*) and *A*/g_s under full (linear regression: $R^2=0.205$; $F_{1,4}=1.032$; $P=0.367$) and water-stressed (linear regression: $R^2=0.185$; $F_{1,4}=0.909$; $P=0.394$) conditions; c) relationship between yield and g_m:g_s (linear regression: $R^2=0.456$; $F_{1,10}=8.379$; $P=0.0160$), and; d) relationship between *HI* and g_m:g_s (linear regression: $R^2=0.359$; $F_{1,10}=5.610$; $P=0.0394$). Error bars as in Figure 1. Numbers next to data points indicate *Oryza sativa* variety as in Figure 1.

grain yields than those that accumulated biomass less efficiently under both well-watered and drought conditions. This may be due to allometric relationships between biomass and yield [53], or the positive influence of increased photosynthetic area on 'whole-plant' carbon uptake [54]. Leaf-level photosynthetic rates were also closely related to yield in all varieties under both well-watered and drought conditions (Fig. 5g). Photosynthesis rates were in turn determined by diffusive limitations to CO_2-uptake (Fig. 5b; 5d; 5f). The most significant of these resistances to CO_2-uptake in *O. sativa* occurs at the interface between the internal sub-stomatal air-space and the mesophyll layer [15,55]. Consistent with other studies, g_m in *O. sativa* correlated most closely with *A* and yield (Fig. 5d) [7,9], and represents a critical 'rate-limiting' step in the determination of grain productivity and assessment of the photosynthetic response to drought stress.

The 'bulk' carbon isotope discrimination values of the *O. sativa* varieties decreased following drought treatment; indicative of water-deficits over the life-span of the flag-leaf reducing discrimination against the heavier ^{13}C isotope [56,57]. However, this pattern of reduced carbon isotope discrimination is not apparent in the carbon isotopic composition of recently synthesised sugars (Table 1). These findings may reflect 'shorter-term' variations in *WUE* in the plants experiencing drought stress, a modification of the photosynthetic pathway affecting sugar metabolism [28] or the breakdown of previously stored carbohydrates with a different isotopic signature into more simple sugars over the course of the stress treatment [58,59]. Differences in the diffusional uptake of CO_2 likely affected carbon isotopic composition [60]; however, this was not apparent from our g_s, g_m (variable J) and g_{tot} datasets (Fig. 5), suggesting that other factors affected carbon isotopic discrimination and possibly the effectiveness of the Δ13C of recently synthesised sugars method of estimating g_m in varieties IR64 and PS80 under control conditions (Fig. 2). It is noteworthy that under water-deficit conditions, g_m estimates based on the variable J method closely correlated with those derived from the Δ13C of recently synthesised sugars (Fig. 2). These findings do not support the suggestion that reductions in C_i may affect estimation of the total photosynthetic electron transport rate associated with enhanced photorespiration, thus rendering the variable J method unreliable in the investigation of g_m in plant responses to environmental stress [34].

Selective breeding has improved the productivity of *O. sativa* varieties through improved growth form and the increased proportional allocation of biomass as grain [4]; yet these increases in yield have been constrained by the relatively low leaf-level photosynthetic rate of *O. sativa* in comparison to other C3 crops [5,61]. The results of this study indicate that under well-watered and water-deficit conditions, *A* is an important determinant of yield in *O. sativa* and is largely related to the rate of transport of CO_2 across the mesophyll (Fig. 5d). Further improvements in *O. sativa* grain yield may likely be achieved through the enhancement of g_m to CO_2 inducing enhanced *A* [6,9,16,62]. Those *O. sativa* varieties that showed the greatest photosynthetic performance and yield under control conditions were not necessarily the most effective varieties under drought conditions. Varieties that maintained transport of CO_2 during water-stress sustained A and yield to a greater extent than varieties that exhibited more pronounced diffusive limitations to photosynthetic CO_2-uptake (Fig. 5). Drought stress induced a relatively consistent mean reduction in g_s in the *O. sativa* varieties (Fig. 3), yet the C_i:C_a ratio was unaffected (Fig. 4), indicating that g_s was not limiting to A under water-deficit, or that the gas-exchange data were possibly biased by the occurrence of non-uniform stomatal closure over the leaf lamina and consequent overestimation of C_i [27,63,64]. However, the C_i/C_a ratio in the *O. sativa* varieties was reduced by 16.8 to 33.8% after drought, consistent with the impairment of A through mesophyll diffusive limitations [7,11,12,65,66]. Mesophyll conductance to CO_2 declined during the drought treatment, constraining A and growth (Fig. 5; 6). However, significant variability in the extent of reductions in g_m were observed in the *O. sativa* cultivars ranging from −61.7% (IR71) to −89.3% (PS80). Furthermore, those varieties that showed the smallest reductions in g_m also exhibited greater A; demonstrating that g_m largely determines A in *O. sativa* under differing degrees of water-deficit [7,9]. These differences in the ability to sustain g_m values under drought stress are likely related to the maintenance of biochemical function in the mesophyll such as the activity of cooporin and carbonic anhydrase [eg. 25,67].

The conductance of the mesophyll layer to the transport of CO_2 determined A and yield in the seven *O. sativa* varieties studied. Under conditions of water-stress it has been suggested that a low g_s (to reduce transpirative water-loss) and a high g_m (to maintain the uptake of CO_2) are beneficial [11,68]. However, in the case of *O. sativa*, yield and HI were only weakly related to the ratio of g_m:g_s (R^2<0.460) (Fig. 7). The reduction in g_s under water-deficit is consistent with active stomatal control in the optimisation of WUE [69,70]; nevertheless, the lack of a strong correlation to the g_m:g_s ratio may be indicative of the lack of a stomatal limitation to A under drought stress in the *O. sativa* varieties apparent in the C_i:C_a ratios (Fig. 4a). The photosynthetic and yield responses of *O. sativa* under well-watered and drought conditions were more dependent upon g_m than g_s (Fig. 5b and 5d). This is consistent with earlier studies reporting that modification and improvement of g_m is more effective in the enhancement of *O. sativa* productivity [16] and likely drought tolerance [7,9]. The strong dependence of A and yield on g_m is also evident in the lack of a relationship between A/g_s and yield or HI (Fig. 7a and 7c). This may suggest that stomatal control is either less effective [70,71], or plays a reduced role, in respect to mesophyll limitations to CO_2 in a drought intolerant crop such as *O. sativa* in comparison to crops adapted to more water-limited environments [72].

The disparity in A values observed between the cultivars raises the possibility of crossing different varieties to enhance leaf-level A and yield in subsequent generations [6,8,16]. Selection of drought tolerance on the basis of the ability of a particular genotype to maintain g_m function under drought stress would be viable [11,73]. However, genetic sequencing and identification of quantitative trait loci permit the possibility of extracting genetic material from a *O. sativa* cultivar with a valuable characteristic (such as high g_m and/or stable g_m under water deficit) and in the creation of a new variety with a suite of desired attributes [17,19,74–76]. For example, from the results of this study it would be desirable to combine the high g_m and yield of IR55 under well-watered conditions with the capacity to sustain g_m under water-stress exhibited by IR71 and water-stressed yield of Van (Fig. 3). In this context it may be possible to phenotypically screen a large number of *O. sativa* genotypes to identify quantitative trait loci responsible for g_m values and g_m response to drought. Those varieties developed to be resistant to drought may be suited to more marginal land-types or in scenarios of insufficient water availability. Nonetheless, given the sporadic and unpredictable nature of sustained and persistent drought events it is unlikely that large numbers of farmers would select a drought tolerant variety such as Van over a more productive cultivar such as IR55, and thus limit their harvest to accommodate the risk of interruption to water supplies [73,77]. An ideal *O. sativa* variety would therefore possess both values of g_m capable of sustaining high A and the ability to maintain g_m in the event of water-deficits [8,78]. However, the results of this study suggest that those attributes may be mutually exclusive in the species analysed. Furthermore, no single parameter analysed in this study exclusively accounted for yield under either well-watered or drought conditions; therefore attempts to improve the productivity of *O. sativa* cultivars may also involve selection for a suite of characteristics that improve productivity under specific conditions of water availability. Nonetheless, future efforts to improve *O. sativa* productivity should focus on biochemical [eg. 25] and physical modification [eg. 16] of the interface between the mesophyll layer and sub-stomatal air-space to enhance CO_2 uptake and A.

Conclusions

Physiological analysis of *O. sativa* has great potential in identifying those photosynthetic and leaf gas-exchange characteristics that confer attributes such as high yield and tolerance to drought. The analysis of the seven *O. sativa* varieties indicates that under both well-watered and drought stressed conditions yield and A are determined by diffusional constraints to CO_2 uptake and not by biochemical limitations. The varieties showed significant variation in leaf-level photosynthetic rates, symptomatic of the photosynthetic constraints to *O. sativa* production that have previously been recorded. Selection of those *O. sativa* genotypes that exhibit higher rates of leaf-level A would likely be beneficial for grain yields. Photosynthesis was most closely related to g_m, and the tolerance of the *O. sativa* varieties to drought was largely determined by the ability to maintain g_m values under water-deficits. However, the highest g_m and yield under well-watered conditions (IR55) and the most pronounced ability to sustain g_m under water-deficit (IR71) were observed in different varieties. This may suggest that the desired attributes for enhanced yield during high water availability and the ability to maintain g_m, and thus sustain grain production under water-deficits, may be mutually exclusive. Nonetheless, the results of studies such as this could make it possible through selective breeding, or the identification of a particular genetic expression associated with a desired attribute, to increase leaf-level photosynthetic rates and drought tolerance in the same variety of *O. sativa*. These findings indicate that the interface between the internal sub-stomatal air-space and the mesophyll layer offers the greatest potential in terms of possible modification to enhance the performance of *O. sativa* under optimal and stressed growth conditions in the future.

Acknowledgments

The following are thanked for scientific discussion and technical assistance: L. Spaccino for IRMS analyses (CNR-IBAF) and R. Torres G. Dimayuga and the drought physiology team at IRRI for technical support (IRRI, Philippines). The comments of two anonymous referees significantly improved this manuscript.

Author Contributions

Conceived and designed the experiments: MC ML RS. Performed the experiments: MC ML MCM. Analyzed the data: MH MC. Contributed reagents/materials/analysis tools: MC ML RS. Wrote the paper: MH MC ML.

References

1. Pandey S, Bhandari H, Ding S, Prapertchob P, Sharan R, et al. (2007) Coping with drought in rice farming in Asia: insights from a cross-country comparative study. Agricultural Economics 37: 213–224.
2. Serraj R, McNally KL, Slamet-Loedin I, Kohli A, Haefele SM, et al. (2011) Drought resistance improvement in rice: an integrated genetic and resource management strategy. Plant Production Science 14: 1–14.
3. Parent B, Suard B, Serraj R, Tardieu F (2010) Rice leaf growth and water potential are resilient to evaporative demand and soil water deficit once the effects of root system are neutralized. Plant, Cell & Environment 33: 1256–1267.
4. Peng S, Cassman KG, Virmani SS, Sheehy J, Khush GS (1999) Yield potential trends of tropical rice since the release of IR8 and the challenge of increasing rice yield potential. Crop Science 39: 1552–1559.
5. Hubbart S, Peng S, Horton P, Chen Y, Murchie EH (2007) Trends in leaf photosynthesis in historical rice varieties developed in the Philippines since 1966. Journal of Experimental Botany 58: 3429–3438.
6. Gu J, Yin X, Stomph TJ, Struik PC (2014) Can exploiting natural genetic variation in leaf photosynthesis contribute to increasing rice productivity? A simulation analysis. Plant, cell & environment 37: 22–34.
7. Centritto M, Lauteri M, Monteverdi MC, Serraj R (2009) Leaf gas exchange, carbon isotope discrimination, and grain yield in contrasting rice genotypes subjected to water deficits during the reproductive stage. Journal of Experimental Botany 60: 2325–2339.
8. Long SP (2014) We need winners in the race to increase photosynthesis in rice, whether from conventional breeding, biotechnology or both. Plant, Cell & Environment 37: 19–21.
9. Gu J, Yin X, Stomph T-J, Wang H, Struik PC (2012) Physiological basis of genetic variation in leaf photosynthesis among rice (*Oryza sativa* L.) introgression lines under drought and well-watered conditions. Journal of Experimental Botany 63: 5137–5153.
10. Niinemets Ü, Diaz-Espejo A, Flexas J, Galmes J, Warren CR (2009) Role of mesophyll diffusion conductance in constraining potential photosynthetic productivity in the field. Journal of Experimental Botany 60: 2249–2270.
11. Flexas J, Ribas-Carbó M, Diaz-Espejo A, Galmēs J, Medrano H (2008) Mesophyll conductance to CO_2: current knowledge and future prospects. Plant, Cell & Environment 31: 602–621.
12. Centritto M, Loreto F, Chartzoulakis K (2003) The use of low [CO_2] to estimate diffusional and non-diffusional limitations of photosynthetic capacity of salt-stressed olive saplings. Plant, Cell & Environment 26: 585–594.
13. Pallozzi E, Fortunati A, Marino G, Loreto F, Agati G, et al. (2013) BVOC emission from *Populus* × *canadensis* saplings in response to acute UV-A radiation. Physiologia Plantarum 148: 51–61.
14. Hu L, Wang Z, Huang B (2010) Diffusion limitations and metabolic factors associated with inhibition and recovery of photosynthesis from drought stress in a C3 perennial grass species. Physiologia Plantarum 139: 93–106.
15. Scafaro AP, Von Caemmerer S, Evans JR, Atwell BJ (2011) Temperature response of mesophyll conductance in cultivated and wild *Oryza* species with contrasting mesophyll cell wall thickness. Plant Cell and Environment 34: 1999–2008.
16. Adachi S, Nakae T, Uchida M, Soda K, Takai T, et al. (2013) The mesophyll anatomy enhancing CO_2 diffusion is a key trait for improving rice photosynthesis. Journal of Experimental Botany 64: 1061–1072.
17. Manickavelu A, Nadarajan N, Ganesh SK, Gnanamalar RP, Chandra Babu R (2006) Drought tolerance in rice: morphological and molecular genetic consideration. Plant Growth Regulation 50: 121–138.
18. Huang X-Y, Chao D-Y, Gao J-P, Zhu M-Z, Shi M, et al. (2009) A previously unknown zinc finger protein, DST, regulates drought and salt tolerance in rice via stomatal aperture control. Genes & Development 23: 1805–1817.
19. Hu SP, Zhou Y, Zhang L, Zhu XD, Li L, et al. (2009) Correlation and quantitative trait loci analyses of total chlorophyll content and photosynthetic rate of rice (*Oryza sativa*) under water stress and well-watered conditions. Journal of Integrative Plant Biology 51: 879–888.
20. Gu J, Yin X, Struik PC, Stomph TJ, Wang H (2012) Using chromosome introgression lines to map quantitative trait loci for photosynthesis parameters in rice (*Oryza sativa* L.) leaves under drought and well-watered field conditions. Journal of Experimental Botany 63: 455–469.
21. Flexas J, Niinemets Ü, Gallé A, Barbour M, Centritto M, et al. (2013) Diffusional conductances to CO_2 as a target for increasing photosynthesis and photosynthetic water-use efficiency. Photosynthesis Research 117: 45–59.
22. Farquhar GD, Caemmerer S, Berry JA (1980) A biochemical model of photosynthetic CO_2 assimilation in leaves of C3 species. Planta 149: 78–90.
23. Giuliani R, Koteyeva N, Voznesenskaya E, Evans MA, Cousins AB, et al. (2013) Coordination of leaf photosynthesis, transpiration, and structural traits in rice and wild relatives (Genus *Oryza*). Plant Physiology 162: 1632–1651.
24. Kaldenhoff R (2012) Mechanisms underlying CO_2 diffusion in leaves. Current Opinion in Plant Biology 15: 276–281.
25. Hanba YT, Shibasaka M, Hayashi Y, Hayakawa T, Kasamo K, et al. (2004) Overexpression of the barley aquaporin HvPIP2;1 increases internal CO_2 conductance and CO_2 assimilation in the leaves of transgenic rice plants. Plant and Cell Physiology 45: 521–529.
26. Tyerman SD, Niemietz CM, Bramley H (2002) Plant aquaporins: multifunctional water and solute channels with expanding roles. Plant, Cell & Environment 25: 173–194.
27. Loreto F, Centritto M (2008) Leaf carbon assimilation in a water-limited world. Plant Biosystems 142: 154–161.
28. Chaves M, Flexas J, Pinheiro C (2009) Photosynthesis under drought and salt stress: regulation mechanisms from whole plant to cell. Annals of Botany 103: 551–560.
29. Cornic G (2000) Drought stress inhibits photosynthesis by decreasing stomatal aperture – not by affecting ATP synthesis. Trends in Plant Science 5: 187–188.
30. Pinheiro C, Chaves MM (2011) Photosynthesis and drought: can we make metabolic connections from available data? Journal of Experimental Botany 62: 869–882.
31. Kamoshita A, Rodriguez R, Yamauchi A, Wade LJ (2004) Genotypic variation in response of rainfed lowland rice to prolonged drought and rewatering. Plant Production Science 7: 406–420.
32. Boonjung H, Fukai S (1996) Effects of soil water deficit at different growth stages on rice growth and yield under upland conditions. 2. Phenology, biomass production and yield. Field Crops Research 48: 47–55.
33. Pons TL, Flexas J, von Caemmerer S, Evans JR, Genty B, et al. (2009) Estimating mesophyll conductance to CO_2: methodology, potential errors, and recommendations. Journal of Experimental Botany 60: 2217–2234.
34. Gilbert ME, Pou A, Zwieniecki MA, Holbrook NM (2012) On measuring the response of mesophyll conductance to carbon dioxide with the variable J method. Journal of Experimental Botany 63: 413–425.
35. Harley PC, Loreto F, Dimarco G, Sharkey TD (1992) Theoretical considerations when estimating the mesophyll conductance to CO_2 flux by analysis of the response of photosynthesis to CO_2. Plant Physiology 98: 1429–1436.
36. Bois JF, Couchat PH, Lasceve G (1985) Relationships between transpiration and photosynthesis during a water stress. Acta Horticulturae 171: 297–314.
37. Richter A, Wanek W, Werner RA, Ghashghaie J, Jaeggi M, et al. (2009) Preparation of starch and soluble sugars of plant material for the analysis of carbon isotope composition: a comparison of methods. Rapid Communications in Mass Spectrometry 23: 2476–2488.
38. Farquhar GD, Ehleringer JR, Hubick KT (1989) Carbon isotope discrimination and photosynthesis. Annual Review of Plant Physiology and Plant Molecular Biology 40: 503–537.
39. Rodeghiero M, Niinemets Ü, Cescatti A (2007) Major diffusion leaks of clamp-on leaf cuvettes still unaccounted: how erroneous are the estimates of Farquhar et al. model parameters? Plant, cell & environment 30: 1006–1022.
40. Loreto F, Harley PC, Dimarco G, Sharkey TD (1992) Estimation of mesophyll conductance to CO_2 flux by three different methods. Plant Physiology 98: 1437–1443.
41. Genty B, Briantais J-M, Baker NR (1989) The relationship between the quantum yield of photosynthetic electron transport and quenching of chlorophyll fluorescence. Biochimica et Biophysica Acta (BBA) - General Subjects 990: 87–92.
42. Laisk A, Loreto F (1996) Determining photosynthetic parameters from leaf CO_2 exchange and chlorophyll fluorescence (ribulose-1, 5-bisphosphate carboxylase/oxygenase specificity factor, dark respiration in the light, excitation distribution between photosystems, alternative electron transport rate, and mesophyll diffusion resistance. Plant Physiology 110: 903–912.
43. Galmes J, Flexas J, Keys AJ, Cifre J, Mitchell RAC, et al. (2005) Rubisco specificity factor tends to be larger in plant species from drier habitats and in species with persistent leaves. Plant Cell and Environment 28: 571–579.
44. Furbank RT, von Caemmerer S, Sheehy J, Edwards G (2009) C4 rice: a challenge for plant phenomics. Functional Plant Biology 36: 845–856.
45. Pinelli P, Loreto F (2003) $^{12}CO_2$ emission from different metabolic pathways measured in illuminated and darkened C-3 and C-4 leaves at low, atmospheric and elevated CO_2 concentration. Journal of Experimental Botany 54: 1761–1769.

46. Loreto F, Dimarco G, Tricoli D, Sharkey TD (1994) Measurements of mesophyll conductance, photosynthetic electron transport and alternative electron sinks of field grown wheat leaves. Photosynthesis Research 41: 397–403.
47. Lauteri M, Scartazza A, Guido MC, Brugnoli E (1997) Genetic variation in photosynthetic capacity, carbon isotope discrimination and mesophyll conductance in provenances of *Castanea sativa* adapted to different environments. Functional Ecology 11: 675–683.
48. Scartazza A, Lauteri M, Guido MC, Brugnoli E (1998) Carbon isotope discrimination in leaf and stem sugars, water-use efficiency and mesophyll conductance during different developmental stages in rice subjected to drought. Australian Journal of Plant Physiology 25: 489–498.
49. Evans JR, Sharkey TD, Berry JA, Farquhar GD (1986) Carbon isotope discrimination measured concurrently with gas exchange to investigate CO_2 diffusion in leaves of higher plants. Australian Journal of Plant Physiology 13: 281–292.
50. Lloyd J, Syvertsen JP, Kriedemann PE, Farquhar GD (1992) Low conductances for CO_2 diffusion from stomata to the sites of carboxylation in leaves of woody species. Plant Cell and Environment 15: 873–899.
51. von Caemmerer S, Evans JR (1991) Determination of the average partial pressure of CO_2 in chloroplasts from leaves of several C3 plants. Australian Journal of Plant Physiology 18: 287–305.
52. Pantuwan G, Fukai S, Cooper M, Rajatasereekul S, O'Toole JC (2002) Yield response of rice (*Oryza sativa* L.) genotypes to different types of drought under rainfed lowlands: Part 1. Grain yield and yield components. Field Crops Research 73: 153–168.
53. Siddique K, Kirby E, Perry M (1989) Ear: stem ratio in old and modern wheat varieties; relationship with improvement in number of grains per ear and yield. Field Crops Research 21: 59–78.
54. Koyama K, Kikuzawa K (2009) Is whole-plant photosynthetic rate proportional to leaf area? A test of scalings and a logistic equation by leaf demography census. The American Naturalist 173: 640–649.
55. Tomás M, Flexas J, Copolovici L, Galmés J, Hallik L, et al. (2013) Importance of leaf anatomy in determining mesophyll diffusion conductance to CO_2 across species: quantitative limitations and scaling up by models. Journal of experimental botany 64: 2269–2281.
56. Zhao B, Kondo M, Maeda M, Ozaki Y, Zhang J (2004) Water-use efficiency and carbon isotope discrimination in two cultivars of upland rice during different developmental stages under three water regimes. Plant and Soil 261: 61–75.
57. Tipple BJ, Berke MA, Doman CE, Khachaturyan S, Ehleringer JR (2013) Leaf-wax n-alkanes record the plant–water environment at leaf flush. Proceedings of the National Academy of Sciences 110: 2659–2664.
58. Quick W, Chaves M, Wendler R, David M, Rodrigues M, et al. (1992) The effect of water stress on photosynthetic carbon metabolism in four species grown under field conditions. Plant, Cell and Environment 15: 25–35.
59. Yang J, Zhang J, Wang Z, Zhu Q (2001) Activities of starch hydrolytic enzymes and sucrose-phosphate synthase in the stems of rice subjected to water stress during grain filling. Journal of Experimental Botany 52: 2169–2179.
60. Scheidegger Y, Saurer M, Bahn M, Siegwolf R (2000) Linking stable oxygen and carbon isotopes with stomatal conductance and photosynthetic capacity: a conceptual model. Oecologia 125: 350–357.
61. Schmitt MR, Edwards GE (1981) Photosynthetic capacity and nitrogen use efficiency of maize, wheat, and rice - a comparison between C-3 and C-4 Photosynthesis. Journal of Experimental Botany 32: 459–466.
62. Kiran TV, Rao YV, Subrahmanyam D, Rani NS, Bhadana VP, et al. (2013) Variation in leaf photosynthetic characteristics in wild rice species. Photosynthetica 51: 350–358.
63. Terashima I (1992) Anatomy of non-uniform leaf photosynthesis. Photosynthesis research 31: 195–212.
64. Mott KA, Buckley TN (2000) Patchy stomatal conductance: emergent collective behaviour of stomata. Trends in Plant Science 5: 258–262.
65. Sun P, Wahbi S, Tsonev T, Haworth M, Liu S, et al. (2014) On the use of leaf spectral indices to assess water status and photosynthetic limitations in *Olea europaea* L. during water-stress and recovery. Plos One 9: e105165.
66. Tsonev T, Wahbi S, Sun P, Sorrentino G, Centritto M (2014) Gas exchange, water relations and their relationships with photochemical reflectance index in *Quercus ilex* plants during water stress and recovery. International Journal of Agriculture and Biology 16: 335–341.
67. Warren CR. Stand aside stomata, another actor deserves centre stage: the forgotten role of the internal conductance to CO_2 transfer; 2007 Jul 22–27; Glasgow, Scotland. Oxford University Press. pp.1475–1487.
68. Centritto M, Tognetti R, Leitgeb E, Střelcová K, Cohen S (2011) Above ground processes - anticipating climate change influences. In: Bredemeier M, Cohen S, Godbold DL, Lode E, Pichler V et al., editors. Forest Management and the Water Cycle: An Ecosystem-Based Approach. London: Springer. pp.31–64.
69. Haworth M, Elliott-Kingston C, McElwain J (2013) Co-ordination of physiological and morphological responses of stomata to elevated $[CO_2]$ in vascular plants. Oecologia 171: 71–82.
70. Haworth M, Elliott-Kingston C, McElwain JC (2011) Stomatal control as a driver of plant evolution. Journal of Experimental Botany 62: 2419–2423.
71. Haworth M, Gallagher A, Elliott-Kingston C, Raschi A, Marandola D, et al. (2010) Stomatal index responses of *Agrostis canina* to carbon dioxide and sulphur dioxide: implications for palaeo-$[CO_2]$ using the stomatal proxy. New Phytologist 188: 845–855.
72. Kubiskem ME, Abrams MD (1993) Stomatal and nonstomatal limitations of photosynthesis in 19 temperate tree species on contrasting sites during wet and dry years. Plant, Cell and Environment 16: 1123–1129.
73. Venuprasad R, Lafitte HR, Atlin GN (2007) Response to direct selection for grain yield under drought stress in rice. Crop Sci 47: 285–293.
74. Karaba A, Dixit S, Greco R, Aharoni A, Trijatmiko KR, et al. (2007) Improvement of water use efficiency in rice by expression of HARDY, an *Arabidopsis* drought and salt tolerance gene. Proceedings of the National Academy of Sciences of the United States of America 104: 15270–15275.
75. Lanceras JC, Pantuwan G, Jongdee B, Toojinda T (2004) Quantitative trait loci associated with drought tolerance at reproductive stage in rice. Plant Physiology 135: 384–399.
76. Xu Y, This D, Pausch RC, Vonhof WM, Coburn JR, et al. (2009) Leaf-level water use efficiency determined by carbon isotope discrimination in rice seedlings: genetic variation associated with population structure and QTL mapping. Theoretical and Applied Genetics 118: 1065–1081.
77. Wassmann R, Jagadish SVK, Heuer S, Ismail A, Redona E, et al. (2009) Climate change affecting rice production: the physiological and agronomic basis for possible adaptation strategies. In: Sparks DL, editor. Advances in Agronomy, Vol 101. pp.59–122.
78. Blum A (2005) Drought resistance, water-use efficiency, and yield potential–are they compatible, dissonant, or mutually exclusive? Australian Journal of Agricultural Research 56: 1159–1168.

12

Transcriptome-Wide Profiling and Expression Analysis of Diploid and Autotetraploid *Paulownia tomentosa* × *Paulownia fortunei* under Drought Stress

Enkai Xu[1,2], Guoqiang Fan[1,2*], Suyan Niu[1,2], Zhenli Zhao[1,2], Minjie Deng[1,2], Yanpeng Dong[1,2]

1 Institute of Paulownia, Henan Agricultural University, Jinshui Area, Zhengzhou, Henan, P.R. China, **2** College of Forestry, Henan Agricultural University, Jinshui Area, Zhengzhou, Henan, P.R. China

Abstract

Paulownia is a fast-growing deciduous hardwood species native to China, which has high ecological and economic value. In an earlier study, we reported ploidy-dependent differences in Paulownia drought tolerance by the microscopic observations of the leaves. Autotetraploid Paulownia has a higher resistance to drought stress than their diploid relatives. In order to obtain genetic information on molecular mechanisms responses of Paulownia plants to drought, Illumina/Solexa Genome sequencing platform was used to *de novo* assemble the transcriptomes of leaves from diploid and autotetraploid *Paulownia tomentosa* × *Paulownia fortunei* seedlings (PTF2 and PTF4 respectively) grown under control conditions and under drought stress and obtained 98,671 nonredundant unigenes. A comparative transcriptome analysis revealed that hundreds of unigenes were predicted to be involved mainly in ROS-scavenging system, amino acid and carbohydrate metabolism, plant hormone biosynthesis and signal transduction, while these unigenes exhibited differential transcript alteration of the two accessions. This study provides a comprehensive map of how *P. tomentosa* × *P. fortunei* responds to drought stress at physiological and molecular levels, which may help in understanding the mechanisms involve in water-deficit response and will be useful for further study of drought tolerance in woody plants.

Editor: Keqiang Wu, National Taiwan University, Taiwan

Funding: This work was supported by GQF, National Natural Science Foundation of China (Grant no. 30271082, 30571496,U1204309, http://www.nsfc.gov.cn/), GQF, Outstanding Talents Project of Henan Province (Grant no. 122101110700), GQF, Transformation Project of the National Agricultural Scientific and Technological Achievement of China (Grant no. 2012GB2D000271), and GQF, Science and Technology Innovation Team Project of Zhengzhou City (Grant no. 121PCXTD515). The funders had no role in study design, data collection and analysis, decision to publish, or preparation of the manuscript.

Competing Interests: The authors have declared that no competing interests exist.

* Email: zlxx64@126.com

Introduction

Plants are constantly exposed to a wide range of environmental stresses such as drought, high salt, heat, and extremes of temperature. Water scarcity is one of the major environmental stresses that causes substantial losses in many plants [1]. It affects every aspect of plant growth, modifying anatomy, morphology, physiology, and biochemistry, which negatively affects plant growth and productivity [2].

Global climate change will likely reduce water availability further increasing the need for drought tolerant plants [3]. To adapt to drought stress, trees have evolved complex mechanisms based on modifications in metabolites, gene expression, and proteins. Clearly, drought tolerance is a complex trait that is regulated by multiple genes. Based on their putative biological functions in the drought response process, drought-responsive genes have been classified into two groups: (1) genes encoding proteins that defend the cells from the effects of water-deficit, for example, proteins that combat the accumulation of solutes (e.g., enzymes for biosynthesis of osmolytes like proline, betaine, sugars), passive and active transport systems across membranes (e.g., water channel proteins and membrane transporters), proteins that protect and stabilize cell structures from damage by reactive oxygen species (e.g., detoxification enzymes such as catalase and superoxide dismutase), and other proteins that protect the macromolecules (e.g., late embryogenesis-abundant (LEA) proteins and osmotin) [4–6]; and (2) genes encoding proteins that regulate stress signal transduction and alter gene expression (e.g., transcription factors, protein kinases, protein phosphatases, proteinases, and phytohormones) [7,8].

The underlying mechanisms of the drought stress response in plants have been investigated using various genomic and genetic tools. Cohen et al. [9] assessed the genotype-dependent component of a drought-induced transcriptome response in two poplar genotypes with different drought tolerances using a multi-species designed microarray. Zheng et al. [10] compared genome-wide gene expression profiles between drought-tolerant and drought-sensitive lines of maize during drought stress by microarray.

High-throughput sequencing is a powerful and cost-efficient tool that has been used effectively in many areas, including the study of the transcriptomes of plants under biotic and abiotic stresses. Over 24,000 drought-responsive genes were identified from *Populus euphratica* leaves using a Roche/454 platform [11]. The transcriptomes of 22 different developmental stage tissues from

Cicer arietinum L. were sequenced using a Illumina/Solexa system, which revealed thousands of drought-responsive genes [12]. Using Illumina/Solexa sequencing, Peng et al. [13] reported differences in the transcriptome levels between male and female *Populus yunnanensis* in response to drought stress.

Paulownia is a species of tree that is native to China, where it has been cultivated for over 2000 years. This species has also been planted in many other countries and can now be found throughout the world, except Antarctica [14]. Paulownia is an adaptable and fast-growing deciduous tree, which provides raw material for manufacturing furniture, plywood, and musical instruments. It has a unique biological characteristic of late leaf emergence and fall, a deep root system, and sparse branching; therefore, exemplary areas of agroforestry have been built on the broad plains in North-Central China [15]. Paulownia has a very large leaf area with a high transpiration rate; therefore, sufficient moisture is very important for Paulownia growth. *Paulownia tomentosa* × *Paulownia fortunei* is grown mainly in hilly loess areas or in salinized soil [16] where the annual precipitation is below 500 mm. *P. tomentosa* × *P. fortunei* grow quite well without artificial irrigation because of their inherent ability to withstand drought. Therefore, *P. tomentosa* × *P. fortunei* is an ideal species in which to study the mechanisms of drought tolerance.

Polyploidization of chromosomes is thought to be one of the most important mechanisms for species evolution [17]. Polyploidy usually changes plant morphological and anatomical characteristics introducing, for example, superior breast-height diameter, volume, leaf, fruit, and stoma [18–20]. Polyploidy also changes the physiology, biochemistry, and gene expression in plants [17,21,22]. Several studies have reported that polyploids exhibit higher stress resistance compared with their corresponding diploid relatives [23–25].

To benefit from the possible value of polyploidy, in 2010, an autotetraploid *P. tomentosa* × *P. fortunei* ($2n = 4x = 80$) was induced from the leaves of diploid *P. tomentosa* × *P. fortunei* ($2n = 2x = 40$) using colchicines (an established system of *in vitro* plantlet regeneration) and a detailed physiological study of diploid and autotetraploid plants was conducted [26].

Drought tolerance varies considerably between diploid and autotetraploid Paulownia. In an earlier paper, we studied diploid and autotetraploid Paulownia plants for their drought and cold tolerance by microscopic observations of the leaves [27]. However, little is known about the molecular mechanisms of diploid and autotetraploid Paulownia under drought conditions.

In the present study, to reveal the transcriptome-wide response to drought stress in *P. tomentosa* × *P. fortunei* and to investigate diploid and autotetraploid differences in leaf transcriptome remodeling during drought stress, we analyzed the gene expression profiles of diploid and autotetraploid seedlings using an Illumina/Solexa Genome Analyzer IIx (GAIIx) platform. To the best of our knowledge, this is the first comprehensive analysis of differentially expressed genes in drought-stressed *P. tomentosa* × *P. fortunei*. A comparison of the diploid and autotetraploid transcriptomes revealed differences and similarities, which helped identify stress-related candidate genes and gene regulatory networks for future screening. The acquired information will provide new insights into the molecular mechanisms involved in the *P. tomentosa* × *P. fortunei* response to drought stress.

Materials and Methods

Plant material

Uniformly grown tissue culture seedlings of diploid *P. tomentosa* × *P. fortunei* (PTF2) and tetraploid *P. tomentosa* × *P. fortunei* (PTF4) were collected from the Institute of Paulownia, Henan Agricultural University, Zhengzhou, Henan Province, China and cultivated in nutrition blocks containing ordinary garden soil for 30 days. Samples with the same crown size and height were then planted in plastic pots (20 cm in diameter at the bottom and 20 cm deep) containing ordinary garden soil, one for each plant. The plants were housed randomly in an outdoor nursery for 50 days before being subjected to stress treatment. Twelve plants from each accession (PTF2 and PTF4) were obtained, including three controls and nine plants for water-deficit treatment (divided into three groups, each group contained three individual biological replicates).

The seedlings were subjected to drought conditions in a water-controlled experiment according to the method of Moumeni et al. [28]; the controls were watered to field capacity every day. For drought treatment, the seedlings of the two accessions were subjected to mild (5 days without irrigation, 55% relative soil water content), moderate (15 days without irrigation, 40% relative soil water content), and severe (20 days without irrigation, 25% relative soil water content) drought stress. The mature, fully expanded leaves (second leaf from the apex) were sampled and mixed in equal numbers from each plant in each of the replicate groups. The leaf tissue was frozen in liquid nitrogen and then stored at −80°C for later use.

Physiological measurements

Phenotypic and physiological characteristics of the PTF2 and PTF4 seedlings were measured under the control, mild, moderate, and severe drought stresses.

Physiological measurements were made between 8.30 am and 11.00 am on single, fully expanded leaves (third and fourth leaves from the terminal bud of a twig) immediately after excision. The measurements were taken on three replicates of the leaves from the same plants.

The percent relative water content of the leaf was calculated as (fresh weight - dry weight)/(turgid weight - dry weight) × 100 [29]. Chlorophyll content was measured according to a described method previously [30] and calculated as described by Arnon [31]. The malondialdehyde concentration was measured using the thiobarbituric acid method as described by Hodges et al. [32]. Relative electrical conductivity was measured as described previously [33]. Superoxide dismutase activity was determined using the nitroblue tetrazolium method reported by Dhindsa et al. [34]. Values were expressed in units/gmHb [35]. Total soluble protein content was measured following the procedure of Guy et al. [36], and the protein concentration was measured as described by Bradford [37] using bovine serum albumin (Sigma, USA) as the standard. Proline content was measured using the acid ninhydrin method of Bates et al. [38]. Soluble sugar content was measured using the anthrone colorimetry method described by Irigoyen et al. [39].

Statistical analysis

Analyses of variance (ANOVA) of the parameters were performed using the MSTAT-C computer program (MSTAT Development Team, Michigan State University, 1989) and SPSS (v. 16.5). The mean values of each of the measured traits were compared using the least significance difference (LSD) test at a 5% level of probability according to Snedecor and Cochran [40].

RNA extraction and double-stranded cDNA synthesis

The PTF2 and PTF4 seedlings from the severe drought-stressed group were chosen for sequencing based on the physiological measurements (see the results section for details).

Approximately 8 mg of leaves from the diploid control (well-watered PTF2) and the severe drought-treated PTF2, and 8 mg of leaves from the tetraploid control (well-watered PTF4 control) and the severe drought-treated PTF4 in three replications were used for total RNA extraction.

Total RNA was extracted from the four samples with TRIzol reagent (Invitrogen, Carlsbad, CA). RNA was purified using RNeasy MiniElute Cleanup Kit (Qiagen, Valencia, CA) according to the manufacturer's protocol. Magnetic beads with Oligo (dT) were used to isolate poly-(A) mRNA from the total RNA. The mRNA was mixed with fragmentation buffer and fragmented into short fragments. The first-strand cDNA was synthesized with SuperScript II reverse transcriptase (Life Technologies, Carlsbad, CA), and DNA polymerase I and RNase H were then used to synthesize the second-strand cDNA. The short cDNA fragments were purified and dissolved in EB buffer for end reparation and addition of a single nucleotide A (adenine). The short fragments were then connected to adapter sequences. Suitable fragments were selected as templates for PCR amplification. The cDNA libraries were qualified using a 2100 Bioanalyzer (Agilent Technologies, Inc., Santa Clara, CA), then sequenced on a Illumina Genome Analyzer IIx (GAIIx) platform (Illumina, San Diego, CA) by Beijing Genomics Institute (BGI)-Shenzhen, Shenzhen, China, following the manufacturer's standard cBot and sequencing protocols.

Assembly and annotation

The raw image data were transformed by base calling into sequence data (raw reads). Raw reads that contained adaptor sequences, ambiguous reads, or low-quality reads were removed using DynamicTrim in the SolexaQA package (http://solexaqa.sourceforge.net/) to create clean reads. The clean reads used in this publication have been deposited in the NIH Short Read Archive (SRA) database (http://www.ncbi.nlm.nih.gov/sra) under accession number SRP041407. The quality sequences were assembled into unigenes using the Trinity program [41]. Unigenes from multiple samples from the same accession were further processed by removing redundancy and splicing, and then non-redundant unigenes without gaps were clustered using the TIGR Gene Indices clustering tools (TGICL) [42]. The unigene sequences were annotated by aligning them against the NCBI non-redundant protein sequence database (Nr) (http://www.ncbi.nlm.nih.gov/), Swiss-Prot (http://www.expasy.ch/sprot/), Cluster of Orthologous Groups (COG) (http://www.ncbi.nlm.nih.gov/cog/), and the Kyoto Encyclopedia of Genes and Genomes (KEGG) (http://www.genome.jp/kegg/) using BLASTX (E-value <1.0E-5), and then against the NCBI nucleotide sequence database (Nt) using BLASTN (E-value <1.0E-5).

The best alignments were used to annotate the unigenes and to decide the coding region and sequence direction of the unigenes. Non-annotated unigenes with no matches in any of the searched databases were scanned using ESTScan [43] to predict the sequence (5′–3′) direction and to predict the coding region to obtain the translated amino sequence. Gene Ontology (GO) functional classification of the unigenes was performed using Blast2GO [44] and WEGO software [45]. Pathway annotations were obtained by aligning the unigenes with the KEGG database [46].

Unigene expression difference analysis

Unigene expression in each library was calculated using the FPKM (fragments per kb per million fragments) method [47] as:

$$FPKM = \frac{10^6 C}{NL/10^3}$$

where *FPKM* is the expression of unigene A, C is the number of fragments that aligned uniquely to unigene A, *N* is the total number of fragments that aligned uniquely to all the unigenes, and *L* is the number of bases in the coding sequence of unigene A.

Differentially expressed unigenes (DEUs) between the PTF2 and PTF4 transcriptomes were identified using the rigorous algorithm described by Audic et al. [48].

To screen the DEUs, the p-value threshold in multiple hypothesis testing and analyses was determined by manipulating the false discovery rate (FDR) value [49]. A FDR ≤ 0.001 and absolute value of the $|\log2\text{Ratio}| > 2$ were used as the threshold to determine the significance of the differences in gene expression [49]. The DEUs were then subjected to GO functional and KEGG Pathway analyses.

GO and pathway enrichment analysis

GO functional enrichment analysis was performed to classify the GO functional annotations for the DEUs. First, all the DEUs were mapped to terms in the GO database (http://www.geneontology.org/) and the gene numbers for each GO term were calculated to obtain a gene list and the gene numbers for every mapped GO term. The GO enrichment analysis was performed by applying the hypergeometric test to find significantly enriched GO terms in the DEUs compared with the whole transcriptome background of *P. tomentosa* × *P. fortunei*. The p-value was calculated as:

$$p = 1 - \sum_{i=0}^{m-1} \frac{\binom{M}{i}\binom{N-M}{n-i}}{\binom{N}{n}}$$

where N is the total number of genes with GO annotation; n is the number of DEUs in N; M is the total number of all genes that were annotated to a certain GO term; and m is the number of DEUs in M. After applying the Bonferroni correction to the calculated p-value we chose a corrected p-value of ≤ 0.05 as the threshold, and GO terms with $p \leq 0.05$ were defined as significantly enriched GO terms for the DEUs.

Similarly, a pathway enrichment analysis to identify significantly enriched biochemical or signal transduction pathways associated with the DEUs was performed by comparing the pathway annotations with the whole transcriptome background of *P. tomentosa* × *P. fortunei*. A Q-value, defined as the FDR, was used to correct the p-value. After multiple test corrections, we chose a Q-value of ≤ 0.05 to identify significantly enriched pathways for the DEUs.

Real-time quantitative PCR analysis of potential drought response DEUs

A real-time quantitative PCR (qRT-PCR) analysis was conducted on selected DEUs to validate the Illumina paired-end sequencing results. *P. tomentosa* × *P. fortunei* leaves collected from the severe drought and control PTF2 and PTF4 plants were used to extract total RNA. The first-strand cDNA was synthesized from 2 mg total RNA with the iScriptc DNA synthesis kit (Bio-Rad, Hercules, CA) according to the manufacturer's instructions.

The specific primers used for the real-time PCR assay were designed using Beacon Designer, version 7.7 (Premier Biosoft International, Ltd., Palo Alto, CA) and are listed in Table S1. The cDNA was amplified using a CFX96TM Real-Time System (Bio-Rad) with So Fast Eva Green Supermix (Bio-Rad) according to the manufacturer's instructions. PCR mixtures with a final volume of 20 μL contained 10 μL SYBR Green PCR mix, 1 μL cDNA, 0.4 μM forward primer, 0.4 μM reverse primer, and 7.0 μL SteriLe ddH_2O. The PCR cycles were as follows: 95°C for 1 min, followed by 40 cycles of 95°C for 10 s, and 55°C for 15 s. Relative expression levels of the genes were calculated using the $2^{-\Delta\Delta Ct}$ method and normalized with 18S rRNA from *P. tomentosa* × *P. fortunei*.

Results

Physiological characterization of P. tomentosa × P. fortunei under drought stress

The phenotypic and physiological responses of PTF2 and PTF4 to water deficit were assessed using seedlings exposed to progressive water loss for about three weeks.

There were no significant differences in stress phenotype between the PTF2 and PTF4 under the well-watered conditions (Figure 1a and 1b). The leaves of the PTF2 seedlings drooped after five days of drought stress (Figure 1c), whereas the leaves of the PTF4 seedlings remained green and turgid (Figure 1d). After 15 days, the PTF2 plants showed stunted growth and had leaves that were paler and more wilted compared with the PTF4 plants (Figure 1e and 1f); after 20 days, the PTF2 and PTF4 plants both showed distinct effects of drought stress. However, the PTF4 plants showed much better development, less wilting, and higher biomass compared with the PTF2 plants (Figure 1 g and 1 h).

In general, the PTF4 seedlings maintained markedly higher growth vigor under the drought stress conditions than the PTF2 seedlings.

Changes in physiological parameters in PTF2 and PTF4 plants in response to drought

Drought stress significantly inhibited the relative water content (RWC) and chlorophyll content of the plants. The RWC of PTF2 was similar to that of PTF4 under well-watered conditions. The RWC of both accessions declined with the increase of the drought, as characterized by their significantly lower RWC compared with the corresponding controls (well-watered), particularly under the severe drought treatment. However, after 20 days of drought, the PTF2 plants showed a sharp decrease in RWC (13.25%) compared with the RWC (12.23%) of the PTF4 plants (Figure 2a). PTF2 and PTF4 were recorded the highest chlorophyll content under well-watered conditions while, during the drought treatment, the chlorophyll content significantly decreased in both accessions. The chlorophyll content also obviously showed difference between PTF2 and PTF4 plants, and the PTF4 seedlings always contained higher levels of total chlorophyll (chlorophyll a and b) than the PTF2 seedlings under drought stress (Figure 2b).

Relative electrical conductivity (REC) and malondialdehyde (MDA) content showed overall increased tendencies in the PTF2 and PTF4 plants under drought stress. However, after 20 days of drought stress the REC and MDA values increased significantly in both accessions, whereas the REC and MDA in the PTF2 plants were always at much higher levels than in the PTF4 plants over the entire treatment process (Figure 2c and 2d).

Compared to the control conditions, both the PTF2 and PTF4 plants under mild and moderate drought conditions showed a substantial increase in superoxide dismutase (SOD) activity and soluble proteins content, and then decreased dramatically under severe drought stress, whereas the values of these two traits were higher in both PTF2 and PTF4 plants under severe drought than under well-watered conditions. Besides, SOD activity and soluble proteins content measurements showed striking differences between the two accessions under the same drought treatments. In particular, PTF4 showed higher SOD activity and soluble proteins content than PTF2 at the end of the 20-day experiment (Figure 2e and 2f).

Additionally, the proline and soluble sugar contents tended to increase as the drought strengthened in the two accessions. During the different drought treatments, the proline and soluble sugar contents in PTF4 increased significantly, but in PTF2 they showed relatively little changes. Overall, the PTF4 plants showed much higher levels of the proline and soluble sugar contents compared with the PTF2 plants (Figure 2 g and 2 h).

Paired-end sequencing and de novo assembly of the P. tomentosa × P. fortunei reads

To understand the molecular mechanisms of drought tolerance in *P. tomentosa* × *P. fortunei*, four leaf cDNA libraries, two for the well-watered controls (PTF2W, PTF4W) and two for the severe drought-treated plants (PTF2T, PTF4T), were constructed and sequenced on the GAIIx platform. Over 291 million raw reads with a mean length of 110 nt, were generated from the four libraries; approximately 73 million reads from each sample. After removing adapter sequences, duplication sequences, and ambiguous and low-quality reads, 264 (23.78 Gb) million clean reads

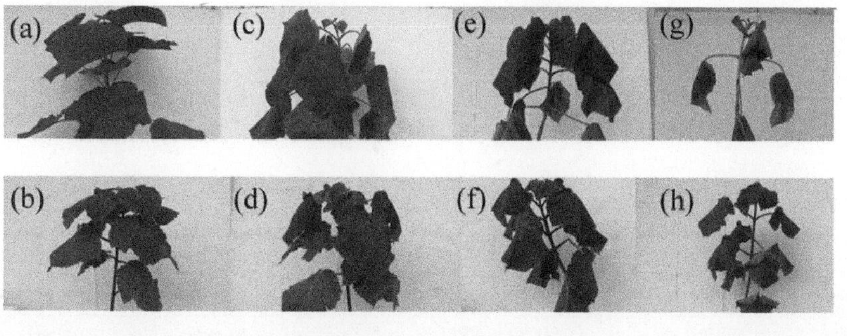

Figure 1. Physiological feature of diploids and autotetraploid in response to drought. (a) PTF2W, well-watered diploid, 75% relative soil water content; (b) PTF4W, well-watered tetraploid, 75% relative soil water content; (c) PTF2T-1, 5-day drought treated diploid, 55% relative soil water content; (d) PTF4T-1, 5-day drought treated tetraploid, 55% relative soil water content; (e) PTF2T-2, 15-day drought treated diploid, 40% relative soil water content; (f) PTF4T-2, 15-day drought treated tetraploid, 40% relative soil water content; (g) PTF2T-3, 20-day drought treated diploid, 25% relative soil water content; (h) PTF4T-3, 20-day drought treated tetraploid, 25% relative soil water content.

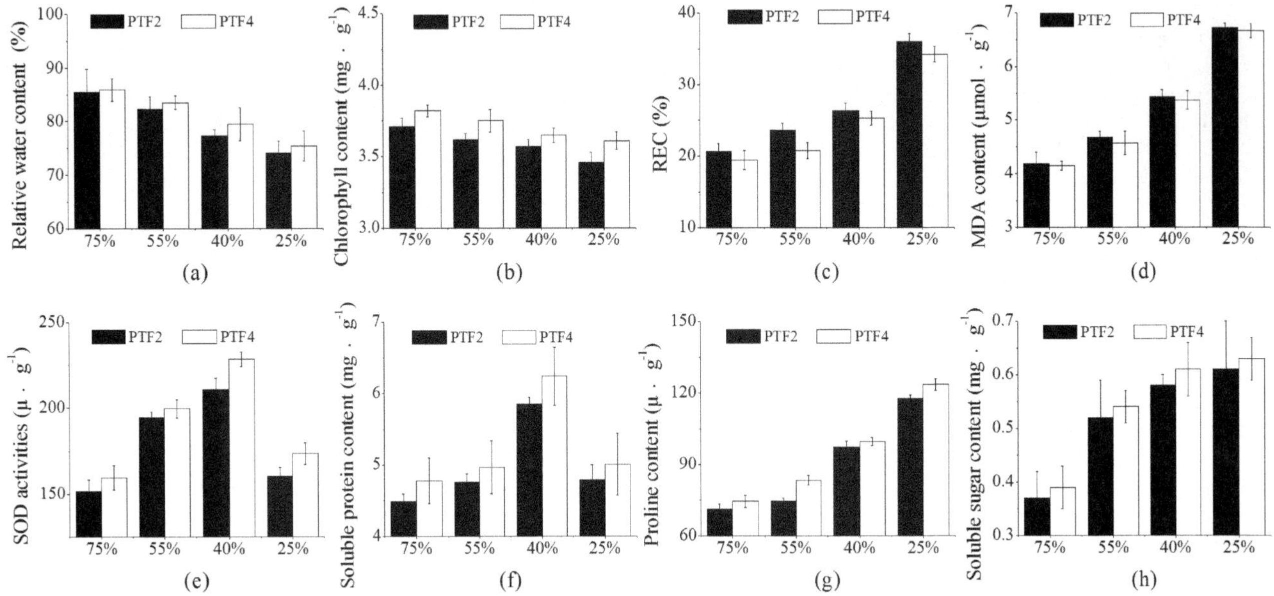

Figure 2. Effects of drought stress on *P. tomentosa* × *P. fortunei* physiology. PTF2 represents diploid *P. tomentosa* × *P. fortunei*, PTF4represents autotetraploid *P. tomentosa* × *P. fortunei*, 75%, 55%, 40% and 25% relative soil water contents were used. (a) Effect of drought stress on leaf relative water content;(b) Effect of drought stress on leaf chlorophyll content; (c) Effect of drought stress on leaf relative electrical conductivity; (d) Effect of drought stress on leaf malondialdehyde (MDA) content; (e) Effect of drought stress on leaf superoxide dismutase (SOD) activities; (f) Effect of drought stress on leaf soluble protein content; (g) Effect of drought stress on leaf proline content; (h) Effect of drought stress on leaf soluble sugar content.

remained. The Q20 (sequencing error rate, 1%) percentage, N percentage, and GC content were 97.43%, 0%, and 49.04%, respectively (Table 1). The *de novo* assembly of the clean reads using Trinity generated 599,908 contigs without gaps, with an average length of 304 nt and an N50 of 471 nt. After clustering using the TGICL software, a total of 98,671 nonredundant unigenes were produced, including 47,582 clusters and 51,089 singletons. The total length of the unigenes was 96,940,229 nt, with a mean length of 982 nt and an N50 of 1626 nt (Table 2); 63,015 (63.86%) unigenes were longer than 500 nt and 38,134 (38.65%) were longer than 1,000 nt. The length distribution of the unigenes is shown in Figure S1. None of the unigenes had gaps, indicating that the sequences were of high quality. All the assembled unigenes were aligned against the Nr, Swiss-Prot, KEGG, and COG protein databases, and a total of 70,524 coding sequences (CDSs) were predicted; 67,605 CDSs were inferred using BLASTX hits and 2,919 were assigned using ESTScan.

All the statistics showed that the sequencing of the *P. tomentosa* × *P. fortunei* transcriptomes was of high depth and that high quality reads were produced.

Annotation and functional analysis of the unigenes

The BLAST searches against the various databases yielded a total of 67,326 (68.23%) unigenes that found matches against the Nr database, followed by Nt (58,508, 59.30%), GO (55,413, 56.16%), Swiss-Prot (43,851, 44.44%), KEGG (41,206, 41.76%), and COG (28,165, 28.55%) (Table S2).

Overall, 70,100 (71.04%) unigenes shared similarities with entries in one or more of the six databases. No matches were found for 28,571 of the unigene sequences in any of these databases (E-value >10-5), suggesting that some of these unigenes may serve as a source for novel gene discovery.

Based on the Nr annotations, we determined that 34.0% of the mapped unigenes had very significant homology to known sequences (E-value <10–100), 35.9% showed significant homology (10–100<E-value <10–30), and 30.1% showed weak homology (E-value 10–30 to 10–5) (Figure 3A); 25.1% of unigenes shared >80% similarity with known sequences, 67.8% shared between 40% and 80% similarity, and 7.2% shared 19% to 40% similarity (Figure 3B). Further, the unigenes shared strong homology with sequences from *Vitis vinifera* (46.0%), *Ricinus communis* (12.8%), *Populus trichocarpa* (10.8%), *Glycine max* (5.9%), and *Nicotiana tabacum* (2.0%) (Figure 3C).

Table 1. Summary of the sequence assembly after Illumina sequencing.

Samples	Total Raw Reads	Total Clean Reads	Total Clean Nucleotides (nt)	Q20 percentage	N percentage	GC percentage
PTF2W	72,147,152	65,657,050	5,909,134,500	97.50%	0.00%	46.42%
PTF2 T	72,176,810	65,227,994	5,870,519,460	97.42%	0.00%	46.07%
PTF4W	76,705,742	69,169,960	6,225,296,400	97.43%	0.00%	46.90%
PTF4 T	70,571,228	64,168,586	5,775,172,740	97.37%	0.00%	47.29%

Table 2. Summary statistics of contig and unigene in four libraries from *P. tomentosa* × *P. fortune.*

	Sample	Total Number	Total Length(nt)	Mean Length(nt)	N50	Total Consensus Sequences	Distinct Clusters	Distinct Singletons
Contig	PTF2W	144,755	45,259,057	313	508	–	–	–
	PTF2 T	161,799	47,608,475	294	441	–	–	–
	PTF4W	143,358	45,001,926	314	494	–	–	–
	PTF4 T	149,996	44,356,767	296	442	–	–	–
Unigene	PTF2W	78,883	59,470,771	754	1332	78,883	32,443	46,440
	PTF2 T	89,241	69,693,869	781	1429	89,241	36,185	53,056
	PTF4W	81,816	57,290,184	700	1209	81,816	31,763	50,053
	PTF4 T	79,025	55,450,884	702	1240	79,025	30,515	48,510
	All	98,671	96,940,229	982	1626	98,671	47,582	51,089

COG functional categories

When the translated protein sequences of the unigenes were searched against the COG database, 28,165 unigenes (28.55%) were assigned to 25 function categories. The largest category was "General function prediction only" (9,316, 9.44%), followed by "Transcription" (5,232, 5.30%) and "Replication, recombination and repair" (4,299, 4.35%), while "Extracellular structures" (44, 0.045%) and "Nuclear structure" (8, 0.0081%) were the smallest categories (Figure 4).

GO annotations

The GO classification system was used to assign possible functions to the unigenes based on the annotations in Nr. A total of 55,413 unigenes (56.16%) were categorized into the three main GO categories (biological process, cellular component, and molecular function) and 44 functional groups. The largest groups under biological process were "cellular process" (36,028, 36.51%), "metabolic process" (34,472, 34.94%), and "response to stimulus" (19,059, 19.32%). Under cellular component, "cell" and "cell part" were the two largest groups, both containing 44,582 unigenes (45.18%). Under molecular function, the top three groups were "catalytic activity" (27,131, 27.50%), "binding" (26,550, 26.90%), and "transporter activity" (4,071, 4.12%) (Figure 5).

KEGG pathways

To identify the different biochemical pathway that the annotated unigenes were associated with, we assigned the Enzyme Commission numbers (EC numbers) in the KEGG pathways. A total of 41,206 unigenes (approximately 41.76%) were mapped to 128 KEGG pathways, the top four of which were "Metabolic pathways"(ko01100, 10,041 unigenes, 24.37%), "Biosynthesis of secondary metabolites" (ko01110, 4,681 unigenes, 11.36%), "Plant-pathogen interaction" (ko04626, 2,284 unigenes, 5.54%), and "Plant hormone signal transduction" (ko04075, 2,089 unigenes, 5.07%), while "Betalain biosynthesis" (ko00965, 3 unigenes, 0.01%) was the least represented (Table S3).

Transcriptional profiling of PTF2 and PTF4 under drought stress

To identify transcriptional changes under drought stress, we performed four pairwise comparisons of the transcriptomes as follows: drought-treated PTF2 plants versus well-watered PTF2 plants (PTF2T vs. PTF2W), and drought-treated PTF4 versus well-watered PTF4 plants (PTF4T vs. PTF4W), well-watered PTF4 plants versus well-watered PTF2 plants (PTF4W vs. PTF2W), and drought-treated PTF4 versus drought-treated PTF2 plants (PTF4T vs. PTF2T). Large numbers of DEUs were detected (Table S4, S5, S6 and S7). The results of the pairwise comparisons of the DEUs are shown in Figure 6. Co-regulated DEUs between the PTF2T vs. PTF2W and PTF4T vs. PTF4W (the drought-treated vs. the well-watered) libraries were identified; 1,616 DEUs were up-regulated and 3,457 DEUs were down-regulated in the two comparisons. Similarly, co-regulated DEUs between the PTF4W vs. PTF2W and PTF4T vs. PTF2T (the tetraploid vs. the diploid) libraries were also identified; 173 DEUs were up-regulated and 98 DEUs were down-regulated in the two comparisons. Cell wall genes, photosynthesis-related genes, and genes encoding drought stress-related kinases, reactive oxygen species (ROS) scavenging enzymes, and osmotic adjustment proteins were among the most abundant transcripts. Some of the DEUs had no matches to known proteins; however, they may

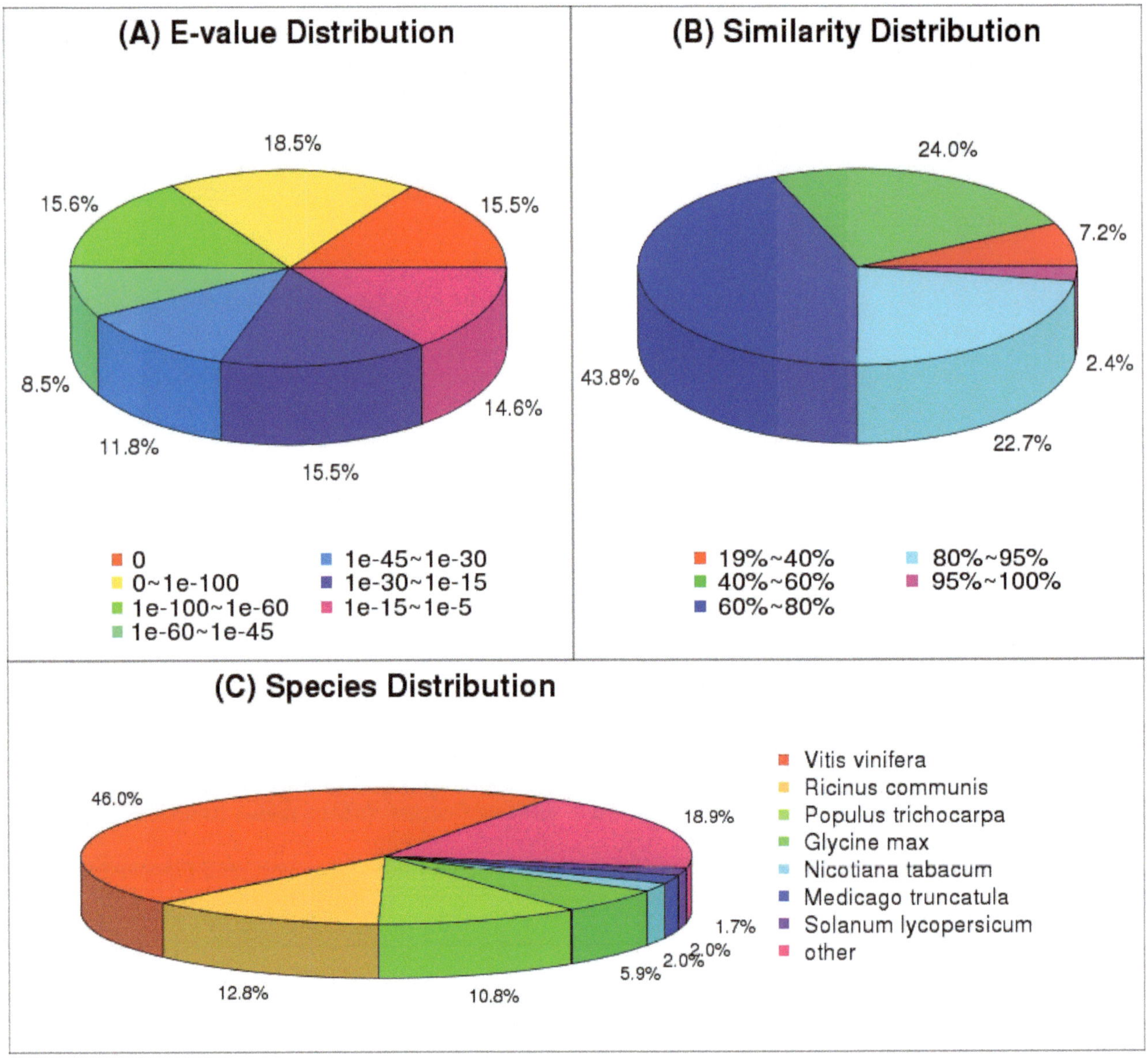

Figure 3. E-value Distribution. (A) similarity Distribution (B) and species distribution (C) of the BLAST matches of the transcriptome unigenes. This figure shows the distributions of unigenes BLASTX matches against the nr protein database (cutoff value E <1.0E-5) and the proportions for each species.

be found to contribute to drought tolerance when studied further (Table S8 and S9).

GO assignments allowed us to determine the major biological process, cellular component, and molecular function with which DEUs were associated (Figure 7 and 8). We then performed a GO enrichment analysis to identify the functional categories that were significantly enriched among the DEUs. Under biological process, stress-responsive processes (e.g., response to water deprivation, response to biotic stimulus, response to hormone stimulus, response to abiotic stimulus, and) were more represented in the up-regulated than in down-regulated DEUs, suggesting an increase of the related activities during drought treatment. Additionally, as reported previously in drought-stressed *P. euphratica* [50], GO terms like signal transduction, response to osmotic stress, small molecule metabolic process, carbohydrate catabolic process, arginine and proline metabolism, maltose metabolic process, disaccharide metabolic process, glucose catabolic process, hexose catabolic process, monosaccharide catabolic process, oligosaccharide metabolic process, and the related MAP kinase kinase activity were enriched. These findings may suggest that the osmotic stress-activated MAPK pathway and osmotic adjustment were crucial in the responses of PTF2 and PTF4 plants to water deprivation. Under cellular component, cell wall and plant-type cell wall were significantly enriched. Under molecular function, catalytic activity and binding were highly enriched for the drought-treated vs. the well-watered and the tetraploid vs. the diploid expression profiles, suggesting that the activities of many genes may be modulated in response to drought stress [51].

The co-regulated DEUs were also mapped to the KEGG metabolic pathways. Thirty-nine and seven metabolic pathways were significantly enriched (Q-value ≤ 0.05) in the drought-treated

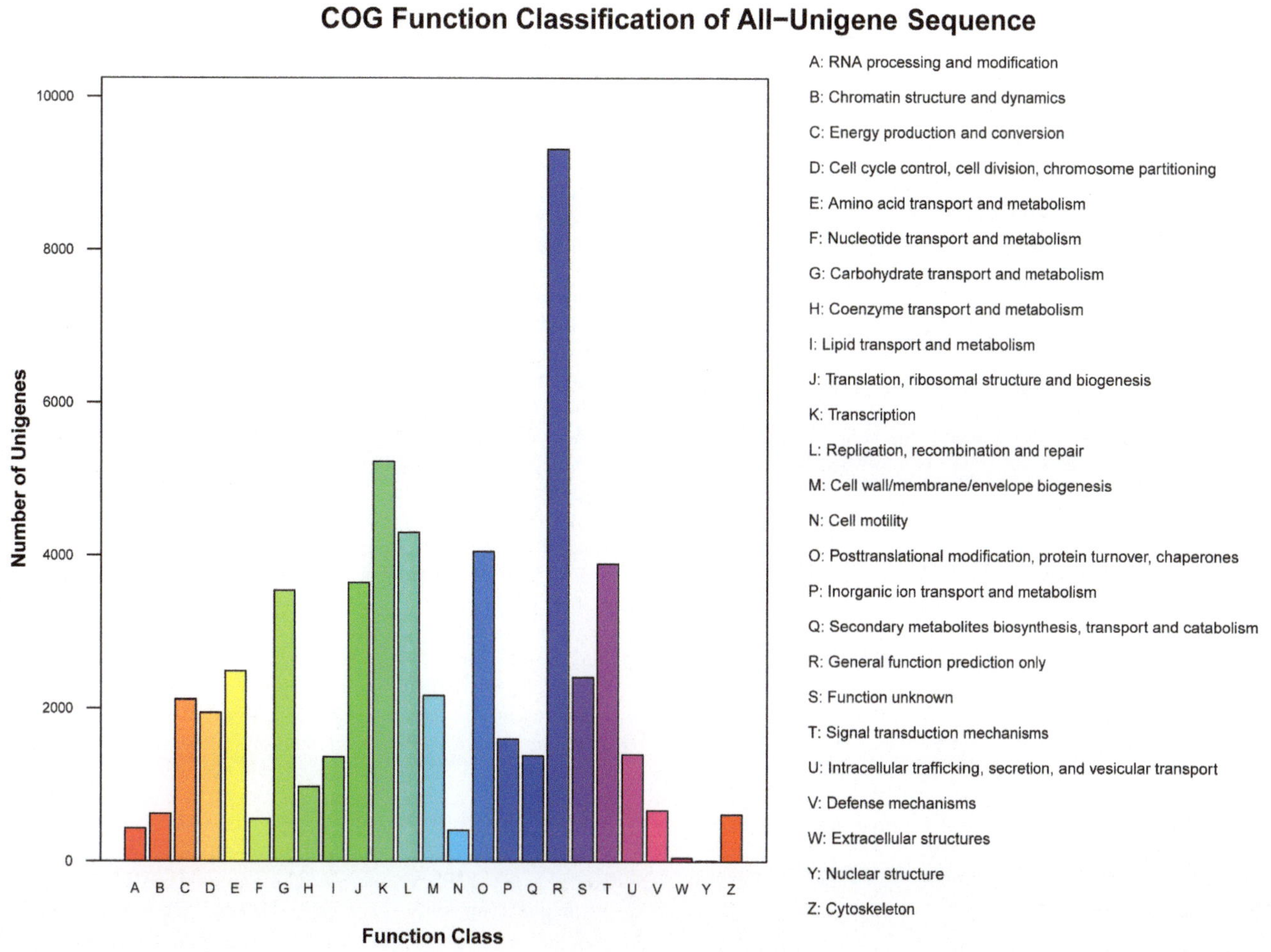

Figure 4. Classification of the clusters of orthologous groups (COG) for the transcriptome of *P. tomentosa* × *P. fortune.*

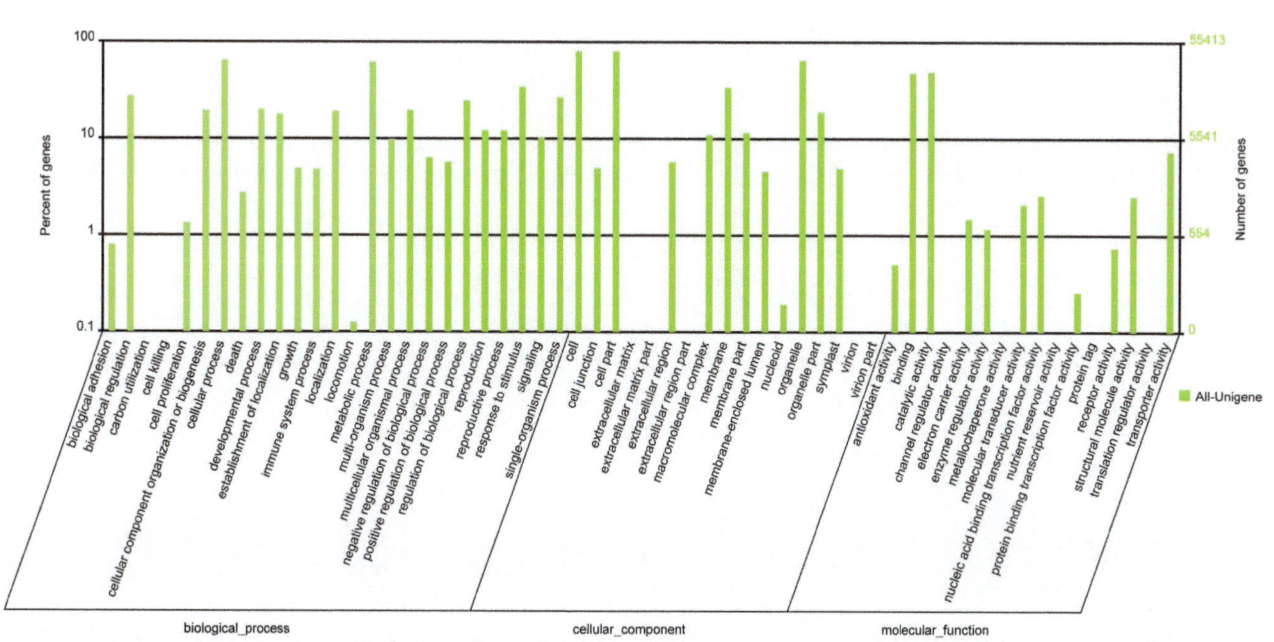

Figure 5. Classification of the gene ontology (GO) for the transcriptome of *P. tomentosa* × *P. fortune.*

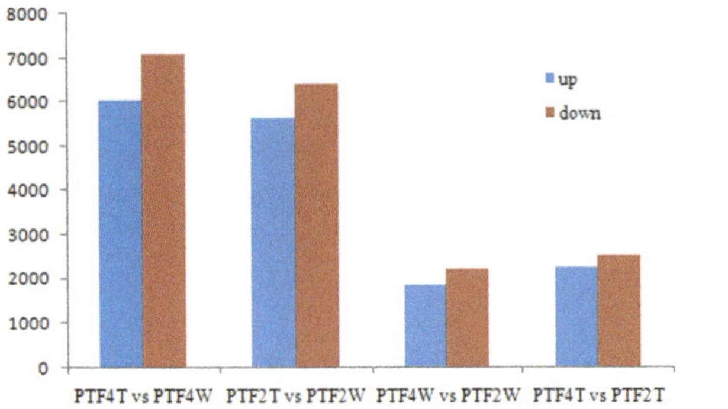

Figure 6. Statistics of differentially expressed genes in each pairwise comparison. Red bars represent the up-regulated genes while blue bars represent the down-regulated ones. PTF4T, 20 days drought treated tetraploid. PTF4W, well-watered tetraploid. PF2FT, 20 days drought treated diploid. PTF2W, well-watered diploid.

vs. the well-watered and the tetraploid vs. the diploid, respectively (Table S10 and S11).

qRT-PCR verification of genes related to drought response in Paulownia leaves

The expression profiles of 13 DEUs from leaves of the drought-treated and the well-watered plants were assessed by qRT-PCR. The results showed that except for the mannose-1-phosphate guanyl transferase beta gene in the PTF2 plants, the expressions of all the other genes were consistent with the expressions from the transcriptome analyses (Figure 9).

Discussion

Water deficiency significantly decreased plant growth in terms of height and basal stem diameter in the PTF2 and PTF4 plants. Paulownia is regarded as a relatively drought-tolerant species, and *P. tomentosa* × *P. fortune* plants, which were established in an ecosystem characterized by drought, have evolved special mechanisms that allow them to deal with insufficient water supply. In this study, the PTF4 plants exhibited more significant growth than the PTF2 plants in both the drought-treated and the well-watered samples. We also observed that PTF2 and PTF4 plants showed substantial differences in several physiological parameters in response to drought stress. The PTF4 seedlings exhibited higher relative water content (RWC), chlorophyll content, superoxide dismutase (SOD) activity, proline content and soluble protein content than the PTF2 seedlings under drought stress, this may indicate that PTF4 has stronger water absorption and cell turgor, and a more efficient enzymatic antioxidant system under water deficit conditions. Additionally, PTF4 seedlings showed lower malondialdehyde (MDA) content and relative electrical conductivity (REC) than the PTF2 seedlings in response to drought stress, suggesting more damage to membranes in PTF2 plants. These physiological parameters imply that PTF4 plants may be better capable of surviving drought conditions, which is consistent with the findings of our previous study [27]. In the present study, the transcriptome-wide gene expression profiling revealed differences in the abundance of drought-responsive genes between the PTF2 and PTF4 plants (Table S4, S5, S6, and S7). Under control conditions, transcript abundances of the majority of DEUs co-up-regulated in PTF2 plants were higher than those in PTF4 plants. However, under water-stress conditions, transcript abundances of co-up-regulated DEUs significantly increased in PTF4, but they maintained a relatively stable level in PTF2 plants, especially DEUs involved in ROS-scavenging system, amino acid and carbohydrate metabolism, and plant hormone biosynthesis and transduction (Table S8 and S9). This implies that PTF4 plants had a greater capability to tackle drought stress by increasing abundance of metabolism and defense-related unigenes. GO terms and pathways analysis showed that diverse metabolic processes were altered and some DEUs indicated the possibility

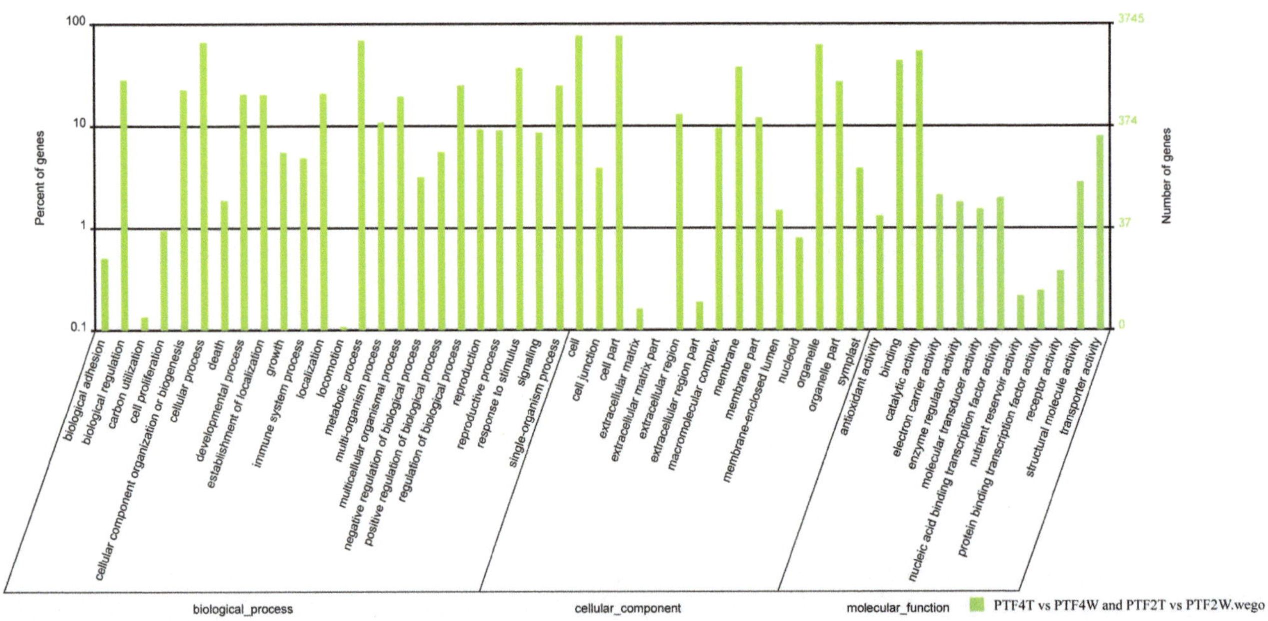

Figure 7. GO function analysis results for the consistently differentially expressed genes in both the PTF4T vs. PT4W and PTF2T vs. PT2W comparisons. PTF4T, 20 days drought treated tetraploid. PTF4W, well-watered tetraploid. PF2FT, 20 days drought treated diploid. PTF2W, well-watered diploid.

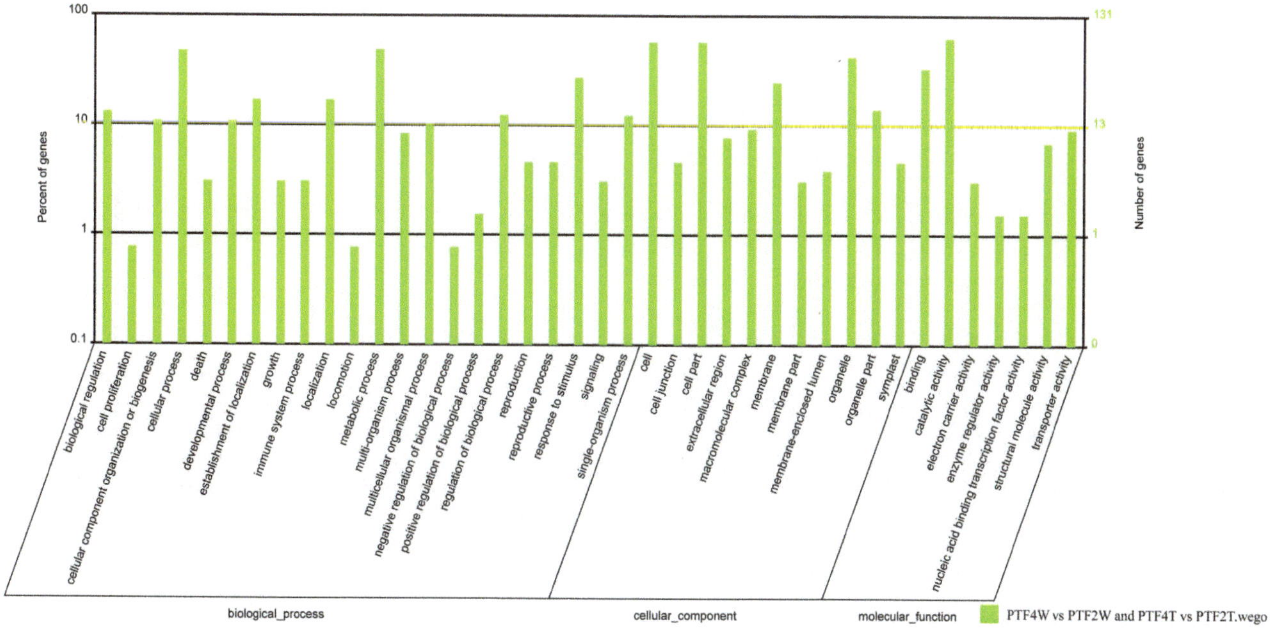

Figure 8. GO function analysis results for the consistently differentially expressed genes in both the PTF4W vs. PTF2W and PTF4T vs. PTF2T comparisons. PTF4W, well-watered tetraploid. PTF2W, well-watered diploid. PTF4T, 20 days drought treated tetraploid. PF2FT, 20 days drought treated diploid.

of crosstalk between pathways in the plants during drought stress. These changes might help the PTF2 and PTF4 plants survive under drought conditions.

Protection from oxidative damage

ROS accumulate in chloroplasts and produce a leakage of electrons towards oxygen under water deficit conditions. ROS serve as integral cellular signal molecules in stress sensing/signal transduction processes; however, excessive levels of ROS may induce oxidative damage to proteins, DNA, and lipids [52]. To protect tissues from oxidative damage, plants have developed an enzymatic antioxidant defense system to remove excess ROS. The ROS-scavenging system of plants includes several important antioxidant enzymes such as superoxide dismutase and ascorbate peroxidase [53]. In the present study, major ROS-scavenging enzymes and antioxidants including superoxide dismutase, ascorbate peroxidase, glutathione peroxidase, and lactoylglutathione lyase were up-regulated by drought stress (Table S8). However, differences in the expressions of these antioxidant genes were also observed between the two accessions, and the fold changes of these unigenes in the SOD, APX, and CAT families were almost always greater in the PTF4 plants than that in the PTF2 plants (Table S9). This is supported by the observation that both PTF2 and PTF4 plants showed increasing SOD activity, and the values of this trait in the PTF4 plants were always at much higher levels than in the PTF2 plants over the entire treatment process.

A senescence-associated protein was up-regulated in the PTF2 and PTF4 plants under drought stress, and its homologs, senescence associated protein 1 and senescence-associated gene 21, were over-expression in Arabidopsis to protect the plant from oxidative damage [54]. The Unigene22025 transcript was strongly enriched and up-regulated in the PTF2 and PTF4 plants during drought stress; interestingly, this transcript was also reported to be significantly differentially expressed in a previous study [55]. Additionally, the PTF2 and PTF4 plants were found to accumulate of high amounts of galactinol and raffinose during drought stress (discussed below), which may also help improve ROS scavenging in the drought-stressed plants [56].

Amino acid and carbohydrate metabolism

Low-molecular weight compatible osmolytes (osmoprotectants), such as proline, glycine, betaine, and sugar alcohols have been reported to accumulate in a variety of plant species in response to environmental stresses. This accumulation allows plants to maintain their water absorption and cell turgor under water deficit conditions by allowing rapid water intake from rather dry soil, thereby, mitigating the effects of water stress [57–59]. In the present study, various genes that may be involved in proline and carbohydrate metabolism were found.

Proline is a common osmoprotectant in water-stressed plants and its dramatic accumulation has been shown to confer osmotic tolerance to plants during drought stress [60]. In our study, unigenes encoding aldehyde dehydrogenases and glutamate synthase, which play important roles in catalyzing the conversion of Δ^1-pyrroline 5-carboxylate synthetase (P5C) to glutamate, were found to be down-regulated in the PTF2T and PTF4T plants; however, the expression levels of these DEUs were higher in PTF2T than in PTF4T. Conversely, a unigene encoding proline dehydrogenase (ProDH) was up-regulated in PTF2T, while its expression was unaltered in PTF4T plants (Table S8 and S9).

In plants, glutamate is catalyzed to P5C by Δ^1-pyrroline 5-carboxylate synthetase (P5CS) and P5C is then converted to proline by P5C reductase (P5CR). However, ProDH catalyzes the conversion of proline to P5C [61]. Therefore, the accumulation of proline in dehydrated plants has been linked to the up-regulation of P5CS and the down-regulation of ProDH [62]. In our study, unigenes encoding P5CS and P5CR were not among the selected DEUs (Table S8 and S9). However, the values of physiological parameters showed that the proline content in the PTF4 plants increased significantly with the increase of the drought stress, we

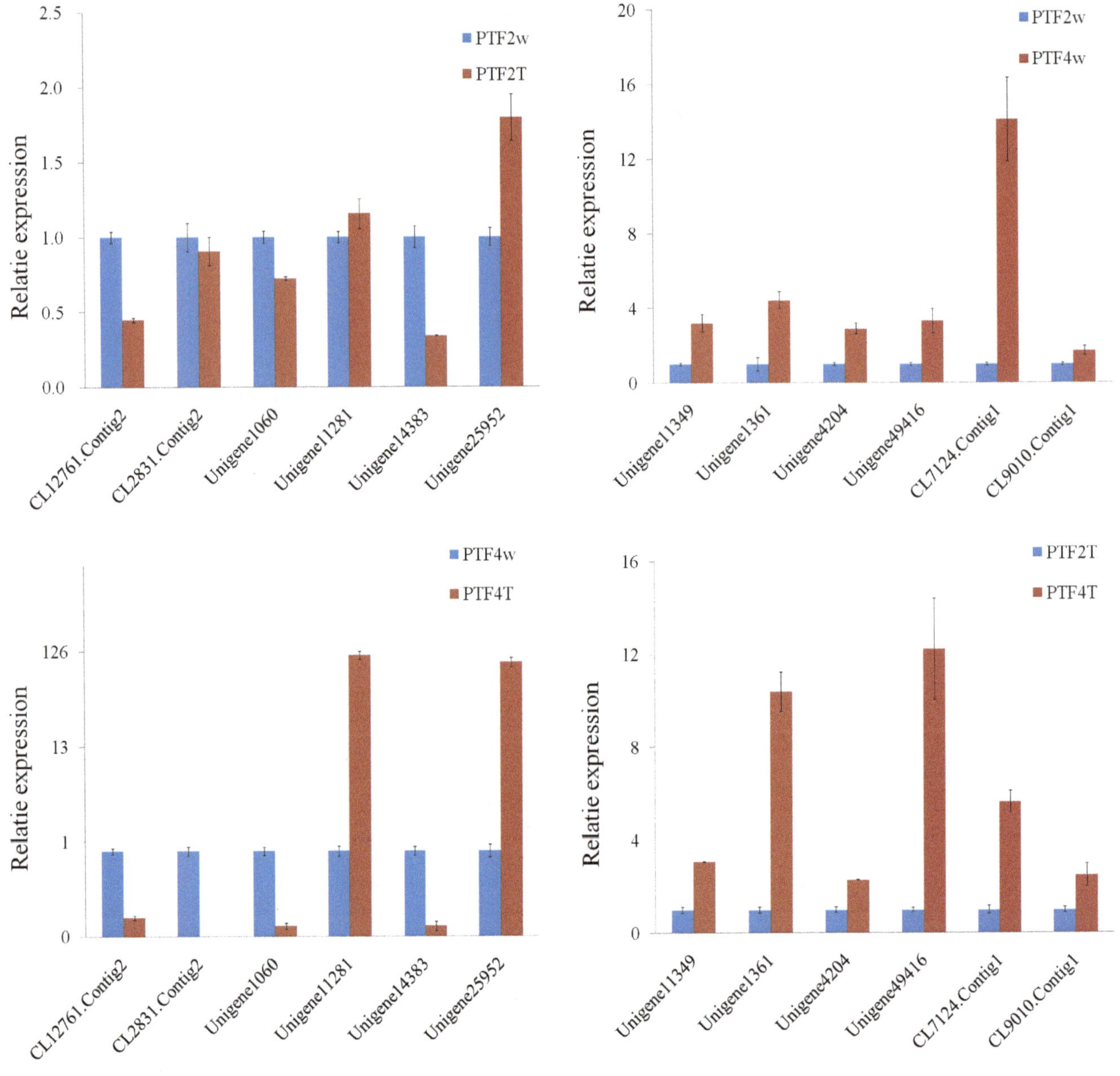

Figure 9. qRT-PCR analysis of candidate drought response genes. A, PTF2W, well-watered diploid; PTF2T, drought treated diploid. B, PTF4W, well-watered diploid; PTF4T, drought treated diploid.

therefore speculate that processes other than transcriptional regulation might play a role in influencing the accumulation of proline in PTF4 plants.

Unigenes encoding proteins involved in carbohydrate metabolism (e.g., galactinol, sucrose, fructose, and mannose metabolism) accounted for a large number of the DEUs. It has been reported that the accumulation of soluble carbohydrates is strongly correlated with stress tolerance in plants [63]. Under drought stress, we observed that most of these DEUs involved in carbohydrate metabolism were co-up-regulated in the PTF4 and PTF2 plants; however, the fold changes of these unigenes were greater in PTF4 than in PTF2 plants (Table S8 and S9). The expression of the unigene encoding galactinol synthase (GolS), which is involved in galactinol and raffinose family oligosaccharides (RFO) metabolism, was increased in PTF2T and PTF4T plants. In Arabidopsis, the homologs of *GolS*, *AtGolS1* and/or *AtGolS2*, were reported to be over-expressed, which increased galactinol and raffinose accumulation and enhanced the plant's tolerance to drought stress and oxidative damage [56]. The GolS gene in *Populus balsamifera* was found to be significantly differentially expressed in response to drought stress [64]. GolS catalyzes the first step in the biosynthetic pathway of galactinol and raffinose family oligosaccharides using galactose and myo-inositol as substrates. Several unigenes encoding aldose 1-epimerase, which are involved in galactose and myo-inositol synthesis, were found to be elevated in both PTF4T and PTF2T plants (Table S8 and S9). The expressions of several trehalose synthase encoding unigenes were found to be significant enhanced in response to drought stress in both accessions. The trehalose-6-phosphate synthase 1 (TPS1) encoding gene, which was homologous to *AtTPS1,* was of particular interest because *AtTPS1* plays an important role in abscisic acid (ABA) and sugar signaling and its over-expression resulted in improved drought tolerance in Arabidopsis [65]. Further, the over-expression of *AtTPS1* in rice was reported to elevate the sucrose, fructose, and glucose content and enhance the drought stress response [66]. Several DEUs with homology to glucose-6-phosphate/phosphate translocator 2 (GPT2), which was found to be involved in glucose 6-phosphate transport in Arabidopsis [67], were up-regulated in PTF4T and PTF2T, implying that the sugar signal triggered by GPT2 may be important in regulating drought stress tolerance in *P. tomentosa* × *P. fortunei* plants. We also found a very robust increase in soluble sugar contents that was more pronounced in PTF4 than PTF2 plants in our physiological measurements. The accumulation of these osmoprotectants in the leaves could help enhance the osmotic stress tolerance of *P. tomentosa* × *P. fortunei.*

Hormone biosynthesis

Plant hormones are closely related to plant growth and stress tolerance [68,69]. ABA plays crucial roles in various stress responses and the accumulation of endogenous ABA is known to be associated with drought tolerance [70,71]. Recently, several studies have revealed the molecular basis of ABA biosynthesis and its response mechanisms [72–74].

The over-expression of 9-cis-epoxycarotenoid dioxygenase (NCED), Zeaxanthin epoxidase (ABA2), and phytoene synthase were reported to be associated with the conversion of zeaxanthin to ABA, which can be metabolized and catalyzed by the ABA hydroxylase CYP707A1 [75]. In transgenic Arabidopsis plants, *AtNCED3* and the phytoene synthase gene were induced by drought and their up-regulation was reported to enhance drought tolerance by increasing the endogenous ABA level [76,77]. Furthermore, ABA2 was found to be a negative regulator of ABA in Arabidopsis [78,79]. Therefore, ABA biosynthesis can be enhanced by positive regulators (NCED3, phytoene synthase) and inhibited by negative regulators (ABA2). In our study, transcript abundances of hormone biosynthesis-related DEUs were significantly different between PTF2 and PTF4. The unigenes that were most similar to the Arabidopsis *NCED3* and phytoene synthase gene sequences were up-regulated, and the unigenes encoding the ABA2 enzymes were down-regulated in the PTF2T and PTF4T plants, transcript abundances of these co-up-regulated and co-down-regulated DEUs have higher fold change in PTF4 than in PTF2 (Table S8 and S9). Our data suggest that these genes may play important roles in regulating the biosynthesis of ABA in the two accessions.

ABA signaling pathways have been linked to phospholipase D (PLD), protein phosphatase 2C (PP2C), SNF1-related kinases (SnPKs), and CBL-interacting protein kinases (CIPKs), and unigenes encoding these enzymes were significantly up- or down-regulation more in the PTF4T than that in the PTF2T plants under drought stress (Table S8 and S9). PP2C was found previously to negatively regulate ABA signaling in Arabidopsis [80] and links between PP2C and SnRK2 have been reported [81–83]. Several members of the SnRK2 family in Arabidopsis and rice were shown to be activated in response to ABA or osmotic stress [84,85], and in fava bean guard cells SnRK2 was shown to play an important role in ABA-dependent stomatal closure [86]. Together these results suggest that SnRK2 is an important regulator of ABA signal transduction during drought stress [83]. CIPKs and their interacting partner calcineurin B-like (CBL) (also known as SnRK3) form a complex that was reported to be a central hub of the ABA-mediated signaling pathway under drought, cold, or salt stress [87,88]. Although several genes encoding potential factors of the ABA signaling pathway were identified in the PTF2T and PTF4T plants, the molecular mechanisms that control this complex signaling pathway in *P. tomentosa* × *P. fortunei* need to be investigated further.

Recent studied have revealed that the plant hormone indol-3yl-acetic acid (IAA) plays a critical role in regulating plant growth and cell enlargement [89]. The auxin transporter protein AUX1 has been found to be a major influx carrier [90]. Aux/IAA proteins and the ARF transcription factors are important regulators of IAA-modulated gene expression, and auxin-responsive promoter (GH3) and auxin-responsive protein (SAUR) have been reported to regulate plant growth and cell enlargement [89,91]. In the present study, several increased transcript abundance of the DEUs in both PTF2T and PTF4T were annotated as being involved in crosstalk among the IAA encoding unigenes, whereas we observed that the expression level of these DEUs were higher in PTF4 under drought stress (Table S8 and S9), suggesting that different levels of IAA accumulation might explain the cell enlargement and plant growth differences observed between the PTF2 and PTF4 plants.

Multi-stress responsive genes in *P. tomentosa* × *P. fortunei*

Transcription factors (TFs) have been shown to be critical upstream regulators of plant responses to various biotic and abiotic stresses [92]. In the present study, a large number of DEUs encoded TFs belonging to the MYB/MYC, ABI3VP1, AP2-EREBP, DREB, ARF, bzip, ABA responsive, NAC, WRKY, bHLH, ZF-HD, AREB/ABF, and ZFP families. Under control conditions, most of the DEUs encoded TFs had higher transcript abundance in PTF2 than in PTF4, while under drought stress, expression changes of these DEUs were significantly higher in PTF4 than in PTF2 plants (Table S8 and S9). These TFs were divided into ABA-dependent signaling systems and ABA-independent signaling pathways. Over-expression of the MYB/MYC, bHLH, NAC, WRKY, ZFP, bzip, and ZF-HD families has been shown to play significant roles in controlling the expression of specific stress-related genes and plant stress signal transduction, thereby greatly enhancing the tolerance of plants to drought and high salinity in Arabidopsis and rice [93–97]. Recently, transcriptional activators that up-regulate stress-responsive genes have been used widely to impart drought tolerance to transgenic plants [98]. The over-expression of the AP2-type TF and ethylene-responsive transcription factors (SHN1–3) in transgenic plants were found to regulate several stress inducible target genes and increase drought tolerance [99]. Increased expression of the NAC transcription factor 45 (OsNAC045) in transgenic rice plants was found to enhance drought tolerance, which may serve as a useful tool for improving drought stress response in genetically engineered rice [94]. Recent studies have shown that ABRE-binding bZIP and OsABF2 were detectable in transgenic rice where they positively regulated drought tolerance and ABA signaling [100]. Transcriptional repressors that have the ability to repress the expression of genes under drought stress have been shown to be important in engineering drought tolerance in plants [101]. AtMYB60 was reported to adjust stomatal movement and function as a transcriptional repressor in the plants response to dehydration; specifically, the expression of this gene in Arabidopsis guard cells was negatively regulated during drought stress [102]. A null T-DNA insertion mutation in the AtMYB60 gene led to reduced stomatal opening and minimized wilting during drought stress, this mutation resulted in defective guard cell but had no significant effect on other physiological processes and agricultural traits [102]. In our study, several signal transduction pathways that independently respond to drought stress (in both an ABA dependent- and independent- manner) were identified in the PTF2T and PTF4T plants. ABA-dependent signaling pathway have been illustrated that they could mediate stress adaptation by the action of ABA-inducible TFs controlling the expression of genes containing cis-acting ABA response elements (e.g., TFs such as MYB/MYC and AREB/ABF), while ABA-independent stress signaling pathway leading to gene expression involve different TFs such as (1) NAC and ZF-HD, and (2) CBF/DREB (Figure S2) [103–106], suggesting that drought stress tolerance at the transcriptional level was controlled by an extremely intricate gene regulatory network. Thus, functional analysis of drought inducible TFs will help in the development of drought tolerant plants that will survive well in acute field condition.

Different sets of genes are known to be associated with specific stress-responsive signal transduction. Mitogen-activated protein kinases (MAPKs) are a highly conserved family of serine/threonine protein kinases that are involved in a variety of fundamental cellular processes such as proliferation, differentiation, stress response, and survival. The MAPK cascade has been found to lead to accumulated production of osmolytes by activating, for example, the protein tyrosine kinases and G-protein receptors [107,108]. The MAPK pathway phosphorylates a three-tiered kinase cascade: MAPKKK→MAPKK→MAPK, which propagates downstream stress signals [109]. Activated MAPK can migrate to the nucleus to activate the TFs directly. These findings indicate that modulating MAPK may lead to altered expression of specific regulatory TFs and protein kinases towards enhanced drought tolerance in plants. In this study, many DEUs in the PTF2T and PTF4T plants were identified as MAPKs including MAPK8, MAPK9, MAPK14, MAPKK2, MAPKK6, MAPKKK1, and MAPKKK2, and a relatively large proportion of DEUs encoding MAPKs were up-regulated both in the PTF2 and PTF4 plants under drought stress, but the fold changes of these DEUs were greater in PTF4 than in PTF2. We also identified a great number of calcium-dependent protein kinases (CDPKs) that were activated and the expression levels of these unigenes were induced strongly in PTF4 plants under drought treatments (Table S8 and S9). Previous studies have indicated that CDPKs play a crucial role in Ca^{2+}-mediated signal transduction in plants in response to drought and salinity stresses, suggesting that signal transduction may play a critical role in the *P. tomentosa* × *P. fortunei* drought response [107,110].

Plants have been found to survive different stresses using multifunctional genes that enable crosstalk among the various abiotic stress resistance systems [111]. We also found that in PTF4, several stress-regulated unigenes like lipid transfer protein, late embryogenesis abundant protein (LEA), osmotin, small heat shock protein, HSP70, HSP90, antiporter, and dehydrins were expressed at higher transcript abundance than in PTF2 plants under drought stress. Several study reported that these genes were generally involved in cell rescue and defense mechanisms and it was suggested that their overexpression may play an important role in conferring dehydration and drought stress by causing stomatal closure and osmotic adjustment [112–116].

Conclusions

PTF4 trees have high ecological and economic value. In the present work, we performed large-scale transcriptome sequencing of diploid and autotetraploid *P. tomentosa* × *P. fortune* plants under drought stress using an Illumina/Solexa sequencing platform. More than 291 million reads were generated from four libraries and *de novo* assembled into 98,671 nonredundant unigenes that will provide a useful resource for this species. GO terms enrichment and pathways analyses of the annotated DEUs revealed changes in several metabolic processes as a result of drought stress. These results may shed some light on the biology of the drought response and tolerance in Paulownia species. In addition to the previously reported drought stress related genes detected in the libraries, a large number of unigene sequences could not be mapped to known sequences in the protein databases. Some of the unidentified DEU sequences may be novel and/or unknown Paulownia genes associated with drought stress. Functional analysis of these genes may contribute to a deeper understanding of the response of Paulownia plants to drought. Our findings will help enrich our understanding of the drought response in diploid and autotetraploid *P. tomentosa* × *P. fortune* and may provide new insights into the responses of woody plants to drought stress.

Supporting Information

Figure S1 Distribution of unigene lengths in the transcriptome of *P. tomentosa* × *P. fortunei*. The sizes of all unigenes were calculated.

Figure S2 A schematic representation of transcriptional regulatory networks of cis-acting elements and transcription factors involved in abiotic-stress-responses. Transcription factors are shown in ellipses; cis-acting elements are shown in boxes.

Table S1 Primers of quantitative RT-PCR analysis of candidate drought response genes. -f represents forward primers and -r represents reverse primers.

Table S2 Unigene annotations based on their BLASTX hits in the nr database.

Table S3 KEGG annotation of unigenes.

Table S4 Differetially expressed genes in the PTF2T vs. PTF2W comparison. PTF2T, 20 days drought-treated diploid; PTF2W, well-watered diploid.

Table S5 Differetially expressed genes in the PTF4T vs. PTF4W comparison. PTF4T, 12 days drought-treated tetraploid; PTF4W, well-watered tetraploid.

Table S6 Differetially expressed genes in the PTF4W vs. PTF2W comparison. PTF4W, well-watered tetraploid; PTF2W, well-watered diploid.

Table S7 Differetially expressed genes in the PTF4T vs. PTF2T comparison. PTF4T, 20 days drought-treated tetraploid; PTF2T, 20 days drought-treated diploid.

Table S8 Consistently differentially expressed genes in both the PTF2T vs. PTF2W and PTF4T vs. PTF4W comparisons. PTF2T, 20 days drought-treated diploid; PTF2W, well-watered diploid; PTF4T, 20 days drought-treated tetraploid; PTF4W, well-watered tetraploid.

Table S9 Consistently differentially expressed genes in both the PTF4W vs. PTF2W and PTF4T vs. PTF2T comparisons. PTF4W, well-watered tetraploid; PTF2W, well-watered diploid; PTF4T, 20 days drought-treated tetraploid; PTF2T, 20 days drought-treated diploid.

Table S10 KEGG pathway analysis results for the consistently differentially expressed genes in both the PTF2T vs. PTF2W and PTF4T vs. PTF4W comparisons. PTF2T, 20 days drought-treated diploid; PTF2W, well-watered diploid; PTF4T, 20 days drought-treated tetraploid; PTF4W, well-watered tetraploid.

Table S11 KEGG pathway analysis results for the consistently differentially expressed genes in both the PTF4W vs. PTF2W and PTF4T vs. PTF2T comparisons. PTF4W, well-watered tetraploid; PTF2W, well-watered diploid; PTF4T, 20 days drought-treated tetraploid; PTF2T, 20 days drought-treated diploid.

Author Contributions

Conceived and designed the experiments: EKX GQF SYN ZLZ MJD YPD. Performed the experiments: EKX GQF SYN ZLZ MJD YPD. Analyzed the data: EKX GQF SYN ZLZ MJD YPD. Contributed reagents/materials/analysis tools: EKX GQF SYN ZLZ MJD YPD. Contributed to the writing of the manuscript: EKX GQF SYN.

References

1. Valliyodan B, Nguyen HT (2006) Understanding regulatory networks and engineering for enhanced drought tolerance in plants. Current opinion in plant biology 9: 189–195.
2. Wang W, Vinocur B, Shoseyov O, Altman A (2000) Biotechnology of plant osmotic stress tolerance physiological and molecular considerations. pp. 285–292.
3. Hamdy A, Ragab R, Scarascia-Mugnozza E (2003) Coping with water scarcity: water saving and increasing water productivity. Irrigation and drainage 52: 3–20.
4. Bohnert HJ, Ayoubi P, Borchert C, Bressan RA, Burnap RL, et al. (2001) A genomics approach towards salt stress tolerance. Plant Physiology and Biochemistry 39: 295–311.
5. Seki M, Narusaka M, Ishida J, Nanjo T, Fujita M, et al. (2002) Monitoring the expression profiles of 7000 Arabidopsis genes under drought, cold and high-salinity stresses using a full-length cDNA microarray. The Plant Journal 31: 279–292.
6. Fowler S, Thomashow MF (2002) Arabidopsis transcriptome profiling indicates that multiple regulatory pathways are activated during cold acclimation in addition to the CBF cold response pathway. The Plant Cell Online 14: 1675–1690.
7. Seki M, Kamei A, Satou M, Sakurai T, Fujita M, et al. (2004) Transcriptome analysis in abiotic stress conditions in higher plants. Plant Responses to Abiotic Stress: Springer. pp. 271–308.
8. Shinozaki K, Yamaguchi-Shinozaki K (2007) Gene networks involved in drought stress response and tolerance. Journal of experimental botany 58: 221–227.
9. Cohen D, Bogeat-Triboulot M-B, Tisserant E, Balzergue S, Martin-Magniette M-L, et al. (2010) Comparative transcriptomics of drought responses in Populus: a meta-analysis of genome-wide expression profiling in mature leaves and root apices across two genotypes. BMC genomics 11: 630.
10. Zheng J, Fu J, Gou M, Huai J, Liu Y, et al. (2010) Genome-wide transcriptome analysis of two maize inbred lines under drought stress. Plant molecular biology 72: 407–421.
11. Tang S, Liang H, Yan D, Zhao Y, Han X, et al. (2013) Populus euphratica: the transcriptomic response to drought stress. Plant molecular biology 83: 539–557.
12. Hiremath PJ, Farmer A, Cannon SB, Woodward J, Kudapa H, et al. (2011) Large-scale transcriptome analysis in chickpea (Cicer arietinum L.), an orphan legume crop of the semi-arid tropics of Asia and Africa. Plant biotechnology journal 9: 922–931.
13. Peng S, Jiang H, Zhang S, Chen L, Li X, et al. (2012) Transcriptional profiling reveals sexual differences of the leaf transcriptomes in response to drought stress in Populus yunnanensis. Tree physiology 32: 1541–1555.
14. Hall T (2008) Paulownia: An Agroforestry Gem. Trees for Life Journal 3: 3.
15. Zhu Z, Chao C, Lu X, Xiong Y (1986) Paulownia in China: cultivation and utilization: International Development Research Centre, beijing, china.
16. Ipekci Z, Gozukirmizi N (2003) Direct somatic embryogenesis and synthetic seed production from Paulownia elongata. Plant Cell Reports 22: 16–24.
17. Masterson J (1994) Stomatal size in fossil plants: evidence for polyploidy in majority of angiosperms. Science (New York, NY) 264: 421–424.
18. Dwivedi N, Suryanarayana N, Sikdar A, Susheelamma B, Jolly M (1989) Cytomorphological studies in triploid mulberry evolved by diploidization of female gamete cells. Cytologia 54: 13–19.
19. Leitch A, Leitch I (2008) Genomic plasticity and the diversity of polyploid plants. Science 320: 481–483.
20. Zhang ZY, Yu XS, Zhu ZT (1997) Advances in the triploid breeding programme of Populus tomentosa in China. Journal of Beijing Forestry University, English edition 6: 1–8.
21. Stupar RM, Bhaskar PB, Yandell BS, Rensink WA, Hart AL, et al. (2007) Phenotypic and transcriptomic changes associated with potato autopolyploidization. Genetics 176: 2055–2067.
22. Riddle NC, Jiang H, An L, Doerge R, Birchler JA (2010) Gene expression analysis at the intersection of ploidy and hybridity in maize. Theoretical and applied genetics 120: 341–353.
23. Fock II, Collonnier C, Purwito A, Luisetti J, Souvannavong VV, et al. (2000) Resistance to bacterial wilt in somatic hybrids between Solanum tuberosum and Solanum phureja. Plant Sci 160: 165–176.
24. Huang S, Sirikhachornkit A, Su X, Faris J, Gill B, et al. (2002) Genes encoding plastid acetyl-CoA carboxylase and 3-phosphoglycerate kinase of the Triticum/Aegilops complex and the evolutionary history of polyploid wheat. Proc Natl Acad Sci U S A 99: 8133–8138.
25. Zhang XY, Hu CG, Yao JL (2010) Tetraploidization of diploid Dioscorea results in activation of the antioxidant defense system and increased heat tolerance. J Plant Physiol 167: 88–94.
26. Fan GQ, Zhai XQ, Wei ZZ, Yang ZQ (2010) Induction of autotetraploid from the somatic cell of *Paulownia tomentosa* × *P. fortunei* and its in vitro plantlet regeneration. Journal of Northwest Forestry University 38: 22–26.
27. Zhang XS, Zhai XQ, Fan GQ, Deng MJ, Zhao ZL (2013) Observation on microstructure of leaves and stress tolerance analysis of different tetraploid Paulownia. Journal of Henan Agricultural University 46: 646–650.
28. Moumeni A, Satoh K, Kondoh H, Asano T, Hosaka A, et al. (2011) Comparative analysis of root transcriptome profiles of two pairs of drought-tolerant and susceptible rice near-isogenic lines under different drought stress. BMC plant biology 11: 174.
29. Barrs HD WP (1962) A re-examination of the relative turgidity technique for estimating water deficit in leaves. Aust J Biol Sci 15: 413–428.
30. Bojović BM, Stojanović J (2005) Chlorophyll and carotenoid content in wheat cultivars as a function of mineral nutrition. Archives of Biological Sciences 57: 283–290.
31. Arnon DI (1949) Copper enzymes in isolated chloroplasts. Polyphenoloxidase in Beta vulgaris. Plant physiology 24: 1–15.
32. Hodges DM, DeLong JM, Forney CF, Prange RK (1999) Improving the thiobarbituric acid-reactive-substances assay for estimating lipid peroxidation in plant tissues containing anthocyanin and other interfering compounds. Planta 207: 604–611.
33. Liu FH, Guo Y, Gu DM, Xiao G, Chen ZH, Chen SY (1996) Salt tolerance of transgenic plants with BADH cDNA. Acta Genetica Sinica 24: 54–58.
34. Dhindsa RS, Plumb-Dhindsa P, Thorpe TA (1981) Leaf senescence: correlated with increased levels of membrane permeability and lipid peroxidation, and decreased levels of superoxide dismutase and catalase. Journal of Experimental botany 32: 93–101.
35. Sun Y, Oberley LW, Li Y (1988) A simple method for clinical assay of superoxide dismutase. Clinical Chemistry 34: 497–500.
36. Guy C, Haskell D, Neven L, Klein P, Smelser C (1992) Hydration-state-responsive proteins link cold and drought stress in spinach. Planta 188: 265–270.
37. Bradford MM (1976) A rapid and sensitive method for the quantitation of microgram quantities of protein utilizing the principle of protein-dye binding. Analytical biochemistry 72: 248–254.
38. Bates L, Waldren R, Teare I (1973) Rapid determination of free proline for water-stress studies. Plant and soil 39: 205–207.
39. Irigoyen J, Einerich D, Sánchez-Díaz M (1992) Water stress induced changes in concentrations of proline and total soluble sugars in nodulated alfalfa (Medicago sativd) plants. Physiologia Plantarum 84: 55–60.
40. SnedecorGW., WG C (1980). Statistical methods7th Edition: Iowa State University Press, Ames, IOWA. pp. 507.
41. Grabherr MG, Haas BJ, Yassour M, Levin JZ, Thompson DA, et al. (2011) Full-length transcriptome assembly from RNA-Seq data without a reference genome. Nat Biotechnol 29: 644–652.
42. Pertea G, Huang X, Liang F, Antonescu V, Sultana R, et al. (2003) TIGR Gene Indices clustering tools (TGICL): a software system for fast clustering of large EST datasets. Bioinformatics 19: 651–652.

43. Iseli C, Jongeneel CV, Bucher P (1999) ESTScan: a program for detecting, evaluating, and reconstructing potential coding regions in EST sequences. Proc Int Conf Intell Syst Mol Biol: 138–148.
44. Conesa A, Gotz S, Garcia-Gomez JM, Terol J, Talon M, et al. (2005) Blast2GO: a universal tool for annotation, visualization and analysis in functional genomics research. Bioinformatics 21: 3674–3676.
45. Ye J, Fang L, Zheng H, Zhang Y, Chen J, et al. (2006) WEGO: a web tool for plotting GO annotations. Nucleic Acids Res 34: W293–297.
46. Kanehisa M, Araki M, Goto S, Hattori M, Hirakawa M, et al. (2008) KEGG for linking genomes to life and the environment. Nucleic Acids Res 36: D480–484.
47. Ali Mortazavi BAW, Kenneth McCue, et al. (2008) Lorian Schaeffer and Barbara Wold Mapping and quantifying mammalian transcriptomes by RNA-Seq. Nature Methods 5: 621–628.
48. Audic S, Claverie JM (1997) The significance of digital gene expression profiles. Genome Res 7: 986–995.
49. Benjamini Y, Yekutieli D (2001) The control of the false discovery rate in multiple testing under dependency. Annals of statistics: 1165–1188.
50. Tang S, Liang H, Yan D, Zhao Y, Han X, et al. (2013) Populus euphratica: the transcriptomic response to drought stress. Plant Mol Biol 83: 539–557.
51. Cohen D, Bogeat-Triboulot MB, Tisserant E, Balzergue S, Martin-Magniette ML, et al. (2010) Comparative transcriptomics of drought responses in Populus: a meta-analysis of genome-wide expression profiling in mature leaves and root apices across two genotypes. BMC Genomics 11: 630.
52. Mittler R, Zilinskas BA (1994) Regulation of pea cytosolic ascorbate peroxidase and other antioxidant enzymes during the progression of drought stress and following recovery from drought. Plant J 5: 397–405.
53. Bowler C, Slooten L, Vandenbranden S, De Rycke R, Botterman J, et al. (1991) Manganese superoxide dismutase can reduce cellular damage mediated by oxygen radicals in transgenic plants. The EMBO journal 10: 1723.
54. Woo HR, Kim JH, Nam HG, Lim PO (2004) The delayed leaf senescence mutants of Arabidopsis, ore1, ore3, and ore9 are tolerant to oxidative stress. Plant Cell Physiol 45: 923–932.
55. Yan DH, Fenning T, Tang S, Xia X, Yin W (2012) Genome-wide transcriptional response of Populus euphratica to long-term drought stress. Plant Sci 195: 24–35.
56. Nishizawa A, Yabuta Y, Shigeoka S (2008) Galactinol and raffinose constitute a novel function to protect plants from oxidative damage. Plant Physiol 147: 1251–1263.
57. Ashraf M, Foolad M (2007) Roles of glycine betaine and proline in improving plant abiotic stress resistance. Environmental and Experimental Botany 59: 206–216.
58. Csonka L, Gelvin S, Goodner B, Orser C, Siemieniak D, et al. (1988) Nucleotide sequence of a mutation in the proB gene of Escherichia coli that confers proline overproduction and enhanced tolerance to osmotic stress. Gene 64: 199–205.
59. Delauney AJ, Verma DPS (1993) Proline biosynthesis and osmoregulation in plants. The plant journal 4: 215–223.
60. Verma DPS (1999) Osmotic Stress Tolerance in Plants: Role of Proline and Sulfur Metabolisms. Molecular Responses to Cold, Drought, Heat and Salt Stress in Higher Plants: Landes, Austin, pp. 153–168.
61. Kishor PK, Sangam S, Amrutha R, Laxmi PS, Naidu K, et al. (2005) Regulation of proline biosynthesis, degradation, uptake and transport in higher plants: its implications in plant growth and abiotic stress tolerance. Curr Sci 88: 424–438.
62. Evers D, Lefèvre I, Legay S, Lamoureux D, Hausman J-F, et al. (2010) Identification of drought-responsive compounds in potato through a combined transcriptomic and targeted metabolite approach. Journal of experimental botany 61: 2327–2343.
63. Oliver MJ, Cushman JC, Koster KL (2010) Dehydration tolerance in plants. Methods Mol Biol 639: 3–24.
64. Wilkins O, Waldron L, Nahal H, Provart NJ, Campbell MM (2009) Genotype and time of day shape the Populus drought response. Plant J 60: 703–715.
65. Kolukisaoglu U, Weinl S, Blazevic D, Batistic O, Kudla J (2004) Calcium sensors and their interacting protein kinases: genomics of the Arabidopsis and rice CBL-CIPK signaling networks. Plant Physiol 134: 43–58.
66. Yoshida R, Umezawa T, Mizoguchi T, Takahashi S, Takahashi F, et al. (2006) The regulatory domain of SRK2E/OST1/SnRK2.6 interacts with ABI1 and integrates abscisic acid (ABA) and osmotic stress signals controlling stomatal closure in Arabidopsis. J Biol Chem 281: 5310–5318.
67. Niewiadomski P, Knappe S, Geimer S, Fischer K, Schulz B, et al. (2005) The Arabidopsis plastidic glucose 6-phosphate/phosphate translocator GPT1 is essential for pollen maturation and embryo sac development. Plant Cell 17: 760–775.
68. He XJ, Mu RL, Cao WH, Zhang ZG, Zhang JS, et al. (2005) AtNAC2, a transcription factor downstream of ethylene and auxin signaling pathways, is involved in salt stress response and lateral root development. The Plant Journal 44: 903–916.
69. Wolters H, Jürgens G (2009) Survival of the flexible: hormonal growth control and adaptation in plant development. Nature Reviews Genetics 10: 305–317.
70. Wasilewska A, Vlad F, Sirichandra C, Redko Y, Jammes F, et al. (2008) An update on abscisic acid signaling in plants and more.... Molecular Plant 1: 198–217.
71. Zeevaart J, Creelman R (1988) Metabolism and physiology of abscisic acid. Annual review of plant physiology and plant molecular biology 39: 439–473.
72. Shinozaki K, Yamaguchi-Shinozaki K (1997) Gene expression and signal transduction in water-stress response. Plant Physiology 115: 327–334.
73. Zhu JK (2002) Salt and drought stress signal transduction in plants. Annual review of plant biology 53: 247–273.
74. Nambara E, Marion-Poll A (2005) Abscisic acid biosynthesis and catabolism. Annu Rev Plant Biol 56: 165–185.
75. Nambara E, Marion-Poll A (2005) Abscisic acid biosynthesis and catabolism. Annu Rev Plant Biol 56: 165–185.
76. Iuchi S, Kobayashi M, Taji T, Naramoto M, Seki M, et al. (2001) Regulation of drought tolerance by gene manipulation of 9-cis-epoxycarotenoid dioxygenase, a key enzyme in abscisic acid biosynthesis in Arabidopsis. Plant J 27: 325–333.
77. Thompson AJ, Mulholland BJ, Jackson AC, McKee JM, Hilton HW, et al. (2007) Regulation and manipulation of ABA biosynthesis in roots. Plant Cell Environ 30: 67–78.
78. Cheng WH, Endo A, Zhou L, Penney J, Chen HC, et al. (2002) A unique short-chain dehydrogenase/reductase in Arabidopsis glucose signaling and abscisic acid biosynthesis and functions. Plant Cell 14: 2723–2743.
79. Hoth S, Morgante M, Sanchez JP, Hanafey MK, Tingey SV, et al. (2002) Genome-wide gene expression profiling in Arabidopsis thaliana reveals new targets of abscisic acid and largely impaired gene regulation in the abi1-1 mutant. J Cell Sci 115: 4891–4900.
80. Bertauche N, Leung J, Giraudat J (1996) Protein phosphatase activity of abscisic acid insensitive 1 (ABI1) protein from Arabidopsis thaliana. Eur J Biochem 241: 193–200.
81. Mustilli AC, Merlot S, Vavasseur A, Fenzi F, Giraudat J (2002) Arabidopsis OST1 protein kinase mediates the regulation of stomatal aperture by abscisic acid and acts upstream of reactive oxygen species production. Plant Cell 14: 3089–3099.
82. Yoshida R, Hobo T, Ichimura K, Mizoguchi T, Takahashi F, et al. (2002) ABA-activated SnRK2 protein kinase is required for dehydration stress signaling in Arabidopsis. Plant Cell Physiol 43: 1473–1483.
83. Fujii H, Verslues PE, Zhu JK (2007) Identification of two protein kinases required for abscisic acid regulation of seed germination, root growth, and gene expression in Arabidopsis. Plant Cell 19: 485–494.
84. Wang Y, Ying J, Kuzma M, Chalifoux M, Sample A, et al. (2005) Molecular tailoring of farnesylation for plant drought tolerance and yield protection. Plant J 43: 413–424.
85. Mikolajczyk M, Awotunde OS, Muszynska G, Klessig DF, Dobrowolska G (2000) Osmotic stress induces rapid activation of a salicylic acid-induced protein kinase and a homolog of protein kinase ASK1 in tobacco cells. Plant Cell 12: 165–178.
86. Li J, Wang XQ, Watson MB, Assmann SM (2000) Regulation of abscisic acid-induced stomatal closure and anion channels by guard cell AAPK kinase. Science 287: 300–303.
87. Cheong YH, Kim KN, Pandey GK, Gupta R, Grant JJ, et al. (2003) CBL1, a calcium sensor that differentially regulates salt, drought, and cold responses in Arabidopsis. Plant Cell 15: 1833–1845.
88. Hrabak EM, Chan CW, Gribskov M, Harper JF, Choi JH, et al. (2003) The Arabidopsis CDPK-SnRK superfamily of protein kinases. Plant Physiol 132: 666–680.
89. Jain M, Kaur N, Tyagi AK, Khurana JP (2006) The auxin-responsive GH3 gene family in rice (Oryza sativa). Functional & integrative genomics 6: 36–46.
90. Vandenbussche F, Petrášek J, Žádníková P, Hoyerová K, Pešek B, et al. (2010) The auxin influx carriers AUX1 and LAX3 are involved in auxin-ethylene interactions during apical hook development in Arabidopsis thaliana seedlings. Development 137: 597–606.
91. Liscum E, Reed J (2002) Genetics of Aux/IAA and ARF action in plant growth and development. Auxin Molecular Biology: Springer. pp. 387–400.
92. Baloglu M OM, Oktem H, Yucel M (2012) Expression analysis of TaNAC69-1 and TtNAMB-2, wheat NAC family transcription factor genes under abiotic stress conditions in durum wheat (Triticum turgidum). Plant Mol Biol Report 30: 1246–1252.
93. Abe H, Urao T, Ito T, Seki M, Shinozaki K, et al. (2003) Arabidopsis AtMYC2 (bHLH) and AtMYB2 (MYB) function as transcriptional activators in abscisic acid signaling. Plant Cell 15: 63–78.
94. Zheng X, Chen B, Lu G, Han B (2009) Overexpression of a NAC transcription factor enhances rice drought and salt tolerance. Biochem Biophys Res Commun 379: 985–989.
95. Sugano S, Kaminaka H, Rybka Z, Catala R, Salinas J, et al. (2003) Stress-responsive zinc finger gene ZPT2-3 plays a role in drought tolerance in petunia. Plant J 36: 830–841.
96. Choi H, Hong J, Ha J, Kang J, Kim SY (2000) ABFs, a family of ABA-responsive element binding factors. J Biol Chem 275: 1723–1730.
97. Tran LS, Nakashima K, Sakuma Y, Osakabe Y, Qin F, et al. (2007) Co-expression of the stress-inducible zinc finger homeodomain ZFHD1 and NAC transcription factors enhances expression of the ERD1 gene in Arabidopsis. Plant J 49: 46–63.
98. Muoma J, Ombori O, Jesse M (2011) Signal Transduction One of the Current Molecular Approaches hi the Management of Drought Stress hi the Sub-Saharan Region. Biotechnology 10: 469–476.

99. Aharoni A, Dixit S, Jetter R, Thoenes E, van Arkel G, et al. (2004) The SHINE clade of AP2 domain transcription factors activates wax biosynthesis, alters cuticle properties, and confers drought tolerance when overexpressed in Arabidopsis. Plant Cell 16: 2463–2480.
100. Hossain MA, Cho JI, Han M, Ahn CH, Jeon JS, et al. (2010) The ABRE-binding bZIP transcription factor OsABF2 is a positive regulator of abiotic stress and ABA signaling in rice. J Plant Physiol 167: 1512–1520.
101. Zhang JZ (2003) Overexpression analysis of plant transcription factors. Curr Opin Plant Biol 6: 430–440.
102. Cominelli E, Galbiati M, Vavasseur A, Conti L, Sala T, et al. (2005) A guard-cell-specific MYB transcription factor regulates stomatal movements and plant drought tolerance. Curr Biol 15: 1196–1200.
103. Chen JH, Song YP, Zhang H, Zhang DQ (2013) Genome-Wide Analysis of Gene Expression in Response to Drought Stress in Populus simonii. Plant Molecular Biology Reporter 31: 946–962.
104. Tardieu F, Tuberosa R (2010) Dissection and modelling of abiotic stress tolerance in plants. Curr Opin Plant Biol 13: 206–212.
105. Hadiarto T, Tran LSP (2011) Progress studies of drought-responsive genes in rice. Plant Cell Reports 30: 297–310.
106. Umezawa T, Fujita M, Fujita Y, Yamaguchi-Shinozaki K, Shinozaki K (2006) Engineering drought tolerance in plants: discovering and tailoring genes to unlock the future. Curr Opin Biotechnol 17: 113–122.
107. Tuteja N, Mahajan S (2007) Calcium signaling network in plants: an overview. Plant signaling & behavior 2: 79–85.
108. Hasegawa PM, Bressan RA, Zhu J-K, Bohnert HJ (2000) Plant cellular and molecular responses to high salinity. Annual review of plant biology 51: 463–499.
109. Nadarajah K, Sidek HM (2010) The Green MAPKs. Asian Journal of Plant Sciences 9: 1–10.
110. Wan B, Lin Y, Mou T (2007) Expression of rice Ca2+-dependent protein kinases(CDPKs) genes under different environmental stresses. FEBS letters 581: 1179–1189.
111. Hu X, Zhang Z, Xu P, Fu Z, Hu S, et al. (2010) Multifunctional genes: the cross-talk among the regulation networks of abiotic stress responses. Biologia plantarum 54: 213–223.
112. Watkinson JI, Sioson AA, Vasquez-Robinet C, Shukla M, Kumar D, et al. (2003) Photosynthetic acclimation is reflected in specific patterns of gene expression in drought-stressed loblolly pine. Plant Physiology 133: 1702–1716.
113. Queitsch C, Sangster TA, Lindquist S (2002) Hsp90 as a capacitor of phenotypic variation. Nature 417: 618–624.
114. Alvim FC, Carolino SM, Cascardo JC, Nunes CC, Martinez CA, et al. (2001) Enhanced accumulation of BiP in transgenic plants confers tolerance to water stress. Plant Physiology 126: 1042–1054.
115. Cho EK, Hong CB (2006) Over-expression of tobacco NtHSP70-1 contributes to drought-stress tolerance in plants. Plant cell reports 25: 349–358.
116. Wittenmayer L, Merbach W (2005) Plant responses to drought and phosphorus deficiency: contribution of phytohormones in root-related processes. Journal of Plant Nutrition and Soil Science 168: 531–540.

13

Estimated Dietary Intake of Radionuclides and Health Risks for the Citizens of Fukushima City, Tokyo, and Osaka after the 2011 Nuclear Accident

Michio Murakami[1,2]*, Taikan Oki[1,2]

1 Institute of Industrial Science, The University of Tokyo, Meguro, Tokyo, Japan, **2** Japan Science and Technology Agency, Core Research for Evolutionary Science and Technology (CREST), Chiyoda, Tokyo, Japan

Abstract

The radionuclides released from the Fukushima Daiichi nuclear power plant in 2011 pose a health risk. In this study, we estimated the 1st-year average doses resulting from the intake of iodine 131 (^{131}I) and cesium 134 and 137 (^{134}Cs and ^{137}Cs) in drinking water and food ingested by citizens of Fukushima City (~50 km from the nuclear power plant; outside the evacuation zone), Tokyo (~230 km), and Osaka (~580 km) after the accident. For citizens in Fukushima City, we considered two scenarios: Case 1, citizens consumed vegetables bought from markets; Case 2, citizens consumed vegetables grown locally (conservative scenario). The estimated effective doses of ^{134}Cs and ^{137}Cs agreed well with those estimated through market basket and food-duplicate surveys. The average thyroid equivalent doses due to ingestion of ^{131}I for adults were 840 μSv (Case 1) and 2700 μSv (Case 2) in Fukushima City, 370 μSv in Tokyo, and 16 μSv in Osaka. The average effective doses due to ^{134}Cs and ^{137}Cs were 19, 120, 6.1, and 1.9 μSv, respectively. The doses estimated in this study were much lower than values reported by the World Health Organization and the United Nations Scientific Committee on the Effects of Atomic Radiation, whose assessments lacked validation and full consideration of regional trade in foods, highlighting the importance of including regional trade. The 95th percentile effective doses were 2–3 times the average values. Lifetime attributable risks (LARs) of thyroid cancers due to ingestion were 2.3–39×10^{-6} (Case 1) and 10–98×10^{-6} (Case 2) in Fukushima City, 0.95–14×10^{-6} in Tokyo, and 0.11–1.3×10^{-6} in Osaka. The contributions of LARs of thyroid cancers due to ingestion were 7.5%–12% of all exposure (Case 1) and 12%–30% (Case 2) in Fukushima City.

Editor: Bart O. Williams, Van Andel Institute, United States of America

Funding: This study was supported by a "Core Research for Evolutionary Science and Technology (CREST)" grant from the Japan Science and Technology Agency, http://www.jst.go.jp/kisoken/crest/en/index.html. MM and TO receive the funding. The funder had no role in study design, data collection and analysis, discussion to publish, or preparation of the manuscript.

Competing Interests: The authors have declared that no competing interests exist.

* Email: michio@iis.u-tokyo.ac.jp

Introduction

Radionuclides were released from the Tokyo Electric Power Company's Fukushima Daiichi nuclear power plant, mainly on 15 March 2011, after the Great East Japan Earthquake and tsunami on 11 March. They were diffused into the atmosphere, deposited mainly through precipitation, and incorporated into surface waters, drinking waters, agricultural crops, and aquatic organisms. Radionuclides have been detected in drinking water and foods in Fukushima and other prefectures in Japan, including Tokyo [1,2,3]. Radioactive iodine 131 (^{131}I), which has a short half-life (8.04 d [4]), can contaminate drinking water and foods immediately after its release, whereas radioactive cesium 134 and 137 (^{134}Cs and ^{137}Cs), which have longer half-lives (2.06 and 30 y, respectively [4]), cause contamination over longer periods of time. Although other radionuclides such as strontium 90 (^{90}Sr), ruthenium 106 (^{106}Ru), and plutonium have half-lives of >1 y, in the year after the accident ^{134}Cs and ^{137}Cs contributed an estimated 84% to 88% of the total radiation dose from radionuclides with half-lives of >1 y in the diet [5].

In response to the accident, the Japanese government announced provisional "indices relating to limits on food and drink ingestion" on 17 March 2011 [1]. The government began releasing monitoring data on radionuclide concentrations in foods on 19 March and restricted the distribution of foods collected in some municipalities from 21 March, as the foods exceeded the limits. Some municipalities in Fukushima prefecture voluntarily imposed more stringent limits. In addition, some local governments, including the Tokyo Metropolitan Government, distributed bottled water to protect the health of citizens, particularly infants, against the high concentrations of ^{131}I detected in drinking water.

Among the exposure pathways, internal exposure from dietary intake has not been fully elucidated. External exposure can be determined from personal dosimeters [6], calculated from ambient dose monitoring and shielding factors [7], or estimated by using numerical dispersion models [8]. Internal exposure from inhalation can be estimated by air monitoring [9] or by numerical dispersion models [8,10]. Internal exposure to Cs can be measured by whole-body-counter surveys; however, the Cs burdens in most inhabitants of Fukushima have been below the detection limit

[11,12]. Internal exposure from the intake of water and food can be determined from market basket surveys (i.e., foods are purchased from the market according to consumption patterns) or food-duplicate surveys (i.e., cooked foods are collected from home) [1,13], but exposure immediately after the accident could not be evaluated by these methods because of the lack of samples. On the other hand, individual foods were monitored just after the accident for restriction of the distribution of foods, as described above. On the basis of the monitoring data, working groups of the Ministry of Health, Labour and Welfare [14] and the World Health Organization (WHO) [15] approximated the nationwide average internal exposure to radionuclides from foods. The United Nations Scientific Committee on the Effects of Atomic Radiation (UNSCEAR) also recently reported that an increased risk of thyroid cancer for infants and children in particular is of concern, that most of the absorbed dose to the thyroid was received during the first year after the accident, and that the major route of exposure was ingestion; e.g., absorbed dose to thyroid in the first year for 1-year-olds in Fukushima Prefecture (districts not evacuated): external + inhalation, 0.2–19 mGy; ingestion, 33 mGy [16]. However, the regional trade in foods was not fully considered in these assessments despite the low food self-sufficiency rates in Japan (54% for major grains, 39% on basis of total energy [17]), and no validations based on the results of market basket, food-duplicate, or whole-body-counter surveys were done. More detailed estimates and validation are required on account of regional variations in concentrations in drinking water and foods, the regional trade in foods, and variations in individual exposure. Furthermore, recent screening of all children in Fukushima Prefecture identified thyroid cancer in 90 of more than 280 000 children (an incidence of 313 per 1 million) by March 2014 [18]. Although this incidence might be attributable to increased screening by advanced ultrasound techniques [18], an increase incidence of cancer in future, especially thyroid cancer in children, is a matter of concern. The the contribution of cancer risk from ingestion needs to be understood.

To provide more detail, we estimated the 1st-year average doses due to intake of drinking water and foods after the accident from radionuclide concentrations measured in water and foods and the effects of countermeasures (i.e., restriction of food distribution, voluntary withholding of rice, and distribution of bottled water for infants). We previously reported the average thyroid doses resulting from the intake of ^{131}I by citizens of Tokyo and the effects of countermeasures [19]. Here, we estimated the intake of ^{131}I, ^{134}Cs, and ^{137}Cs by the citizens of Fukushima City, Tokyo, and Osaka, which differ in the regional trade in foods. The results were validated against observations in market basket, food-duplicate, and whole-body-counter surveys. By comparing the results with reports by WHO and UNSCEAR, whose assessments lacked validation and full consideration of regional trade in foods, we evaluated the role of regional trade in foods. We also estimated variations in intakes by using a Monte Carlo simulation. Finally, we assessed the cancer risks due to intake. We considered all sources of exposure for citizens in Fukushima City, and evaluated the contributions of cancer risks due to ingestion to the overall risk.This information will help to address current concerns of citizens and be useful for establishing the regulation of foods and drinking water.

Methods

Intake pathway

To assess intake via the ingestion of drinking water and foods, we estimated the thyroid equivalent dose and effective dose of ^{131}I and the effective dose of ^{134}Cs and ^{137}Cs among citizens of Fukushima City (~50 km from the nuclear power plant; capital of prefecture), Tokyo (~230 km), and Osaka (~580 km). All three cities, including Fukushima City, were outside the evacuation zone. Citizens were classified into ten groups based on age, sex, and pregnancy: <1-y-old infants, males and females aged 1–6, 7–12, 13–18, and ≥19 y old, and pregnant females.

Monitoring data on drinking water and foods were obtained from ref [1,2,3,20]. Drinking water and foods were classified into 18 categories [21]: drinking water, rice, other grains, potato, leafy vegetables, root crops, beans, fruit vegetables, milk, dairy products, formula milk, beef, pork, chicken, chicken eggs, fresh fisheries products, marine products, and others. "Fruit vegetables" included vegetable-like fruits such as tomatoes and fruits such as persimmons and apples. They were then further classified into sub-categories based on governmental restrictions [1]: leafy vegetables were divided into "spinach", "garland chrysanthemum and ging-geng-cai", "mustard spinach and non-heading lettuce", "heading leafy vegetables", "broccoli and cauliflower", and "naganegi onion, chive, and asparagus"; root crops into "turnip", "bamboo shoots", and "other root crops"; fruit vegetables into "kiwifruit", "chestnut", and "other fruit vegetables"; fresh fisheries products into "wild *ayu*, wild Japanese dace, and wild landlocked *masu* salmon" and "other fresh fisheries products"; and other into "tea", "shiitake mushroom (virgin wood)", and "other mushrooms".

Drinking water

The average doses (thyroid equivalent dose of ^{131}I and the effective dose of ^{131}I, ^{134}Cs, and ^{137}Cs) from drinking water were estimated as:

$$Dose_k(\mu Sv) = A_k \times B \times \sum_t C_{kt} \quad (1)$$

where k is the radionuclide, A_k is the dose coefficient for radionuclide k (μSv/Bq) (Table S1), B is the daily consumption of drinking water per person (g/d) (Table S2), t is the number of days after the accident (consumption date), and C_{kt} is the concentration of radionuclide k in drinking water at t days after the accident (Bq/g).

So as not to underestimate the dose, we set A_k for ^{131}I on the assumption that the fractional thyroid uptake from blood is 0.3, as used in the determination of control index levels by the Nuclear Safety Commission of Japan [22]. All the dose coefficients are reference values taken from the publications of the International Commission on Radiological Protection (ICRP) (Table S1).

B was set at 710 g/d for <1 y, 1000 g/d for 1–6 y, and 1650 g/d for >6 y [22]. The average daily consumption of water in soup and rice was then added [19]: The consumption of water in soup was calculated from the consumption of soup by adults [23] and the ratio of miso (soybean paste) consumption by each age group to that by adults [24]. The consumption of water in rice was calculated from the consumption of rice [21] and the ratio of cooking water to rice [25].

The radionuclide concentrations in drinking water were the values monitored in tap water in Fukushima City [3]; Shinjuku, Tokyo [2]; and Osaka (not detected) [20]. As details of ^{134}Cs and ^{137}Cs concentrations in tap water in Fukushima City were not available, we estimated them from the ^{131}I concentrations and the ratios of ^{134}Cs and ^{137}Cs to ^{131}I in tap water in Tokyo between 19 March and 8 April. Intake was estimated from the date when radionuclides were first detected (16 March in Fukushima City, 18 March in Tokyo). As they were not detected in tap water after 4

May (except on 2 July in Tokyo), intake after then was considered to be negligible.

Foods

The average doses from foods were estimated by the following equation [19], which is modified to include ^{134}Cs and ^{137}Cs, product origins, and food categories:

$$Dose_k(\mu Sv) = A_k \times \sum_i \sum_j \sum_t (B_i \times C_{kijt} \times D_{ij}/100) \quad (2)$$

where k is the radionuclide, A_k is the dose coefficient for radionuclide k (μSv/Bq) (Table S1), i is the individual food category, j is the individual area (source prefecture or prefectural sub-group), t is the number of days after the accident (consumption date), B_i is the daily consumption of food i per person (g/d) (Table S2), C_{kijt} is the arithmetic mean concentration of radionuclide k in food i in area j at t days after the accident (Bq/g), and D_{ij} is the arrival share in an area (the fraction of food i in the market that comes from prefecture or area j) (%).

Daily consumption. For the daily consumption of each food recently reported detailed values were adopted [21]. The daily consumption of the sub-categorized foods was estimated from the daily consumption of the foods in the main categories and the ratio of the amount of each sub-category arriving at the Tokyo Metropolitan Central Wholesale Market to the total in 2010 [26]. The national fishery yield was used to apportion the daily consumption of fresh fisheries products [27]. The daily consumption of mushrooms followed reported values [24]. The daily consumption of tea by ≥19-y-old females was 439 g/d [28], and that by other groups was calculated from that and the ratio of the daily consumption of drinking water excluding water in soup and rice in each group to that in ≥19-y-old females. We assumed that 10 g of leaf is used to make 300 g of tea and that 60% of radionuclides in the leaf enter the tea [14]. The ratio of virgin wood to other substrates used to grow shiitake was used to estimate the daily consumption of "shiitake mushroom (virgin wood)" [29].

Radionuclide concentrations. The radionuclide concentrations in food were drawn from >130 000 food monitoring data [1] and rice monitoring data [3]. Details of the number of samples analyzed each month and in each prefecture are given in the Supporting Information and Tables S3 and S4. Foods were washed with water and the edible parts were then analyzed for radionuclides by Ge, NaI or CsI detectors. We considered the intake of radionuclides via the consumption of foods collected in 16 prefectures: Hokkaido, Aomori, Iwate, Miyagi, Akita, Yamagata, Fukushima, Ibaraki, Tochigi, Gunma, Saitama, Chiba, Kanagawa, Tokyo (Metropolis), Yamanashi, and Shizuoka (Figure S1). Analysis of the variations in radionuclide concentrations among foods in Fukushima Prefecture showed that spinach, broccoli and cauliflower, and marine products were heavily contaminated (Figures S2 and S3). Analysis of the regional differences in radionuclide concentrations in spinach showed that foods collected in Fukushima Prefecture and in prefectures in the Kanto region (Ibaraki, Tochigi, Gunma, Saitama, Chiba, Kanagawa, and Tokyo) had much higher levels than those in the other 8 prefectures (Figure S4). The amounts of radionuclides originating from other prefectures or from overseas were regarded as negligible. When radionuclides were not detected in a food, the contribution of radionuclides from that food to the dose was considered to be negligible. The concentrations in each food were classified by source prefecture and date. The prefectures where the distribution of the food was restricted (for example, rice in Fukushima Prefecture) were further classified into prefectural sub-groups. The arithmetic mean concentration in each area on each day was calculated. When only the sum of ^{134}Cs and ^{137}Cs concentrations was available, individual concentrations were estimated from the average ratio of ^{134}Cs to the sum of ^{134}Cs and ^{137}Cs, namely 0.49 (March–June 2011), 0.46 (July–September 2011), 0.45 (October–December 2011), or 0.42 (January–March 2012), as calculated from foods in which both the ^{134}Cs and the ^{137}Cs concentration exceeded 25 Bq/kg [1]. The radionuclide concentrations in crude tea and manufactured tea were converted to those in tea leaf by multiplying by the weight ratio of crude tea to tea leaf (0.22) [30]. The concentrations of ^{131}I in foods on days before monitoring began or for which there were no data were approximated from the concentrations measured on the closest day and from the half-life of ^{131}I (8.04 d [4]). The concentrations of ^{134}Cs and ^{137}Cs on days before monitoring began or for which there were no data were regarded as the same as the concentrations measured on the closest day. The consumption dates and concentrations in foods were assumed to be as reported [1,3].

Arrival shares. Since citizens generally purchase food from markets, which are supplied from the whole country, the arrival share is an important parameter for estimating the doses due to ingestion. A few citizens in Fukushima City consumed vegetables produced in local fields or home gardens (4.5% of workers in Fukushima City were agricultural workers [31]). We therefore considered two scenarios: Case 1, citizens consumed foods from markets; Case 2, citizens consumed vegetables produced in Fukushima Prefecture and other foods from markets. Case 2 means the arrival shares of vegetables from Fukushima Prefecture are 100%, so this scenario can be regarded as conservative. For citizens in Tokyo and Osaka, only Case 1 was considered. The proportions of targeted foods produced in each area for the markets in each city were calculated as follows. The arrival shares of vegetables and fisheries products were based on the amounts arriving from each prefecture at each central wholesale market in 2010 [26,32,33]. We used the 2010 data so as to evaluate the effects of countermeasures. Since data on the importation of fisheries products from overseas were not available, we used the Japanese food self-sufficiency rates based on quantities of production, imports, and exports [17]. The production of marine products in Fukushima Prefecture was regarded to be nil owing to the tsunami damage and the suspension of fishing. The arrival shares of beef, pork, and chicken eggs were based on the quantities transported from each prefecture to Fukushima, Tokyo, and Osaka prefectures [34,35]. The arrival shares of beef and pork were corrected to the food self-sufficiency rates of Japan. The arrival shares of milk and dairy products were based on the production and transport of raw milk in each prefecture, the amount of milk transported from each prefecture to Fukushima, Tokyo, and Osaka prefectures [36], and the food self-sufficiency rates. For rice, other grains, tea, soybeans, and chicken, we used the nationwide production rates [30,35,37,38] and the food self-sufficiency rates of Japan. The use of these sources is reasonable, especially for dried foods. To estimate the ratios of restricted food production among the prefectural sub-groups, we used the number of cows in each municipality for milk and dairy products [39], the cultivated areas of vegetables and tea [40], the areas of rice paddies [40] and the number of rice farm households [3], and the production [41] or the number of management bodies [42] for shiitake mushrooms.

Calculation periods. It is likely that rainfall on the night of 15–16 March 2011 [43] caused heavy contamination of leafy vegetables and of milk and dairy products by radionuclides, so we estimated intake via these products before the first release of data

(19 March 2011): from 17 March 2011 in Fukushima City, on the assumption that delivery to citizens takes 1 day; and from 18 March 2011 in Tokyo and Osaka, on the assumption that it takes 2 days. The intake of other foods was estimated from 21 March 2011, when the Japanese government restricted the distribution of foods. Since radionuclide contamination in other foods was not attributable to direct deposition [22], the intake before 21 March was not included. This was supported by the lower contamination of other foods by ^{131}I and ^{134}Cs and ^{137}Cs than of leafy vegetables (Figures S2, S3). The intake was estimated from these days to 20 March 2012.

Effects of countermeasures

We evaluated the effects of countermeasures (restrictions on the distribution of foods, voluntary withholding of rice, and distribution of bottled water for infants) on reducing intake. We assumed the consumption of foods harvested in the areas where distribution was restricted or rice was withheld to be nil: citizens ate alternative foods containing negligible radionuclides. In this regard, however, for Case 2, the conservative scenario, we considered the effects of restrictions on the distribution of other foods except vegetables, because we assumed that some citizens consumed vegetables grown locally. Since the Tokyo Metropolitan Government distributed bottled water for infants on the morning of 24 March, we evaluated the effect of this countermeasure on 24 and 25 March. (Fukushima City distributed bottled water on the night of 11 March following the earthquake, but this was too early to consider).

Variations in individual doses

The variations in ingestion doses for ≥19-y-old males (with countermeasures in place) were estimated by Monte Carlo simulation using the commercially available software Crystal Ball (Oracle, California, USA). The drinking water and food variation data input into the Monte Carlo simulation were obtained as below.

We took into account variations in daily consumption of drinking water [44] (Table S2). The drinking water in Tokyo is derived from several sources: 38.0% of tap water comes from the Edogawa River, 41.3% from the Arakawa River, 14.4% from the Tamagawa River, and 6.3% from other sources such as ground water [45,46]. Therefore, we also took into account the radionuclide concentrations in each source. We assumed that citizens drank water from a single source. The radionuclide concentration factors were estimated from the ratios of ^{131}I concentrations in representative purification plants (Kanamachi purification plant for the Edogawa River, Asaka purification plant for the Arakawa River, Ozaku purification plant for the Tamagawa River) to those in tap water in Shinjuku [2,46]. We used the arithmetic mean concentrations on 22–28 March in the purification plants and on 23–29 March in Shinjuku, on the assumption that delivery to the tap takes 1 day. Concentrations were corrected for half-life. Concentration factors were 3.51 for the Edogawa River, 1.31 for the Arakawa River, and 0.46 for the Tamagawa River (i.e., the ^{131}I concentration in the Ozaku purification plant on the Tamagawa River was lower than that in tap water in Shinjuku). The ^{131}I concentration in other water systems was regarded as negligible. Fukushima City has only one drinking water treatment plant, and radionuclides were not detected in the drinking water in Osaka. We therefore did not need to take into account variations in the water resources in these two cities.

For input into the Monte Carlo simulation, variations in the dose from drinking water in Fukushima City were estimated from the variation in daily consumption under the assumption of a log-normal distribution. Variations in Tokyo were estimated from variations in daily consumption, the dependence on water source, and the concentration factor.

Variations in the dose from foods for input into the model were estimated under the assumption that citizens bought each food once a week, and the production area was randomly selected according to the distribution of the arrival shares. The weekly average concentrations of radionuclides in each food were calculated. The variations in the daily consumption of each food were as reported [28,47]. Since these values are based on 1-day surveys, not long-term surveys, their use could have caused us to overestimate the results. The variations in doses obtained by ingesting each food from the same area in the same week at the same daily rate of consumption were considered to indicate variations in the radionuclide concentrations. The relative standard deviations of the variations were estimated from the effective doses calculated from data measured on the same day in Tochigi Prefecture for "ging-geng-cai," in Ibaraki Prefecture for "marine products," in Shizuoka Prefecture for "tea," and in Fukushima Prefecture for other foods ($n \geq 4$ for each food; Table S2). Daily consumption of foods and the radionuclide concentrations in foods were assumed to follow a log-normal distribution. The variations in the dose from each food were estimated from the input data for the Monte Carlo simulation, namely, the variation in daily consumption, the distribution of arrival shares, and the variation in doses received by ingesting the foods in the same areas. The simulation was performed 10 000 times for drinking water and for each food.

Estimation of cancer risk

The lifetime attributable risks (LARs) of cancer incidences up to the age of 89 y were estimated in accordance with the method described in a WHO report [48] and Harada et al. [49]. We estimated those due only to the ingestion of ^{131}I, ^{134}Cs, and ^{137}Cs for citizens in each city and those due to three pathways–ingestion, inhalation in the radioactive cloud, and external exposure to material deposited on the ground and in the cloud–for citizens in Fukushima City. Estimation of doses due to inhalation and external exposure followed previous references [15,50] as described in the Supporting Information. The risk models were based on data from survivors of the atomic bombing in Japan; the validity of the thyroid cancer incidence was also confirmed by a cohort study performed after the Chernobyl accident [51] in the WHO report [48]. A linear-quadratic dose-response model was used for leukemia [52], and linear non-threshold (LNT) models were used for all solid cancers, breast cancer, and thyroid cancer [53]. LAR was calculated from cancer-free survival rates ($S(a, g)$), an excess absolute risk (EAR) model, and an excess relative risk (ERR) model, as follows:

$$LAR(D,e,g) = \int_{e+L}^{89} [w \times EAR(D,e,a,g) + (1-w) \times ERR(D,e,a,g) \times m(a,g)] \frac{S(a,g)}{S(e,g)} da \quad (3)$$

where D is the annual dose (Sv), e is the age at exposure, g is sex, L is the minimum latency period, w is the weight of the EAR and the ERR models combined, a is the age attained, and $m(a, g)$ is the baseline cancer incidence rate in the unexposed population.

The organ (colon, bone marrow, breast, and thyroid) doses were calculated from the effective doses and the organ dose-to-effective dose ratios for three age groups (<1 y, 10 y, and 20 y in the first year) and for the two sexes in the three cities [48]. Doses in the second and subsequent years were calculated from the doses of ^{134}Cs and ^{137}Cs in March 2012 and from the physical decay of ^{134}Cs and ^{137}Cs (half-lives of 2.06 and 30 y, respectively [4]).

The cancer-free survival rates of males and females were derived from the age- and sex-stratified all-cause mortality in Japan in 2010 [54] plus the difference between the all-cancer incidence in Japan in 2008 [55,56] and the all-cancer mortality in Japan in 2010 [54]. Cancer incidences in Japan in 2008 [55,56] were used as baseline incidence rates for all solid cancers, leukemia, breast cancer, and thyroid cancer.

The minimum latency period was set at 2 y for leukemia, 3 y for thyroid cancer, and 5 y for breast and all solid cancers. The weights of the EAR model and the ERR model combined were set at 0.5 for all solid cancers, leukemia, and thyroid cancer, and 1 for breast cancer.

The details of the EAR and ERR models and their parameters for leukemia, all solid cancers, breast cancer, and thyroid cancer are described in the Supporting Information.

The relationship between cancer risk and dose is still uncertain [57]. However, for the same dose, the risks at low dose rates are known to be lower than those at high dose rates [58]. In particular, coordination of DNA repair processes plays a critical role in allowing proper development and survival of organisms [59].To correct the risk at low dose levels, the International Commission on Radiological Protection (ICRP) and the BEIR VII committee adopted "dose and dose-rate effectiveness factors" of 2 [60] and 1.5 [61]. We followed the ICRP lead and set a factor of 2 for LNT models from the perspective of radiological protection.

Sources of uncertainty

This study included sources of uncertainty: ^{134}Cs and ^{137}Cs concentrations in tap water in Fukushima City, limited data on foods in the early stages, data that were less than detection limits, individual behaviors (such as purchasing of bottled water or not purchasing products from Fukushima Prefecture), and assessment of doses in the second and subsequent years for cancer estimation.

^{134}Cs and ^{137}Cs concentrations in tap water in Fukushima City were estimated from the ^{131}I concentrations and the ratios of ^{134}Cs and ^{137}Cs to ^{131}I in tap water in Tokyo by assuming that the ratios of ^{134}Cs and ^{137}Cs deposition to ^{131}I deposition and the removal efficiencies in drinking water treatment plants were similar between Fukushima City and Tokyo. The deposition ratios are known to be similar in the two cities [62], and drinking water treatment plants in both cities use sedimentation and rapid sand filtration, which are effective for removing ^{134}Cs and ^{137}Cs [63]. This uncertainty did not have a large effect, because the contributions of ^{134}Cs and ^{137}Cs from drinking water were minor (see "Comparison of doses among ages, effects of countermeasures, and change of intake over time).

In the early stages of monitoring, data were not available for some foods in some prefectures in the Kanto region, but these contributions were judged to be small (see Supporting Information). The number of samples was limited for some foods in the early stages even in Fukushima Prefecture. We therefore used a Monte Carlo simulation and show the variations in dose (e.g., 95th percentile value), including those resulting from radionuclide concentrations, in foods in the early stages.

We regarded radionuclide concentrations that were less than detection limits as nil. There were differences in detection limits among periods and institutions surveyed. We therefore confirmed the results through validation against observations in market basket, food-duplicate, and whole-body-counter surveys.

Differences in individual behaviors, such as not purchasing products from Fukushima Prefecture, could have influenced the variation in estimates. However, as the volume of major crops shipped from Fukushima Prefecture did not decrease after the accident [64,65], we ignored differences in behaviors. The nationwide average of bottled soft drink consumption was 410 g/d in 2011 [66], and some citizens also purchased bottled water so as to avoid tap water. We ignored the consumption of bottled water and soft drinks: the assumption that people drank only tap water would conservatively overestimate the dose from drinking water.

The doses due to ingestion in the second and subsequent years for cancer estimation were calculated from the physical decay, although the actual doses might be lower because of tighter regulations. The doses due to external exposure from September 2014 were also calculated from the physical decay. These assumptions can be regarded as conservative.

Results and Discussion

Validation of results against market basket and food-duplicate surveys

The effective doses due to ingestion of ^{134}Cs and ^{137}Cs in the diet in Fukushima City (Case 1), Tokyo, and Osaka agreed within a factor of 2, in general, with those calculated in the market basket [1] and food-duplicate surveys [1,13,67] in five periods from July 2011 to March 2012 (Table 1). The market basket survey includes drinking water as well as foods. This good agreement supports the accuracy and reliability of our results. The doses in the diet in Fukushima City (Case 2) were higher than those in the market basket and food-duplicate surveys. This result is reasonable because Case 2 is conservative. Validation for the period from March to June, including for the thyroid equivalent doses due to ^{131}I, was not done, because no data from market basket and food-duplicate surveys were available.

Comparison of doses among cities

The average thyroid equivalent doses (and effective doses) due to ingestion of ^{131}I for a ≥19-y-old male were 840 μSv (43 μSv) in Fukushima City (Case 1), 2700 μSv (140 μSv) in Fukushima City (Case 2), 370 μSv (19 μSv) in Tokyo, and 16 μSv (0.82 μSv) in Osaka (Table 2). The average effective doses due to ^{134}Cs and ^{137}Cs were 19, 120, 6.1, and 1.9 μSv, respectively. The average effective doses due to total radionuclides were 62, 260, 25, and 2.7 μSv, respectively. There were no large differences between the average and median (within a factor of <1.5, with the exception of the thyroid equivalent dose due to ^{131}I in Osaka). The slightly lower thyroid equivalent dose due to ^{131}I for citizens of Tokyo in this study (370 μSv) than in our previous study (410 μSv) [19] is attributable to the modification of daily consumption rates.

The dose due to ingestion of ^{131}I in Case 2 was 3 times that in Case 1 and the dose due to ^{134}Cs and ^{137}Cs was 6 times that in Case 1. The doses in Fukushima City (Case 1) were lower than the values reported in the WHO preliminary assessment (adults in Fukushima City, thyroid equivalent dose, 800–8000 μSv; effective dose due to total radionuclides, 500–5000 μSv) [15] and the UNSCEAR report (absorbed dose to thyroid, 7800 μGy; effective dose due to total radionuclides, 900 μSv) [16]. Regional trade in foods was considered in this study, whereas the WHO preliminary assessment and UNSCEAR report assumed that consumers consumed mainly food produced in Fukushima and neighboring prefectures. Inclusion of regional trade in foods is a key to accurate dose assessment.

In both doses, Fukushima City had the highest values and Osaka the lowest. This was consistent with the distances from the nuclear power plant (i.e. Fukushima City, ~50 km; Tokyo,

Table 1. Comparison of effective doses of ^{134}Cs and ^{137}Cs in the diet between this study and the market basket and food-duplicate surveys (μSv/month).

	Jul.2011	Sep.-Nov. 2011	Dec. 2011	Feb.-Mar. 2012	Mar. 2012
Fukushima City (Case 1; this study)[a]	2.36	1.18	0.96	0.51	0.44
Fukushima City (Case 2; this study)[a]	4.15	2.85	3.08	6.84	2.60
Fukushima Prefecture (ref[67])[b]	0.53 ± 1.04[f]	-	-	-	-
Fukushima Prefecture (ref[1])[c]	-	1.6	-	0.33-0.55	-
Fukushima Prefecture (ref[1])[b]	-	-	-	-	0.18[g]
Fukushima Prefecture (ref[13])[b]	-	-	2.17± 1.67[f]	-	-
Tokyo (this study) [a]	-	0.44	0.24	0.18	0.19
Tokyo (ref[1])[c]	-	0.22	-	-	-
Kanto (ref[1])[c, d]	-	-	-	0.28-0.33	-
Kanto (ref[1])[b, d]	-	-	-	-	0.15[g]
Kanto (ref[13])[b, e]	-	-	0.92± 1.42[f]	-	-
Osaka (this study)[a]	-	0.15	0.09	0.07	0.07
Osaka (ref[1])[c]	-	-	-	0.13	-
Osaka (ref[1])[b]	-	-	-	-	0.1[g]

Case 1, citizens consumed vegetables bought from markets. Case 2, citizens consumed vegetables grown locally.
[a]with countermeasure (≥19 y old male).
[b]food-duplicate survey.
[c]market basket survey.
[d]including Saitama, and Kanagawa prefectures.
[e]including Tochigi, Gunma, Ibaraki, Saitama, Chiba, Tokyo, Kanagawa, and Nagano prefectures.
[f]arithmetic mean ± standard deviation.
[g]Mar.-May, 2012.

~230 km; Osaka, ~580 km). The contribution of effective dose from ^{131}I to total radionuclides was higher than that from ^{134}Cs and ^{137}Cs in Fukushima City and Tokyo, and lower in Osaka. Except in Case 2, drinking water contributed the highest thyroid equivalent dose due to ingestion of ^{131}I in Fukushima City and Tokyo, followed by vegetables (Figures 1, S5, and S6). ^{131}I was not

Table 2. Average doses and variations due to countermeasures resulting from the intake of ^{131}I, ^{134}Cs and ^{137}Cs from drinking water and foods by a ≥19-y-old male from 16 March 2011 to 20 March 2012.

	Average	5th percentile	25th percentile	Median	75th percentile	95th percentile
^{131}I (thyroid equivalent dose: μSv)						
Fukushima City (Case 1)	840	360	520	690	940	1700
Fukushima City (Case 2)	2700	940	1500	2200	3300	6200
Tokyo	370	140	290	470	780	1400
Osaka	16	<1	1	4	13	74
^{134}Cs and ^{137}Cs (effective dose: μSv)						
Fukushima City (Case 1)	19	8.1	11	15	22	43
Fukushima City (Case 2)	120	60	86	110	150	220
Tokyo	6.1	3.0	4.4	5.8	7.7	12
Osaka	1.9	0.49	0.88	1.4	2.1	4.5
Total (effective dose: μSv)						
Fukushima City (Case 1)	62	29	39	50	66	110
Fukushima City (Case 2)	260	120	170	230	310	530
Tokyo	25	11	21	30	47	80
Osaka	2.7	0.59	1.2	1.8	3.0	7.1

Case 1, citizens consumed vegetables bought from markets. Case 2, citizens consumed vegetables grown locally.

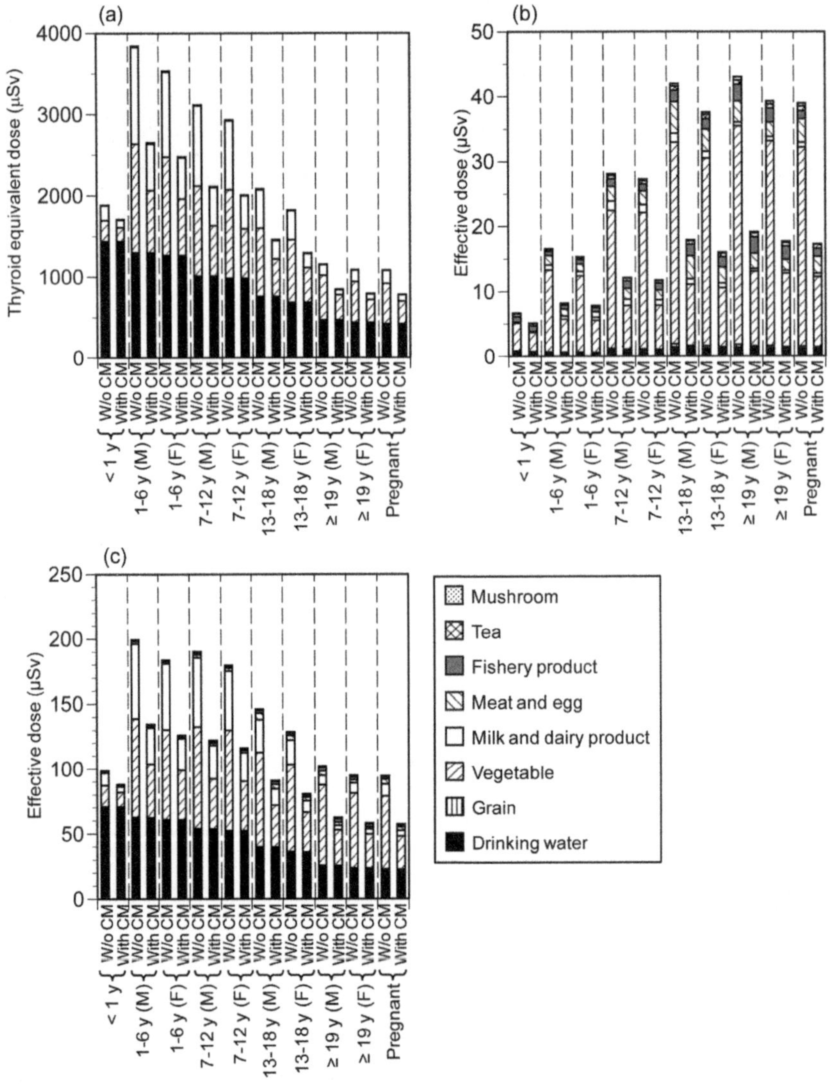

Figure 1. Average doses with and without countermeasures in Fukushima City (Case 1): (a) ^{131}I, (b) ^{134}Cs and ^{137}Cs, (c) total. CM, countermeasures; M, male; F, female. Case 1, citizens consumed vegetables bought from markets.

detected in drinking water in Osaka, and vegetables contributed the highest effective dose due to ^{134}Cs and ^{137}Cs (Figure S7). These results indicate that local contamination of drinking water caused higher intake of ^{131}I by citizens in Fukushima City and Tokyo than in Osaka, and that regional trade in foods played an important role in the intake of ^{134}Cs and ^{137}Cs by citizens in Osaka.

The 95th percentile effective dose due to ingestion of ^{134}Cs and ^{137}Cs in Fukushima City (Case 1) was 43 μSv; this was 4 times the 97th percentile internal exposure to ^{134}Cs and ^{137}Cs (i.e., 10 μSv, the value below which the exposures of 328 out of 337 individuals fell) in inhabitants of the village of Kawauchi in Fukushima Prefecture, as monitored by whole-body-counter surveys in November 2011 [12]. The effects of differences in monitoring time were small, because our average effective dose due to ^{134}Cs and ^{137}Cs in November 2011 (1.3 μSv/month) was similar to the annual average (19 μSv/y = 1.6 μSv/month). Note that the whole-body-counter surveys had sampling bias and may have underestimated the doses received by the general population [11]. The 95th percentile effective doses due to total radionuclides were 110 μSv in Fukushima City (Case 1), 530 μSv in Fukushima City (Case 2), 80 μSv in Tokyo, and 7.1 μSv in Osaka, 2 to 3 times the average values (Table 2). These values were much lower than the annual effective dose due to other natural radionuclides in the diet; e.g., potassium 40 (^{40}K), 130–217 μSv [68]; polonium 210 (^{210}Po), 730 μSv [69]. The provisional limits of radionuclides in drinking water and foods, which were announced by the Japanese government just after the accident, were determined from the intervention levels of 50 000 μSv of an annual thyroid equivalent dose due to ^{131}I and 5000 μSv of an annual effective dose due to ^{134}Cs, ^{137}Cs, ^{89}Sr, and ^{90}Sr [22]. Standards released in April 2012 were determined from the intervention level of 1000 μSv of annual effective dose [21]. The 95th percentile doses due to ingestion of individual and total radionuclides in Fukushima City (Cases 1 and 2) and Tokyo were greater than or equal to 1 order of magnitude lower than the provisional limits and also lower than the new limits. Those in Osaka were greater than 2 orders of magnitude lower.

Comparison of doses among ages, effects of countermeasures, and change of intake over time

The thyroid equivalent dose due to ingestion of ^{131}I in Fukushima City (Case 1) decreased with increasing age from 1 y (Figure 1, Tables S5, S7), although the daily consumption of drinking water and most foods increases with age (Table S2). This discrepancy between dose and daily consumption is attributable to the thyroid ingestion dose coefficient, which depends largely on age. In contrast, the effective doses due to ^{134}Cs and ^{137}Cs increased with age. Overall, the effective doses due to total radionuclides were higher for younger people except for <1-y-old infants (Table S6, S8). The difference in dose between sexes was small. Because the dose coefficients are the same for males and females (Table S1), the differences in doses can be attributed to differences in consumption patterns. Similar results were found in Case 2, Tokyo and Osaka, with the exception that there were no large differences in effective dose due to total radionuclides among ages in Osaka (Figures S5–S7, Tables S9–S20).

The countermeasures reduced the intake of ^{131}I by >1-y-olds by 27%–32% in Fukushima City (Case 1), by 11%–15% in Tokyo, and by 33%–37% in Osaka, and reduced the intake of ^{134}Cs and ^{137}Cs by 49%–57% in Fukushima City (Case 1), by 18%–25% in Tokyo, and by 16%–24% in Osaka (Table S21). The reductions of intake were generally lower for <1-y-old infants because the contributions of drinking water were higher for them.

The cumulative intakes of ^{131}I for ≥19-y-old males in Fukushima City (Case 1) in the presence of countermeasures were 72% of the total in week 1 and 87% in week 2 (Figure 2) and stopped increasing within 1 month. This was consistent with our previous result [19] and is attributable to the short half-life of ^{131}I and the absence of new serious releases from the nuclear power plant to the atmosphere. The intake of ^{131}I was dominant within the first 2 weeks: rapid countermeasures are therefore important in reducing intake. On the other hand, the cumulative intakes of ^{134}Cs and ^{137}Cs were 25% of the total in week 1 and 27% in week 2. The longer-term intake of ^{134}Cs and ^{137}Cs is due to the longer half-lives. In particular, the intake via consumption of vegetables and fisheries products was continuous. Similar results were found in Case 2, Tokyo and Osaka (Figures S8–S10).

LARs of cancer incidence

The LARs of cancer from ingestion are summarized in Table 3. These values do not include risks from external exposure and inhalation. The LARs of all the solid cancer risks were 8.4–12×10^{-6} in Fukushima City (Case 1), 45–71×10^{-6} in Fukushima City (Case 2), 3.0–5.0×10^{-6} in Tokyo, and 1.1–2.0×10^{-6} in Osaka, whereas the LARs of leukemia were >1 order of magnitude lower than those for all the solid cancers (i.e., 0.34–0.59×10^{-6} in Fukushima City (Case 1), 2.0–3.5×10^{-6} in Fukushima City (Case 2), 0.14–0.25×10^{-6} in Tokyo, and 0.05–0.09×10^{-6} in Osaka). The risk of all solid cancers combined together with leukemia is intended to provide an overall indication of the lifetime risk of cancer; however, in circumstances where the tissue doses are highly heterogeneous, such as with doses of ^{131}I to the thyroid, the risk of all solid cancers underestimates the cancer risk in specific tissues [48]. This is because the LARs of thyroid cancer in some young groups (<1 y and 10 y in the first year) in Fukushima City and Tokyo exceeded those of all solid cancers. The LARs of all solid cancers estimated from the colon dose are not suitable in the case of ingestion, although they may be still useful as indicators of the risk of deadly cancers. The LARs of thyroid cancers were 2.3–39×10^{-6} in Fukushima City (Case 1), 10–98×10^{-6} in Fukushima City (Case 2), 0.95–14×10^{-6} in Tokyo, and 0.11–1.3×10^{-6} in Osaka, and the maximum LAR in Fukushima City (Case 1) was found in females <1 y old. These values were calculated from the average doses. The 95th percentile doses were 2× the average values for Fukushima City (Cases 1 and 2) and 3× those for Tokyo and Osaka. Because of the variation of doses, a factor of at least 2–3 could be applied to the 95th percentiles of LARs. In addition, these values cannot be directly compared with the ongoing results of diagnosis of thyroid cancers in Fukushima Prefecture because of the screening effects of diagnosis and the use of advanced ultrasound techniques [18]. The contributions of ^{131}I to the lifetime-exposure thyroid doses were 92%–96% in Fukushima City (Case 1), 84%–92% in Fukushima

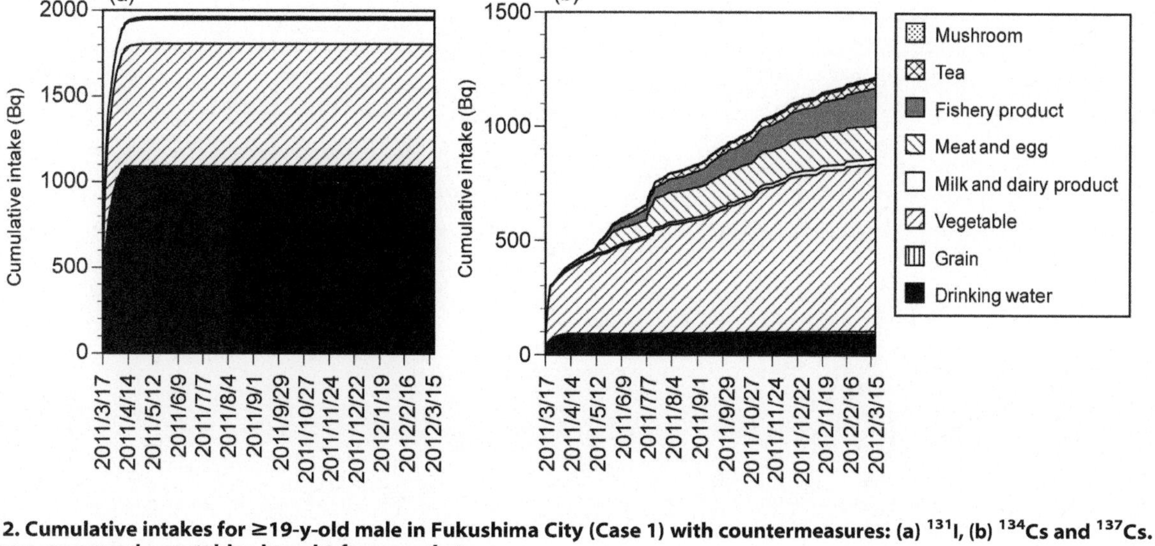

Figure 2. Cumulative intakes for ≥19-y-old male in Fukushima City (Case 1) with countermeasures: (a) ^{131}I, (b) ^{134}Cs and ^{137}Cs. Case 1, citizens consumed vegetables bought from markets.

Table 3. LARs for all solid cancers, leukemia, breast cancer, and thyroid cancer ($\times 10^{-6}$) due to ingestion.

	All solid cancers	Leukemia	Breast cancer	Thyroid cancer
Fukushima City (Case 1)				
<1 y (M)	8.6	0.57	-	9.5
<1 y (F)	12	0.36	1.5	39
10 y (M)	8.4	0.59	-	7.6
10 y (F)	11	0.36	1.5	30
20 y (M)	7.4	0.56	-	2.3
20 y (F)	9.8	0.34	1.1	9.0
Fukushima City (Case 2)				
<1 y (M)	47	3.1	-	15
<1 y (F)	66	2.0	8.4	62
10 y (M)	51	3.5	-	25
10 y (F)	71	2.2	9.3	98
20 y (M)	45	3.4	-	10
20 y (F)	61	2.1	7.1	38
Tokyo				
<1 y (M)	3.6	0.24	-	3.4
<1 y (F)	5.0	0.15	0.63	14
10 y (M)	3.6	0.25	-	3.1
10 y (F)	4.7	0.15	0.61	12
20 y (M)	3.0	0.23	-	0.95
20 y (F)	3.9	0.14	0.44	3.6
Osaka				
<1 y (M)	1.4	0.09	-	0.15
<1 y (F)	2.0	0.06	0.25	0.62
10 y (M)	1.3	0.09	-	0.20
10 y (F)	1.8	0.06	0.23	1.3
20 y (M)	1.1	0.08	-	0.11
20 y (F)	1.4	0.05	0.16	0.42

Ages represent ones in the first year. Case 1, citizens consumed vegetables bought from markets. Case 2, citizens consumed vegetables grown locally.

Table 4. LARs for all solid cancers, leukemia, breast cancer, and thyroid cancer ($\times 10^{-6}$) due to external exposure, inhalation, and ingestion.

	All solid cancers	Leukemia	Breast cancer	Thyroid cancer
Fukushima City (Case 1)				
<1 y (M)	2700	300	-	120
<1 y (F)	4100	210	1200	490
10 y (M)	2000	170	-	66
10 y (F)	3000	110	690	270
20 y (M)	1400	120	-	29
20 y (F)	2100	79	390	120
Fukushima City (Case 2)				
<1 y (M)	2700	310	-	130
<1 y (F)	4200	210	1200	520
10 y (M)	2100	170	-	83
10 y (F)	3100	120	700	340
20 y (M)	1400	120	-	37
20 y (F)	2100	81	390	150

Ages are those in the first year. Case 1, citizens consumed vegetables bought from markets. Case 2, citizens consumed vegetables grown locally.

City (Case 2), 91%–96% in Tokyo, and 62%–78% in Osaka. As described above, most of the ^{131}I was taken up within the first 2 weeks. The LARs of thyroid cancers were attributed mainly to ^{131}I in the first 2 weeks, highlighting again the fact that rapid implementation of countermeasures after a nuclear accident is important. On the other hand, Japanese dietary exposure gives lifetime cancer risks of 310×10^{-6} from inorganic arsenic (for the sum of skin, liver, and lung cancers) [70] and 1400×10^{-6} from acrylamide [71]. Cancer risks from ingestion–especially in the case of the thyroid cancer risk in young females in Fukushima City–might not be negligible, but they are 1 to 2 orders of magnitude lower than the cancer risks from ubiquitous carcinogens in the daily diet.

The total effective doses due to ingestion, inhalation, and external exposure in the first year and over a lifetime (up to 89 y) for Fukushima City are summarized in Table S22. The LARs of all solid cancers due to the three pathways were $1400–4100\times10^{-6}$ in Fukushima City (Case 1) and $1400–4200\times10^{-6}$ in Fukushima City (Case 2); those of thyroid cancer were $29–490\times10^{-6}$ in Fukushima City (Case 1) and $37–520\times10^{-6}$ in Fukushima City (Case 2) (Table 4). The contributions of LARs of all solid cancer due to ingestion were 0.3%–0.5% (Case 1) and 1.6%–3.1% (Case 2); those of thyroid cancer were 7.5%–12% (Case 1) and 12%–30% (Case 2). Even when the 95th percentile values were 2× the average, the contributions of LARs of all solid cancers due to ingestion were minor. However, for the 95th percentile values in Case 2, the contribution of LARs of thyroid cancer due to ingestion was approximately half of the total. Except in such extreme situations, the contributions of LARs of thyroid cancer due to ingestion cannot be regarded as either negligible or predominant but as minor.

As described above, the contributions of ^{131}I to the lifetime-exposure thyroid doses were >84% for Fukushima City, and the contributions of LARs from foods in the second and subsequent years were negligible. However, in Japan and other countries, some people are still concerned about artificial radionuclides in the diet. Although a greater understanding of risks alone may not lead to reasonable countermeasures, the doses and cancer risks calculated in this study are essential information.

Supporting Information

Figure S1 Locations of each prefecture in Japan.

Figure S2 ^{131}I concentrations in leafy vegetables, fruit vegetables, milk and dairy products, meat and eggs, and marine products in Fukushima Prefecture in March 2011. a: Gunma; b: April.

Figure S3 ^{134}Cs and ^{137}Cs concentrations in foods in Fukushima Prefecture in 2011. a: August; b: July; c: June; d: March; e: Gunma in March; f: March–April; g: May; h: April; i: October; j: September; k: Tochigi; December; l: April–May; m: Ibaraki in May.

Figure S4 Radionuclide concentrations in each prefecture in March 2011. (a) ^{131}I; (b) ^{134}Cs and ^{137}Cs.

Figure S5 Average doses with and without countermeasures in Fukushima City (Case 2): (a) ^{131}I, (b) ^{134}Cs and ^{137}Cs, (c) total. CM, countermeasures; M, male; F, female. Case 2, citizens consumed vegetables grown locally.

Figure S6 Average doses with and without countermeasures in Tokyo: (a) ^{131}I, (b) ^{134}Cs and ^{137}Cs, (c) total. CM, countermeasures; M, male; F, female.

Figure S7 Average doses with and without countermeasures in Osaka: (a) ^{131}I, (b) ^{134}Cs and ^{137}Cs, (c) total. CM, countermeasures; M, male; F, female.

Figure S8 Cumulative intakes for ≥ 19 y old male in Fukushima City (Case 2) with countermeasures: (a) ^{131}I, (b) ^{134}Cs and ^{137}Cs. Case 2, citizens consumed vegetables grown locally.

Figure S9 Cumulative intakes for ≥ 19 y old male in Tokyo with countermeasures: (a) ^{131}I, (b) ^{134}Cs and ^{137}Cs.

Figure S10 Cumulative intakes for ≥ 19 y old male in Osaka with countermeasures: (a) ^{131}I, (b) ^{134}Cs and ^{137}Cs.

Methods S1 Numbers of samples analyzed; Dose estimation for inhalation and external exposure; Risk models and their parameters.

Table S1 Thyroid equivalent dose coefficients for ingestion of ^{131}I and the effective dose coefficients for ingestion of ^{131}I, ^{134}Cs and ^{137}Cs (μSv/Bq).

Table S2 Daily consumption rates of drinking water and foods, and relative standard deviations (RSDs) of doses received by ingesting the foods in the same areas.

Table S3 Numbers of samples analyzed each month.

Table S4 Numbers of samples (leafy vegetables, fruit vegetables, milk and dairy products, meat and eggs, and marine products) analyzed in each prefecture in March and April 2011.

Table S5 Average thyroid equivalent doses of ^{131}I without countermeasures in Fukushima City (Case 1) in the first year after the accident (μSv). M, male; F, female. Case 1, citizens consumed vegetables bought from markets.

Table S6 Average effective doses of ^{134}Cs and ^{137}Cs without countermeasures in Fukushima City (Case 1) in the first year after the accident (μSv). M, male; F, female. Case 1, citizens consumed vegetables bought from markets.

Table S7 Average thyroid equivalent doses of ^{131}I with countermeasures in Fukushima City (Case 1) in the first year after the accident (μSv). M, male; F, female. Case 1, citizens consumed vegetables bought from markets.

Table S8 Average effective doses of ^{134}Cs and ^{137}Cs with countermeasures in Fukushima City (Case 1) in the first

year after the accident (μSv). M, male; F, female. Case 1, citizens consumed vegetables bought from markets.

Table S9 Average thyroid equivalent doses of ^{131}I without countermeasures in Fukushima City (Case 2) in the first year after the accident (μSv). M, male; F, female. Case 2, citizens consumed vegetables grown locally.

Table S10 Average effective doses of ^{134}Cs and ^{137}Cs without countermeasures in Fukushima City (Case 2) in the first year after the accident (μSv). M, male; F, female. Case 2, citizens consumed vegetables grown locally.

Table S11 Average thyroid equivalent doses of ^{131}I with countermeasures in Fukushima City (Case 2) in the first year after the accident (μSv). M, male; F, female. Case 2, citizens consumed vegetables grown locally.

Table S12 Average effective doses of ^{134}Cs and ^{137}Cs with countermeasures in Fukushima City (Case 2) in the first year after the accident (μSv). M, male; F, female. Case 2, citizens consumed vegetables grown locally.

Table S13 Average thyroid equivalent doses of ^{131}I without countermeasures in Tokyo in the first year after the accident (μSv). M, male; F, female.

Table S14 Average effective doses of ^{134}Cs and ^{137}Cs without countermeasures in Tokyo in the first year after the accident (μSv). M, male; F, female.

Table S15 Average thyroid equivalent doses of ^{131}I with countermeasures in Tokyo in the first year after the accident (μSv). M, male; F, female.

Table S16 Average effective doses of ^{134}Cs and ^{137}Cs with countermeasures in Tokyo in the first year after the accident (μSv). M, male; F, female.

Table S17 Average thyroid equivalent doses of ^{131}I without countermeasures in Osaka in the first year after the accident (μSv). M, male; F, female.

Table S18 Average effective doses of ^{134}Cs and ^{137}Cs without countermeasures in Osaka in the first year after the accident (μSv). M, male; F, female.

Table S19 Average thyroid equivalent doses of ^{131}I with countermeasures in Osaka in the first year after the accident (μSv). M, male; F, female.

Table S20 Average effective doses of ^{134}Cs and ^{137}Cs with countermeasures in Osaka in the first year after the accident (μSv). M, male; F, female.

Table S21 Reductions of doses by countermeasures (%). M, male; F, female. Case 1, citizens consumed vegetables bought from markets.

Table S22 Effective doses in the first year and over the total lifetime (up to 89 y) due to three pathways (μSv). Ages are those in the first year. M, male; F, female. Case 1, citizens consumed vegetables bought from markets. Case 2, citizens consumed vegetables grown locally.

Author Contributions

Conceived and designed the experiments: MM TO. Performed the experiments: MM. Analyzed the data: MM. Contributed reagents/materials/analysis tools: MM. Wrote the paper: MM TO.

References

1. Ministry of Health Labour and Welfare (2012) Available: http://www.mhlw.go.jp/shinsai_jouhou/shokuhin.html. Accessed: 7 Oct. 2013. [in Japanese]
2. Tokyo Metropolitan Institute of Public Health (2011) Available: http://monitoring.tokyo-eiken.go.jp/monitoring/w-past_data.html. Accessed: 7 Oct. 2013.
3. Fukushima Prefecture (2013) Available: http://www.new-fukushima.jp/monitoring/. Accessed: 13 Sep. 2013. [in Japanese]
4. ICRP (1983) Radionuclide transformations - Energy and intensity of emissions. ICRP Publication 38, Ann. ICRP 11–13.
5. Working Group on Radionuclides Pharmaceutical Affairs and Food Sanitation Council (2011) Estimation of contribution of radiocesium to internal exposure from dietary intake. [in Japanese]
6. Yoshida-Ohuchi H, Hirasawa N, Kobayashi I, Yoshizawa T (2013) Evaluation of personal dose equivalent using optically stimulated luminescent dosemeters in Marumori after the Fukushima nuclear accident. Radiat Prot Dosim 154: 385–390.
7. Yasutaka T, Iwasaki Y, Hashimoto S, Naito W, Ono K, et al. (2013) A GIS-based evaluation of the effect of decontamination on effective doses due to long-term external exposures in Fukushima. Chemosphere 93: 1222–1229.
8. Evangeliou N, Balkanski Y, Cozic A, Moller AP (2013) Global transport and deposition of Cs-137 following the Fukushima nuclear power plant accident in Japan: Emphasis on Europe and Asia using high-resolution model versions and radiological impact assessment of the human population and the environment using interactive tools. Environ Sci Technol 47: 5803–5812.
9. Amano H, Akiyama M, Bi C, Kawamura T, Kishimoto T, et al. (2012) Radiation measurements in the Chiba Metropolitan Area and radiological aspects of fallout from the Fukushima Dai-ichi nuclear power plants accident. J Environ Radioactiv 111: 42–52.
10. Christoudias T, Lelieveld J (2013) Modelling the global atmospheric transport and deposition of radionuclides from the Fukushima Dai-ichi nuclear accident. Atmos Chem Phys 13: 1425–1438.
11. Hayano RS, Tsubokura M, Miyazaki M, Satou H, Sato K, et al. (2013) Internal radiocesium contamination of adults and children in Fukushima 7 to 20 months after the Fukushima NPP accident as measured by extensive whole-body-counter surveys. P Jpn Acad B-Phys 89: 157–163.
12. Tsubokura M, Kato S, Nihei M, Sakuma Y, Furutani T, et al. (2013) Limited internal radiation exposure associated with resettlements to a radiation-contaminated homeland after the Fukushima Daiichi Nuclear Disaster. PLOS ONE 8: e81909.
13. Harada KH, Fujii Y, Adachi A, Tsukidate A, Asai F, et al. (2013) Dietary intake of radiocesium in adult residents in Fukushima prefecture and neighboring regions after the Fukushima nuclear power plant accident: 24-h food-duplicate survey in December 2011. Environ Sci Technol 47: 2520–2526.
14. Ministry of Health Labour and Welfare (2011) Available: http://www.mhlw.go.jp/stf/shingi/2r98520000001ip01-att/2r98520000001ipae.pdf. Accessed: 7 Oct. 2013. [in Japanese]
15. World Health Organization (2012) Preliminary dose estimation from the nuclear accident after the 2011 Great East Japan Earthquake and Tsunami. World Health Organization, Geneva.
16. United Nations Scientific Committee on the Effects of Atomic Radiation (2014) Sources, effects and risks of ionizing radiation. UNSCEAR 2013 Report to the General Assembly with Scientific Annexes. United Nations, New York.
17. Ministry of Agriculture Forestry and Fisheries (2013) Food demand and supply 2010: http://www.maff.go.jp/j/tokei/kouhyou/zyukyu/index.html. Accessed: 13 Sep. 2013. [in Japanese]
18. Nagataki S, Takamura N (2014) A review of the Fukushima nuclear reactor accident: radiation effects on the thyroid and strategies for prevention. Curr Opin Endocrinol Diabetes Obes 21: 384–393.
19. Murakami M, Oki T (2012) Estimation of thyroid doses and health risks resulting from the intake of radioactive iodine in foods and drinking water by the citizens of Tokyo after the Fukushima nuclear accident. Chemosphere 87: 1355–1360.

20. Osaka Water Supply Authority (2013) Available: http://www.wsa-osaka.jp/gaiyou/jigyoukanri/keikakuka/housyanou.html. Accessed: 13 Sep. 2013. [in Japanese]
21. Working Group on Radionuclides Pharmaceutical Affairs and Food Sanitation Council (2011) Revision of new standard for radionuclides in foods. [in Japanese]
22. Nuclear Safety Commission of Japan (1998) Control index levels on the ingestion of food and water. [in Japanese]
23. Kuroda M, Ohta M, Okufuji T, Takigami C, Eguchi M, et al. (2011) Frequency of soup intake and amount of dietary fiber intake are inversely associated with plasma leptin concentrations in Japanese adults. Appetite 54: 538–543.
24. Izumo Y (2001) Ministry of Health Labour and Welfare Grants 200000078A. [in Japanese]
25. Nakamura T, Yurimoto M, Matsumoto K, Kawasaki N, Tanada S (1996) Sensory taste evaluation of milled rice cooked with water different in hardness. Jpn J Food Chem 3: 141–144.
26. Tokyo Metropolitan Central Wholesale Market (2011) Available: http://www.shijou-tokei.metro.tokyo.jp/index.html. Accessed: 7 Oct. 2013. [in Japanese]
27. Ministry of Agriculture Forestry and Fisheries (2013) Fishery and aquaculture industrial production statistics 2011: http://www.e-stat.go.jp/SG1/estat/List.do?lid=000001104479. Accessed: 13 Sep. 2013. [in Japanese]
28. Suzuki K, Okuda T, HIgasine Y (2006) The Daily intakes of drinks and folate from the drinks by young women. Mem Osaka Kyoiku Univ Ser 2 Soc Sci Home Econ 54: 27–34. [in Japanese]
29. Ministry of Agriculture Forestry and Fisheries (2011) Basic data on specific forest products industry: http://www.e-stat.go.jp/SG1/estat/List.do?lid=000001085178. Accessed: 13 Sep. 2013. [in Japanese]
30. Ministry of Agriculture Forestry and Fisheries (2011) Crop production statistics 2009: http://www.e-stat.go.jp/SG1/estat/List.do?lid=000001069768. Accessed: 7 Oct. 2013. [in Japanese]
31. Statistics Bureau of Japan (2011) National Population Census 2010.
32. Fukushima City Central Wholesale Market (2013) Available: http://www.city.fukushima.fukushima.jp/soshiki/24/1030.html. Accessed: 13 Sep. 2013. [in Japanese]
33. Osaka Central Wholesale Market (2013) Available: http://www.pref.osaka.jp/fuichiba/toukeijouhou/nenpou22.html. Accessed: 13 Sep. 2013. [in Japanese]
34. Ministry of Agriculture Forestry and Fisheries (2011) Statistics of animal product distribution 2009: http://www.e-stat.go.jp/SG1/estat/List.do?lid=000001072695. Accessed: 13 Sep. 2013. [in Japanese]
35. Ministry of Agriculture Forestry and Fisheries (2011) Statistics of animal product distribution 2010: http://www.e-stat.go.jp/SG1/estat/List.do?lid=000001076341. Accessed: 7 Oct. 2013. [in Japanese]
36. Ministry of Agriculture Forestry and Fisheries (2010) Statistical survey on milk and dairy products 2008: http://www.e-stat.go.jp/SG1/estat/List.do?lid=000001063504. Accessed: 13 Sep. 2013. [in Japanese]
37. Ministry of Agriculture Forestry and Fisheries (2013) Crop statistics 2011: http://www.e-stat.go.jp/SG1/estat/List.do?lid=000001087011. Accessed: 13 Sep. 2013. [in Japanese]
38. Ministry of Agriculture Forestry and Fisheries (2012) Crop production statistics: http://www.e-stat.go.jp/SG1/estat/List.do?lid=000001081820. Accessed: 13 Sep. 2013. [in Japanese]
39. Ministry of Agriculture Forestry and Fisheries (2007) Statistical survey on livestock 2006: http://www.e-stat.go.jp/SG1/estat/List.do?lid=000001060567. Accessed: 7 Oct. 2013. [in Japanese]
40. Ministry of Agriculture Forestry and Fisheries (2011) Statistical survey on cultivated acreage (FY 2009): http://www.e-stat.go.jp/SG1/estat/List.do?lid=000001061777. Accessed: 7 Oct. 2013. [in Japanese]
41. Fukushima Prefecture (2010) Statistics of forest industry in Fukushima 2009. [in Japanese]
42. Ministry of Agriculture Forestry and Fisheries (2008) Census of agriculture and forestry 2005: http://www.maff.go.jp/j/tokei/census/afc/2010/report05_archives.html. Accessed: 13 Sep. 2013. [in Japanese]
43. Japan Meteorological Agency (2011) Available: http://www.data.jma.go.jp/obd/stats/etrn/index.php. Accessed: 7 Oct. 2013. [in Japanese]
44. Murakami M, Takeda H, Okaneya M, Kobayashi Y, Oki T (2012) Classification and motives of drink intakes based on the situations using behavioral record. SEISAM-KENKYU 64: 359–366. [in Japanese]
45. Japan Water Works Association (2011) Statistics of tap water (2009 FY). [in Japanese]
46. Bureau of Waterworks Tokyo Metropolitan Government (2013) Available: https://www.waterworks.metro.tokyo.jp/press/shinsai22/press01.html. Accessed: 13 Sep. 2013.
47. Ministry of Health and Welfare (2012) The National Health and Nutrition Survey in Japan, 2010. [in Japanese]
48. World Health Organization (2013) Health risk assessment from the nuclear accident after the 2011 Great East Japan Earthquake and Tsunami based on a preliminary dose estimation.
49. Harada KH, Niisoe T, Imanaka M, Takahashi T, Amako K, et al. (2014) Radiation dose rates now and in the future for residents neighboring restricted areas of the Fukushima Daiichi nuclear power plant. P Natl Acad Sci USA 111: E914–E923.
50. Akahane K, Yonai S, Fukuda S, Miyahara N, Yasuda H, et al. (2013) NIRS external dose estimation system for Fukushima residents after the Fukushima Dai-ichi NPP accident. Sci Rep 3: 1670.
51. Brenner AV, Tronko MD, Hatch M, Bogdanova TI, Oliynik VA, et al. (2011) I-131 dose response for incident thyroid cancers in Ukraine related to the Chornobyl accident. Environ Health Persp 119: 933–939.
52. United Nations Scientific Committee on the Effects of Atomic Radiation (2008) Annex A: Epidemiological studies of radiation and cancer. UNSCEAR 2006 Report to the General Assembly: Effects of Ionizing Radiation (United Nations Scientific Committee on the Effects of Atomic Radiation, Vienna), Vol. 1.
53. Preston DL, Ron E, Tokuoka S, Funamoto S, Nishi N, et al. (2007) Solid cancer incidence in atomic bomb survivors: 1958–1998. Radiat Res 168: 1–64.
54. Statistics Bureau of Japan (2011) General mortality in 2010: http://www.e-stat.go.jp/SG1/estat/ListE.do?lid=000001101825. Accessed: 5 Mar. 2014.
55. Center for Cancer Control and Information Services National Cancer Center Japan (2013) Cancer incidence (1975–2008): http://ganjoho.jp/professional/statistics/statistics.html. Accessed: 5 Mar. 2014. [in Japanese]
56. Matsuda A, Matsuda T, Shibata A, Katanoda K, Sobue T, et al. (2013) Cancer incidence and incidence rates in Japan in 2007: A study of 21 population-based cancer registries for the monitoring of cancer incidence in Japan (MCIJ) project. Jpn J Clin Oncol 43: 328–336.
57. Mullenders L, Atkinson M, Paretzke H, Sabatier L, Bouffler S (2009) Assessing cancer risks of low-dose radiation. Nat Rev Cancer 9: 596–604.
58. Russell WL, Kelly EM (1982) Mutation frequencies in male-mice and the estimation of genetic hazards of radiation in men. P Natl Acad Sci–Biol 79: 542–544.
59. Ciccia A, Elledge SJ (2010) The DNA damage response: Making it safe to play with knives. Mol Cell 40: 179–204.
60. ICRP (2007) The 2007 recommendations of the international commission on radiological protection. ICRP Publication 103, Ann. ICRP. 37 (2–4).
61. National Research Council of the National Academies (2006) Health risks from exposure to low levels of ionizing radiation: BEIR VII Phase 2. Board on Radiation Effects Research.
62. Kinoshita N, Sueki K, Sasa K, Kitagawa J-i, Ikarashi S, et al. (2011) Assessment of individual radionuclide distributions from the Fukushima nuclear accident covering central-east Japan. P Natl Acad Sci USA 108: 19526–19529.
63. Kosaka K, Asami M, Kobashigawa N, Ohkubo K, Teracia H, et al. (2011) Removal of radioactive iodine and cesium in water purification processes after an explosion at a nuclear power plant due to the Great East Japan Earthquake. Water Res 46: 4397–4404.
64. Diamond Online (2013) 8th July. http://diamond.jp/articles/-/38458. [in Japanese]
65. Hangui S (2012) Enormous influences on agriculture in Fukushima prefecture caused by the Great East Japan Earthquake and nuclear disaster: Focused on distribution and consumption stage. Jpn J Farm Manag 49: 93–96. [in Japanese]
66. Japan Soft Drink Association (2010) Available: http://j-sda.or.jp/statistically-information/stati01.php. Accessed: 5 Sep. 2013.
67. Koizumi A, Harada KH, Niisoe T, Adachi A, Fujii Y, et al. (2012) Preliminary assessment of ecological exposure of adult residents in Fukushima Prefecture to radioactive cesium through ingestion and inhalation. Environ Health Prev Med 17: 292–298.
68. Sugiyama H, Terada H, Takahashi M, Iijima I, Isomura K (2007) Contents and daily intakes of gamma-ray emitting nuclides, ^{90}Sr, and ^{238}U using market-basket studies in Japan. J Health Sci 53: 107–118.
69. Ota T, Sanada T, Kashiwara Y, Morimoto T, Sato K (2009) Evaluation for committed effective dose due to dietary foods by the intake for Japanese adults. Jpn J Health Phys 44: 80–88.
70. Oguri T, Yoshinaga J, Tao H, Nakazato T (2012) Daily intake of inorganic arsenic and some organic arsenic species of Japanese subjects. Food Chem Toxicol 50: 2663–2667.
71. Murakami M, Nagai T (2013) Cancer risks from micro-pollutants and their acceptable levels. J Jpn Soc Water Environ 36(A): 322–326. [in Japanese]

14

Cloning of *Gossypium hirsutum* Sucrose Non-Fermenting 1-Related Protein Kinase 2 Gene (*GhSnRK2*) and Its Overexpression in Transgenic *Arabidopsis* Escalates Drought and Low Temperature Tolerance

Babatunde Bello, Xueyan Zhang, Chuanliang Liu, Zhaoen Yang, Zuoren Yang, Qianhua Wang, Ge Zhao, Fuguang Li*

State Key Laboratory of Cotton Biology, Cotton Research Institute, Chinese Academy of Agricultural Sciences, Beijing, China

Abstract

The molecular mechanisms of stress tolerance and the use of modern genetics approaches for the improvement of drought stress tolerance have been major focuses of plant molecular biologists. In the present study, we cloned the *Gossypium hirsutum* sucrose non-fermenting 1-related protein kinase 2 (*GhSnRK2*) gene and investigated its functions in transgenic Arabidopsis. We further elucidated the function of this gene in transgenic cotton using virus-induced gene silencing (VIGS) techniques. We hypothesized that *GhSnRK2* participates in the stress signaling pathway and elucidated its role in enhancing stress tolerance in plants via various stress-related pathways and stress-responsive genes. We determined that the subcellular localization of the *GhSnRK2*-green fluorescent protein (GFP) was localized in the nuclei and cytoplasm. In contrast to wild-type plants, transgenic plants overexpressing *GhSnRK2* exhibited increased tolerance to drought, cold, abscisic acid and salt stresses, suggesting that *GhSnRK2* acts as a positive regulator in response to cold and drought stresses. Plants overexpressing *GhSnRK2* displayed evidence of reduced water loss, turgor regulation, elevated relative water content, biomass, and proline accumulation. qRT-PCR analysis of *GhSnRK2* expression suggested that this gene may function in diverse tissues. Under normal and stress conditions, the expression levels of stress-inducible genes, such as *AtRD29A, AtRD29B, AtP5CS1, AtABI3, AtCBF1*, and *AtABI5*, were increased in the *GhSnRK2*-overexpressing plants compared to the wild-type plants. *GhSnRK2* gene silencing alleviated drought tolerance in cotton plants, indicating that VIGS technique can certainly be used as an effective means to examine gene function by knocking down the expression of distinctly expressed genes. The results of this study suggested that the *GhSnRK2* gene, when incorporated into Arabidopsis, functions in positive responses to drought stress and in low temperature tolerance.

Editor: Girdhar K. Pandey, University of Delhi South Campus, India

Funding: This work was supported by Major Program of Joint Funds (Sinkiang) of the National Natural Science Foundation of China (Grant No. U1303282). The funders had no role in study design, data collection and analysis, decision to publish, or preparation of the manuscript.

Competing Interests: The authors have declared that no competing interests exist.

* Email: aylifug@163.com

Introduction

Plants have developed complex signaling pathways in response to various environmental stresses, such as salt, drought, and cold, and have acquired metabolic functions and developmental methods to survive changing environmental conditions [1]. Improving crop resistance to drought stress would be the most valuable means to improve agricultural productivity and to reduce crop loss caused by environmental stress. As a result, understanding the mechanisms of drought tolerance and developing drought-resistant crop plants have been major targets of plant molecular biologists and biotechnologists. Low-temperature constraints have been progressively overcome by the recognition of cold-tolerant genes for applications in transgenic plants. Transgenic approach has reveal many possibilities to improve cold stress in plants by incorporating or deleting genes that regulate a particular characteristic [2]. These approaches also provide unique opportunities to improve the genetic quality of plants via the development of particular crop varieties that exhibit enhanced resistance to biotic and cold stresses and improved nutritional quality. The plant response to salt stress typically results in osmotic alterations, which play a major role in ensuring osmotic balance in plant cells. During plant stress responses, the regulation of gene expression involves both universal and unique changes in the transcript levels of certain plant genes [3]. Plants directly or indirectly respond to stresses by initiating signal transduction pathways. Various abiotic stresses result in both general and specific effects on plant growth and development. For example, drought limits plant growth due to difficulties in maintaining turgor pressure, photosynthetic decline, osmotic stress-imposed constraints on plant processes and interference with nutrient availability as the soil dries [4].

Protein kinases and phosphatases are major elements of stress signals which are transmitted to different cellular regions via specialized signaling pathways. Some of the protein kinases involved in stress signal transduction in plants, such as mitogen-activated protein kinases (*MAPKs*) [5,6,7,8], glycogen synthase kinase 3 (*GSK3*) [9,10], and S6 kinase (*S6K*) [11], are similar among all eukaryotic organisms, whereas others, including calcium-dependent protein kinases (*CDPKs*) [12,13,14] and *SNF1*-related kinases (*SnRKs*), are plant-specific.

SnRK2 is an important stress-related protein kinase in plants that has been implicated in stress and abscisic acid-mediated signaling pathways [15]. Previously, it was reported that *SRK2C/SnRK2.8*, a subclass II member, was strongly activated by drought stress and that plants overexpressing *SRK2C* exhibited improved drought tolerance as a result of the up-regulation of many stress-inducible genes [16]. *SnRK2s* have a molecular weight of approximately 40 kDa and are monomeric serine/threonine protein kinases [17,18]. Based on phylogenetic analysis, three groups of *SnRK2* family members have been identified. Group 1 consists of kinases not activated by ABA, group 2 consists of kinases not activated or activated very weakly by ABA, and group 3 consists of kinases strongly activated by ABA. The amino acid sequences of all *SnRK2s* can be separated into two regions, the highly conserved N-terminal kinase domain and the regulatory C-terminal domain, which contains stretches of acidic amino acids. Furthermore, the C-terminal domain consists of two subdomains, Domain I and Domain II. Domain I is characteristic of all *SnRK2* family members and is required for activation by osmotic stress. Based on our investigation, we found that the expression of *GhSnRK2* was induced by PEG. PEG was quite commonly used in physiological experiments to induce controlled drought stress. In the present study, we generated a gene construct containing *GhSnRK2* driven by the constitutive cauliflower mosaic virus (CaMV) 35S promoter and transformed this construct into *Arabidopsis* in order to investigate the functional analysis of *GhSnRK2* gene. We monitored the activities of this gene with respect to drought and low temperature tolerance in transgenic plants. We further elucided the function of this gene in transgenic cotton using virus-induced gene silencing (VIGS) techniques. We showed that *GhSnRK2*, the cotton *SnRK2* gene, is involved in multi-stress responses. Our results promote the analysis of gene function in *G. hirsutum* to facilitate the exploitation of desirable genes from this species. The elucidation of *GhSnRK2* gene function contributes to our understanding of the mechanism by which this plant adapts to abiotic stress and provides a valuable gene resource for plant breeders.

Materials and Methods

Cloning of theGhSnRK2 gene

Total RNA was extracted from CCRI24 cotton tissues using Trizol (Sigma-Aldrich) according to the manufacturer's instructions. Reverse transcription (RT) was performed using total RNA extracted from seedlings oligo(dT)16 primer, and SuperScript II reverse transcriptase (Promega). The RT product was used in PCRs to amplify the predicted *GhSnRK2* open reading frame using primer star polymerase (enzymes). The cDNA regions of *GhSnRK2* were cloned into the T-simple vector. All of the clones were confirmed via sequencing. Primers specific to the sequence of *GhSnRK2* were designed, synthesized, and used to clone the *SnRK2* gene. For plant transformation, *GhSnRK2* cDNAs were introduced into the modified pCAMBIA2301 plant transformation vector under the control of the CaMV 35S promoter.

Localization of the *GhSnRK2*-GFP fusion protein

The method of [19] was adopted to perform the subcellular localization assay. The *GhSnRK2* coding sequence was cloned and ligated into the XbaI and SpeI sites of the PCAMBIA2301-GFP vector to generate PCAMBIA2301-*GhSnRK2*-GFP, which expressed the *GhSnRK2*-GFP fusion protein under the control of the CaMV35S promoter. The construct was used for transient transformation of onion. Onion epidermal peels were bombarded with DNA-coated gold particles, and GFP expression was visualized 24 h later. Transformed onion cells were observed under a confocal microscope.

Effect of polyethylene glycol (PEG) treatment on *GhSnRK2* gene expression

The effect of PEG on the expression level of the *GhSnRK2* gene was evaluated via qRT- PCR. The root of a three-week-old upland cotton plant was submerged in 10% PEG solution, and samples were collected from the root at one hour intervals for 6 hours. RNA was extracted from the samples, and RT was performed as described above to generate cDNA for qRT- PCR analysis.

Arabidopsis transformation and screening of transgenic plants

pCAMBIA2301 carrying *GhSnRK2* was introduced into *Agrobacterium tumefaciens* strain GV3101. The transgenic *Arabidopsis* plants were generated using the flower dipping method [20], and transgenic plants were selected based on their growth in 0.8% agar containing half-strength MS salts and kanamycin. The transformants were transferred to soil and allowed to set seed. The T3 generation of the transgenic plants was used for all experiments.

Plant materials and growth conditions

Arabidopsis thaliana seeds and ecotype Col-0 wild-type, mutant, and transgenic seeds were surface-sterilized using bleach (5% Sodium hypochlorite) and 0.1% Triton X-100. After cold exposure at 4°C for 2 d, the seeds were germinated and cultured in plates containing 0.5 × MS medium, 0.8% agar, and 1% sucrose under continuous light at 22°C. For plants grown in soil, 7-day-old seedlings were transferred from the MS plates to soil and cultured under 80–100 μmol $m^{-2}s^{-1}$ photoperiodic cycles of 16 h light and 8 h dark at 22°C in a growth chamber under fluorescent light using one cool-light and one warm-light tube, each of which was suspended several inches above the plants. Mature seeds obtained from transformed plants (T_0 generation) were cultured in MS medium containing the antibiotic kanamycin in order to screen for transformed plants. The seeds were allowed to grow in the medium for two weeks before transfer to soil for growth continuation. Cotton seeds were planted in moist soil and grown in a growth chamber with a 14 h photoperiod at a 20/30°C night/day temperature cycle, with a light intensity of 400 μmol$m^{-2}s^{-2}$ and at 60% relative humidity.

Drought tolerance assay

Drought stress tolerance was measured by transferring 2-week-old plants cultured in Petri dishes to pots (10-cm diameter) filled with a 1:1 (v/v) vermiculite:perlite mixture. The seedlings were cultured for 2 weeks with constant watering before the water was withheld. After 9 days without water, all of the pots were re-watered simultaneously, and plant re-growth was scored 4 days later. Thirty-five plants for each individual transgenic line were used in each repeated experiment, and one representative image is shown. Plants were scored as survivors if there were healthy green young leaves after re-watering treatment. The survival rate was calculated as the ratio of number of surviving plants to the total number of treated plants in the flower pot. Representative transgenic and WT plants of were photographed before and after drought treatment. Each stress assay was repeated at least three times. The water use efficiency (WUE) of the plants was calculated as the amount of water needed to maintain the weight of each experimental pot containing the plants. To determine the soil water content (SWC), the pots were immediately weighed after saturation with water prior to the initiation of drought stress and then periodically weighed during the drought stress period. The SWC was calculated as (final fresh weight-dry weight)/(initial weight–dry weight)x100. The plants were exposed to water stress by reducing the water content to 25 to 30% of the SWC. For plant tugor pressure assay, drought stressed plants were removed from the growth chamber and transferred to a dark laboratory cupboard for 4 h before turgor measurements using the pressure chamber [21]. To measure the drought tolerance of *GhSnRK2* gene silenced and non-silenced plants, water was withheld for 5 days, approximately two weeks post-inoculation, and the survival rate was recorded as the percentage of plants that survived after re-watering for 3 days. Each of the ten treated groups consisted of five plants.

Freezing and cold stress treatment

Transgenic plants overexpressing the *GhSnRK2* gene and wild-type *Arabidopsis* plants were cultured in pots filled with a 1:1 mixture of perlite and vermiculite under a long-day photoperiod at 22°C. Freezing stress was performed by transferring 4-week-old plants to a chamber in which the temperature was adjusted to −4°C and −8°C for 10 h, respectively, and then returned to the typical standard growth conditions. For cold stress, 4-week-old plants were transferred to 4°C for three weeks. The plants were analyzed after recovery for 7 days under normal growth conditions. The survival rate (%) under freezing and cold conditions was calculated as the percentage of clearly green plants after returning to normal conditions, and plants exhibiting clear sign of wilting were denoted as dead. Thirty-five plants for each individual line were used in each repeated experiment, and one representative image is shown.

Transpirational water loss assay

The leaves of mutant and wild-type seedlings at the rosette stage were detached and placed in a weighing boat, and the changes in fresh weight over time were monitored using an electronic balance. The rate of water loss was calculated as the loss in fresh weight of the samples. Six plants of each transgenic and WT line were analyzed in this assay, and this experiment was replicated three times [22].

Relative water content

The relative water content of the leaves was measured according to the method of [23]. Fully expanded leaves were cut from the plants, and the fresh weight (FW) was recorded immediately. Then, the fresh portions were immersed in distilled water for 4 h and the turgid weight (TW) was recorded. Finally, the dry weight (DW) was recorded after drying for 48 h at 80°C in an oven. The RWC was calculated according to the following formula: $RWC\ (\%) = (FW-DW)/(TW-DW)\times 100$.

Measurement of stomatal closure in response to ABA treatment

Stomatal closure assays were conducted as described by [24]. Leaves from *GhSnRK2* transgenic and WT plants were immersed in a solution containing 50 μM $CaCl_2$, 10 mM KCl, and 10 mM 2-(N-morpholino) ethanesulfonic acid (MES)-Tris, pH 6.15, and were exposed to light for 2 h. Subsequently, various concentrations of ABA were added to the solution. The stomatal apertures were measured after 2 h of ABA treatment.

Determination of proline content

The free proline content was measured using the method described by [25]. Leaf segments were homogenized in 3% sulfosalicylic acid, and the homogenates were centrifuged at 3000 g for 20 min. Mixtures containing 2 ml of sample supernatant, 2 ml of acetic acid, and 2 ml of 2.5% acid ninhydrin solution were boiled for 30 min, and the absorbance at 520 nm (A520) was measured.

Chlorophyll content assay

The determination of chlorophyll content was performed according to the method of [26]. Extracts were obtained from 0.1 g (fresh weight) leaf samples from 4-week-old plants and were homogenized in 1 ml of 80% acetone to quantify the chlorophyll content via spectrophotometric analysis.

Seedling growth in response to ABA and NaCl treatment

The sensitivity of seedling germination to ABA and NaCl was assessed on MS agar plates containing various concentrations of ABA and NaCl solution [27]. Seedlings from WT and transgenic plants were placed on MS agar plates supplemented with distilled water or different concentrations of ABA or NaCl and were placed in the growth chamber at 22°C.

Germination assay

To evaluate seed germination, the method of [27] was employed. Seeds from wild-type and transgenic plants were surface-sterilized using sodium hypochlorite and were placed on Murashige and Skoog (MS) solid medium containing various concentrations of ABA (0 μm, 0.3 μm, or 0.5 μm). For the evaluation of seed germination under salt stress, the MS medium was supplemented with various concentrations of NaCl (0 mM, 50 mM, or 100 mM). The percentage of germinating seeds was recorded after 7 days.

Biomass accumulation

Plants were harvested for biomass measurements after 4 weeks of germination. The fresh weight of each individual plant shoot and root was measured immediately after harvesting. For fresh weight biomass, the fresh weight of the root and the shoot was measured immediately after harvesting, and the dry weight was recorded in (g) after drying in an oven to a constant weight at 70° for 48 h.

qRT-PCR

Total RNA was extracted from *Arabidopsis* seedlings using Trizol reagent (Takara) and was treated with RNase-free DNase (Promega). cDNA was synthesized using a Promega kit according to the manufacturer's protocol. qPCR was performed using a SYBR green PCR master mix kit. After PCR, the data was quantified using the comparative CT method ($2^{-\Delta\Delta CT}$ method) based on the CT values [28]. *AtACT2* (gene accession At3G18780) and cotton *histone3* (gene accession AF024716) were used as internal controls, and the relative expression level of each target gene was quantified. Each RT-PCR measurement was performed in triplicate. All RT-PCR experiments were reproduced at least three times using independent cDNA samples.

Silencing construct development and cotton transformation

The method described by [29] was employed to construct the VIGS vector. Specific fragments (approximately 300 bp) were amplified via PCR and cloned into the T-simple vector according to the manufacturer's specifications (Promega). The resulting clones were sequenced. The plasmids were digested using the restriction enzymes Xba1 and BamH1 and were subsequently ligated into the VIGS vector (*PYL156:(pTRV-RNA2)-GhSnRK2*). The resulting constructs were named according to the putative function of the stress-related gene. For transformation, *Agrobacterium* (GV3101) carrying *GhSnRK2* derivatives were cultured at 28°C in LB medium containing appropriate antibiotics. Cells were harvested from cultures grown overnight, re-suspended in inoculation buffer (200 μM acetosyringone, 10 mM MES, pH 5.5), and incubated for 2 h at room temperature in a shaker. *Agrobacterium* strains containing the *GhSnRK2* derivative *PYL156:(pTRV-RNA2)-GhSnRk2*, *PYL192:(pTRV-RNA1)*, *PYL156:(pTRV-RNA2)-GrCLA1*(positive control), or*PYL156:(pTRV-RNA2)* vector (OD 1.5) were mixed at a 1 : 1 ratio in 5 mM MES buffer (pH 5.5) and inoculated into *Gossypium hirsutum* leaves using a needleless syringe. The inoculated plants were maintained in a green house at 23±2°C for effective viral infection and spreading.

Experimental design and statistical analysis

The data obtained were subjected to statistical analysis and were expressed as the means and standard error (SE) of at least three replicates. All experimental data are presented as the means of at least three independent replicates, and the analysis for significance was performed using Student's T-test.

The sequences of primers used in this study are listed in Table S2.

Results

Phylogenetic analysis, subcellular localization, and expression pattern of *GhSnRK2* in cotton plant

The phylogenetic tree constructed using the full length amino acid sequence of selected *SnRK2* genes to analyze the evolutionary relationship between *GhSnRK2* and other *SnRK2* family genes is shown in the neighbor-joining tree developed based on an alignment of the complete protein sequences. The bootstrap values are shown on the branches. *GhSnRK2* clustered with the known stress-related genes *AtSnRK2.1*, AED91326.1; *AtSnRK2.4*, AEE28666.1; *AtSnRK*2.5, AED97781.1; *AtSnRK2.10*, AEE33751.1; and *Oryza sativa* (*RK1*), ABB89146.1. *SnRK2.10* and *SnRK2.4* are closely related to *GhSnRK2*, suggesting that *GhSnRK2* belongs to the *SnRK2* family (Figure 1A). Alignment of the *GhSnRK2* amino acid sequence with that of other *SnRK2s* revealed that the *GhSnRK2* protein is highly similar to other *SnRK2s*; the relatively conserved motif is underlined. The deduced amino acid sequence displays relatively high homology with the monocot *SnRK2* family members *Oryza sativa* (*RK1*), ABB89146 and with the dicot species *AtSnRK2.10*, AEE33751.1 (Figure S1A). The *GhSnRK2*-GFP fusion protein driven by the CaMV 35S promoter was transiently expressed in onion epidermal cells, and the green fluorescent *GhSnRK2* protein signals were localized to the nuclei and the cytoplasm, whereas GFP alone was detected throughout the cell (Figure1B). qRT-PCR was performed to quantify the expression level of the *GhSnRK2* gene. Treatment of 3-week-old CCRI24 upland cotton cultivar plants with 10% PEG for different periods induced the expression of the *GhSnRK2* gene for 3 h, after which the expression level declined (n = 3) (Figure 1C). To investigate the distribution of *GhSnRK2* in different tissues, samples from root (RT), stem (ST), cauline leaves (CL), rosette leaves (RL) and flowers (FL) were analyzed. *GhSnRK2* gene expression was detected in all of the examined tissues in distinct expression patterns and was highest in the root and lowest in the flower, suggesting that the *GhSnRK2* gene may most actively function in the root (Figure 1D). Cotton *histone*3 (gene accession AF024716) was used as internal control to normalize expression data. The values are presented as the means of three replicates, and the error bars denote the SE. Different letters denote the means ± standard deviation displaying significant difference at P≤0.05.

Plant transformation vector and the expression pattern of the *GhSnRK2* gene in the transgenic lines

The schematic representation of the 35S *GhSnRK2* construct used for *Arabidopsis* transformation (Figure 2A). The full-length *GhSnRK2* cDNA was introduced into the *Arabidopsis* genome via the floral dip method using *Agrobacterium* GV3101; the XbaI–BamHI region of *GhSnRK2* was inserted between the CaMV 35S promoter and the nopaline synthase gene terminator in the pCAMBIA2301 vector to generate the recombinant plasmid *35S-GhSnRK2-NOST*. The neomycin phosphotransferase II (*NPTII*) gene, driven by the nopaline synthase gene promoter (NOSP), was carried by the *pCAMBIA2301* vector for transgenic cell selection in kanamycin-containing LB medium. *GhSnRK2* in transgenic *Arabidopsis* plants was confirmed by PCR analysis. Genomic DNA from the first generation of the plants (T1) was extracted and used as a template for gene-specific primers. Of the 43 plants evaluated, 35 were positive for *GhSnRK2* amplification. No product was formed by the amplification reaction using untransformed plant DNA as the template (Figure S2B). At T1 generation, the ratio of dead to surviving plants was approximately 1:3 in kanamycin-containing LB medium. However, at T3 generation, all of the plants became homozygous, as their survival rate was 100% in kanamycin-containing LB medium (Figure S2A, Table S1). The values are expressed as the mean germination rate (%) of approximately 200 seeds. qRT-PCR analysis was performed to evaluate the differences in transgene expression between all of the independently generated transgenic lines. Various transgene expression patterns were detected in the transgenic lines overexpressing the *GhSnRK2* gene, but no expression was detected in the WT line, indicating that the *GhSnRK2* genes were successfully introduced into the transformed plants (Figure 2B). The values are presented as the means of three experimental replicates; the error bars indicate the standard deviations. *AtACT2* (gene accession At3G18780) was used as internal control to normalize expression data. Three transgenic lines (L1, L2, and L4) were selected for further investigation based on their seed availability and relative expression pattern.

Overexpression of *GhSnRK2 in transgenic plants improves their tolerance to drought and cold stresses*

To investigate the roles of *GhSnRK2* in stress response pathways, the responses of the three selected transgenic *GhSnRK2*-overexpressing *Arabidopsis* lines to drought and low temperature stresses were analyzed. Constitutive overexpression of

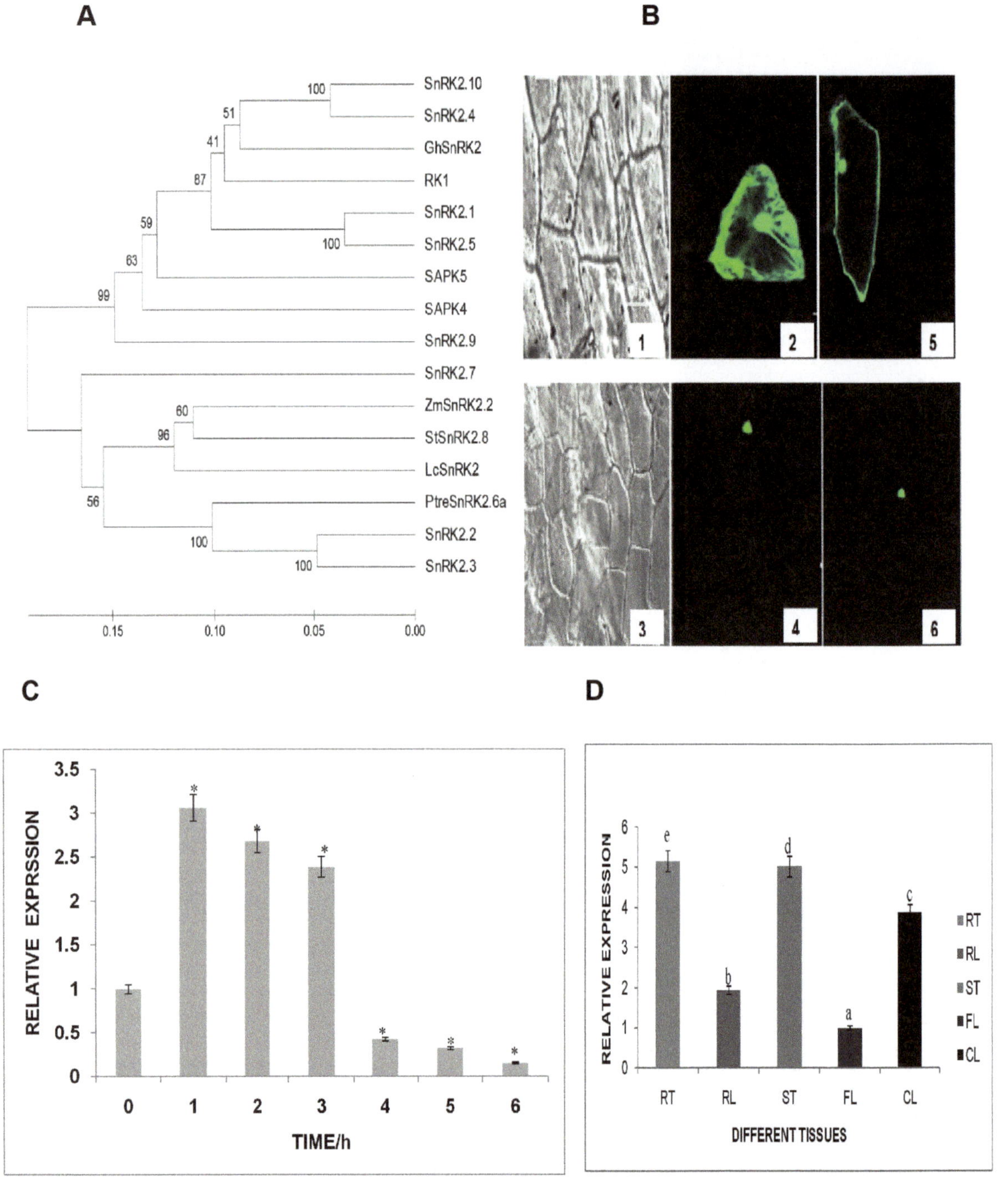

Figure 1. Phylogenetic analysis, subcellular localization, and expression pattern of *GhSnRK2* in cotton plant. (A) A phylogenetic tree of *GhSnRK2* and other *SnRK2* proteins from different plants was constructed using the neighbor-joining method with MEGA 5. The sequences used for analysis are listed by accession number: *Litchi chinensis* (*LcSnRK2*), AFX72761.1; *Arabidopsis thaliana* (*AtSNRK2.2*), CP002686.1; *Arabidopsis thaliana* (*AtSnRK2.3*), AED98274.1; *AtSnRK2.1*, AED91326.1; *AtSnRK2.9*, AEC07398.1; *AtSnRK2.10*, AEE33751.1; *AtSnRK2.4*, AEE28666.1; *AtSnRK2.5*, AED97781.1; *AtSnRK2.7*, AEE87152.1; *Populus tremula* (*PtreSnRK2.6a*), AGW51610.1; *Zea mays* (*ZmSnRK2.2*), NM_001137717.1; *Solanum tuberosum* (*StSnRK2.8*), AFR68945.1; *Oryza sativa* (*RK1*), ABB89146.1; *Oryza sativa* (*SAPK4*), BAD18000.1; *Sorghum bicolor* (*SAPK4*), AGM39623.1; and *Zea mays* (*SAPK5*), ACG42286.1. The bootstrap values are shown on the tree branches. (B) Subcellular localization of the *GhSnRK2-GFP* protein. (2 and 5) GFP alone; (4 and 6) *GhSnRK2-GFP* in onion epidermal cells; (1 and 3) corresponding bright-field images. (C) The expression pattern of the *GhSnRK2* gene in cotton plants subjected to 10% PEG stress. The gene expression data were normalized to that of the cotton histone 3 gene. The values are presented as the means of three experimental replicates. The vertical axis represents the relative expression level. The values from1 to 6 indicate the time (h) of PEG treatment. Asterisk denotes a significant difference ($P<0.05$) compared with the control (0 h). (D) Relative expression levels of the *GhSnRK2* gene in various cotton plant tissues. Samples from root (RT) stem (ST), cauline leaves (CL), rosettes leaves (RL) and flowers (FL) were analyzed. The vertical axis represents the relative expression level. The letters denote significant differences ($P<0.05$) based on Duncan's multiple range tests. The cotton histone 3 gene was used as an internal control for normalization of the gene expression data.

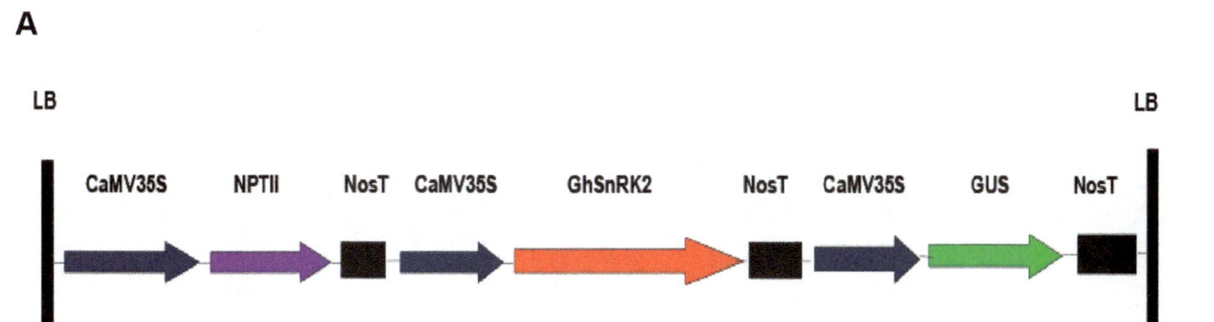

Figure 2. Plant transformation vector and the expression pattern of the *GhSnRK2* gene in the transgenic lines. (A) Schematic representation of the T-DNA region of the binary vector *pCAMBIA2301-GhSnRK2*. (B) Expression pattern of the *GhSnRK2* gene in the transgenic plants. Various upregulated expression patterns of the *GhSnRK2* gene in transgenic lines were detected, as indicated by the vertical axis. The values are presented as the means of three experimental replicates; the error bars indicate the standard deviations. The *AtACT2* gene was used as an internal control for normalization of gene expression.

the 35S-*GhSnRK2* gene resulted in an increase in the drought tolerance of the transgenic plants. After withholding water for 9 days, some of the WT plants grew slowly, wilted severely and died, whereas few of the *GhSnRK2* transgenic plants wilted. The recovery rate of the transgenic plants was higher than that of the WT plants after re-watering. Indeed, the survival rate was significantly different between the *GhSnRK2* transgenic and wild-type plants: L2 (100%), L4 (94.4%), L1 (88.9%), and WT (4.4%); Student's T-test ($n = 3$) ($p < 0.05$) (Figure 3A, B). These results suggested that the *GhSnRK2* gene is involved in the response to drought tolerance.

Under low temperature stress conditions, the survival rate of the transgenic lines was significantly higher than that of the WT line. Most of the transgenic plants were intact and recovered from this stress, whereas most WT plants were found to have died and could not recover after transferring them to the normal growth conditions. The survival rates at 4°C were: WT (16.67%), L2 (80.55%), L1 (77.78%), and L4 (78.33%). The survival rates at −4°C were: WT (5.56%), L1 (54.44%), L2 (58.33%), and L4 (56.11%). At −8°C, only 3.3% of the WT plants survived, whereas the survival rates of transgenic lines ranged between 30% and 33.33% after 7 days under normal growth conditions (Figure 3C, D). However, the survival rate under different low temperatures clearly indicates the difference between the transgenic and WT plants, suggesting that the *GhSnRK2* overexpression alleviates low temperature stress in transgenic Arabidopsis. Our findings revealed that the water use efficiency was higher in *GhSnRK2*-overexpressing plants compared to the corresponding WT plants, suggesting that *GhSnRK2*-overexpressing plants exhibit greater photosynthetic potential during drought stress treatment (Figure 3E). Normal turgor pressure which is regulated by the amount of water in the plant's cells is required for healthy growth of a plant and is a powerful determinant of the plant's drought tolerance. We determined the turgor of the *GhSnRK2* transgenic plant to understand the role of turgor pressure in cellular signaling during water deficit condition. Turgor values of *GhSnRK2* transgenic plant was notably higher compared to the corresponding WT plant and this may be attributable to osmotic adjustment, like stomatal closure, and the maintenance of water content which prevent plants from desiccation and turgor loss. Student's T-test ($n = 3$) ($p < 0.05$) (Figure 3F).

Biochemical and physiological assays of plants overexpressing the *GhSnRK2* gene

The rate of water loss and the RWC were investigated to further understand the tolerance of *GhSnRK2*-overexpressing plants to

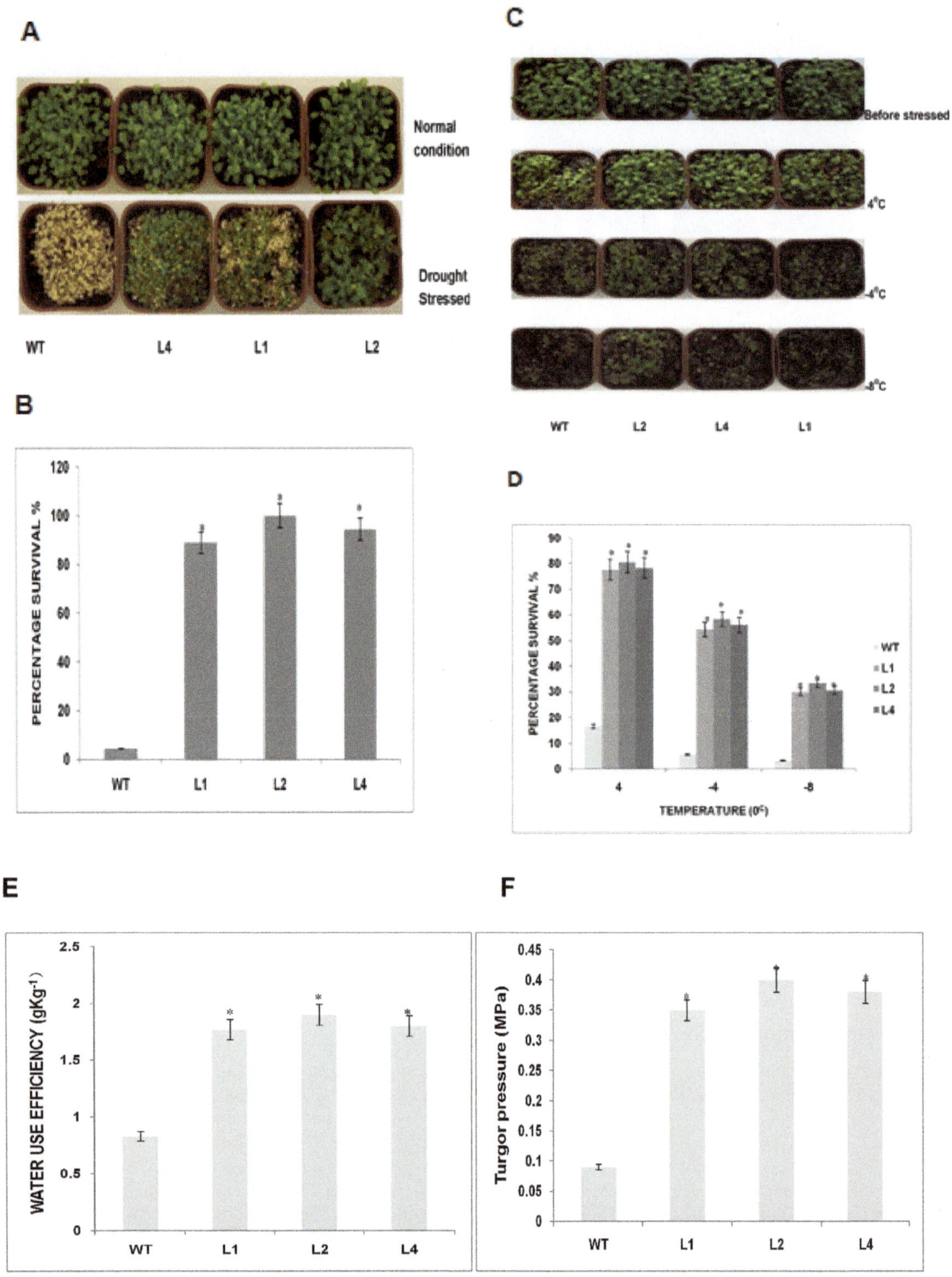

Figure 3. Survival rates of *GhSnRK2* transgenic plants under drought and low temperature stresses. (**A**) WT and *GhSnRK2* transgenic plants before and after drought stress. (**B**) Seedlings were cultured for 2 weeks with constant watering before the withholding of water. After 9 days without water, all of the plants were irrigated, and plant re-growth was scored 4 days later. The plants were scored as survivors if there were healthy green young leaves after re-watering. The survival rate was calculated as the ratio of the number of surviving plants to the total number of treated plants in the flower pot. Asterisk denotes a significant difference ($P<0.05$). (**C**) WT and *GhSnRK2* transgenic plants before and after low temperature stress. (**D**) Survival rates under low temperature stress conditions. The survival rate after transferring 4-week-old *GhSnRK2* transgenic and WT plants to a low temperature chamber at 4°C, −4°C or −8°C for 10 h, followed by returning the plants to normal growth conditions. Clearly green plants after returning to the normal growth condition were scored as survivors, and plants exhibiting clear signs of wilting were denoted as dead. The mean survival rates of the WT line were compared with those of the transgenic lines using Student's T-test. (E) The water use efficiency of the WT and *GhSnRK2* transgenic plants. Asterisk denotes a significant difference ($P<0.05$).

water stress via the maintenance of a higher RWC and a reduced rate of water loss. Our results revealed that due to their smaller stomata aperture, the transgenic *Arabidopsis* plants overexpressing the *GhSnRK2* gene lost water more slowly than the WT plant in the same period under normal conditions, which may underlie the capacity of the transgenic plants to maintain a higher leaf RWC and to tolerate water stress conditions; "tstat">t Critical two-tail (Figure 4A). The *GhSnRK2*-overexpressing plants exhibited a higher RWC than the WT plants, which is most likely due to the reduction in transpiration-mediated water loss. The RWCs in each line were L2(74.56%), L4(72.09%), L1 (70.04%), and WT (53.98%); Student's T-test (n = 3) ($p<0.05$) (Figure 4B). It is known that during the plant response to abiotic stress, stomata typically close to reduce the rate of water loss due to transpiration. Moreover, abscisic acid (ABA) regulates the stomatal aperture under water-deficient conditions. We found that the stomata apertures of the *GhSnRK2* transgenic plants in response to exogenous ABA treatment were smaller than those of the WT plants, indicating the protective effect of the *GhSnRK2* protein in response to ABA. At 0 μM ABA, the stomata aperture of the WT plants was observed slightly wider than that of the *GhSnRK2* transgenic plants. However, when the concentration of ABA was increased to 1 μM, the aperture size of the transgenic plants was significantly reduced. A similar pattern was detected at 5 μM ABA (n = 5) (Figure 4C, D). Free proline is an osmoprotective molecule that accumulates under stress conditions. Our results revealed that the proline contents of the *GhSnRK2* transgenic lines were significantly higher than those of the WT line, indicating that the transgenic plants accumulate a higher amount of proline than the WT plants. Moreover, the upregulation of Arabidopsis pyrroline-5-carboxylate synthetase 1, a proline biosynthesis gene, in *GhSnRK2*-overexpressing plants suggested that proline production may be improved by *GhSnRK2* via the regulation of proline biosynthesis genes; Student's T-test (n = 3) ($p<0.05$) (Figure 4E). Greater leaf chlorophyll content at all stages of plant development has been associated with improved transpiration efficiency under drought stress, and this trait may indicate the presence of drought avoidance mechanisms. Based on our results, the chlorophyll content of the *GhSnRK2* transgenic lines was higher than that of the WT line, suggesting that *GhSnRK2*-overexpressing plants exhibit greater photosynthetic potential ($p<0.05$) (Figure 4F).

Overexpression of *GhSnRK2* in transgenic plants enhances seedling growth in response to NaCl and exogenous ABA treatment

To investigate the effect of *GhSnRK2* overexpression in response to NaCl and exogenous ABA treatment, a root growth experiment was conducted to elucidate the physiological differences between the transgenic and WT plants. The transgenic plants overexpressing *GhSnRK2* exhibited enhanced seedling growth under NaCl and ABA stress treatments, suggesting that *GhSnRK2* may be involved in the oxidative stress response pathway. Seedlings of WT and *GhSnRK2* transgenic lines grew normally in 0 mM NaCl. In 100 mM and 150 mMNaCl, the transgenic seedlings formed longer roots and displayed significantly larger growth than the WT seedlings. When the NaCl concentration was increased to 200 mM, the growth of the WT seedlings was completely inhibited, and the color of these seedlings was found to have turned brownish, whereas the transgenic seedlings remained green and continued to grow, although at a slower rate (n = 3) (Figure 5A, B).

The seedlings from the WT and *GhSnRK2*-overexpressing lines cultured under normal conditions displayed no significant difference. However, when various concentrations of ABA (0.5 μM, 1 μM, or 2 μM) were introduced to the MS medium and the plants were allowed to grow vertically, the root growth of the *GhSnRK2*-overexpressing lines, although slightly inhibited, displayed greater elongation than the WT line, the root growth of which was severely inhibited (n = 3). These results indicated that *GhSnRK2*-overexpressing plants were more tolerant to ABA than the WT plants (Figure 5C, D).

GhSnRK2 overexpression in transgenic plants enhances seed germination in response to NaCl and exogenous ABA treatment

We investigated the germination of *GhSnRK2*-overexpressing plants under exogenous ABA and salt stresses to determine whether *GhSnRK2* is involved in stress response pathways. The germination rates of WT and *GhSnRK2*-overexpressing seeds were similar at 0 mM NaCl and ABA. However, when different concentrations of ABA or NaCl were introduced, the germination of the WT and *GhSnRK2*-overexpressing seeds was inhibited. When 50 mM NaCl was introduced to the "MS" medium, the germination rate was 47% for the WT line and 85%, 89%, and 86% for the L1, L2, and L4 lines, respectively. At100 mM NaCl, the germination rates were: WT (20.5%), L1(52%), L2 (56%), and L4 (54%)(Figure 6A, B). These results suggested that the *GhSnRK2*-overexpressing plants were more tolerant to salt stress than the corresponding WT plants. ABA plays a prominent role in the regulation of germinating and post-germinating growth arrest and mediates the adaptation of the plant to stress. To investigate this hypothesis, we measured the response of *GhSnRK2*-overexpressing plants to different concentrations of ABA during the germination stage. At 0.3 μm ABA the seed germination rates were WT (45%), L1 (69%), L2 (73%), and L4 (70%), respectively. When the concentration of ABA was increased to 0.5 μm, the germination rate of the WT line was clearly reduced (27%), whereas the germination rate of the transgenic lines ranged from 43% to 48% (Figure 6A, C). In summary, our findings revealed that the *GhSnRK2*-overexpressing mutants were more tolerant to NaCl and exogenous ABA stresses than the corresponding WT plants. Student's t-test was used to compare the mean germination rates. Asterisks denote a significant difference compared to the control ($P\leq0.05$).

We further investigated the biomass accumulation of *GhSnRK2* transgenic and WT plants. The transgenic plants exhibited highly significant increases in both the fresh and dry weight biomasses compared with the corresponding WT plants (Figure 6D, E). These increases were all statistically significant; suggesting that overexpression of the *GhSnRk2* gene increases biomass accumulation in plants. The increased biomass accumulation may be attributable to the increased chlorophyll content in *GhSnRK2* overexpressing plant, as chlorophyll in plant cells carries out the bulk of energy fixation in the process of photosynthesis and is probably the most-often used estimator of plant biomass.

Expression analysis of stress-responsive marker genes

To further elucidate the biological function and molecular mechanisms of the *GhSnRK2* gene, we determined the transcript levels of several stress-associated genes in *GhSnRK2*-overexpressing lines. Seedlings from 10-day-old *GhSnRK2* transgenic and WT *Arabidopsis* lines cultured in MS medium were in the presence or absence of NaCl (250 mM) for 2 h. qRT-PCR was performed as described in the Materials and Methods section. Our findings revealed that *GhSnRK2* regulates the expression levels of ABA and stress-responsive marker genes, suggesting that *GhSnRK2* may positively affect ABA signaling and plant stress

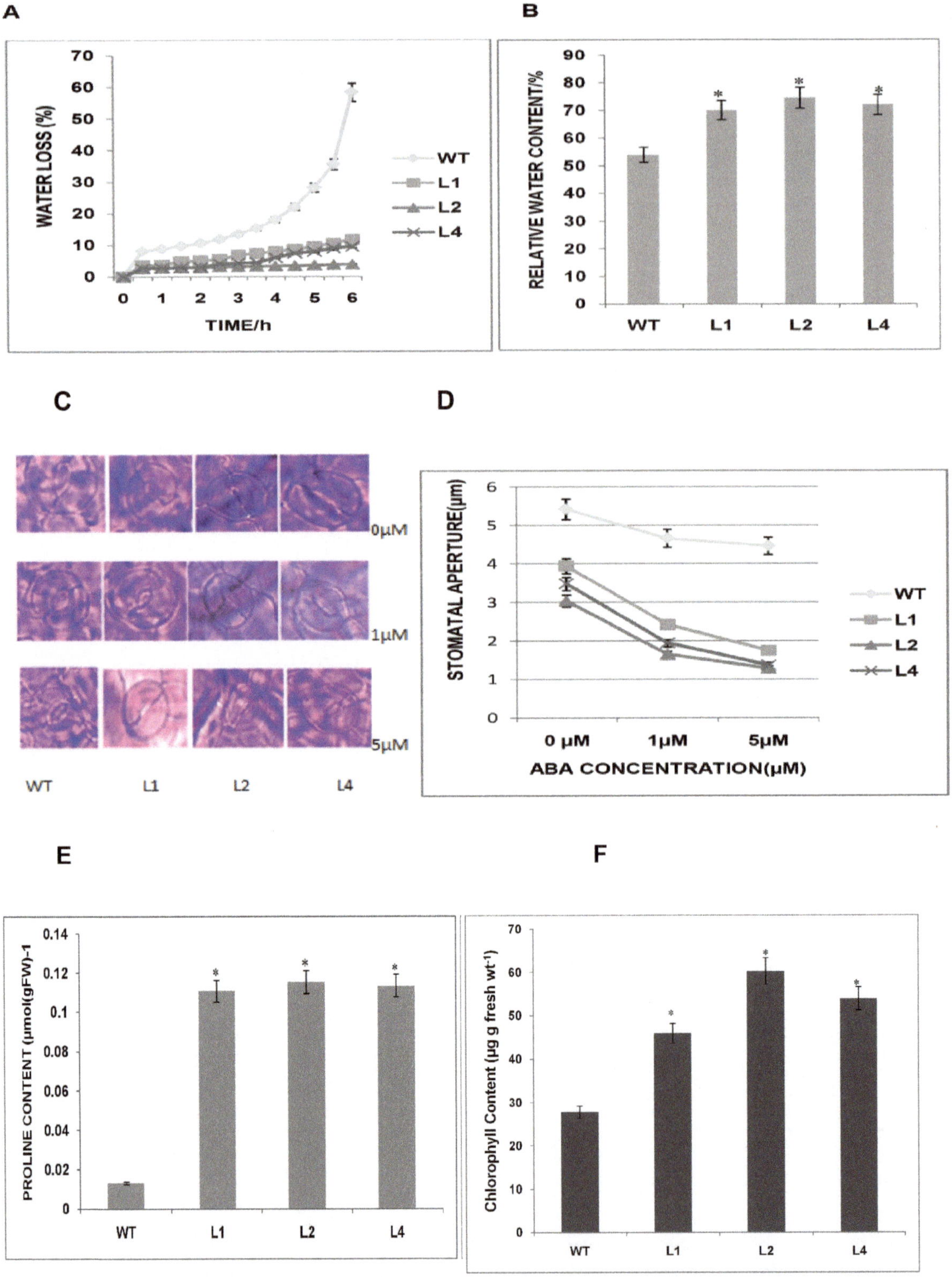

Figure 4. Biochemical and physiological assays of plants overexpressing the *GhSnRK2* gene. (A) Rate of water loss from *GhSnRK2* transgenic plants. Six plants of each transgenic and WT line were analyzed. Three biological replicates produced similar results. Asterisk denotes a significant difference ($P<0.05$). (B) The RWC of *GhSnRK2* transgenic plants. (C) Stomata aperture size of WT and *GhSnRK2* transgenic plants treated with different concentrations of ABA (D) The stomata apertures were measured after 2 h of treatment with different concentrations of ABA, and the mean values of the WT and transgenic lines at each ABA concentration were compared using Student's T-test ($P<0.05$). Asterisk denotes a significant difference ($P<0.05$). (E) Proline accumulation in WT and *GhSnRK2* transgenic plants. The proline content of *GhSnRK2* transgenic lines was consistently higher than that of the WT line. Student's T-test revealed a significant difference ($p<0.05$) between the transgenic and WT lines ($n=3$). (F) The chlorophyll content of the WT and *GhSnRK2* transgenic plants. The mean values were compared using Student's "t-test". Asterisk denotes a significant difference ($P<0.05$).

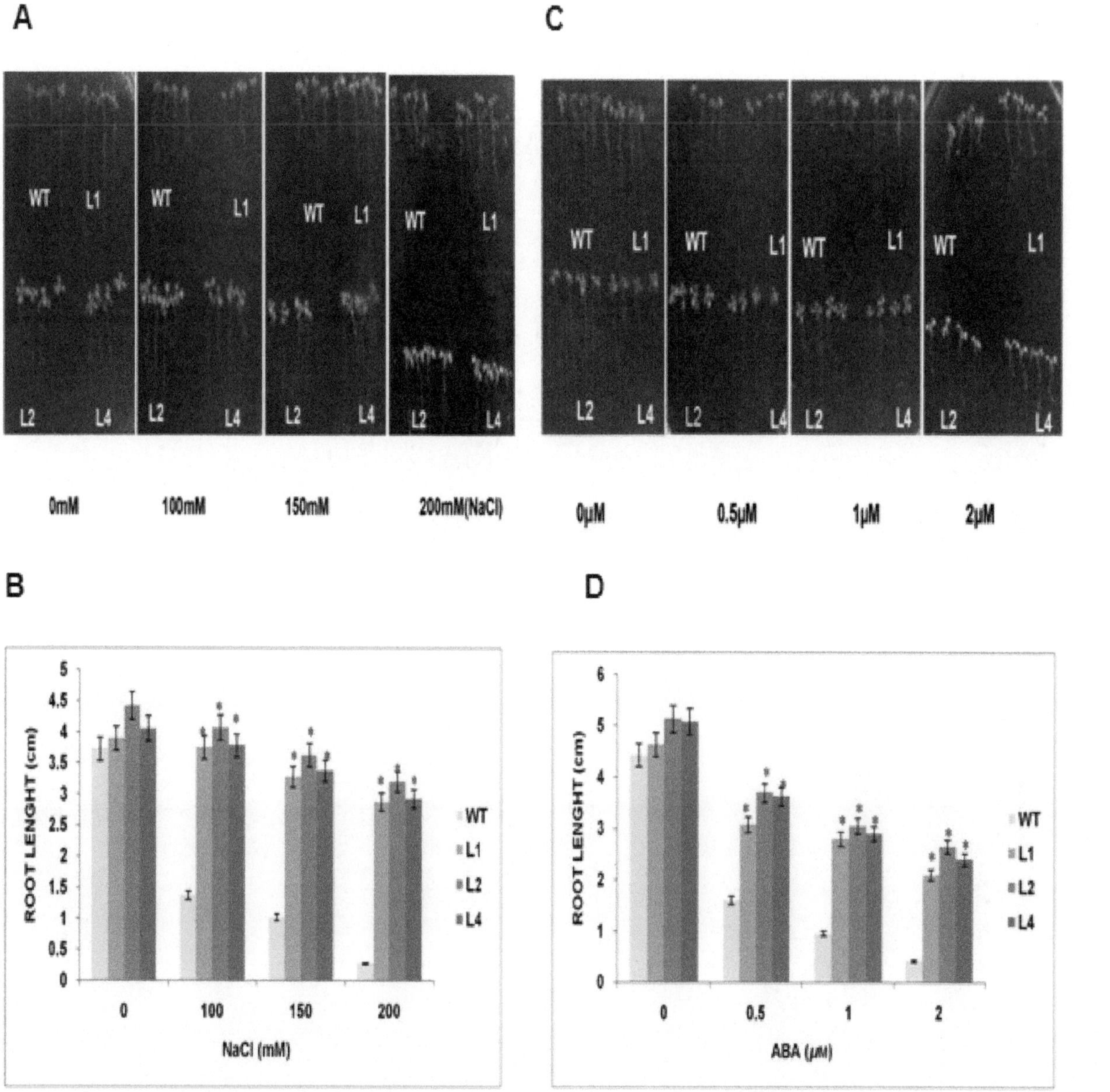

Figure 5. Seedling growth of the WT and *GhSnRK2*-overexpressing lines in response to NaCl and ABA treatment. (A) WT and *GhSnRK2* transgenic plants subjected to different concentrations of NaCl. (B) Approximately one-week-old seedlings were transferred to 1/2MS medium supplemented with different concentrations of NaCl; the root length was measured after 7 days. Each of the three biological replicates consisted of 16 plants. The mean values were compared using Student's T-test ($p<0.05$). (C) The root growth of the WT and *GhSnRK2* transgenic plants treated with different concentrations of ABA. (D) The plants were cultured vertically on MS-containing medium treated with different concentrations of ABA (0.5 μM, 1 μM, or 2 μM); the root length was measured after 7 days. Each of the three biological replicates consisted of 16 plants. Student's T-test was used to compare the mean values of the WT and transgenic lines. Asterisk denotes a significant difference ($P<0.05$).

responses. The expression levels of C-repeat binding factor-1 (*AtCBF1*; accession: ABV27087), delta-1-pyrroline-5-carboxylate-synthase-1 (*AtP5CS1*; accession:NP001189715), protein abscisic acid-insensitive-5 (*AtABI5*; accession:NP565840), protein abscisic acid-insensitive-3 (*AtABI3*; accession:CAA48241), desiccation-responsive genes (*Atrd29B*; accession: BAA02375 and *Atrd29A*; accession:BAA02376) in the *GhSnRK2*-overexpressing lines were significantly higher than those of the WT lines ($p \leq 0.05$), revealing that *GhSnRK2* is actively involved in stress signaling pathways (Figure 7). Three experimental replicates produced similar results ($n = 3$); Student's T-test.

VIGS efficiency and the transcript level of *GhSnRK2* in gene silenced plants

We employed the VIGS technique to further dissect the function of the *GhSnRK2* gene in transgenic cotton. The TRV VIGS vectors were modified based on a pTRV1 containing RNA-dependent RNA polymerase (RdRp), movement protein (MP), a 16 kDa cysteine-rich protein (16K), CaMV 35S promoters (2X35S) and a NOS terminator (NOSt) T-DNA vector. pTRV2 contains the coat protein (CP), multiple cloning sites (MCSs), CaMV 35S promoters (2X35S) and a NOS terminator (NOSt) T-DNA vector. Both vectors contain Rz, which is designated as a

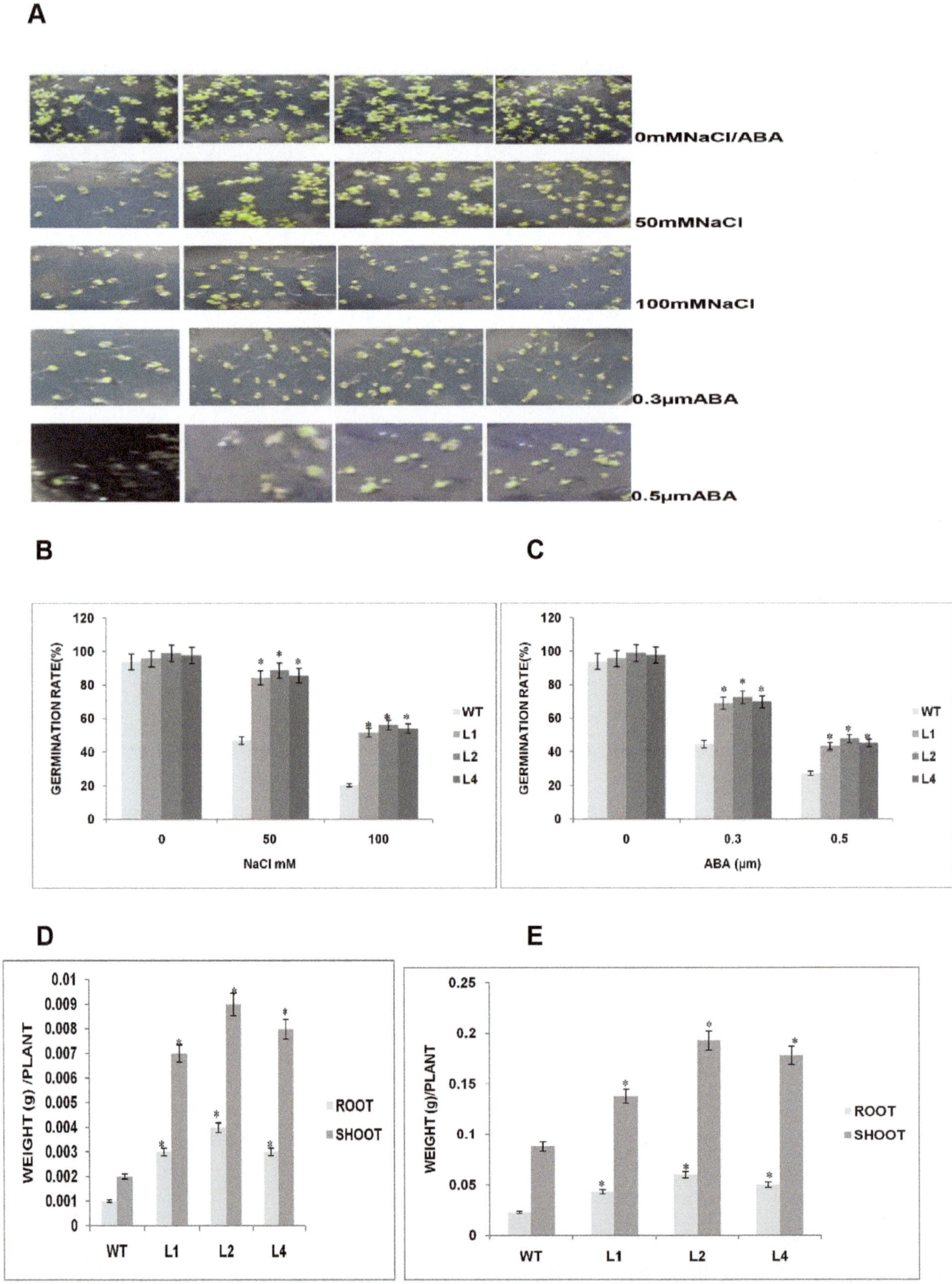

Figure 6. Seed germination of the WT and *GhSnRK2* plants subjected to NaCl and exogenous ABA treatment and biomass accumulation of these plants. (A) Seed germination frequency of the WT and *GhSnRK2* transgenic plants cultured on MS medium supplemented with different concentrations of NaCl (50 mM or 100 mM) or ABA (0 μM, 0.3 μM, or 0.5 μM). (B) The survival rate of the WT and *GhSnRK2* transgenic plants cultured in MS medium containing 50 mM or100 mMNaCl. The mean values were compared using Student's T-test ($p<0.05$). (C) The

germination rate in MS medium supplemented with 0.3 μM or 0.5 μM ABA. The values are presented as the mean germination rates (%) of approximately 200 seeds. Asterisk denotes a significant difference (P<0.05). (D) Biomass accumulation of the *GhSnRK2* transgenic and WT plants. For dry weight biomass, the dry weight in the roots and shoots was recorded after drying in an oven to a constant weight at 70°C for 48 h. (E) Fresh weight biomass of *GhSnRK2* transgenic plants and corresponding WT plants. The fresh weight of the roots and shoots was measured immediately after harvesting. Each of the three biological replicates consisted of 12 plants. Student's T-test was performed. Asterisk denotes a significant difference (P<0.05).

self-cleaving ribozyme, LB (left border) and RB (right border) of the T-DNA (Figure S1B). After inoculating *Agrobacterium* strains containing *GhSnRK2* derivatives (PYL156: (pTRV-RNA2)-*GhSnRk2,* PYL192:(pTRV-RNA1), PYL156:(pTRV-RNA2)-*GhCLA1* (positive control), or PYL156:(pTRV-RNA2) vector) into *Gossypium hirsutum* leaves, the phenotype of both silenced and non-silenced plants was monitored daily for gene silencing efficiency. Loss of normal green coloration in plant leaves and albino phenotype was detected in PYL156:(pTRV-RNA2)-*GrCLA1* silenced plants, which served as the positive control indicating gene silencing efficiency. The *GrCLA1* gene represents the ideal visual marker for gene silencing efficiency due to its involvement in chloroplast development, as its loss of function results in an albino phenotype in true leaves (Figure 8A). The transcript levels of *GhSnRK2* in gene silenced and non-silenced plants were analyzed via qRT-PCR. The transcript level of *GhSnRK2* in the gene silenced two plant cultivars were downregulated compared to the non-silence plants inoculated with empty vector and WT plants, which exhibited a higher expression level. The downregulation of *GhSnRK2* indicated that the gene was successfully knocked down in the silenced plants (Figure 8B). qRT-PCR analysis of the distribution of the TRV-construct in the silenced plants revealed that its infection results in gene-specific transcript degradation, that pTRV-*GhSnRK2* functions in diverse tissues, and that pTRV-*GhSnRK2* is abundantly expressed in the root. The roots of gene silenced plants exhibited further down regulated expression levels compared to the stems and the leaves. The TRV-construct displays increased infectivity and meristem invasion, both of which are key requirements for efficient VIGS-based functional characterization of genes in root tissues (Figure 8C).

Physiological assay and responses of *GhSnRK2* gene silenced plants to various stresses

To further elucidate the function of the *GhSnRK2* gene in stress tolerance, we investigated the response of *GhSnRK2*gene silenced plants to various stresses. Our findings validated the importance of this technique for stress tolerance studies. Under water-deficient conditions, the silenced plants wilted and drooped, regardless of the cultivar (Figure 9A). The phenotype of the non-silenced plants inoculated with empty vector (TRV) was indistinguishable from that of the WT plants. The two cotton cultivars (CRI99668 and CRI409) inoculated with the target gene displayed similar phenotypes and symptoms. The WT and control vector-treated plant survival rates were 80% and 77.3%, respectively, whereas the survival rate of the two gene silenced cotton cultivars (CRI99668 and CRI409) were 34.66% and 38.67%, respectively (Figure 9B). Student's T-test revealed a significant difference ($p < 0.05$) between the survival rate of the gene silenced and non-silenced plants. The rate of water loss from the detached leaves was determined in both the gene silenced and non-silenced plants (Figure 9C). After monitoring the water loss for 6-h, we found that the *GhSnRK2* gene silenced plants inoculated lost more water than the non-silenced plants inoculated with empty vector and the WT plants. The rate of water loss was slightly higher in the CRI99668 cultivars than in the CRI409 cultivars. Under similar conditions, the gene silenced plants, regardless of the cultivar, exhibited a substantially reduced water content compared with the non-silenced plants inoculated with empty vector and the WT plants (Figure 9D). Oxidative stress, including that mediated by salinity, alters the physiological and morphological responses of plants. To investigate the effect of salt stress on *GhSnRK2* gene silenced plants, the gene silenced and non-silenced plants were treated with 150 mM NaCl for 7 days, and the effect of NaCl was measured following this stress. Our findings revealed that the chlorophyll content of the non-silenced plants was higher than that of the gene silenced plants, indicating that the *GhSnRK2* gene may be involved in the oxidative stress response (Figure 9E). Growth retardation was detected in the *GhSnRK2* silenced plants, and the effect of salt accumulation in the plant cells was detected as elevated blisters, which were visibly detectable on the leaf surface of *GhSnRK2* gene silenced plants (Figure 9F).

Discussion

The ability of plants to withstand water shortage while sustaining proper physiological activities can be associated with drought tolerance. In this study, we generated a gene construct containing *GhSnRK2* driven by the CaMV 35S promoter and functionally characterized it via gene overexpression in Arabidopsis. We further elucidated the function of this gene in transgenic cotton plants using VIGS techniques. We found that the amino acid sequence of *GhSnRK2* is highly similar to that of other *SnRK2* proteins, and a neighbor-joining tree developed based on an alignment of the complete protein sequences revealed that *GhSnRK2* clustered with known stress-related genes from other plants. The conserved motif detected demonstrated that *GhSnRK2* is a functional protein. The conserved domain (CD) region in these proteins plays a crucial role in protein interactions, DNA binding, enzyme activity, and other important cellular processes. Furthermore, we found that *GhSnRK2* was localized to the cytoplasm and the nucleus, suggesting that *GhSnRK2* might perform diverse functions in cotton plants. Our findings revealed that *GhSnRK2* is abundantly expressed in the root and is widely distributed throughout plant tissues, suggesting that it may function in diverse tissues. Transgenic *Arabidopsis* plants overexpressing *GhSnRK2* exhibited enhanced tolerance to various abiotic stresses. Consistent with previous studies, *SnRK2* genes have been implicated in the response to multi-environmental stresses [16,30,31]. The improved drought tolerance of plants overexpressing the *GhSnRK2* gene may be due to the abundance of gene expression in roots. In addition, protein kinases that are specifically found in roots, such as *SnRK2C,* may perform certain important functions in root tissues as a sensor of water and nutrients in soil. The root tip plays a vital role in the response to water and nutrient detection via appropriate signal transduction [32]. Turgor regulation of the *GhSnRK2* transgenic plant under water deficit condition may be attributable to the dynamic process of cell wall adjustment of the transgenic plant due to their enhanced drought-tolerant. Plants can maintain turgor by solute accumulation, i.e. by osmotic adjustment, and possibly by elastic adjustment of their cell walls [33].

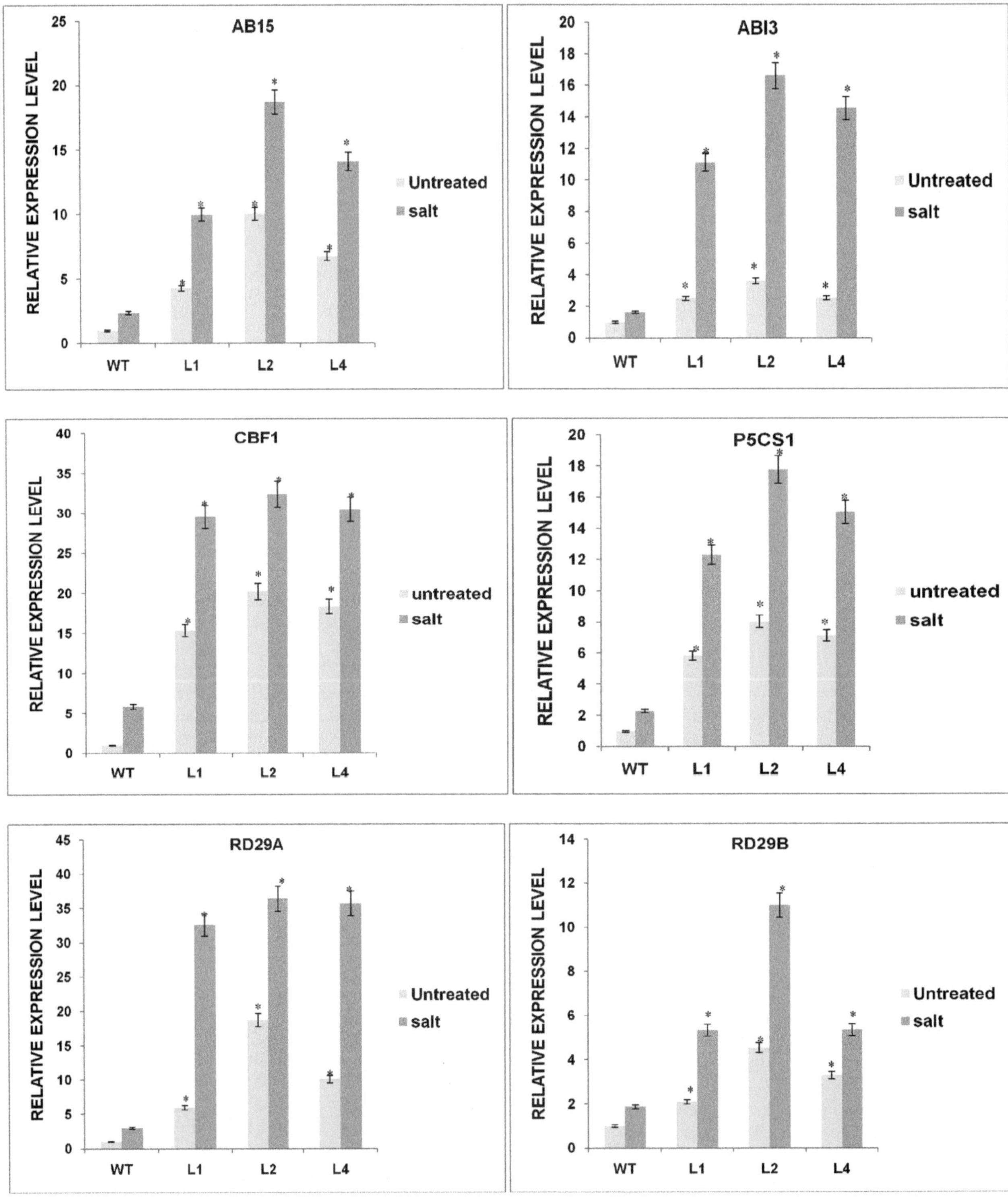

Figure 7. Expression analysis of stress-responsive marker genes. The relative transcript levels of the stress-responsive genes *AtABI5, AtABI3, AtP5CS1, AtRD29A, AtCBF1*, and *AtRD29B* in the *GhSnRK2*-overexpressing and WT lines. qRTPCR was performed for gene expression analysis. The vertical axis displays the expression pattern. Three biological replicates produced similar results.

Based on our results, the transgenic plants overexpressing *GhSnRK2* exhibited enhanced low temperature tolerance compared to WT plants. This result may be attributable to the resistance of the membrane system of the transgenic plant cells to cold stress. These findings corroborated the results of [30], who reported enhanced multi-stress tolerance in *Arabidopsis* plants overexpressing *TaSnRK2.8*. Numerous studies have demonstrated that the membrane systems of the cell are the primary site of low temperature contusion in plants [34,35]. Indeed, it is well confirmed that low temperature-mediated membrane injury

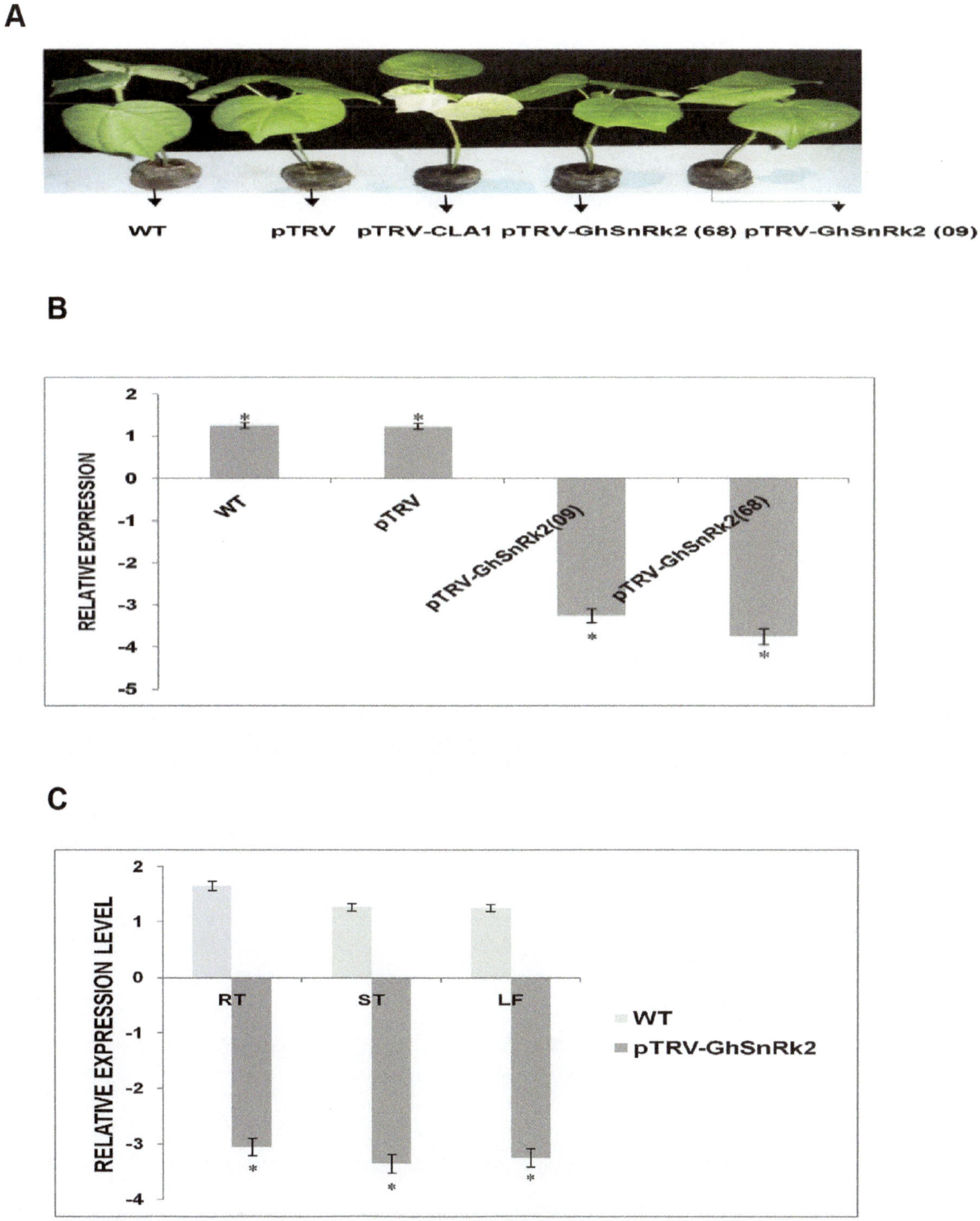

Figure 8. Silencing efficiency and transcript level of the *GhSnRK2* gene silenced plants. (A) Phenotype of gene silenced and non-silenced plants; 14 dpi. Wild-type (WT), negative control Empty vector (pTRV), positive control (pTRV-GrCLA1), gene silenced CRI409 cotton cultivars [pTRV-GhSnRk2 (09)], and gene silenced CRI99668 cotton cultivars [pTRV-*GhSnRK2* (68)]. (B) The silencing efficiency as determined by the expression pattern of WT, vector control (pTRV) and *GhSnRK2* gene silenced plants based on qRT-PCR. (C) The distribution of the TRV- construct in the gene silenced plants. Samples from the root, the stem and the leaves of gene silenced and non-silenced plants were analyzed via qRT-PCR. The values are presented as the means of three biological replicates. Asterisk denotes a significant difference ($P<0.05$).

occurs primarily as a result of acute dehydration associated with low temperature [35,36].

Our findings revealed the accumulation of compatible osmolytes, such as free proline, in the *GhSnRK2*-overexpressing plants; therefore, we speculate that the *GhSnRK2* transgenic plants exhibit increased stress tolerance by regulating downstream gene expression and accumulating a larger amount of compatible osmolytes, which may account for the increase in stress tolerance

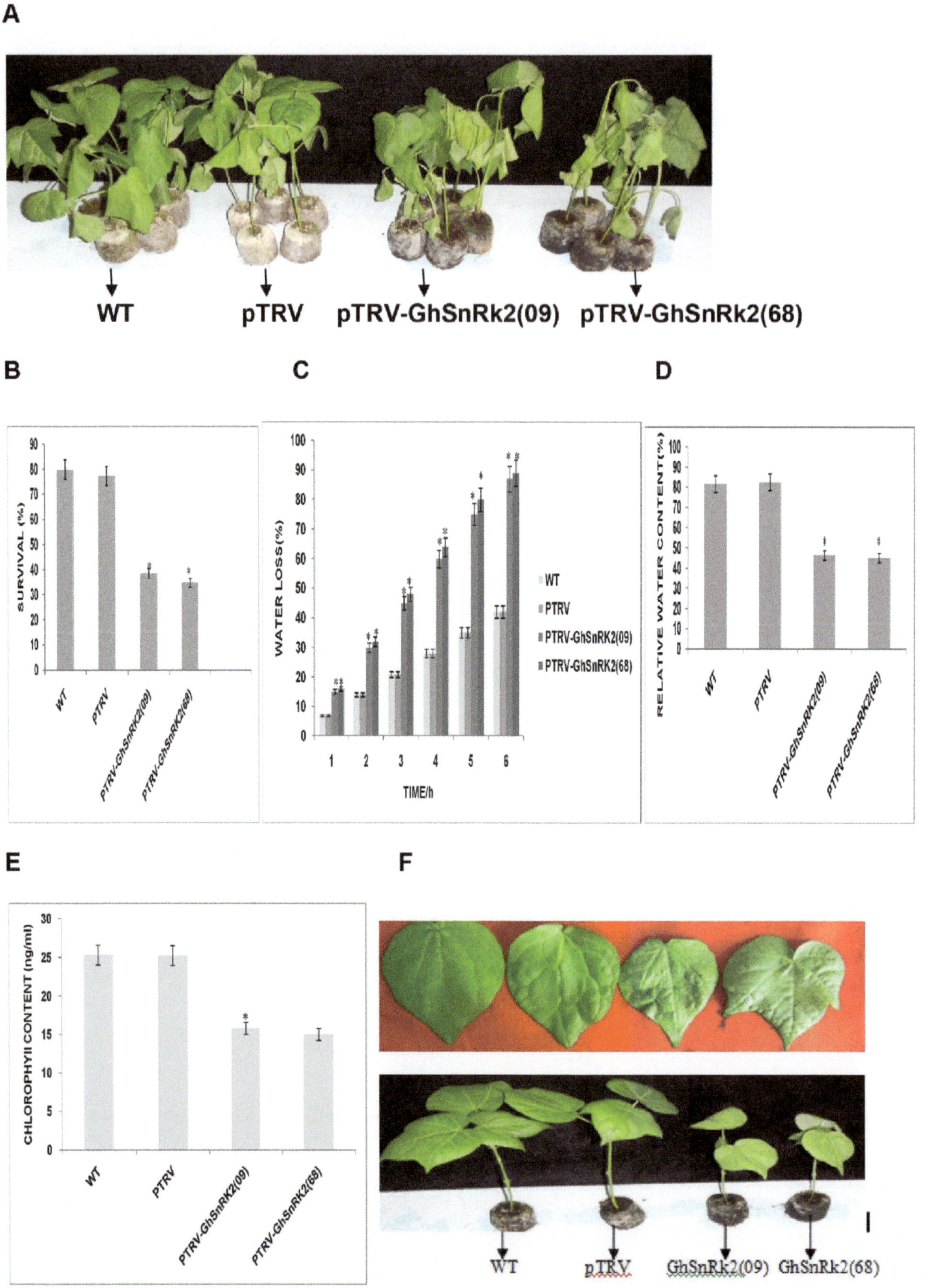
A
WT
pTRV
pTRV-GhSnRk2(09)
pTRV-GhSnRk2(68)
B
SURVIVAL (%)
WT
PTRV
PTRV-GhSnRK2(09)
PTRV-GhSnRK2(68)
C
WATER LOSS(%)
TIME/h
WT
PTRV
PTRV-GhSnRK2(09)
PTRV-GhSnRK2(68)
D
RELATIVE WATER CONTENT(%)
WT
PTRV
PTRV-GhSnRK2(09)
PTRV-GhSnRK2(68)
E
CHLOROPHYII CONTENT (ng/ml)
WT
PTRV
PTRV-GhSnRK2(09)
PTRV-GhSnRK2(68)
F
WT
pTRV
GhSnRk2(09)
GhSnRk2(68)

Figure 9. Physiological assay of *GhSnRK2* gene silenced plants. (A) Phenotype of drought stressed plants. Water was withheld from *GhSnRK2* gene silenced and non-silenced plants for 5 days approximately two weeks post-inoculation. Each of the ten treated groups consisted of five plants. A photograph of each group representative was captured. (B) The survival rate was determined by withholding water from *GhSnRK2* gene silenced and non-silenced plants for 5 days approximately two weeks post-inoculation, and the survival rate was recorded as the percentage of plants that survived after re-watering for 3 days. (C) The relative water loss was determined two weeks after inoculation. The reduction in the fresh weight from the initial weight was determined at the indicated time and represented as the percentage of water loss. The experiment was replicated three times. (D) The mean RWC was analyzed after immersing the fresh detached leaves in distilled water for 4 h and oven drying at 80°C for 48 h. The values are presented as the means of three biological replicates. Asterisk denotes a significant difference ($P<0.05$). (E) The change in the chlorophyll content of *GhSnRK2* gene silenced and non-silenced plants under salt stress. (F) The phenotype of *GhSnRK2* gene silenced and non-silenced plants under salt stress.

of *GhSnRK2* plants. Proline is an osmoprotective molecule that accumulates in response to water stress and salinity [37,38]. Proline is considered as a typical physiological parameter for evaluating abiotic stress tolerance and resistance in plants. Many plants reduce their cellular osmotic potential via the accumulation of intracellular organic osmolytes, such as proline, to maintain a stable intracellular environment when subjected to external environmental stresses [39,40,41].

We detected decreased stomatal apertures in the *GhSnRK2* transgenic plants compared with the corresponding WT plants. When exposed to adverse environmental conditions, the stomata must close for the plant to survive [42]. Stomatal closure reduces water loss, particularly in plants that have been exposed to water stress conditions caused by high solute concentrations in the nutrient medium, such that minimizing the rate of water loss due to transpiration reduces the accumulation of toxic ions [43]. The reduced rate of water loss from *GhSnRK2* transgenic plants may be due to the decrease in the size of the stomata aperture. ABA is involved in stomata movement, enhanced drought tolerance and plant osmoregulation. Under water stress conditions, ABA is released into leaves, inducing the release of potassium salts from the guard cells, resulting in stomatal closure. Plant guard cells modulate the opening and closure of the stomata in response to phytohormones and various environmental signals, such as light and temperature, thereby regulating gas exchange for photosynthesis and water status via transpiration [44]. The *SnRK2* gene *SRK2E*, or *SnRK2.6*, plays a significant role in stomatal closure in *Arabidopsis* leaves [45,46].

The *GhSnRK2* gene was upregulated after treatment with 10% PEG, which may suggest that the *GhSnRK2* gene sufficiently confers drought tolerance, as one strategy that can be employed to mimic the effect of drought on plants is treatment with PEG. PEG was quite commonly used in physiological experiments to induce controlled drought stress.

We speculate that the *GhSnRK2*-overexpressing plants were more tolerant to NaCl than the WT plants, which may be due to the improved accumulation of Na+ in the vacuoles, increased sequestration of Na+ into the vacuoles, improved cellular Na+ exclusion, or their greater capacity for vacuolar osmoregulation. High salt can result in oxidative damage to cell membranes. The ability to avert such damage is consistent with the degree of tolerance exhibited by the plant [47]. Under salt stress, this damage is less extensive in transgenic plants overexpressing the *GhSnRK2* gene than in WT plants because *GhSnRK2* overexpression can tolerate the accumulation of Na+ in the vacuoles, thus preventing the toxicity of excess Na+. Hypersensitivity to ABA treatment during seed germination and early seedling development is typically followed by improved drought tolerance [48,49]. Our finding reveals that the transgenic plants overexpressing *GhSnRK2* exhibited enhanced seedling growth under ABA stress treatment. This may be attributable to ABA-independent signaling pathway involving SnRK2 in root growth. This finding reveals the complexity in ABA-signaling, as different signaling components may function in different cells or tissues in ABA signal transduction. *SnRK2.2* and *SnRK2.3* double mutant showed strong ABA-insensitive phenotypes in seed germination and root growth inhibition.[50]. The decreased stomatal apertures detected in the *GhSnRK2* transgenic plants may be attributable to the role of *SnRK2* in the regulation of ABA-induced activation of plasma membrane anion channels in guard cells and concomitant response to their closure. These results suggest that *SnRK2* functions in ABA signaling of stomatal closure. *SRK2E* knockout mutant lost the ABA-dependent stomatal closure. SRK2E affects ABA-signaling in stomatal closure, but not in germination stage [51]. ABA and PYR/PYLs inhibit protein phosphatase2C (PP2C), which in turn relieves the repression of positive factors, such as SnRK2s [52].

We analyzed the expression of the stress-inducible marker genes *RD29A*, *RD29B*, *P5CS1*, *ABI3*, *CBF1*, and *ABI5* in *GhSnRK2*-overexpressing plants, and our results revealed increases in the transcript levels of these marker genes. Consistent with a previous study, the upregulation of stress response genes, such as *P5CS1*, can contribute to enhanced salt stress tolerance in plants [53]. *ABI3* and *ABI5* activation is necessary to sustain the germination of seedlings during intense drought stress [54]. *ABI5* and *ABI3* regulate numerous ABA responses, such as osmotic water permeability of the plasma membrane, stomatal closure, drought-induced resistance, germination, and inhibition of vegetative growth. We speculated that *GhSnRK2* functions in the transcriptional regulation of ABA-inducible genes in seedlings, suggesting that *SnRK2* protein kinases are involved in several processes of ABA signal transduction, including transcriptional as well as post-transcriptional regulation pathways. Therefore, we suggest that stress signaling pathways may be involved in *GhSnRK2*-mediated stress tolerance.

Our findings revealed the downregulation of *GhSnRK2* in gene silenced plants, indicating that this gene was knocked down in the gene silenced plants. TRV spread vigorously throughout the entire plant, including meristem tissue, and the symptoms induced by TRV are not severe compared with those induced by other viruses [55]. The albino phenotype observed in the *GhCLA1*gene silenced plants in this study is an indicator of effective VIGS function in inoculated plants, which is consistent with previous findings [56]. The results of this study revealed that *GhSnRK2* gene silencing in *Gossypium hirsutum* greatly reduced its tolerance to drought stress, corroborating the findings of [57], who reported that silencing of the function of the *SpMPK1*, *SpMPK2*, and *SpMPK3* genes in tomato plants alleviates their tolerance to drought stress. The decrease in the chlorophyll content of silenced plants in response to salt stress indicates that *GhSnRK2* may be involved in oxidative stress. High-concentration salt stress inhibits plant biochemical processes. Decreased chlorophyll content in salt-stressed pumpkin plants was detected by [58]. The stunted growth observed in salt-stressed *GhSnRK2* gene silenced plants corroborate the findings of [59], who reported that the typical effect of salt stress in plants is growth retardation as a result of cell elongation inhibition.

Conclusions

Despite the significant innovations that have been made in elucidating the genetic mechanisms underlying drought tolerance, considerable challenges remain. The present study revealed that the genetic manipulation of the *GhSnRK2* gene from cotton using a transgenic technique results in enhanced drought and cold stress tolerance in plants. Thus, *GhSnRK2* is hypothesized to participate in the stress signaling pathway, and therefore, overexpression of *GhSnRK2* may alleviate abiotic stress by regulating stress-responsive genes. Moreover, the findings in this study have helped to elucidate that *GhSnRK2* enhances stress tolerance in plants by affecting various stress-related pathways. Thus, the *GhSnRK2* gene represents a candidate gene for future research of abiotic stress signaling pathways and the genetic modification of novel *Gossypium hirsutum* varieties.

Supporting Information

Figure S1 Multiple sequence alignment of *GhSnRK2* and closely related *SnRK2s* from other plants species and VIGS construct. (A) Alignment of *GhSnRK2* and closely related *SnRK2s*. The relatively conserved motif is underlined. The deduced amino acid sequence displays relatively high homology with the monocot *SnRK2* family members *Oryza sativa* (*RK1*), ABB89146 and with the dicot species *AtSnRK2.10*, AEE33751.1. (B) Virus-induced gene silencing construct. The TRV VIGS vectors were modified based on a pTRV1 containing RNA-dependent RNA polymerase (RdRp), movement protein (MP), a 16 kDa cysteine-rich protein (16K), CaMV 35S promoters (2X35S) and a NOS terminator (NOSt) T-DNA vector. pTRV2 contains the coat protein (CP), multiple cloning sites (MCSs), CaMV 35S promoters (2X35S) and a NOS terminator (NOSt) T-DNA vector. Both vectors contain Rz, which is designated as a self-cleaving ribozyme, LB (left border) and RB (right border) of the T-DNA.

Figure S2 Survival rate in kanamycin-containing medium and PCR confirmation of *GhSnRK2* gene expression in transgenic Arabidopsis. (A) The survival rate of WT and *GhSnRK2* transgenic plants in MS medium supplemented with the antibiotic kanamycin. Photograph of a representative plant was captured after 9 days of germination in kanamycin-containing medium. (B) Confirmation of *GhSnRK2* gene in transgenic Arabidopsis. Genomic DNA from the first generation of the plants (T1) was extracted and used as a template for gene-specific primers. Lane 1: DNA molecular marker III standard; lane 2: negative control.

Table S1 Survival (%) of *GhSnRK2* transgenic plants in kanamycin-containing medium. At T1 generation, the ratio of dead to surviving plant was approximately 1:3 in kanamycin-containing LB medium. The survival rate at T3 generation was 100% in kanamycin-containing LB medium. The values are expressed as the mean germination rate (%) of approximately 200 seeds.

Table S2 The lists of primers sequences used in this study.

Acknowledgments

Our appreciation goes to the staff and students of the State Key Laboratory of Cotton Biology, Chinese Academy of Agricultural Sciences, Cotton Research Institute, Anyang, China, for their support during the course of this research. We also thank Jiahe Wu for VIGS technical support and suggestions during manuscript preparation.

Author Contributions

Conceived and designed the experiments: BB FL XZ Zhaoen Yang CL QW GZ Zuoren Yang. Performed the experiments: BB FL XZ Zhaoen Yang CL QW GZ Zuoren Yang. Analyzed the data: BB FL XZ Zhaoen Yang CL QW GZ Zuoren Yang. Contributed reagents/materials/analysis tools: BB FL XZ Zhaoen Yang CL QW GZ Zuoren Yang. Wrote the paper: BB FL XZ Zhaoen Yang CL QW GZ Zuoren Yang.

References

1. Genoud T, Metraux JP (1999) Crosstalk in plant cell signaling: structure and function of the genetic network. Trends in Plant Science 4: 503–507.
2. Kumar N, Bhatt RP (2006) Transgenics: An emerging approach for cold tolerance to enhance vegetables production in high altitude areas. Indian J. Crop Sci 1: 8–12.
3. Shinozaki K, Yamaguchi-Shinozaki K (2000) Molecular responses to dehydration and low temperature: differences and cross-talk between two stress signaling pathways. Curr Opin Plant Biol 3: 217–223
4. Guo-Tao H, Shi-Liang M, Li-Ping B, Li Z, Hui M, et al. (2000) The role of root border cells in plant defense. Trends Plant Sci 5: 128–133
5. Cho K, Agrawal GK, Jwa NS, Kubo A, Rakwal R (2009) Rice OsSIPK and its orthologs: a "central master switch" for stress responses. Biochem Biophys Res Commun 379: 649–653.
6. Mishra NS, Tuteja R, Tuteja N (2006) Signaling through MAP kinase networks in plants. Arch Biochem Biophys 452: 55–68.
7. Pitzschke A, Schikora A, Hirt H (2009) MAPK cascade signalling networks in plant defence. Curr Opin Plant Biol 12: 421–426.
8. Rodriguez MC, Petersen M, Mundy J (2010) Mitogen-activated protein kinase signaling in plants. Annu Rev Plant Biol 61: 621–649.
9. Jonak C, Hirt H (2002) Glycogen synthase kinase 3/SHAGGY-like kinases in plants: an emerging family with novel functions. Trends Plant Sci 7: 457–461.
10. Koh S, Lee SC, Kim MK, Koh JH, Lee S, et al. (2007) T-DNA tagged knockout mutation of rice OsGSK1, an orthologue of Arabidopsis BIN2, with enhanced tolerance to various abiotic stresses. Plant Mol Biol 65: 453–466.
11. Mahfouz MM, Kim S, Delauney AJ, Verma DP (2006) Arabidopsis TARGET OF RAPAMYCIN interacts with RAPTOR, which regulates the activity of S6 kinase in response to osmotic stress signals. Plant Cell 18: 477–490.
12. Das R, Pandey GK (2010) Expressional analysis and role of calcium regulated kinases in abiotic stress signaling. Curr Genomics 11: 2–13.
13. Hrabak EM, Chan CWM, Gribskov M, Harper JF, Choi JH, et al. (2003) The Arabidopsis CDPK-SnRK superfamily of protein kinases. Plant Physiol 132: 666–680.
14. Wurzinger B, Mair A, Pfister B, Teige M (2011) Cross-talk of calcium-dependent protein kinase and MAP kinase signaling. Plant Signal Behav 6: 8–12.
15. Coello P, Hirano E, Hey SJ, Muttucumaru N, Martinez-Barajas E, et al. (2012) Evidence that abscisic acid promotes degradation of SNF1-related protein kinase (SnRK) 1 in wheat and activation of a putative calcium-dependent SnRK2. J Exp Bot 63: 913–924
16. Umezawa T, Yoshida R, Maruyama K, Yamaguchi-Shinozaki K, Shinozaki K (2004) SRK2C, a SNF1-related protein kinase 2,improves drought tolerance by controlling stress-responsive gene expression in Arabidopsis thaliana. Proc Natl Acad Sci. USA 101: 17306–173011.
17. Hardie DG (1999) Plant protein serine/threonine kinases: classification and functions. Annu Rev Plant Physiol Plant Mol Biol 50: 97–131.
18. Mikołajczyk M, Awotunde OS, Muszyńska G, Klessi DF, Dobrowolska G (2000) Osmotic stress induces rapid activation of a salicylic acid-induced protein kinase and a homolog of protein kinase ASK1 in tobacco cells. Plant cell 12: 165–178
19. Wang X, Xu W, Xu Y, Chong K, Xu Z, et al (2004) Wheat RAN1, a nuclear small G protein, is involved in regulation of cell division in yeast. Plant Sci 167: 1183–1190.
20. Clough SJ, Bent AF (1998) Floral dip: a simplified method for Agrobacterium-mediated transformation of Arabidopsis thaliana. Plant J 16(6):735–743
21. Tyree MT, Hammel HT (1972) The measurement of the turgor pressure and the water relations of plants by the pressure-bomb technique. J Exp Bot 23: 267–282
22. Duan J, Zhang M, Zhang H, Xiong H, Liu P, et al. (2012) OsMIOX, a myoinositol oxygenase gene, improves drought tolerance through scavenging of reactive oxygen species in rice (Oryza sativa L.). Plant Sci 196: 143–151.

23. Parida AK, Dasgaonkar VS, Phalak MS, Umalkar GV, Aurangabadkar LP (2007) Alterations in photosynthetic pigments, protein, and osmotic components in cotton genotypes subjected to short-term drought stress followed by recovery. Plant Biotechnology Reports 1: 37–48.
24. Pei ZM, Kuchitsu K, Ward JM, Schwarz M, Schroeder JI (1997 Mar) Differential abscisic acid regulation of guard cell slow anion channels in Arabidopsis wild-type and abi1 and abi2 mutants. Plant cell 9(3):409–423.
25. Bates L, Waldren RP, Teare ID (1973) Rapid determination of free proline for water stress studies. Plant and Soil 39: 205–207
26. Arnon DI (1949) Copper enzymes in isolated chloroplasts, polyphenoxidase in beta vulgaris. Plant physiology 24: 1–15.
27. Xiong L, Ishitani M, Lee H, and Zhu JK (2001b) The *Arabidopsis LOS5/ABA3* locus encodes a molybdenum cofactor sulfurase and modulates cold and osmotic stress responsive gene expression. Plant cell 13: 2063–2083.
28. Livak KJ, Schmittgen TD (2001) Analysis of Relative Gene Expression Data Using Real-Time Quantitative PCR and the 2DDCT Method. METHODS 25: 402–408
29. Gao X, Britt RC, Shan L, He P (2011) Agrobacterium-Mediated Virus-Induced Gene Silencing Assay In Cotton. J. Vis. Exp 54:e2938
30. Kobayashi Y, Murata M, Minami H, Yamamoto S, Kagaya Y (2005) Abscisic acid-activated SNRK2 protein kinases function in the gene-regulation pathway of ABA signal transduction by phosphorylating ABA response element binding factors. Plant J 44: 939–949.
31. Zhang H, Mao X, Wang C, Jing R (2010) Overexpression of a common wheat gene *TaSnRK2.8* enhances tolerance to drought, salt and low temperature in *Arabidopsis*. PLoS ONE 5: e16041.
32. Hawes MC, Gunawardena U, Miyasaka S, Zhao X (2000) The role of root border cells in plant defense. Trends Plant Sci 5: 128–133.
33. Dainty J (1976) Water relations of plant cells. *In* AP Gottingen, MH Zimmermann, eds, Encyclopedia of Plant Physiology, Vol 2. New Series, Part A. Springer-Verlag, Berlin, pp 12–35.
34. Levitt J (1980) Response of plant to Environmental Stress, water, radiation, salt and other stresses. Academic press, New York.
35. Steponkus PL (1984) Role of the plasma membrane in freezing injury and cold acclimation. Annu. Rev. Plant Physiol 35: 543–584.
36. Steponkus PL, Uemura M, Webb MS (1993) A contrast of the cryostability of the plasma membrane of winter rye and spring oat-two species that widely differ in their freezing tolerance and plasma membrane lipid composition. In: Steponkus P L, editor. Adv. Low-Temperature Biol. Vol. 2. London: JAI Press pp. 211–312.
37. Claussen W (2005) Proline as a measure of stress in tomato plants. Plant Sci 168: 241–248.
38. Younis ME, Hasaneen MNA, Tourky MNS (2009) Plant growth, metabolism and adaptation in relation to stress conditions. XXIV. Salinity biofertility interactive effects on proline, glycine and various antioxidants in Lactuca sativa. Plant Omics J 2: 197–205.
39. Zhu JK (2002) Salt and drought stress signal transduction in plants. Annu Rev Plant Biol 53: 247–273
40. Granier C, Tardieu F (1999) Water deficit and spatial pattern of leaf development variability in responses can be simulated using a simple model of leaf development. Plant Physiol 119: 609–620.
41. Wang ZQ, Yuan YZ, Ou JQ, Lin QH, Zhang CF (2007) Glutamine synthetase and glutamate dehydrogenase contribute differentially to proline accumulation in leaves of wheat (*Triticum aestivum*) seedlings exposed to different salinity. Plant Physiol 164: 695–701.
42. Pareek A, Sopory SK, Bohnert HJ, Govindjee EDS (2010) Abiotic Stress Adaptation in Plants: Physiological, Molecular and Genomic Foundation, Springer, Dordrecht, pp. 283e305.
43. Everard JD, Gucci R, Kahn JA, Flore WH (1994). Gas exchange and carbon partitioning in the leaves of celery (Apium graveolens L.) at various levels of root zone salinity, Plant Physiol 106: 281e292.
44. Schroeder JI, Kwak JM, Allen GJ (2001) Guard cell abscisic acid signalling and engineering drought hardiness in plants. Nature 410: 327–330.
45. Yoshida R, Hobo T, Ichimura K, Mizoguchi T, Takahashi F (2002) ABA-activated SnRK2 protein kinase is required for dehydration stress signaling in Arabidopsis Plant Cell Physiol 43: 1473–1483.
46. Mustilli AC, Merlot S, Vavasseur A, Fenzi F, Giraudat J (2002) Arabidopsis OST1 Protein Kinase Mediates the Regulation of Stomatal Aperture by Abscisic Acid and Acts Upstream of Reactive Oxygen Species Production. Plant Cell 14: 3089–3099.
47. Chinnusamy V, Jagendorf A, Zhu JK (2005) Understanding and improving salt tolerance in plants, Crop Sci 45: 437e448.
48. Hu H, Dai M, Yao J, Xiao B, Li X, et al. (2006) Overexpressing a NAM, ATAF, and CUC (NAC) transcription factor enhances drought resistance and salt tolerance in rice. Proc Natl Acad Sci USA 103: 12987–12992.
49. Ko JH, Yang SH, Han KH (2006) Upregulation of an Arabidopsis RING-H2 gene, *XERICO*, confers drought tolerance through increased abscisic acid biosynthesis. Plant J 47 343–355.
50. Hiroaki F, Paul E, Jian-Kang Z (2007) Identification of Two Protein Kinases Required for Abscisic Acid Regulation of Seed Germination, Root Growth, and Gene Expression in *Arabidopsis*. The Plant Cell 19 (2):485–494.
51. Riichiro Y, Tokunori H, Kazuya I, Tsuyoshi M, Fuminori T, et al. (2002) ABA-Activated SnRK2 Protein Kinase is Required for Dehydration Stress Signaling in *Arabidopsis*. *Plant Cell Physiol* 43 (12): 1473–1483.
52. Sang-Youl P, Pauline F, Noriyuki N, Davin RJ, Hiroaki F, et al. (2009) Abscisic Acid Inhibits Type 2C Protein Phosphatases via the PYR/PYL Family of START Proteins. Science 324: 1068
53. Kavi PB, Zonglie H, Cuo-Hua M, Chein-An AH, DeshPal SV (1995) Overexpression of A1-Pyrroline-5-Carboxylate Synthetase Increases Proline Production and Confers Osmotolerance in Transgenic Plants Plant Physiol 108: 1387–1394
54. Lopez-Molina L, Mongrand S, McLachlin DT, Chait BT, Chua NH (2002) ABI5 acts downstream of ABI3 to execute an ABA-dependent growth arrest during germination. Plant J 32: 317–328.
55. Ratcliff F, Martin-Hernandez AM, Baulcombe DC (2001) Tobacco rattle virus as a vector for analysis of gene functions by silencing. Plant J 25: 237–245
56. Liu Y, Schiff M, Dinesh-Kumar S (2002) Virus-induced gene silencing in tomato. Plant J 31: 777–786.
57. Cui L, Jian-Min Y, Yun-Zhou L, Zhen-Cai Z, Qiao-Li W, et al. (2013) Silencing the *SpMPK1*, *SpMPK2*, and *SpMPK3* Genes in Tomato Reduces Abscisic Acid—Mediated Drought Tolerance Int. J. Mol. Sci 14: 21983–21996
58. Senay S, Fikret Y, Sebnem K, Sebnem E (2011) The effect of salt stress on growth, chlorophyll content, lipid peroxidation and antioxidative enzymes of pumpkin seedling. African Journal of Agricultural Research 6(21): 4920–4924
59. Yasar F, Ellialtioglu S, Yildiz K (2008) Effect of salt stress on antioxidant defense systems, lipid peroxidation, and chlorophyll content in green bean. Russian J. Plant Physiol 55(6): 782–786.

15

Isoprene Emissions from Downy Oak under Water Limitation during an Entire Growing Season: What Cost for Growth?

Anne-Cyrielle Genard-Zielinski[1,2], Elena Ormeño[1*], Christophe Boissard[2], Catherine Fernandez[1]

1 Institut Méditerranéen de Biodiversité et d'Ecologie marine et continentale (IMBE) Aix Marseille Université, CNRS, IRD, Avignon Université, Technopôle Arbois-Méditerranée, France, **2** Laboratoire des Sciences du Climat et de l'Environnement (LSCE-IPSL), Unité Mixte CEA-CNRS-UVSQ (Commissariat à l'Energie Atomique, Centre National de la Recherche Scientifique, Université de Versailles Saint-Quentin-en-Yvelines), Gif-sur-Yvette, France

Abstract

Increases in the production of terpene- and phenolic-like compounds in plant species under abiotic stress conditions have been interpreted in physiological studies as a supplementary defense system due to their capacity to limit cell oxidation. From an ecological perspective however, these increases are only expected to confer competitive advantages if they do not imply a significant cost for the plant, that is, growth reduction. We investigated shifts of isoprene emissions, and to a lesser extent phenolic compound concentration, of *Quercus pubescens* Willd. from early leaf development to leaf senescence under optimal watering (control: *C*), mild and severe water stress (*MS*, *SS*). The impact of water stress was concomitantly assessed on plant physiological (chlorophyll fluorescence, stomatal conductance, net photosynthesis, water potential) functional (relative leaf water content, leaf mass per area ratio) and growth (aerial and root biomass) traits. Growth changes allowed to estimate the eventual costs related to the production of isoprene and phenolics. The total phenolic content was not modified under water stress whereas isoprene emissions were promoted under *MS* over the entire growing cycle despite the decline of P_n by 35%. Under *SS*, isoprene emissions remained similar to *C* all over the study despite the decline of *Pn* by 47% and were thereby clearly uncoupled to *Pn* leading to an overestimation of the isoprene emission factor by 44%. Under *SS*, maintenance of isoprene emissions and phenolic compound concentration resulted in very significant costs for the plants as growth rates were very significantly reduced. Under *MS*, increases of isoprene emission and maintenance of phenolic compound concentration resulted in moderate growth reduction. Hence, it is likely that investment in isoprene emissions confers *Q. pubescens* an important competitive advantage during moderate but not severe periods of water scarcity. Consequences of this response for air quality in North Mediterranean areas are also discussed.

Editor: Martin Heil, Centro de Investigación y de Estudios Avanzados, Mexico

Funding: This research was supported by ADEME (Agence de l'environnement et de la maitrise de l'énergie), the region Provence-Alpes Côte d'Azur (PACA) and the European Union and was funded by the French National Research Agency (ANR) through the SecPriMe2 project (ANR-12-BSV7-0016-01). The funders had no role in study design, data collection and analysis, decision to publish, or preparation of the manuscript.

Competing Interests: The authors have declared that no competing interests exist.

* Email: elena.ormeno-lafuente@imbe.fr

Introduction

Water limitation induced by climate change is going to affect particularly the Mediterranean area. Global models show, for the next century, a rise of 4°C [1] and a reduction of 30% of precipitation during summer [2]. In this region, with typical long dry and hot summers, vegetation undergoes an accentuated seasonal water stress. It has been shown that secondary metabolites both, volatile (Biogenic Volatile Organic Compounds or BVOC) and non-volatile (phenolic compounds) could help plants to overcome climate-related stress conditions, due to the protection they confer against thermal stress and [3,4], overall, against any environmental constraint leading to cell oxidation (water stress, air pollution) [5–7]. Physiological studies indicate that phenolic compounds and isoprene cooperate to scavenge reactive oxygen species, thereby resulting in limited cell oxidation [4,6,8,9]. It has been suggested that isoprene primarily participates to the enhancement of phospholipidic interactions between thylakoid membranes which are damaged under environmental constraints, thereby reducing cell oxydation [9]. However, from an ecological perspective, the physiological action of these secondary metabolites is not enough to conclude that increases in leaf phenolic content or isoprene emission rates confer competitive advantages for plants coping environmental stresses. This can be concluded if plant resource allocation for secondary metabolite production does not imply a cost for the plant in terms of growth reduction.

Isoprene (2-methyl, 1–3 butadiene, C_5H_8) represents 50% of total BVOC emissions at the global scale [10]. This volatile secondary metabolite impacts air pollutant formation including tropospheric ozone and Secondary Organic Aerosol (SOA) [11,12]. These processes are particularly important in the Mediterranean region where ozone thresholds are exceeded numerous times every year, especially during summer, and particulate pollution has become a major concern for air quality.

Isoprene biosynthesis is dependent on photosynthesis and enzymatic processes and is therefore mainly controlled by light and temperature [13–16]. By contrast, the impact of drought on isoprene emission is controversial as reviewed by Peñuelas and Staudt (2010) [17], despite the well-known negative impact of drought on photosynthesis. These authors showed that around 50% of the studies relating isoprenoid emissions and drought found an emission decrease while 25% found an increase and the remaining reported no change, such variability being due to water stress intensity, timing and species.

The common point of most previous investigations focused on isoprene emission response to water limitation is that water stress is applied by stopping water supply for days or weeks (not an entire growing season) until a very significant decline of water potential, photosynthetic activity or stomatal conductivity is observed [18–24]. Under mild water stress, during the first days of water withholding, isoprene has been found to be slightly stimulated [22,23] or unchanged [19,24,25] despite a decline of CO_2 assimilation. Thereafter, when water availability becomes highly scarce, isoprene decline continues for a length of time which depends on the species studied, before emissions are fully inhibited [18–20]. Since isoprene emission occurs even when water stress dramatically inhibits photosynthesis due to stomata closure, emissions are uncoupled from photosynthesis during induced water stress. Funk et al. (2004) [23] demonstrated that in water-stressed plants isoprene production was not supported by recently assimilated carbon (as occurred in non-stressed plants), but rather by alternative carbon sources. These results suggest, on the one hand, that temperature and light dependency of isoprene emissions are weakened under this type of abiotic constraint, and, on the second hand, that isoprene confers resistance against water stress.

The objectifs of the present study were to **(i)** test the response of *Quercus pubescens* Willd. - a deciduous drought-resistant species [26] - against mild and severe water stress in terms of secondary metabolism (carbon-based secondary metabolites) with concomitant analysis of growth traits (using above and below–ground biomass as a proxy), physiological traits (photosyntheis, stomatal conductance, chlorophyll fluorescence, stem water potential) and functional traits (leaf mass to area ratio, leaf relative water content). We focused on isoprene emissions as a carbon-based secondary metabolite and, to a lesser extent, on phenolic compound concentrations, since *Q. pubescens* accounts for the major isoprene emitter in Northern Mediterranean regions (10–50 nmol.m^{-2}.s^{-1}) [18,20,27]; **(ii)** assess whether plant investment in isoprene emissions under water stress implies a cost in terms of growth reduction and requires an important fraction of the recently assimilated CO_2; **(iii)** test the adequacy of the Guenther et al., (1993) (G93) algorithm [13] to calculate isoprene emission factors according to light and temperature under the different watering treatments. To attend such objectifs we performed a long-term study under greenhouse conditions consisting of three watering treaments (optimal watering, mild and severe water withholding) applied on *Q. pubescens* saplings over an entire growing season. We hypothesize that mild water stress promotes the production of carbon-based secondary metabolites based on their antioxidant properties and hence, the benefits they confer to plants under moderate stress conditions [6] and previous findings that report a rise. Our hypothesis is also based on the Growth Differentiation Balance Hypothesis (GDBH) [28] which states that moderate resource availability impacts growth rather than photosynthesis, resulting in an increasing carbon pool available to synthesize secondary metabolites. For severe stress however, we expect a decline of isoprene emissions based on both, previous findings performed over days or weeks of full watering withholding and the GDBH. The theory postulates that both growth and photosynthesis are constrained and both baseline maintenance and primary metabolic processes receive priority for carbon allocation compared to carbon-based secondary metabolites. We also hypothesize that, under water scarcity, estimations of isoprene emissions using the G93 algorithm are significantly different from those experimentally obtained considering the uncoupling between isoprene emissions and photosynthesis, and so with light [18–20,29].

Materials and Methods

Ethic statement

The experiment took place under greenhouse conditions in a national nursery in Aix-en-Provence (5°24′31.5″E, 43°30′35.8″N Southern France). The director of this nursery (Patrice Brahic) gave permission to conduct the study on this site. No specific permissions were required for these locations/activities, and studies did not involve endangered or protected species.

Experimental design

Four year-old *Q. pubescens* saplings were grown into 6 L pots, containing natural soil from a national experimental Downy oak forest (https://o3hp.obs-hp.fr/index.php/fr/). Greenhouses were built with transparent polycarbonate plates (Ombrex) which allowed ~90% of the photosynthetic photon flux density (PPDF) to pass through. Environmental greenhouse conditions (air temperature and relative humidity) were recorded with HOBO devices (U12-012 HOBO data logger, temp/RH/light/ext channel).

Water withholding started in April, one week before measurements took place, and stopped late November, during leaf senescence. Saplings were divided into three groups corresponding to three watering treatments: (i) control *(C)*, where saplings were watered to reach and maintain soil field capacity, (ii) mild stress *(MS)*, and (iii) severe stress *(SS)* where saplings received 20% and 10% of soil field capacity respectively. Below 10% of soil field capacity, plants reached their wilting point. The degree of water stress under *MS* and *SS* was monitored by regularly checking sapling physiological activity. As net photosynthesis (P_n) was below 1 μmol_{CO2} m^{-2} h^{-1}, saplings were punctually rewatered (this was only necessary during one day under *SS* just before June measurements, and during another day under *MS* just before July measurements), allowing to follow *Q. pubescens* response to water withholding all over an entire growing season (8 months). A non-destructive set of 15 saplings (5 *C*, 5 *MS* and 5 *SS*) was studied from April to November for gas exchange, isoprene emission and chlorophyll fluorescence measurements. Growth was also studied on this set of plants at the end of the study, at leaf senescence (i.e. in November). A destructive set of 15 different saplings (5 per treatment) was also used each month to measure soil water status, stem water potential and phenolic compound concentration. A total of 105 pots were thus used.

Measurement of water status in soil and plants and leaf mass per area ratio

All parameters described herein were studied in the destructive set of saplings over the entire growing season and, only at the end of the study (November), they were also studied on the non-destructive set of saplings. Soil relative water content (SRWC) was measured using 100 g of soil harvested from each pot. Plant water

Table 1. Indicators of water stress degree.

Parameter	Treatment	April	May	June	July	September	November	Annual mean	Statistics
SRWC ($cm^3.kg^{-1}$)	C	$180.9^{a}\pm21.4$	$205.9^{a}\pm13.9$	$226.2^{a}\pm10.8$	$202.8^{a}\pm7.8$	$192.0^{a}\pm10.0$	$222.0^{a}\pm12.5$	205.0 ± 7.1	$F_{treatment}$: 50.5; df = 2; $P<0.001$ F_{month}: 1.1; df = 5; N.S. $F_{interaction}$: 6.0; df = 10; $P<0.001$
	MS	$174.6^{a}\pm15.9$	$144.9^{b}\pm15.4$	$113.7^{b}\pm13.9$	$143.5^{b}\pm10.9$	$200.3^{a}\pm13.7$	$199.5^{a}\pm12.4$	162.8 ± 14.1	
	SS	$157.5^{a}\pm13.6$	$128.2^{b}\pm4.4$	$135.6^{b}\pm7.3$	$131.8^{b}\pm7.8$	$140.7^{(b)}\pm23.4$	$74.3^{b}\pm11.9$	128.1 ± 11.5	
LRWC (%)	C	71.2 ± 3.2	74.1 ± 3.4	78.4 ± 2.0	78.6 ± 3.5	81.0 ± 2.3	85.0 ± 3.1	$78.1^{A}\pm2.0$	$F_{treatment}$: 3.2; df = 2; $P=0.0469$ F_{month}: 6.1; df = 5; $P=0.001$ $F_{interaction}$: 0.7; df = 10; N.S.
	MS	69.9 ± 4.9	71.6 ± 2.0	76.3 ± 2.3	84.9 ± 1.3	82.3 ± 1.6	88.0 ± 1.5	$78.8^{A}\pm3.0$	
	SS	66.7 ± 5.1	55.5 ± 11.9	69.6 ± 9.8	80.4 ± 1.2	82.3 ± 1.2	78.8 ± 4.1	$72.2^{(B)}\pm4.2$	
ψ_{mid} [MPa]	C	$1.6^{a}\pm0.2$	$2.2^{a}\pm0.4$	$2.3^{a}\pm0.3$	$2.3^{a}\pm0.4$	$1.2^{a}\pm0.1$	$1.0^{a}\pm0.1$	1.8 ± 0.2	$F_{treatment}$: 13.1; df = 2, $P<0.001$ F_{month}: 14.7; df = 5; $P<0.001$ $F_{interaction}$: 6.8;df = 10; $P<0.001$
	MS	$1.8^{a}\pm0.4$	$2.5^{ab}\pm0.3$	$2.8^{a}\pm0.2$	$2.9^{a}\pm0.3$	$1.2^{a}\pm0.1$	$1.2^{a}\pm0.1$	2.1 ± 0.3	
	SS	$1.8^{a}\pm0.3$	$4.8^{b}\pm0.8$	$2.1^{a}\pm0.1$	$2.4^{a}\pm0.2$	$1.3^{a}\pm0.1$	$3.7^{b}\pm0.4$	2.7 ± 0.5	
Climatic parameters	T (°C)[1]	18.7 ± 0.5	23.3 ± 0.3	23.3 ± 0.2	25.2 ± 0.2	24.4 ± 0.2	17.7 ± 0.1		
	RH (%)[2]	63.7 ± 1.5	51.1 ± 0.6	57.8 ± 0.6	52.2 ± 0.5	58.0 ± 0.5	62.0 ± 0.2		

T: mean temperature recorded inside the greenhouses every month; [2] RH: air relative humidity recorded inside the greenhouses every month.
Soil relative water content (SRWC), leaf relative water content (LRWC) and midday water potential (ψ_{mid}) under control, mild and severe water stress (C, MS and SS) all over the growing season. Climatic conditions under the greenhouse are also shown. Variability of these parameters due to water stress, month and their interaction are tested using two-way ANOVA (F) followed by Tukey test. Lower case letters (a>b) represent differences between treatments month by month. Capital letters (A>B) represent differences between treatments all months averaged (Annual mean column). Values are mean ± SE of n = 5; N.S.: not significant. Lower case or capital letters under brackets denote Tukey test differences at 90% confidence level; df: degrees of freedom.

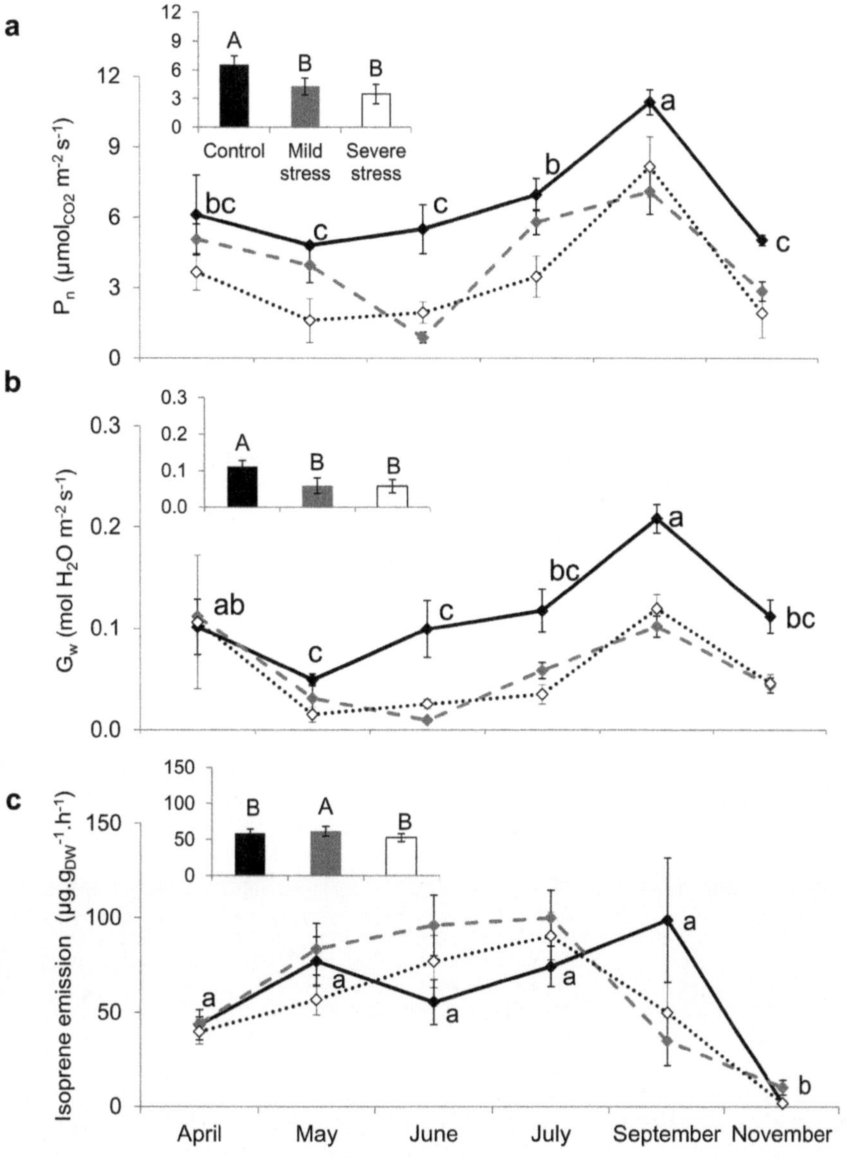

Figure 1

Figure 1. Gas exchanges (photosynthesis, stomatal conductance and isoprene). Seasonal course of the net photosynthesis (P_n, graph a), stomatal conductance to water vapor (G_w, graph b) and isoprene emissions (graph c) under control (–◆–), mild (--◆--) and severe (···◇···) water stressed. Note that 70% of isoprene emissions were mostly measured under standard conditions (refer to materials and method for details). Differences are tested using a two-way ANOVA repeated measures (F) followed by Tukey tests. Since interaction between seasonality and treatment is not significant according to the two-way ANOVA, results of water stress impact are shown in the small graph where data of all months are pooled together and differences between treatments are denoted by capital letters (A>B) while seasonality impact is shown in the main graph (in lower case letters: a>b>c). Values are mean ± SE of n = 5.

status was estimated through two parameters: midday stem water potential (ψ_{mid}) - measured using a pressure chamber (PMS instrument, Co. Oregon USA) - and leaf relative water content (LRWC) calculated following the formula: LRWC = (FM-DM)/(SM-DM) where FM is fresh mass, DM is dry mass and SM is saturated mass. Leaf dry mass was measured by weighing the leaves after they had reached constant mass in an oven at 70°C. Leaf mass to area ratio (LMA), an indicator of leaf sclerophylly [30], was calculated by dividing leaf dry mass by leaf area, which was calculated using a leaf area meter (AM300 Bioscientific).

Chlorophyll fluorescence

Fluorescence was measured on the non-destructive set of plants after isoprene emission collection was performed. Fluorescence of three fully expanded leaves of each sapling was studied and results were averaged. Measurements were performed using a portable Hansatech Fluorescence Monitoring System (FMS 2) allowing computer-assisted data acquisition. The variable to maximum fluorescence ratio (*Fv/Fm*) was measured in the non-energized state after darkness adaptation of about 30 min using a leaf clip holder for adapting leaves to darkness. *Fv/Fm* is a reliable indicator of the maximal potential efficiency of excitation capture

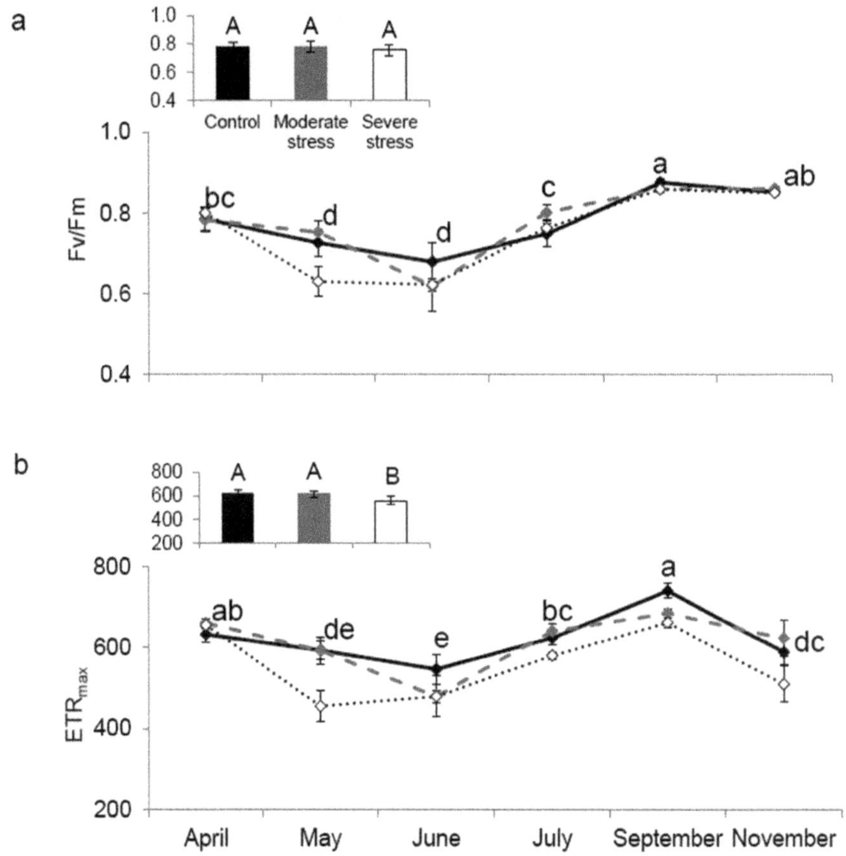

Figure 2. Chlorophyll fluorescence. Seasonal course of the variable to maximum fluorescence ratio (Fv/Fm, graph a) and maximal electron transport rate (ETR_{max}, graph b) under control (–◆–), mild (—◆—) and severe (···◇···) water stress. Differences are tested using a two-way ANOVA repeated measures (F) followed by Tukey tests. Since interaction between seasonality and treatment is not significant according to the two-way ANOVA, results of water stress impact are shown in the small embedded graph where data of all months are pooled together and differences between treatments are denoted by capital letters (A>B) (similar capital letters indicate the absence of water stress influence) while seasonality impact is shown in the main graph (in lower case letters: a>b>c >d>e). Values are mean ± SE of n = 5.

by open PSII in dark-adapted conditions, was also measured to estimate the functional state of the photosynthetic apparatus under the three watering treatments. The apparent maximum photosynthetic electron transport (ETR_{max}) leading to carbon fixation was measured from light-adapted leaves as shown in Ormeño et al. (2009) [31].

Leaf gas exchange

Net photosynthesis (P_n) and stomatal conductance to water vapour (G_w) were measured on the same saplings all over the experimentation, i.e. on the non-destructive set of saplings. The same leaf of each sapling was always studied over the growing season in order to eliminate leaf-to-leaf variability. P_n and G_w were measured using a 6 cm^2 portable leaf LED cuvette (6400-02B LED Light Source Light-Emitting Diode) for PPFD control, coupled to an InfraRed Gas Analyzer (Li-cor 6400 XT system, Li-cor Lincoln, Nebraska, USA). Upstream, flow rate entering the cuvette monitored by the Li-cor 6400 XT device, was fixed at 1 $l.min^{-1}$. Downstream, flow rate was set at 100 $mL.min^{-1}$ using a mass flow controller (Bronckhorst, F-201CV-1K0-AAD-22-V) connected to a pump (E99335; Fisher Scientific). Most leaf gas exchange measurements were studied under standard conditions for isoprene emission (i.e. 30°C temperature and 1000 µmol m^{-2} s^{-1} PPFD) as described by Guenther *et al.* (1993) [14]. Additionally, in May, June and July leaf gas exchanges of some saplings (n = 9, n = 3, n = 8 respectively) were not only measured under standard conditions but also under natural conditions of temperature and PPFD which ranged between 18 and 39°C and 384 and 1800 µmol.m^{-2}.s^{-1} respectively.

Isoprene sampling and analysis

The leaf cuvette previously described was modified in order to sample isoprene emissions from the non-destructive set of plants. Air exiting the cuvette was collected into glass tubes (15 cm long and 3 mm inner diameter) containing 60 mg tenax TA and 140 mg carbotrap. Glass tubes were attached to the cuvette using FPE tubing. For each sapling, isoprene collection took place simultaneously using two different glass tubes for 15 min at 100 $mL.min^{-1}$ using a mass flow controller and a pump. Isoprene emissions collected onto both tubes were averaged. As it was previously described in the "leaf gas exchanges" section, isoprene emissions were mostly measured under standard conditions all over the experimentation excepting May, June and July, where isoprene emissions of some saplings were additionally collected under natural conditions of temperature and light. These additional samplings allowed to compare the experimentally measured isoprene emission factors (i.e. at 30±1°C and 1000±50 µmol m^{-2} s^{-1} of PPFD) to the theoretical isoprene emission factor obtained using the G93 algorithm under the three watering treatments.

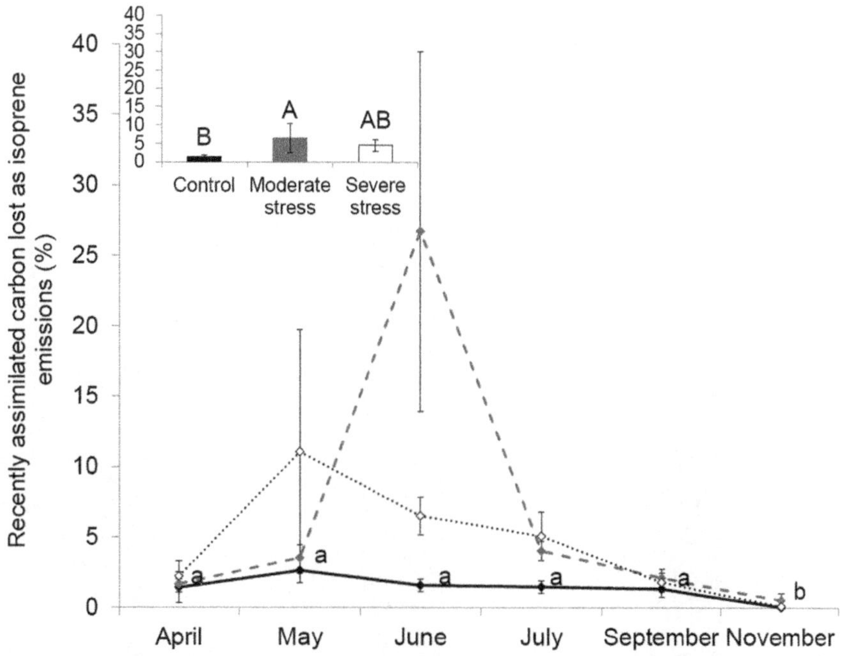

Figure 3. Isoprene emission rate as percentage of carbon re-emitted. Seasonal course of the percentage of photosynthetically assimilated carbon re-emitted as isoprene under control (–◆–), mild (---◆---) and severe (···◇···) water stress. Differences are tested using a two-way ANOVA (F) repeated measurements followed by Tukey tests. Since interaction between seasonality and treatment is not significant according to the two-way ANOVA, results of water stress impact are shown in the small embedded graph where data of all months are pooled together and differences between treatments are denoted by capital letters (A>B) while seasonality impact is shown in the main graph (in lower case letters: a>b). Values are mean ± SE of n = 5 saplings.

Isoprene was analysed with a gas chromatography (HP 6890N), coupled to an injector for thermal desorption (Gerstel TDS3/CIS4) and a mass spectrometer (HP 5973). A capillary column (Al/KCl) was used (30 m long, 250 microns of internal diameter and 5 microns of film thickness). The carrier gas was helium maintained at a constant flow of 1 ml $min^{-1.}$ The desorption program was: pre-cooling to −50°C, 10 min-desorption at 250°C to 50 mL min^{-1}, and then 3 min during injection maintained at 250°C. The analysis program was: 40°C to 200°C at 20°C min^{-1}, maintained for 1 min at 200°C. The transfer line temperature was 200°C. The mass spectrometer fragmentation was done by electron impact at 70 eV, the source temperature was 230°C, the quadrupole temperature was set at 250°C and the potential difference of the electron multiplier was 1400. Data acquisition was done in scan mode from 40 to 150 amu.

Concentration of phenolic compounds

Total water-soluble phenolics of the destructive set of saplings (and in November for the non-destructive set of saplings) were extracted based on the method described by Singleton and Rossi (1965) [32]. Leaf dry mass (0.25 g_{DM}) was extracted with 20 mL of distilled water with 70 % methanol. The mixture was left for 1 h under constant shaking at ambient temperature and light cover. Afterwards, the extract was filtered on Whatman 90 mm paper filter. 1650 µl of distilled water were added to 50 µl of the extract. 200 µl of saturated Na_2CO_3 aqueous solution and 100 µl of Folin-Ciocalteu's reagent were added. After 30 min, the reaction was completed and the extract was measured at 765 nm on a Biomate 3 spectrophotometer (Thermo Electron Scientific Corporation). The quantitative values were expressed in equivalents of gallic acid per leaf dry mass.

Plant growth

Growth of the non-destructive set of saplings was measured at the end of the study period (i.e. November) in terms of plant height, basal diameter, fresh and dry leaf mass, shoot and roots dry mass, root/shoot ratio and main root diameter.

Statistics

Shifts in isoprene emission, gas exchange and chlorophyll fluorescence under the three watering treatments applied over 8 months were tested using a two-way repeated measures ANOVA with treatment (*C*, *MS* and *SS*) and month as factors and the studied traits as dependent variables. Changes in soil and plant water status, LMA as well as concentration of phenolic compounds were analysed using two-way ANOVA, since these parameters were measured on the destructive set of plants. In either case, when these two factors interacted significantly ($P<0.05$), one-way ANOVA was applied for the significant factor. Over the text, simple or repeated ANOVA measurements are named as ANOVA (F). Tukey post hoc tests were used to test differences between months and treatment. Data were log transformed when necessary to achieve normal distribution and homoscedasticity requirements. A Principal Components Analysis (PCA) including the total set of plant traits (growth, emission or concentration of secondary metabolites, physiological and functional traits) in November was performed in order to analyse the adaptation of *Q. pubescens* saplings to the different watering treatments (*C*, *MS* and *SS*). Linear and non-linear regression analyses were used to evaluate isoprene emission dependency on photosynthesis according to water stress level. Differences between the theoretical and the experimental isoprene emission factors of the same sapling were tested for the three watering treatments (*C*, *MS* and *SS*) using

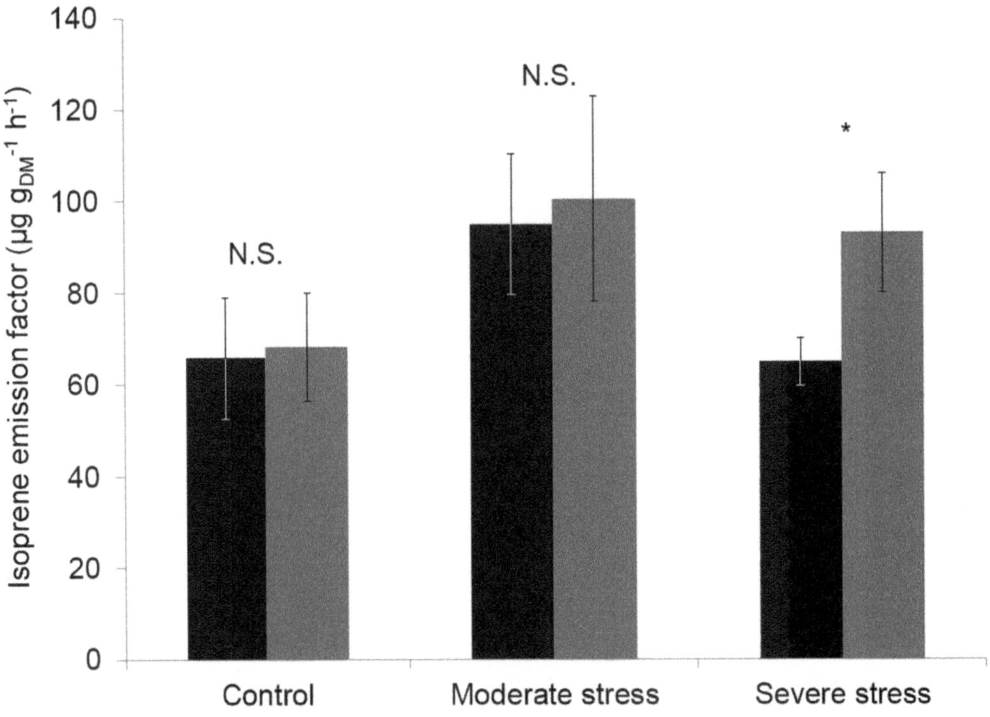

Figure 4. Experimental and calculated isoprene emission factors. Comparison between the isoprene emission factor calculated with Guenther et al. (1993) algorithm (▒) and the experimental isoprene emission factor (■) obtained under standard conditions (30±1°C and 1000±50 µmol m^{-2} s^{-1} of PPFD) for control, mild and severely water stressed sapling. t: value of the paired sample comparison tests. N.S.: not significant, *0.01<*P*<0.05. Values are mean ± SE of n=6–7 saplings.

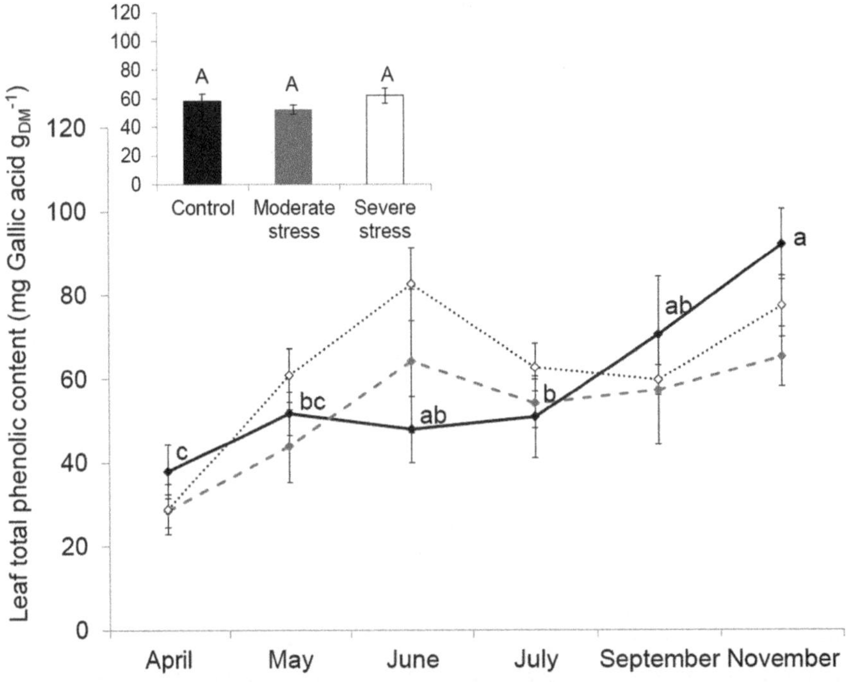

Figure 5. Leaf total phenolic concentration. Seasonal course of leaf total phenolic concentration under control (–◆–), mild (--◆--) and severe (···◇···) water stress. Differences are tested using a two-way ANOVA (F) followed by Tukey tests. Since interaction between seasonality and treatment was not significant, the impact of water stress is shown in the small embedded graph where data of all months are pooled together and differences between treatments are denoted by capital letters (similar capital letters indicate the absence of water stress influence) while seasonality impact is shown in the main graph (in lower case letters: a>b>c). Values are mean ± SE of n=5.

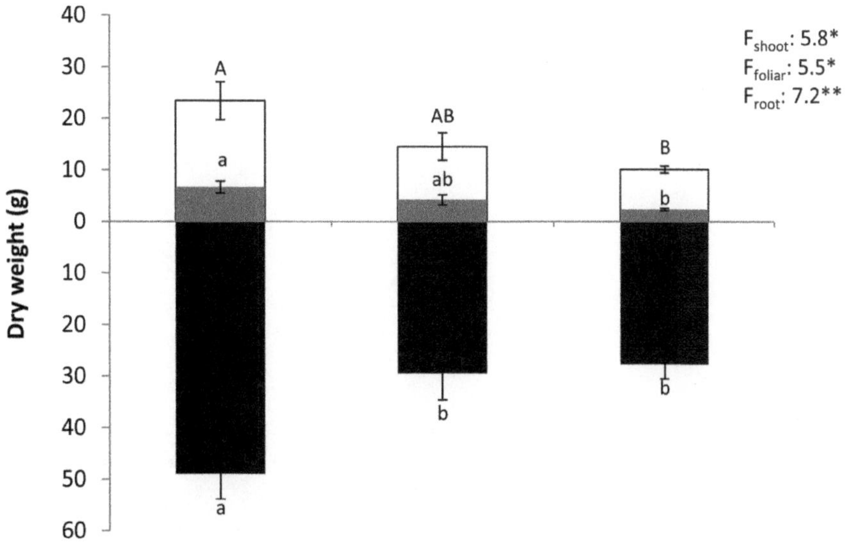

Figure 6. Biomass growth. Shoot, foliar and root biomass (g_{DM} ind^{-1}) during leaf senescence (in November) for control, mild and severe water stress. Differences are tested with one-way ANOVA (F). Capital black lower case black and white lower case letters denote statistical differences for shoot, foliar and root biomasses respectively with a>b or A>B given by Tukey tests. Values are mean ± SE. n = 5.

paired sample comparisons (t). Statistical analyses were conducted using R 3.0.3 and Stagraphics software.

Results

Soil and leaf water status under three different watering treatments

Severe water stress significantly lowered *SRWC* compared to *C* all over the growing season (Table 1, $F_{treatment} = 50.5$, $P<0.001$). Mild water stress reduced *SRWC* excepting in fall (September), when it was similar to *C* pots (One-way ANOVA, $F_{treatment} = 3.82$, $P = 0.054$, Table 1). Despite the soil dryness caused by water shortage, *MS* never lead to a significant decline of leaf water status (no significant differences occurred in either *LRWC* or Ψ_{mid}). Only *SS* tended to decrease *LRWC* (at 90% confidence level; two ways ANOVA, $F_{treatment} = 3.2$, $P = 0.047$), but their Ψ_{mid} only significantly dropped in May and November.

Physiological performances (gas exchange and fluorescence) under water stress

All over the growing season both, saplings under *MS* and *SS* showed a very significant decline of 35% and 47% on average of P_n and 47% and 48% for *Gw* compared to *C* respectively (Fig. 1a,b, two-way ANOVA, $18.8<F_{treatment}<23.7$, $P<0.001$). *Fv/Fm* did not significantly vary between treatments at any month, while ETR_{max} was significantly lowered under *SS*, indicating that photosystem II was only damaged under this level of stress (Fig. 2a: *Fv/Fm* two-way ANOVA, $F_{treatment} = 1.58$, $P = 0.2131$; Fig. 2b: ETR_{max}, two-way ANOVA, $F_{treatment} = 8.23$, $P = 0.0007$).

Physiological status of stressed and non-stressed oaks featured the same seasonal trend with the lowest values of P_n, G_w, *Fv/Fm* and ETR_{max} in May (one month after water stress was applied) and June, and the highest values of these traits in September, before a decline occurred during leaf senescence in November (two-way ANOVA, $16.0<F_{month}<31.0$, $P<0.001$). Since all physiological traits recovered in September, water stress applied in this study did not cause any irreversible damage. All these physiological traits showed a non-significant interaction between month and treatment ($1.05<F_{interaction}<1.46$, $0.178<P<0.417$, Fig. 1 & 2).

Isoprene emissions under water stress

Mean isoprene emissions under *MS* were statistically higher compared to *C*, while not significant increases occurred in *SS* sapling (Fig. 1c, two-way ANOVA, $F_{treatment} = 4.0$, $P = 0.02$; $F_{interaction} = 1.45$, $P = 0.417$). Isoprene emission factors showed a similar trend, although emission rates were slightly higher under *SS* (Fig. S1). In June, saplings under *MS* allocated 27% of the photosynthetically assimilated carbon for isoprene emission - the highest fraction attained over the growing season - while it was only 2% in *C* saplings (Fig. 3). In May, in spite of the very restricted P_n featured by *SS* oaks, isoprene emissions remained similar to *C* saplings. As a result, in May, isoprene emissions under *SS* represented 11% of the photosynthetically assimilated carbon (the highest of all the growing season) while it was only 3% in *C* saplings (Fig. 3).

Isoprene emissions (Fig. 1), as well as isoprene emission factors (Figure S1) remained similar from April to September (70.8 ± 5.1 μg g_{DM}^{-1} h^{-1}) and declined drastically in November (5.9 ± 3.4 μg g_{DM}^{-1} h^{-1}) during leaf senescence. This trend was similar among the three water treatments (two-way ANOVA, $F_{month} = 17.0$, $P<0.001$; $F_{interaction} = 1.45$, $P = 0.417$).

Isoprene emission rates (ER) were strongly linearly correlated to P_n exclusively in *C* saplings ($ER = 5.6 + 8.0 \times P_n$, $r^2 = 18.5$, $F = 19.13$, $P<0.001$). This correlation was weak although significant in *MS* ($ER = 52.1 + 0.5 \times P_n^2$, $r^2 = 6.1$, $F = 3.84$, $P = 0.05$) and was not significant under *SS* ($F = 0.48$, $P = 0.49$). The comparison between the theoretical isoprene emission factor obtained using G93, and the experimental isoprene emission factor of the same trees under the three watering treatments revealed that there were no significant differences between *C* and *MS*, while the G93 algorithm overestimated the isoprene emission factor by 44% under *SS* compared to experimental measurements (Fig. 4, paired sample comparisons, $t = -2.8$, $P = 0.028$).

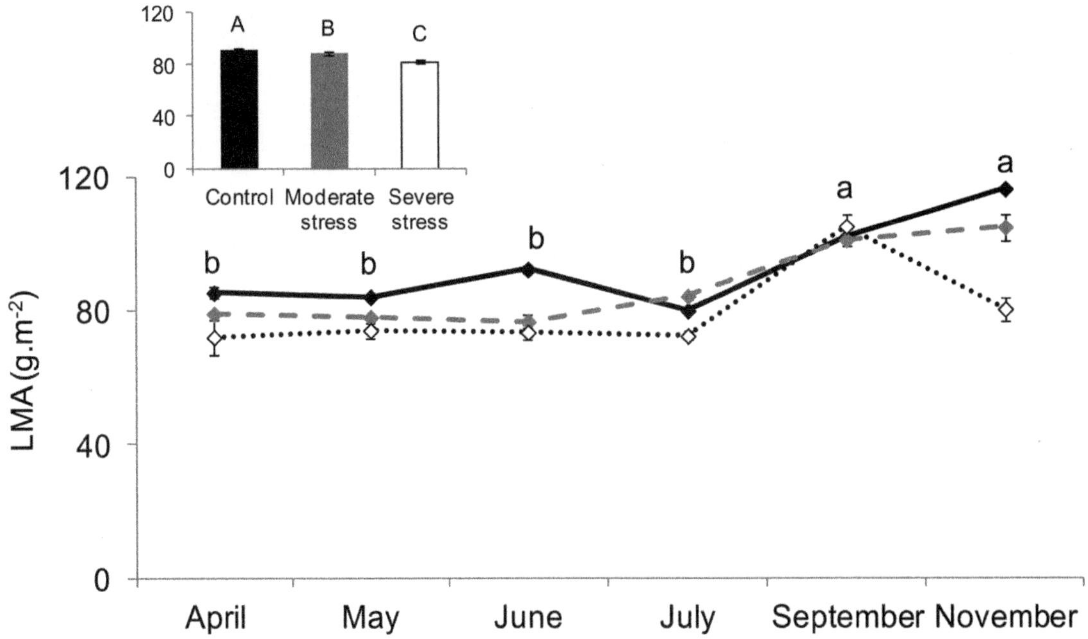

Figure 7. Mapping control, mild stress and severe stress. Two-dimensional mapping of the Principal Component Analysis (PCA) performed for all plant traits measured at the end of the experiment (plant growth, emission or concentration of secondary metabolites, gas exchange and water status). This analysis was performed on n = 5 trees per treatment. Traits shown at the end of some arrows correspond to the most explanatory traits.

Concentration of phenolic compounds under water stress

Unlike isoprene emissions, concentrations of phenolic compounds were not statistically sensitive to water stress (two-way ANOVA; $F_{treatment} = 1.38$, $P = 0.26$, Fig. 5), independently on the month ($F_{interaction} = 1.34$, $P - 0.23$). Their seasonal trend was also different compared to isoprene: *Q. pubescens* produced the lowest phenolic concentrations in April and they increased progressively month by month up to reach the highest values in November (two-way ANOVA, $F_{month} = 8.8$, $P<0.001$, Fig. 5).

Plant growth under water stress

Water scarcity significantly reduced *Q. pubescens* growth. Severely stressed saplings showed a very significant decrease of foliar, shoot and root biomasses and an increase of the root/shoot ratio compared to *C* (one-way ANOVA, $5.8<F_{treatment}<7.2$, $0.008<P<0.0182$, Fig. 6). Under *MS* conditions, only root biomass was significantly reduced (40% lower than *C*), while shoot and foliar biomasses presented intermediate values between *C* and *SS* treatments.

Leaf mass to area ratio under water stress

LMA was significantly reduced by 7% and 14% under *MS* and *SS* all over the growing season (two-way ANOVA, $F_{treatment} = 7.9$, $P<0.0001$, Fig. 7). It also changed over the different months, with the highest values shown at the end of the growing season (November under *MS* and *C* and September under *SS*) and the lowest values from April to July (two-way ANOVA, $F_{month} = 13.7$, $P<0.0001$, $F_{interaction} = 1.93$, $P = 0.05$).

Secondary metabolism, growth, physiological and functional traits under water stress

The PCA - performed with all measured traits in November - revealed that growth, especially root, shoot and foliar biomass, explains sapling repartition on axis 1, while isoprene emissions, followed by Ψ_{mid}, explain sapling reparation on axis 2 (Fig. 8). Control saplings featured a high investment in growth biomass. Only under *MS*, saplings were characterized by high isoprene emissions and their growth was only limited in some of them. Under *SS*, saplings were distinguished by a particularly high $|\psi_{mid}|$ and low growth showing that *Q. pubescens* was unable to maintain the leaf water level and growth rates under such degree of stress. Furthermore, under *SS*, saplings showed a poorer plasticity than *C* and *MS* as reflected by the tight distribution of saplings along both axes.

Discussion

The significant reduction of *Gw* under water stress impeded a significant water status drop under *MS* as evidenced by the maintenance of LRWC and Ψ_{mid}. Nardini and Pitt (1999) [33] showed that *Q. pubescens* is capable to preserve high levels of LRWC under drought due to the high efficiency of its water conduction system through the root system. These responses indicate that *Q. pubescens* adaptation to water scarcity can be attributed to a drought-resistant strategy and, more particularly, to a drought-avoidance strategy, known to confer *Q. pubescens* resistance to arid and semi-arid environments. Thus, there were no irreversible leaf damages (*Pn* and ETR_{max} showed the highest rates in September after the intense summer drought) and photochemical efficiency (*Fv/Fm*) remained unchanged under *C*, *MS* and *SS* all over the study.

Maintenance of similar leaf water levels, especially during *MS*, probably explains why, under this level of water stress, there was a moderate reduction of leaf and shoot growth at the end of the growing season (i.e. at leaf senescence), while there was a marked reduction of root biomass. The ecological advantage of maintaining similar leaf water levels is a low decrease of the leaf area and hence a low drop of CO_2 uptake before leaf senescence [34]. Additionally, this would confer *Q. pubescens,* a deciduous species,

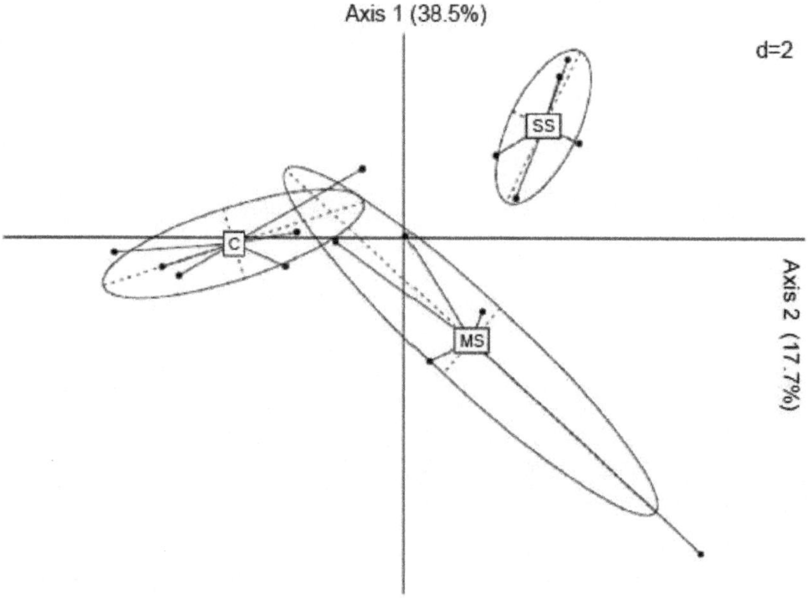

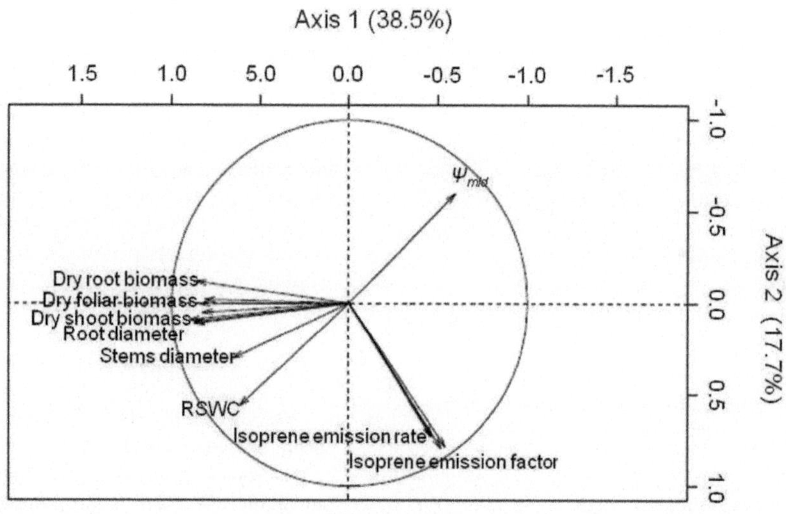

Figure 8. Leaf mass per area ratio. Seasonal course of the leaf mass per area ratio (LMA) under control (–◆–), mild (--◆--) and severe (···◇···) water stress. Differences are tested using a two-way ANOVA (F) repeated measurements followed by Tukey tests. Since interaction between seasonality and treatment is not significant according to the two-way ANOVA, results of water stress impact are shown in the small embedded graph where data of all months are pooled together and differences between treatments are denoted by capital letters (A>B) while seasonality impact is shown in the main graph (in lower case letters: a>b). Values are mean ± SE of n = 5 saplings.

a competitive advantage over the evergreen Mediterranean tolerant species (e.g. *Q. ilex*) during the extended dry periods in summer [35,36].

Our study also showed that LMA increased at the end of the seasonal cycle and decreased with water scarcity as shown for *Q. ilex* [37], indicating that *Q. pubescens* sclerophylly is not strengthened under the drier conditions applied. Indeed, increases of LMA under water scarcity are associated to the degree of sclerophylly (hardness and thickness) [30] known to promote drought resistance ([37]and citations cited therein). The low LMA values observed under *MS* and *SS* conditions can be explained if the plant prioritizes the allocation of the recently assimilated CO_2 to other processes different from growth production, such as isoprene synthesis and so isoprene emissions, rather than to produce leaf biomass.

Unlike all these physiological and functional traits, phenolic compounds as a whole did not respond to water scarcity as already reported for *Capsicum annuum* L. [38]. Phenolic compounds are however a very heterogeneous group of secondary metabolites and, as a result, it is likely that the response of different phenolics (e.g. flavonoids, flavonols, anthocyanins, single phenols) to water scarcity varies.

Nevertheless, our study indicates that isoprene emissions were very sensitive to water stress. Isoprene emissions of *C* saplings –

around 63 μg g_{DM}^{-1} h^{-1} (or 22 nmol m^{-2} s^{-1}) during most part of the growing season (from April to September) – was indeed significantly increased by 38% on average under *MS* independently on the month, while *SS* caused not significant increases. These results are partly consistent with Sharkey and Loreto (1993) [22] who observed a slight increase of isoprene emitted by *Pueraria montana* under *MS*, even if leaf to leaf variation prevented any statistical significance. Funk et al. (2004) [23] described a similar increase of isoprene emission in *Populus deltoides*.

On the contrary, other studies have not reported isoprene emission shifts during water stress [24,25] and previous studies on *Q. pubescens* show a decrease in isoprene emissions under *SS* [18,20]. Differences probably rely on the water stress protocols used. In our study, water stress saplings daily received a lower water supply than *C* saplings over the entire growing season, while the other two cited studies stopped watering in stress sapling while continued watering control saplings, and observed a decrease of isoprene emissions after a period of several weeks of water withholding.

The increase and stability of isoprene under *MS* and *SS* respectively, despite the low *Pn*, reflected an uncoupling between isoprene emissions and *Pn* in agreement with previous investigations [18,19,22,24,29,39]. In our study, this uncoupling was especially clear under *SS* since no correlation was found between both processes and isoprene emissions persisted all over the growing season. In the cited studies, the length of time while isoprene emission persisted varied according to the species studied and the degree of stress. It can be also noted that this length is not always mentioned, as most of the studies stop water stress when physiological processes (e.g. *Pn*, *Gw*) reach their minima. As a result, in our study, as much as 3% of the photosynthetically assimilated carbon in *C* saplings was allocated for isoprene emissions in agreement with pas studies [23] and citations therein). By contrast, this percentage was as high as 27% and 11% under *MS* and *SS* respectively. Some authors show that under extreme water limitation, when *Pn* is highly depressed, isoprene emission represents between 20 and 50% of the recently assimilated carbon from photosynthesis [22,39]. These results suggest that drought resistance - in addition to be conferred by stomatal closure before water potential drops as previously described - is probably implemented by increasing the allocation of the recently assimilated carbon to isoprene emissions in view of the benefits isoprene confers for the plants in terms of cell oxidation reduction and cell stabilization [3,9].

Niinemets (2010) [15] proposed that mild stress decreases stomata openness and therefore, carbon assimilation, but does not inhibit isoprene emissions since sugars, starch reduction, and *ETR* are maintained. This later hypothesis was confirmed in our study where ETR_{max} values under *C* and *MS* conditions were similar. Other authors claim that other carbon sources stored in the plant, rather than recently assimilated carbon, become available to support isoprene biosynthesis under stress conditions [23]. These alternative sources seem hence to be extra-chloroplastic, such as xylem-transported glucose in *Quercus robur* coming from root and stem storage [23,24,40]. In view of the previous results, Rodríguez-Calcerrada et al., (2012) tested if carbohydrate accumulation in leaf cells of *Q. pubescens* were promoted and used for isoprene biosynthesis under water stress. Although authors reported an increase in soluble sugars under water stress, a decrease in isoprene emission was observed. They deduced that water stress reduced the availability of phosphoenolpyruvate. In our study, *MS* could have favoured phosphoenolpyruvate transportation from cytosol to chloroplasts where isoprene is synthesized [41].

The marked uncoupling between isoprene emissions and *Pn* under *SS* also reflects that isoprene emissions are not well explained by light and temperature under such degree of stress as also shown for other isoprene emitters (*Quercus serrata* Thunb., *Quercus crispula* Blume. [19]). Hence, under *SS* there was a significant difference between the calculated isoprene emission factor using G93, which relies on temperature and light conditions, and the experimental isoprene emission factor. Our results indicate that the traditional G93 algorithm for isoprene standardization, overestimates isoprene emissions by 44% under *SS* conditions.

To conclude, the results obtained are partly in agreement with the GDBH as *MS* promoted isoprene emissions as expected but, it was clearly detrimental for *Pn* and only slightly reduced growth (Fig. S2). The opposite underlying mechanism is however stated by the theory, that is, a clear growth reduction and a slight *Pn* limitation, generating an important pool of carbon which remains available for isoprene emission (Fig. S2). Our study indicates that isoprene emission increase under *MS* is rather due to the rise of the CO_2 fraction that is re-emitted as isoprene. Under *SS*, isoprene emissions were as low as under *C* conditions and, as stated by the GDBH, this pattern was associated to an important growth reduction and a dramatic decline of *Pn*.

It is likely that increases of isoprene emissions (as well as the isoprene emission factor) under *MS* (up to 27% of the assimilated CO_2 is reemitted as isoprene) confers a high competitive ability to this species under a climatic scenario moderately dry. Since isoprene emitting biomass (i.e. foliar biomass) was not significantly reduced when *MS* was applied to *Q. pubescens,* it would be interesting to model the consequences on O_3 and SOA formation, especially in the Mediterranean area, where this oak accounts for the main isoprene- emitting species. By contrast, the marked overall growth reduction observed under *SS*, suggests that investment in isoprene emissions under a very dry scenario would imply a poor fitness of *Q. pubescens*. We could hence expect a lower direct implication of this oak in atmospheric pollution processes compared to a moderately dry scenario, if conditions (e.g. temperature and light) driving the photochemical reactions necessary for SOA and O_3 formation were similar under mild and severe drought.

We suggest that future investigations explore the competitive advantages of plant investment in isoprene emissions under natural conditions as their adaptive mechanisms to drought could vary compared to young potted oaks, analyse the response of different phenolic groups to drought, and model the formation of biogenic SOA and O_3 from isoprene under different drought conditions.

Supporting Information

Figure S1 Isoprene emission factor. Isoprene emission factor over the growing season for control (–◆–), mild (--◆--) and severe (···◇···) water stress. Differences are tested using a two-way ANOVA repeated measurements (F) followed by Tukey tests. Since interaction between seasonality and treatment is not significant according to the two-way ANOVA, results of water stress impact are shown in the small graph where data of all months are pooled together and differences between treatments are denoted by capital letters (A>B) while seasonality impact is shown in the main graph (in lower case letters: a>b>c). Values are mean ± SE of n = 5.

Figure S2 Balance between growth and carbon-based secondary metabolism. Response of carbon-based secondary constitutive metabolites, growth and net photosynthesis to resource availability as stated by the Growth Differentiation Balance Hypothesis (graph A) and as observed for isoprene emissions in this research study (graph B).

Acknowledgments

We thank the staff of the administrative plant nursery of Les Milles (Departmental Directorate of Agriculture and Forestry of BDR, France), especially P. Brahic for his useful comments and all the staff for technical assistance. We are also grateful for chemical analysis support provided by S. Greff and C. Lecareux as well as support for physiological and growth measurements provided by S. Dupouyet. Y. Monnier, L. Hardion, M. Santonja and A. Cellier also participated to field work.

Author Contributions

Conceived and designed the experiments: ACGZ EO CB CF. Performed the experiments: ACGZ EO. Analyzed the data: ACGZ EO CF CB. Contributed reagents/materials/analysis tools: ACGZ EO. Contributed to the writing of the manuscript: ACGZ EO CB CF.

References

1. Boberg F, Christensen J (2012) Overestimation of Mediterranean summer temperature projections due to model deficiencies. Nature Climate Change 2: 433–436.
2. Giorgi F, Lionello P (2008) Climate change projections for the Mediterranean region. Global and Planetary Change 63: 90–104.
3. Sharkey TD, Wiberley AE, Donohue AR (2008) Isoprene emission from plants: Why and how. Annals of Botany 101: 5–18.
4. Siwko ME, Marrink SJ, de Vries AH, Kozubek A, Uiterkamp A, et al. (2007) Does isoprene protect plant membranes from thermal shock? A molecular dynamics study. Biochimica et Biophysica Acta-Biomembranes 1768: 198–206.
5. Hernandez I, Alegre L, Munne-Bosch S (2004) Drought-induced changes in flavonoids and other low molecular weight antioxidants in *Cistus clusii* grown under Mediterranean field conditions. Tree Physiology 24: 1303–1311.
6. Vickers CE, Gershenzon J, Lerdau MT, Loreto F (2009) A unified mechanism of action for volatile isoprenoids in plant abiotic stress. Nature Chemical Biology 5: 283–291.
7. Loreto F, Velikova V (2001) Isoprene produced by leaves protects the photosynthetic apparatus against ozone damage, quenches ozone products, and reduces lipid peroxidation of cellular membranes. Plant Physiology 127: 1781–1787.
8. Hollosy F (2002) Effects of ultraviolet radiation on plant cells. Micron 33: 179–197.
9. Velikova V, Sharkey TD, Loreto F (2012) Stabilization of thylakoid membranes in isoprene-emitting plants reduces formation of reactive oxygen species. Plant signaling and behaviour 7: 139–141.
10. Guenther A, Hewitt CN, Erickson D, Fall R, Geron C, et al. (1995) A global-model of natural volatile organic-compound emissions. Journal of Geophysical Research-Atmospheres 100: 8873–8892.
11. Claeys M, Graham B, Vas G, Wang W, Vermeylen R, et al. (2004) Formation of secondary organic aerosols through photooxidation of isoprene. Science 303: 1173–1176.
12. Atkinson R (2000) Atmospheric chemistry of VOCs and NOx. Atmospheric Environment 34: 2063–2101.
13. Guenther AB, Zimmerman PR, Harley PC, Monson RK, Fall R (1993) Isoprene and Monoterpene Emission Rate Variability - Model Evaluations and Sensitivity Analyses. Journal of Geophysical Research-Atmospheres 98: 12609–12617.
14. Owen SM, Harley P, Guenther A, Hewitt CN (2002) Light dependency of VOC emissions from selected Mediterranean plant species. Atmospheric Environment 36: 3147–3159.
15. Niinemets U (2010) Mild versus severe stress and BVOCs: thresholds, priming and consequences. TRENDS in Plant Science 15: 145–153.
16. Kesselmeier J, Staudt M (1999) Biogenic Volatile Organic Compounds (VOC): an overview on emission, physiology and ecology. Journal of Atmospheric Chemistry 33: 23–88.
17. Peñuelas J, Staudt M (2010) BVOCs and global change. Trends in Plant Science 15: 133–144.
18. Bruggemann N, Schnitzler JP (2002) Comparison of isoprene emission, intercellular isoprene concentration and photosynthetic performance in water-limited oak (*Quercus pubescens* Willd. and *Quercus robur* L.) Saplings. Plant Biology 4: 456–463.
19. Tani A, Tozaki D, Okumura M, Nozoe S, Hirano T (2011) Effect of drought stress on isoprene emission from two major Quercus species native to East Asia. Atmospheric Environment 45: 6261–6266.
20. Rodríguez-Calcerrada J, Buatois B, Chiche E, Shahin O, Staudt M (2012) Leaf isoprene emission declines in *Quercus pubescens* seedlings experiencing drought – Any implication of soluble sugars and mitochondrial respiration? Journal of Experimental Botany 85: 36–42.
21. Pegoraro E, Rey A, Greenberg J, Harley P, Grace J, et al. (2004) Effect of drought on isoprene emission rates from leaves of *Quercus virginiana* Mill. Atmospheric Environment 38: 6149–6156.
22. Sharkey TD, Loreto F (1993) Water stress, temperature, and light effects on the capacity for isoprene emission and photosynthesis of kudzu leaves. Oecologia 95: 328–333.
23. Funk JL, Mak JE, Lerdau MT (2004) Stress-induced changes in carbon sources for isoprene production in Populus deltoides. Plant Cell and Environment 27: 747–755.
24. Tingey D, Evans R, Gumpertz M (1981) Effects of environmental conditions on isoprene emission from live oak. Planta 152: 565–570.
25. Pegoraro E, Rey A, Abrell L, Vanharen J, Lin G (2006) Drought effect on isoprene production and consumption in Biosphere 2 tropical rainforest. Global Change Biology 12: 456–469.
26. Contran N, Guenthardt-Goerg MS, Kuster TM, Cerana R, Crosti P, et al. (2013) Physiological and biochemical responses of Quercus pubescens to air warming and drought on acidic and calcareous soils. Plant Biology 15: 157–168.
27. Keenan T, Niinemets U, Sabate S, Gracia C, Peñuelas J (2009) Process based inventory of isoprenoid emissions from European forests: model comparisons, current knowledge and uncertainties. Atmospheric Chemistry and Physics 9: 4053–4076.
28. Herms DA, Mattson WJ (1992) The Dilemma of Plants - to Grow or Defend. Quarterly Review of Biology 67: 283–335.
29. Fang C, Monson RK, Cowling EB (1996) Isoprene emission, photosynthesis, and growth in sweetgum (*Liquidambar styraciflua*) seedlings exposed to short- and long-term drying cycles. Tree Physiology 16: 441–446.
30. Bussotti F, Borghini F, Celesti C, Leonzio C, Bruschi P (2000) Leaf morphology and macronutrients in broadleaved trees in central Italy. Trees 14: 361–368.
31. Ormeño E, Romain O, Mevy J-P, Baldy V, Fernandez C (2009) Compost may affect plant volatiles through changes in nitrogen supply and chlorophyll fluorescence. Chemosphere 77: 94–104.
32. Singleton V, Rossi J (1965) Colorimetry of total phenolics with phosphomolybdic-phosphotungstic acid reagents. American Journal of Enology and Viticulture 16: 144–158.
33. Nardini A, Pitt F (1999) Drought resistance of *Quercus pubescens* as a function of root hydraulic conductance, xylem embolism and hydraulic architecture. New Phytologist 143: 485–493.
34. Goulden M (1996) Carbon assimilation and water-use efficiency by neighboring Mediterranean-climate oaks that differ in water access. Tree Physiology 16: 417–424.
35. Acherar M, Rambal S (1992) Comparative water relations of four Mediterranean oak species. Vegetatio 100: 177–184.
36. Zweifel R, Rigling A, Dobbertin M (2009) Species-specific stomatal response of trees to drought - a link to vegetation dynamics? Journal of Vegetetaion Science 20: 442–454.
37. Ogaya R, Peñuelas J, Martínez-Vilalta J, Mangirón M (2003) Effect of drought on diameter increment of *Quercus ilex*, *Phillyrea latifolia*, and *Arbutus unedo* in a holm oak forest of NE Spain. Forest Ecology and Management 180: 175–184.
38. Estiarte M, Filella I, Serra J, Peñuelas J (1994) Effects of nutrient and water-stress on leaf phenolic content of peppers and susceptibility to generalist herbivore *Helicoverpa-armigera* (Hübner). Oecologia 99: 387–391.
39. Brilli F, Barta C, Fortunati A, Lerdau M, Loreto F, et al. (2007) Response of isoprene emission and carbon metabolism to drought in white poplar (*Populus alba*) saplings. New Phytologist 175: 244–254.
40. Affek H, Yakir D (2003) Natural abundance carbon isotope composition of isoprene reflects incomplete coupling between isoprene synthesis and photosynthetic carbon flow. Plant Physiology 131: 1727–1736.
41. Rosenstiel T, Ebbets A, Khatri W, Fall R, Monson R (2004) Induction of poplar leaf nitrate reductase: A test of extrachloroplastic control of isoprene emission rate. Plant Biology 6: 12–21.

16

Interact to Survive: *Phyllobacterium brassicacearum* Improves Arabidopsis Tolerance to Severe Water Deficit and Growth Recovery

Justine Bresson[1,2], François Vasseur[1,3], Myriam Dauzat[1], Marc Labadie[1,2], Fabrice Varoquaux[2], Bruno Touraine[2], Denis Vile[1]*

1 Laboratoire d'Ecophysiologie des Plantes sous Stress Environnementaux (LEPSE), UMR759, Institut National de la Recherche Agronomique-SupAgro, Montpellier, France, **2** Laboratoire des Symbioses Tropicales et Méditerranéennes (LSTM), UMR113, Université Montpellier 2-IRD-CIRAD-INRA-SupAgro, Montpellier, France, **3** Max Planck Institute for Developmental Biology, Tübingen, Germany

Abstract

Mutualistic bacteria can alter plant phenotypes and confer new abilities to plants. Some plant growth-promoting rhizobacteria (PGPR) are known to improve both plant growth and tolerance to multiple stresses, including drought, but reports on their effects on plant survival under severe water deficits are scarce. We investigated the effect of *Phyllobacterium brassicacearum* STM196 strain, a PGPR isolated from the rhizosphere of oilseed rape, on survival, growth and physiological responses of *Arabidopsis thaliana* to severe water deficits combining destructive and non-destructive high-throughput phenotyping. Soil inoculation with STM196 greatly increased the survival rate of *A. thaliana* under several scenarios of severe water deficit. Photosystem II efficiency, assessed at the whole-plant level by high-throughput fluorescence imaging (F_v/F_m), was related to the probability of survival and revealed that STM196 delayed plant mortality. Inoculated surviving plants tolerated more damages to the photosynthetic tissues through a delayed dehydration and a better tolerance to low water status. Importantly, STM196 allowed a better recovery of plant growth after rewatering and stressed plants reached a similar biomass at flowering than non-stressed plants. Our results highlight the importance of plant-bacteria interactions in plant responses to severe drought and provide a new avenue of investigations to improve drought tolerance in agriculture.

Editor: Jin-Song Zhang, Institute of Genetics and Developmental Biology, Chinese Academy of Sciences, China

Funding: JB was funded by French Ministry of Higher Education and Research. The funders had no role in study design, data collection and analysis, decision to publish, or preparation of the manuscript.

Competing Interests: The authors have declared that no competing interests exist.

* Email: denis.vile@supagro.inra.fr

Introduction

Drought is a global concern and episodes of severe drought will most probably be more frequent with dramatic consequences on agriculture [1]. Severe water stress greatly reduces plant biomass production and can lead to plant mortality [2]. Over the last decade, it has been shown that plants can largely benefit from their interactions with soil microorganisms; especially with plant growth promoting rhizobacteria (PGPR) that colonize the rhizosphere of many plants species [3]. The stimulation of growth by PGPR is often associated with lower plant susceptibility to various biotic and abiotic stresses [4,5] and there is a growing interest in the use of these rhizobacteria in agriculture [6,7].

Survival to drought events is found in plants that are able to maintain key cellular functions under severe water stress and recover similar pre-stress values when conditions become favorable again [8]. The capacity to tolerate low leaf water status, or dehydration tolerance, is widely variable among species [9]. The most spectacular adaptation to severe drought is illustrated by resurrection plants [10]. These plants display rapid physiological responses and metabolic adjustments [11], and tolerate nearly complete tissue dehydration. During mild drought or water stress of limited duration, plants that maintain a good water status can complete their life cycle, although often with reduced performance. However, when stress becomes more drastic or is prolonged the leaf water potential drops and leaf damages occur [12]. Then, dramatic reduction of biomass production and even plant mortality appear [2]. To prevent tissue damages, and survive at low leaf water content, many processes and signaling pathways are involved [13]. Osmotic adjustments and accumulation of specific protective osmolytes such as proline [14], glycine betaine [15] or trehalose [16] allow stabilizing cellular structures. One of the most rapid responses to prevent hydraulic failure is stomatal closure. However under severe water stress, stomatal closure can diminish photosynthetic uptake and induce carbon starvation [17] that can lead to total or partial leaf senescence. Drought-induced senescence of older leaves can contribute to water saving, while allowing the reallocation of nutrient to the younger leaves [18]. However, leaf senescence alters photosynthetic functioning and chlorophyll (Chl) properties [19]. Chl-fluorescence is a powerful, rapid and minimally invasive indicator of plant health [20]. In particular, dark-adapted measurements of the ratio of variable to maximal fluorescence (F_v/F_m) give the potential quantum yield (or efficiency) of the photosystem II (PSII) photochemistry, which varies with plant water status [21,22]. Decrease in F_v/F_m is due to an increase in leaf damages that may to some extent be reversible [22]. After a period of water stress, it has been shown that plants have the capacity to recover progressively, but sometimes incompletely, their photosynthetic [23,24] and growth [25] potential. During stress, plant growth rate is reduced, even stopped, but leaf cells retain their ability to expand when conditions become favorable again [25].

Rhizobacteria can help plants to cope with negative effects of water deficit. Under water stresses of moderate intensity, some PGPR can improve resistance to water deficit through i) modifications in phytohormones content and/or signaling, notably ethylene, auxin, cytokinin, and abscisic acid (*e.g.*, [26,27,28,29]), ii) enhanced cells detoxification by increasing antioxydase activities such as catalase [30] or superoxide dismutase [31], iii) changes in plant functional traits such as photosynthetic capacity through changes in chlorophyll content [31] and in photosynthetic PSII efficiency [32,33], or iv) the formation of a biofilm which enhances soil aggregation and improves water stability in the soil [34]. Even though a rich literature exists on plant responses to rhizobacteria under water stress (for reviews see [4,35]), studies of PGPR effects on plant survival are surprisingly limited.

The aim of this study was to investigate the effects of the free-living PGPR, *Phyllobacterium brassicacearum* strain STM196, on survival, growth and physiological responses of *A. thaliana* during the time-course of severe drought progression. The STM196 strain belongs to the *Phyllobacteriaceae* family in the *Rhizobiales*, order of *α-Proteobacteria* [36]. This strain was the most efficient PGPR isolated from the rhizoplan of field-grown *Brassica napus* roots [37,38]. We have recently shown that STM196 improves *A. thaliana* resistance to moderate water deficit through a reproductive delay and changes in transpiration rate correlated to modifications of leaf ABA content [29]. Moreover, previous *in vitro* studies showed that STM196 modifies root architecture and hormonal signaling [39,40,41,42]. Here, our main experimental goals were (*i*) to determine whether plant-PGPR interaction mitigate the negative consequences of severe drought on plant survival, (*ii*) to assess how biotic interactions with PGPR influence physiological mechanisms of plants (*iii*) to evaluate the benefits of inoculation on growth and productivity of plants after stress. *A. thaliana* plants were subjected to five scenarios of severe soil water deficit, with progressive soil drying and rewatering treatments. The use of the plant phenotyping platform PHENOPSIS allowed fine-tuning of soil water content and daily acquisition of images of plants [43]. The dynamics of physiological changes in plants were investigated independently in surviving and perishing plants under severe drought by estimating survival with non-invasive chlorophyll fluorescence measurements at high throughput levels. This approach is broadly applicable to investigate survival of plants under various stresses affecting chlorophyll properties and leaf functioning.

Materials and Methods

Bacteria material, bacterial inoculum and soil inoculation

The *Phyllobacterium brassicacearum* STM196 strain was grown for three days in Petri dishes on a sterile (20 min at 120°C) 1.5% agar (w/v; Sigma-Aldrich) medium (E′) containing 2.87 mM K_2HPO_4, 0.81 mM $MgSO_4$, 1.71 mM NaCl, 7.91 mM KNO_3, 0.34 mM $CaCl_2$, 30 μM $FeCl_3$, 1% mannitol (w/v) and 0.3% yeast extract (w/v; Sigma-Aldrich), adjusted to pH 6.8. Next, the bacteria were grown aerobically in liquid E′ medium on a rotary shaker (145 rpm) at 25°C for 24 h to reach the exponential phase of growth. Culture of bacteria cells was pelleted by centrifugation (3200 g, 15 min, 20°C) and resuspended in deionized water. To obtain 3.10^7 colony forming units (cfu) per gram of soil, the volume was adjusted based upon a correspondence with the absorbance measured at 595 nm (WPA UV 1101, Biotech Photometer, Cambridge, UK). This inoculum was directly put into the non-sterilized soil substrate (see Table S1 in File S1 for soil chemical properties), which was then manually homogenized.

Plant material, growth conditions and irrigation treatments

All experiments were realized with *A. thaliana* (L.) Heynh accession Col-0. Five seeds were sown at the soil surface in 260 mL culture pots filled with a damped mixture (1:1, v:v) of loamy soil and organic compost (Neuhaus N2; see Table S1 in File S1 for soil chemical properties) inoculated with STM196 or not. Non-inoculated soil was previously damped with deionized water to avoid difference in initial soil humidity with inoculated soil. Soil water content was controlled during pot filling by determining soil fresh weight (FW_{soil}) and soil dry weight (DW_{soil}, after 5 d at 60°C) every ten pots. Initial soil relative water content was determined as $RWC_{soil} = (FW_{soil} - DW_{soil}) \times 100 \times DW_{soil}^{-1}$. The pots were kept in the dark for two days in the PHENOPSIS growth chamber [43] and were damped with sprayed deionized water three times a day until germination. Then, plants were cultivated under 12 h day length (180 μmol m^{-2} s^{-1} photosynthetic photon flux density, PPFD, at plant height). During germination phase (7 d), air temperature was set to 20°C day and night, and air relative humidity was adjusted in order to maintain constant water vapor pressure deficit (VPD) at 0.6 kPa. Then, plants were grown at 20/17°C day/night and 0.8 kPa of VPD. Seedlings with similar sizes and developmental stages were selected and thinned to one to four plants per pot just before the beginning of water stress (see Figure 1A, C and Table 1, for watering scenarios and details on replicate numbers). Soil water content was daily adjusted with a modified one-tenth-strength Hoagland solution [44]. Soil water content was maintained at 0.35 g H_2O g^{-1} dry soil in the well-watered treatment (35%, WW) and it was decreased progressively to the desired RWC_{soil} by stopping irrigation in the water deficit treatments (WD; Table 1). Continuous moderate water deficit ($20\%_c$) was maintained at 0.20 g H_2O g^{-1} dry soil during the whole plant life cycle. In the case of severe punctual stresses, when the soil reached the desired RWC_{soil} level depending on the experiment (*i.e.*, 0.10, 0.07 or 0.06 g H_2O g^{-1} dry soil), irrigation was resumed after 1 day (for $10\%_p$, $7\%_p$ and $6\%_p$ stresses) or after 10 days (for $10\%_{p\text{-}10d}$) to progressively reach the WW soil condition (avoiding no more than 10 ml of the modified Hoagland solution per day to avoid soil leaching). Soil water content was then maintained at WW until final harvests at first flower open (stage 6.00; [45]).

Soil water potential was determined during soil drying (from 0.35 to 0.06 g H_2O g^{-1} dry soil, Table 1 and Figure S1 in File S1; WP4-T dewpoint meter, Decagon Devices, Pullman, WA 99163, USA).

Plant survival

Plants that failed to develop after rewatering and deteriorated until the complete senescence of the rosette were considered as dead. Survival percentage was scored in three consecutive experiments that were carried out following the same experimental procedure (Table 1). In experiments 1 and 2, water stresses ($20\%_c$, $10\%_{p\text{-}10d}$, $10\%_p$ and $7\%_p$) were started at the emergence of the

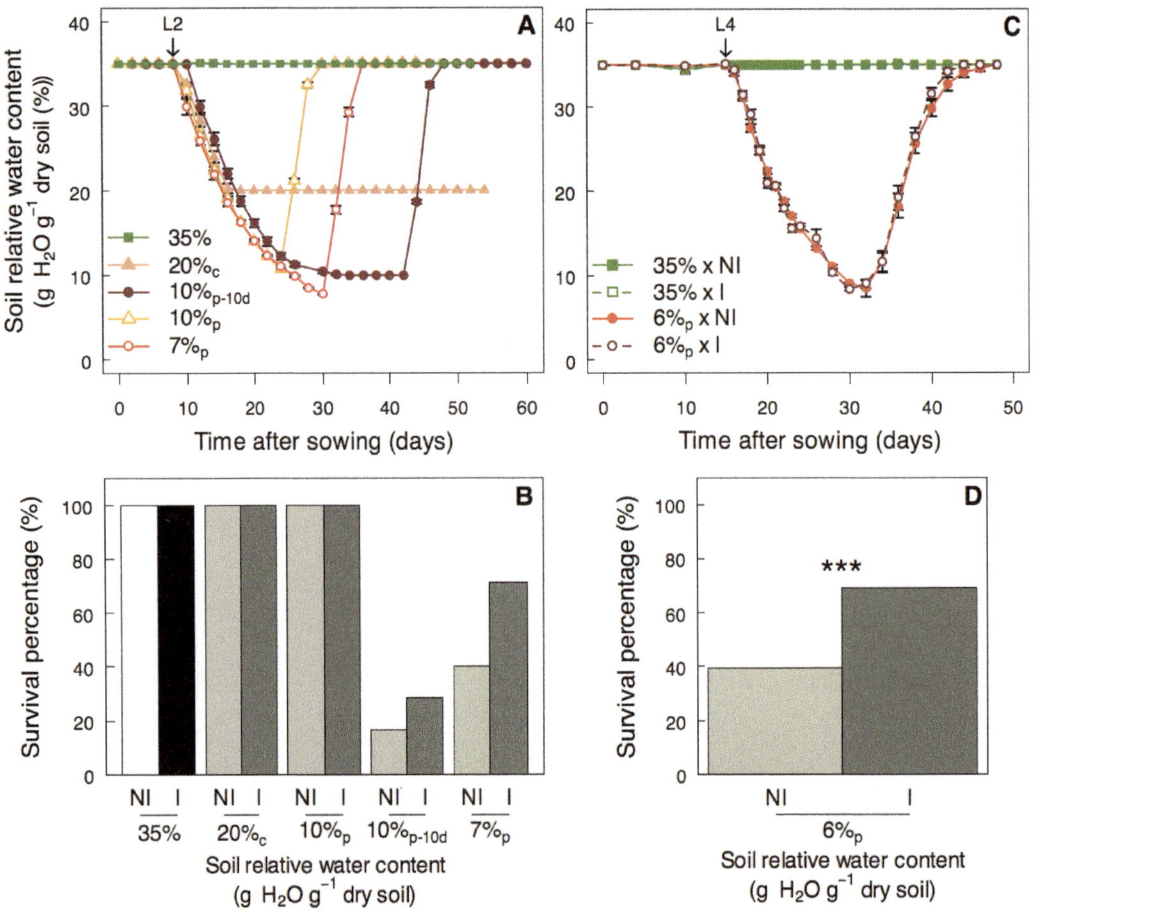

Figure 1. ***Phyllobacterium brassicacearum*** **STM196 enhances** ***A. thaliana*** **survival under severe water deficits. A)** Average soil relative water content and **B)** survival percentage of non-inoculated (NI) and inoculated plants (I) in five watering scenarios including constant well-watered conditions (35% g H_2O^{-1} dry soil; 35%), water withdrawing from the two firsts leaves (L2) followed by constant moderate water deficit (20% g H_2O^{-1} dry soil; 20%$_c$), punctual severe water deficits with rewatering after 10 days at 10% g H_2O^{-1} dry soil (10% g H_2O^{-1} dry soil; 10%$_{p\text{-}10d}$) or after1 day (10%$_p$), and after 1 day at 7% g H_2O^{-1} dry soil (7%$_p$). **C)** Average soil relative water content and **D)** survival percentage of NI (closed symbols) and I (open symbols) plants in two watering scenarios including constant well-watered conditions (35% g H_2O^{-1} dry soil; 35%), and water withdrawing from the four-leaves stage (L4) followed by rewatering after 1 day at 6% g H_2O^{-1} dry soil (6%$_p$). Asterisks indicate significant differences following Chi^2 test between NI and I plants (***: $P<0.001$).

first two leaves (L2; stage 1.02; [45]). In experiment 3, the number of replicates was increased in order to maximize the statistical power, water stress was started at four-leaf stage (L4; stage 1.04; [45]) to allow precise measurements on early developmental stages and RWC_{soil} was decreased to reach 0.06 g H_2O g^{-1} dry soil (6%$_p$ stress) before rewatering to reach well-watered soil condition (RWC_{soil} = 35%).

Measurements of whole-plant traits

Detailed plant phenotyping of growth and physiological status was performed in experiment 3 (6%$_p$ stress) throughout the whole plant cycle.

Measurement of photosynthetic efficiency. Measurement of photosynthetic efficiency was daily performed from early developmental stages to the emergence of the flowering stem (*i.e.*, bolting stage) under WW and water deficit. The maximum quantum yield of PSII was estimated by the ratio of variable to maximal chlorophyll fluorescence (F_v/F_m) on dark-adapted plants, after 8–12 h of dark (IMAGING-PAM; Maxi-version; W-IMAG-K6 camera implemented in PHENOPSIS; Imaging Win software; Walz; Effeltrich, Germany). F_v/F_m is given by $(F_m\text{-}F_0)/F_m$ [46], where F_0 is the basal fluorescence in the dark adapted state and F_m is the maximal fluorescence obtained after saturating light pulse (Si 9, width 800 ms). For unstressed plants, the value of F_v/F_m around 0.83 measured for most plant species and values lower than this indicate that plants are stressed [46]. Whole-rosette F_v/F_m values were extracted by image analyses using ImageJ (ImageJ 1.47V, Rasband, Bethesda, Maryland, USA).

Determination of plant water status. To determinate plant relative water content (RWC_{leaf}), plants were harvested at different levels of RWC_{soil} during establishment of water stress (*i.e.*, at 0.35, 0.20, 0.10 and 0.06 g H_2O g^{-1} dry soil) and after rewatering (*i.e.*, at 0.20r, 0.10r and 0.35r g H_2O g^{-1} dry soil and at first flower open). Rosettes were cut and immediately weighted, after the removal of inflorescence stems for plants harvested at stage 6.00 [45], to determine aboveground vegetative fresh mass (FM). The rosettes were wrapped in moist paper and placed into Petri dishes at 4°C in darkness overnight to achieve complete rehydration. Water-saturated fresh mass (SM) was then determined. The rosettes were oven-dried at 65°C for 48 h, and rosette dry mass (DM) was determined. From these measurements, relative water content (RWC_{leaf} = (FM − DM)×100×(SM −

Table 1. Description of watering scenarios and design of experiments.

			Water stress characteristics				Experiment		
Treatment	**Watering scenario**	**Soil water potential (Mpa)**	**Stage of irrigation withdrawing**	**Days to reach desired RWC_{soil}**	**Days before rewatering**	**Number of plants**	**1**	**2**	**3**
$35\%_c$; WW	RWC_{soil} maintained at 35% by daily irrigation during the whole plant life cycle	0.07	No stress			13–16	x	x	x
$20\%_c$	Irrigation stopped at stage L2 to decrease RWC_{soil} at 20% and RWC_{soil} thereafter maintained constant during the whole plant life cycle	−0.28	L2	6	No rewatering	9–20		x	
$10\%_{c\text{-}10d}$	Irrigation stopped at stage L2 to decrease RWC_{soil} at 10%. RWC_{soil} then maintained constant during 10 days before rewatering to reach well-watered level ($RWC_{soil}=35\%$)	−3.19	L2	15	10	7–12	x		
$10\%_p$	Irrigation stopped at stage L2 to decrease RWC_{soil} at 10%. RWC_{soil} then maintained constant during 1 day before rewatering to reach well-watered level	−3.19	L2	14	1	7–12	x		
$7\%_p$	Irrigation stopped at stage L2 to decrease RWC_{soil} at 7%. RWC_{soil} then maintained constant during 1 day before rewatering to reach well-watered level	−9.52	L2	20	1	7–10	x	x	
$6\%_p$	Irrigation stopped at stage L4 to decrease RWC_{soil} at 6%. RWC_{soil} then maintained constant during 1 day before rewatering to reach well-watered level	–	L4	30	1	48–52			x

WW: well-watered condition (control); RWC_{soil}: soil relative water content (%; g H_2O g^{-1} dry soil); L2: emergence of the first two leaves (stage 1.02; [45]) and L4: emergence of the four-leaf stage (stage 1.04; [45]).
All treatments were performed in soil inoculated or not with *Phyllobacterium brassicacearum* STM196. See Figure 1A, C for a graphical representation of watering scenarios.

$DM)^{-1}$) was calculated at the rosette level. Water content (WC_{leaf}) was calculated as FM × DM^{-1} ratio.

Rosette expansion during time course. Projected area of the rosettes (RA_{proj}) was determined every days from semi-automated analysis (ImageJ 1.43C [47]) of zenithal images of the plants (Sony SSC-DC393P camera). A sigmoid curve was fitted for each plant following $RA_{proj} = a/[1+\exp-[(d-a/2)/b]]$ where a is the maximum area, and d is the number of days after sowing. The maximum rate of leaf expansion (R_{max}, $mm^2\ d^{-1}$) was calculated from the first derivative of this logistic model at d_0 as $R_{max} = a/(4b)$.

Flowering time was determined as the number of days from germination until visualization of the first flower open.

Measurements of leaf morphology at flowering. Surviving individuals were harvested at first flower open. Rosettes were cut and immediately weighted after the removal of inflorescence stems to determine aboveground vegetative FM. SM was then determined as describe above. Total leaf number was determined, and the leaf blades were separated from their petiole in order of leaf emergence and scanned for measurements of individual leaf area (ImageJ 1.43C). Leaf blades, petioles and reproductive structures were then separately oven-dried at 65°C for 48 h, and their dry mass was determined. Rosette DM was calculated as the sum of blades and petioles dry masses and RWC_{leaf} was calculated at the rosette level. All phenotypic data were stored in the PHENOPSIS database [47].

Quantification of bacteria in the soil

To analyze bacterial growth under water stress in soil, a natural mutant of STM196 strain was selected in a selection medium E′ containing 100 µg ml^{-1} of rifampin and then, was transformed using pCH60 vector. The vector pCH60 encodes for tetracycline resistance and contains the *gfp* gene that is constitutively expressed [48]. Bacterial concentration was estimated during soil drying at 0.35, 0.20, 0.10 and 0.06 g $H_2O\ g^{-1}$ dry soil and after rewatering at 0.35r g $H_2O\ g^{-1}$ dry soil. Quantification of bacteria was performed in soil without plant. The concentration of colony-forming units (cfu/mg) was estimated using the most probable number method (MPN; [49]). 100 mg of inoculated soil were put in 1 ml of physiological water (8.5 g l^{-1} de NaCl) on a rotary shaker (145 rpm) at 25°C for 2:30. The solubilized soil samples were serially diluted until 10^{-7}, and 100 µl were spread in Petri dishes on a sterile (20 min at 120°C) 1.5% agar (w/v; Sigma-Aldrich) medium (E′) with addition of 50 µl of rifampin and tetracycline. Bacteria were then counted after 6 days at 25°C.

Statistical analyses and determination of the lethal F_v/F_m threshold

All analyses were performed using R 2.15 [50]. Comparisons of mean trait values between treatments were performed with Kruskal–Wallis non-parametric tests. Plant survival was analyzed by Chi^2 tests. To estimate the survival of harvested plants during water stress (only for 6_p% stress), a 90% lethal threshold was determined just before rewatering (*i.e.*, at 0.06 g $H_2O\ g^{-1}$ dry soil) from plants with known survival, in a dose-response analysis of survival as a function of F_v/F_m values. The relationships between survival probability and whole-rosette F_v/F_m values were modeled using a binomial logistic regression. The effect of inoculation was tested by Chi^2 tests on deviance ratio. The 90%-mortality threshold (*i.e.*, 10% survival probability) of F_v/F_m value was inferred from the regression. Plants with F_v/F_m values above this threshold were considered as able to survive the stress imposed whereas plants with F_v/F_m values below this threshold were considered as perishing plants. Estimated mortality ratios (*i.e.*, proportion of perishing plants) were compared by Chi^2 tests.

Results

Phyllobacterium brassicacearum STM196 strain increases *A. thaliana* survival under multiple scenarios of severe water deficit

Arabidopsis thaliana Col-0 was grown under five scenarios of soil water availability to determine a level of stress that induced plant mortality and then analyze the effects of STM196 strain on plant survival. Soil relative water content was maintained at 0.35 g $H_2O\ g^{-1}$ dry soil in the well-watered (WW) treatment until flowering and it was decreased progressively to the desired RWC_{soil} by stopping irrigation in the water deficit treatments followed by rewatering or not (see Table 1 and Figure 1A, C for a description of the watering treatments). Under WW conditions, all plants survived and reached the reproductive stage (Figure 1A, B). All plants also survived a continuous moderate WD ($20\%_c$; Figure 1A, B), *i.e.* irrigation withdrawn from two first leaves emerged (L2) and RWC_{soil} then maintained at 20% g $H_2O\ g^{-1}$ dry soil until flowering. Decreasing RWC_{soil} punctually to 10% g $H_2O\ g^{-1}$ dry soil ($10\%_p$) did not affect plant survival, but when this RWC_{soil} level was prolonged for 10 days ($10\%_{p\text{-}10d}$) more than 80% of the non-inoculated plants died (Figure 1A, B). Decreasing RWC_{soil} punctually to 7% g $H_2O\ g^{-1}$ dry soil ($7\%_p$) resulted in 40% of non-inoculated plants that survived and reproduced after stepwise rewatering to WW conditions (Figure 1A, B).

To perform accurate measurements of plant development and physiology during soil drying, the beginning of water stress was delayed to four leaves emerged (L4), and RWC_{soil} was punctually decreased to 6% ($6\%_p$; Figure 1C). Under this scenario, plant survival rate of non-inoculated plants was 40%, *i.e.* similar to the rate observed under punctual $7\%_p$ stress (Figure 1B, D). In all watering scenarios causing plant mortality ($10\%_{p\text{-}10d}$, $7\%_p$ and $6\%_p$), soil inoculation by STM196 strain resulted in a great increase in plant survival rate (Figure 1B, D). For instance, 70% of inoculated plants survived against only 40% of non-inoculated plants under $6\%_p$ stress ($P<0.001$). This stress level was reached 1.7 days earlier in inoculated plants than in non-inoculated plants (the mean ± SE number of days to reach 6% RWC_{soil} was 16.8±1.9 ($n = 50$) and 18.5±2.2 ($n = 48$) for inoculated and non-inoculated plants, respectively; $P<0.001$). To decipher the effects of STM196 (only under $6\%_p$ stress) at similar RWC_{soil} levels, the traits of stressed plants were analyzed and presented independently of time but as a function of soil humidity during soil drying and after rewatering. The growth of STM196 in the soil was also analyzed during the WD treatment, without plant. Bacterial growth was not affected by WD and the concentration of bacteria remained constant during the experiment (Figure S2 in File S1).

STM196 strain delays and reduces plant mortality under severe water deficit

Non-destructive measurements of Chl-fluorescence were used as a sensitive indicator of photosynthetic performance (efficiency of PSII) from early developmental stages to the emergence of flowering stem. Under WW conditions, whole-rosette mean F_v/F_m was 0.80 during the entire life cycle and was not affected by soil inoculation with STM196 ($P = 0.57$; see Figure S3 in File S1). As expected, F_v/F_m decreased significantly under severe WD (6_p% stress). Mean F_v/F_m just before rewatering (*i.e.*, $RWC_{soil} = 6\%$ g $H_2O\ g^{-1}$ dry soil) was equal to 0.7 for surviving plants whereas it was equal to 0.3 for the plants that failed to develop and perished after rewatering, for both non-inoculated and inoculated plants

(Figure 2A, C). A 90% lethal threshold was then determined with F_v/F_m values of these latter plants in order to estimate the mortality of harvested plants with unknown survival (Figure 2A, B). There was no difference between the logistic regressions of survival on F_v/F_m performed on non-inoculated and inoculated plants ($P = 0.518$; Figure 2B) and the average fit was therefore used. The 90%-mortality threshold was inferred at $F_v/F_m = 0.398$. In further analyses, plants with F_v/F_m values above this threshold were considered as able to survive the stress imposed and plants with F_v/F_m values below this threshold were considered as perishing. The distinction between surviving and perishing plants was crucial to avoid errors of interpretation of the results due to a higher number of inoculated surviving plants, and could help to differentiate the behavior of plants according to their ability to survive to WD. This threshold showed that estimated mortality rate (*i.e.*, the proportion of perishing plants) tended to increase at 20% g H_2O g^{-1} dry soil in non-inoculated plants and never before 6% g H_2O g^{-1} dry soil in inoculated plants (Figure 2D). From $RWC_{soil} = 6\%$ g H_2O g^{-1} dry soil and after rewatering, the estimated mortality rate of inoculated plants was significantly lower than that of non-inoculated plants (Figure 2D; note that at the end of the experiment most senescing plants were no more detectable because decomposition started, which explains the biased decrease of mortality rate observed).

Delayed dehydration of tissues confers a higher tolerance to photosynthetic damages in STM196-inoculated plants

Whole-rosette Chl-fluorescence was then analyzed independently in surviving and perishing plants inoculated or not with STM196. At the whole-rosette level, the decrease in mean F_v/F_m was not progressive in plants exposed to stress but was dramatically affected beyond 10% RWC_{soil} in both surviving and perishing plants, with a higher magnitude for the latter (Figure 3A, B). At the maximum of stress severity (*i.e.*, 6% RWC_{soil}), lowering of whole-rosette mean F_v/F_m was more pronounced in surviving inoculated plants than in non-inoculated plants ($P<0.05$), and F_v/F_m of inoculated plants was closer to the mortality threshold (see grey points and dashed line in Figure 3A). Upon rewatering, whole-rosette mean F_v/F_m of both non-inoculated and inoculated surviving plants recovered progressively F_v/F_m values to reach initial mean F_v/F_m (0.8), similar to non-stressed plants (Figure 3A and Figure S3 in File S1). Both inoculated and non-inoculated perishing plants reached an equivalent mean F_v/F_m (0.49) at 6% RWC_{soil} (Figure 3B). This result suggests that inoculation by STM196 induced a slight decrease in photosynthetic performance but surviving inoculated plants had higher tolerance to photosynthetic damages under WD.

Severe WD in the soil unequivocally led to reduced water content in plant tissues (Figure 4A). RWC_{leaf} was progressively affected by soil drying and 6% RWC_{soil} resulted in a great decrease causing a RWC_{leaf} as low as 25% in non-inoculated plants compared to 82% in plants grown under WW conditions (Figure 4A). At 10% RWC_{soil}, surviving inoculated plants displayed higher RWC_{leaf} ($P<0.001$) than non-inoculated plants, which suggested that soil inoculation by STM196 slowed the loss of water in the leaves. At 6% RWC_{soil} the effect of inoculation was opposite and the RWC_{leaf} of surviving inoculated plants was lower than that of non-inoculated plants ($P<0.05$). This result showed that soil inoculation by STM196 allowed plants to withstand higher leaf dehydration than non-inoculated plants. RWC_{leaf} and Chl-fluorescence were closely related (Figure 4B). The relationship between F_v/F_m and RWC_{leaf} showed that inoculated plants displayed a lower decline of F_v/F_m for lower values of RWC_{leaf} (*e.g.*, around 20%). Fitting a logistic regression to the relationship between F_v/F_m and WC_{leaf} also showed that the decrease of F_v/F_m in response to WD was delayed in inoculated plants compared to non-inoculated plants and appeared for lower values of WC_{leaf} (see Figure S4 in File S1). Moreover, inoculated plants displayed higher survival probability (estimated from whole-rosette F_v/F_m values; Figure 1B) at very low RWC_{leaf} (*e.g.*, around 20%), and the decline of survival as a function of RWC_{leaf} was delayed in inoculated plants compared to non-inoculated plants (Figure 5). Together these results showed that STM 196 induced a higher plant survival during stress through a good maintenance of photosynthetic efficiency at worst leaf dehydration levels.

STM196 improves growth recovery of surviving plants, and increases biomass production

Establishment of WD (6%$_p$ stress) resulted in reduced leaf growth, and total leaf area declined until rewatering compared to plants under WW conditions (Figure 6A). Upon rewatering, leaf growth of stressed surviving plants resumed and the plants reached the reproductive stage. At flowering, WD induced a decrease by 50% of total leaf area in non-inoculated plants (insert in Fig 6C).

The plant growth promotion effect of STM196 was not effective under WW conditions but strongly occurred under WD (Figure 6). Under WD, soil inoculation by STM196 induced a significant increase in the maximum rate of leaf expansion after rewatering (R_{max}; insert in Figure 6A; $P<0.01$) that led to a larger total leaf area at flowering (insert in Figure 6C; $P<0.01$). This was associated with a significant 45% increase of shoot dry biomass in inoculated plants under WD (Figure S5A in File S1). The increase in total leaf area of inoculated plants under WD was associated with larger individual leaves than non-inoculated plants (Figure 6C). At flowering, inoculated plants displayed also a higher number of leaves (Figure 6C). Flowering time was delayed by 15 d under WD but it was not affected by inoculation (Figure S5B in File S1). At flowering, all surviving plants recovered a complete rehydration of tissues (Figure S5C in File S1). Taken together, all traits demonstrated a better tolerance of inoculated plants to severe WD and an improvement of biomass accumulation upon rewatering.

Discussion

Severe water stress induces dehydration of plant tissues and can cause irreversible cellular damages leading to death [17]. Nonetheless, plants are able to some extent to withstand periods in a dried status and restart their metabolic functions after rehydration (*e.g.*, [25,51,52]). Several genes in Arabidopsis have been shown to be implicated in plant survival to water deficit and transgenic modifications could improve plant survival [53]. In addition, some soil bacteria such as PGPR strains can improve tolerance to water deficit, but reports on their effects on plant survival are scarce [54,55], specifically in response to severe water stress.

We recently showed that the PGPR *Phyllobacterium brassicacearum* strain STM196, previously isolated from the rhizosphere of oilseed rape *Brassica napus* [37,38], improved Arabidopsis resistance to moderate water deficit through delayed developmental transitions and modifications of plant physiology, notably by a decrease of leaf transpiration through an increase of leaf abscisic acid (ABA) content [29]. Here, we show that inoculation by STM196 strain consistently induces a significant increase in survival rate under multiple scenarios of severe water deficit. We highlight that STM196 delayed and reduced mortality rate during water stress establishment through a better tolerance to leaf

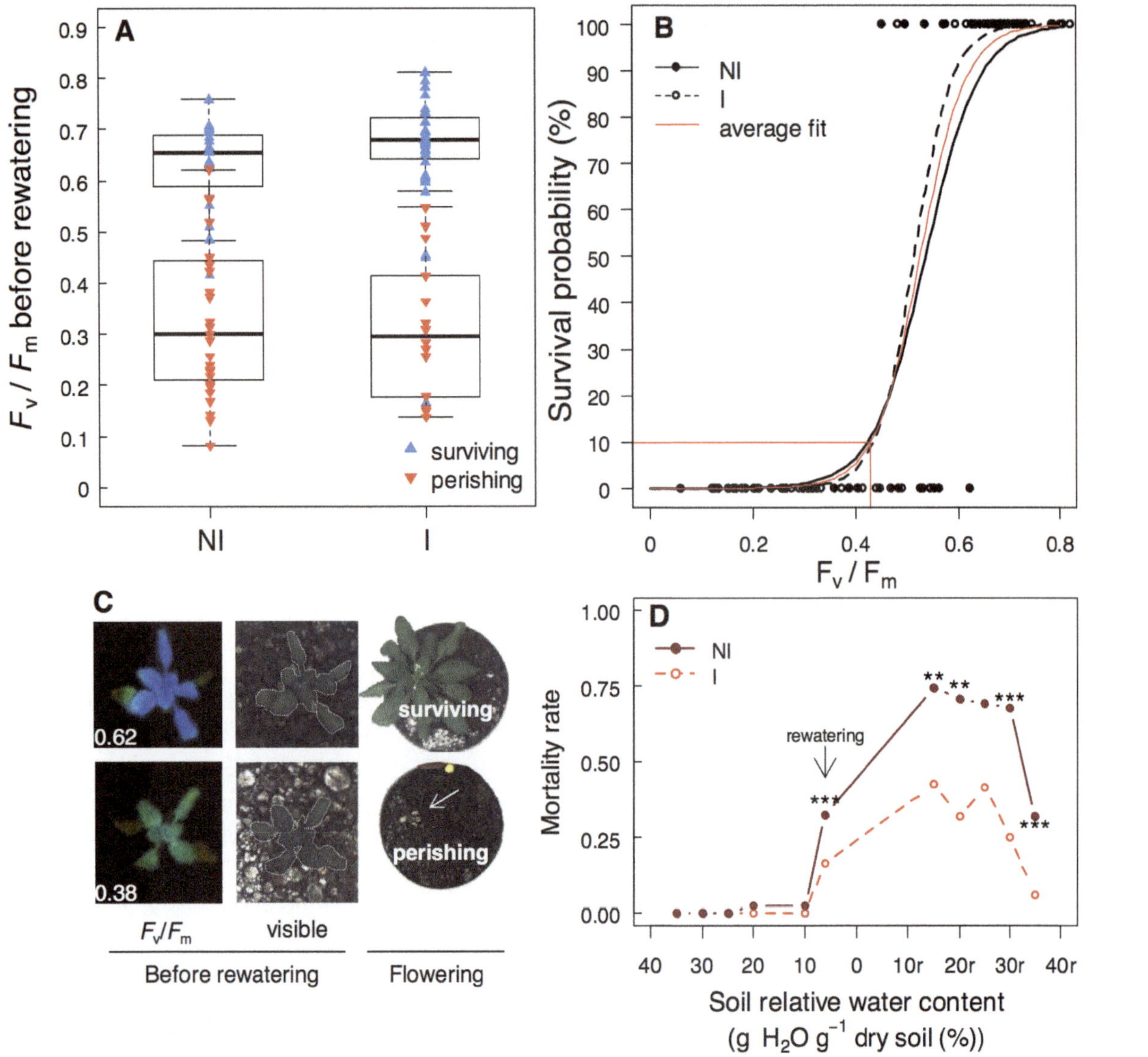

Figure 2. ***P. brassicacearum*** **STM196 does not affect** ***A. thaliana*** **mortality threshold but delays and reduces mortality rate during soil drying. A)** Whole-rosette F_v/F_m just before rewatering (*i.e.*, 0.06 g H_2O^{-1} dry soil) of non-inoculated (NI) and inoculated (I), and surviving (▲; n = 19–36) and perishing (▼; n = 16–29) plants as observed at the end of the experiment conducted with water withdrawing followed by rewatering at 6% g H_2O^{-1} dry soil (6%$_p$ stress in Figure 1). **B)** Relationships between F_v/F_m and survival probability (same data as in A); the 90%-mortality threshold ($F_v/F_m = 0.398$) is shown. **C)** F_v/F_m false-colour images (left) and visible images (middle) of vegetative rosettes before rewatering and of surviving flowering and perishing plant at the end of the experiment (right). **D)** Mortality rate of stressed NI (closed symbols) and I (open symbols) plants during soil drying and rewatering, as estimated from the 90%-mortality threshold. Asterisks indicate significant differences following Chi^2 test between NI (n = 28–242) and I (n = 16–187) plants (**: $P<0.01$; ***: $P<0.001$).

dehydration and leaf photosynthetic damages. Contrary to common findings where rhizobacteria enhance physiological plant status (*e.g.*, leaf water content or photosynthetic performance), here we show that STM196-inoculated plants can survive under stress with more leaf damages. Importantly, STM196 not only increased plant survival but also increased growth recovery in surviving plants and led to a higher biomass production at flowering.

Inoculation by STM196 allows a better tolerance to leaf damages and conservation of leaf water content during stress, and a better growth recovery after rewatering

Although some studies detailed the mechanisms underlying the improvement of plant resistance to water stress by PGPR inoculation, a very few studies have showed that PGPR could improve plant survival under drastic conditions. It has been shown that some rhizobacteria, genetically modified to overproduce trehalose in their cells, can improve survival of plants under severe

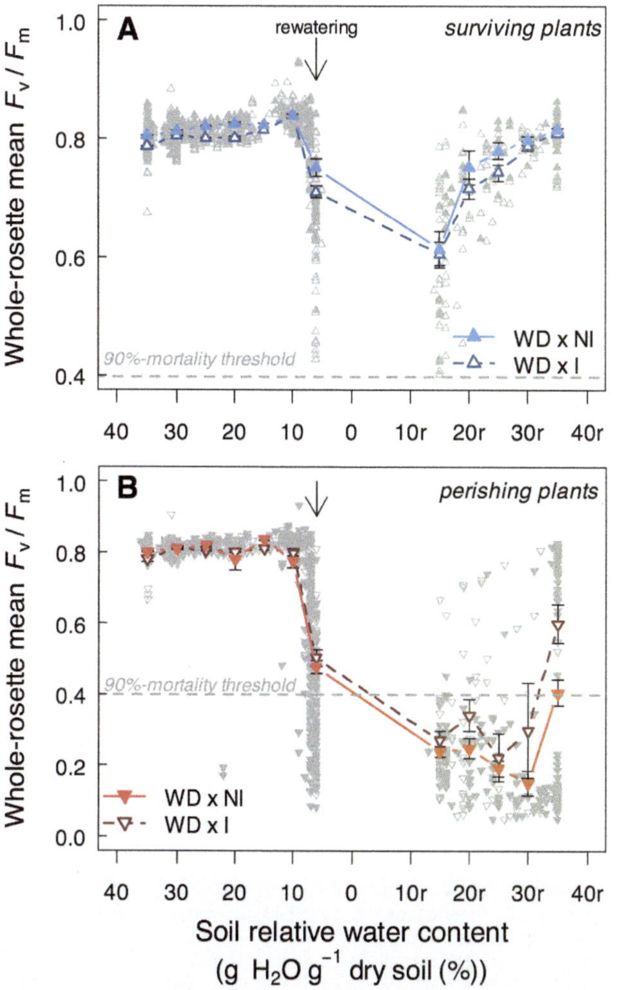

Figure 3. *P. brassicacearum* STM196 improves *A. thaliana* tolerance to higher levels of leaf photosynthetic damages under severe water deficit (WD; 6%$_p$). Whole-rosette F_v/F_m of **A)** surviving and **B)** perishing non-inoculated (NI; $n_{surviving} = 7–147$; $n_{perishing} = 6–137$) and inoculated (I; $n_{surviving} = 10–152$; $n_{perishing} = 6–78$) plants during soil drying and during rewatering. Dashed lines show the 90%-mortality threshold and arrows indicate the beginning of soil rewatering. Grey points represent individuals for each condition (NI; closed symbols and I; open symbols).

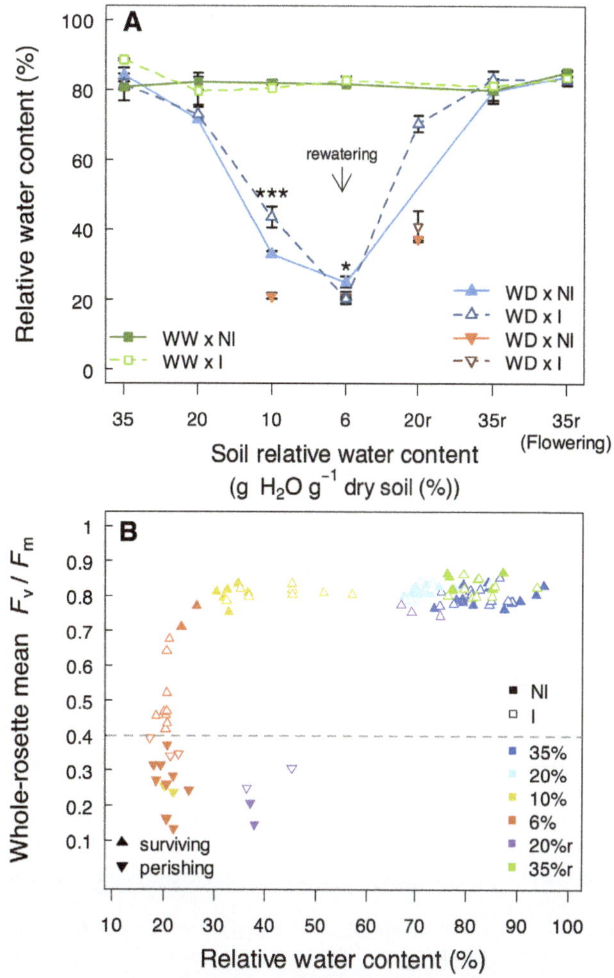

Figure 4. *P. brassicacearum* STM196 induces a delayed dehydration of tissues and increases tolerance to severe water deficit. A) Leaf relative water content and **B)** relationship between whole-rosette F_v/F_m and leaf relative water content of non-inoculated (NI; closed symbols) and STM196-inoculated (I; open symbols) plants under well watered (WW) and water deficit (WD; 6%$_p$) during soil drying (35%, 20%, 10% and 6%) and after rewatering (20%r, 35%r and 35%r at flowering). Arrow in A indicates the beginning of soil rewatering. Dashed line in B represents the 90%-mortality threshold. Surviving plants with mean F_v/F_m values above the threshold, are represented by triangles (▲; n = 3–10 and n = 3–19 for NI and I plants, respectively) and perishing plants, with mean F_v/F_m below the threshold, are represented by upside-down triangles (▼; n = 3–9 and n = 3 for NI and I plants, respectively). Asterisks indicate significant differences following Kruskal-Wallis tests between NI and I plants (*: $P<0.05$ and ***: $P<0.001$).

water-limiting conditions, notably by increasing leaf water content or by inducing the accumulation of trehalose content in the plant [54,55]. Here, we used the automated phenotyping platform PHENOPSIS, that allows the precise control of soil watering [43], to analyze the effects of STM196, a natural PGPR, on the physiology and growth of *A. thaliana* under multiple scenarios of severe water deficit throughout the whole plant cycle. The scenarios of water deficit used in this study induced a large decrease in plant survival from 60 to 83%, which is comparable to a previous report using a similar procedure (water stress/rewatering from stage 1.04) and similar intensities of soil drying [53]. Plants inoculated by STM196 strain consistently presented a higher survival rate in comparison with non-inoculated plants.

It is well established that severe water stress strongly affects plant growth, water status and causes decline of photosynthetic capacity [17], specifically through stomatal closure and leaf senescence. Dedicated measurements require a precise knowledge of the dynamics of stress establishment and are often highly time-consuming. For this reason, non-destructive measurements based on chlorophyll fluorescence imaging have been extensively used to decipher the effects of different stresses on plant physiology (*e.g.*, [56,57,58]) but have rarely been used at high throughput (but see [21]). In this paper, we used chlorophyll fluorescence measurements at high throughput in order to unravel the effects of rhizobacteria on the dynamic plant responses to severe water deficit. Amongst the different photosynthetic parameters existing, dark-adapted F_v/F_m, reflects the maximal efficiency of PSII and is therefore one of the most used parameters for measuring leaf physiological status [20]. Most often the mean F_v/F_m of a photosynthetic organ or a whole-plant is used to characterize the

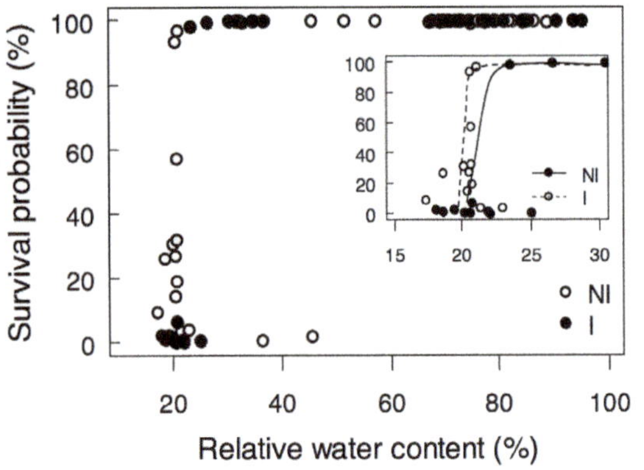

Figure 5. *P. brassicacearum* STM196 increases plant survival to severe leaf dehydration. Relationship between plant survival (estimated from whole-rosette F_v/F_m values; Figure 1B) and leaf relative water content of non-inoculated (NI; closed circle; n = 36) and STM196-inoculated (I; open circle; n = 44) under severe water deficit (6%$_p$). Insert represents fitting of logistic regression at very low leaf relative water content (solid and dashed lines for NI and I plants, respectively).

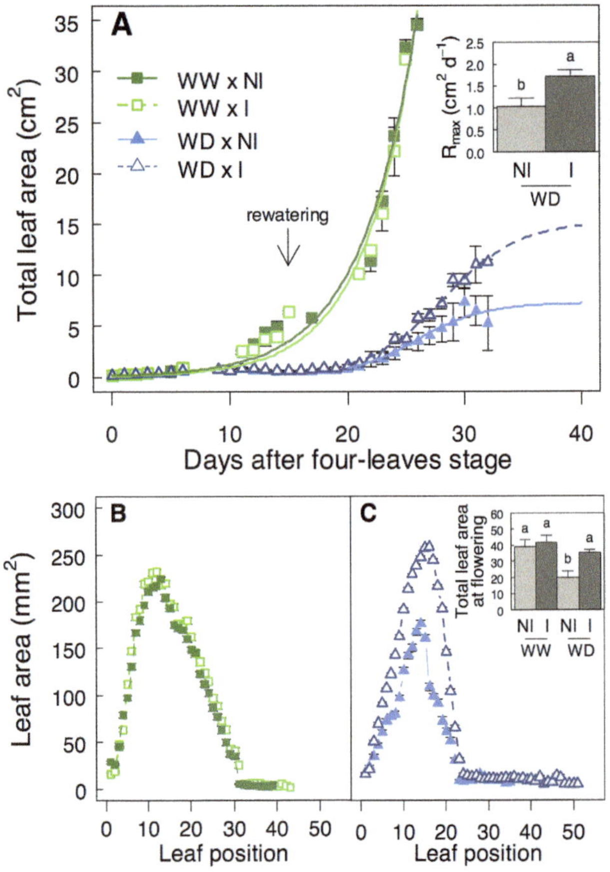

Figure 6. *P. brassicacearum* STM196 increases growth rate of surviving plants after rewatering. A) Total projected leaf area of non-inoculated (NI; closed symbols) and STM196-inoculated (I; open symbols) plants under well-watered condition (WW) and severe water deficit (WD; 6%$_p$ in Figure 1) as a function of days after four-leaves stage. Arrow in A indicate the beginning of soil rewatering. Insert in **A** represents the maximum rate of leaf expansion (R_{max}) after rewatering of surviving stressed plants. Area of individual leaves of I and NI plants under **B)** WW condition and **C)** WD. Insert in **C** shows total leaf area of surviving plants at flowering. Data are means (±SE) of 11-27 plants. Different letters indicate significant differences between means following Kruskal-Wallis tests (P<0.05).

response to a stressor (*e.g.*, [22]). Here, we first showed that the whole-rosette mean F_v/F_m was related to the probability of survival to severe water deficit. The determination of a mortality threshold allowed the estimation of survival of harvested plants and thus, the discrimination between surviving and perishing plants. The mortality threshold also allowed following the variation of plant mortality during time course. Moreover, this method was necessary to decipher the effect of an exogenous treatment that induced differences in sample size. Then, we showed that improvement of plant survival by STM196-inoculation was not related to changes in mortality threshold as determined by whole-rosette F_v/F_m values but was associated to differences in tolerance to WD of surviving plants. During stress, plants are able to some extent to endure leaf photosynthetic damages. Surviving inoculated plants tolerate lower values of whole-rosette mean F_v/F_m just before rewatering. The "Point of no return", the limit point that once passed a plant dies, seemed to appear for lower values of F_v/F_m in inoculated plants. The large decline in mean F_v/F_m during prolonged water deficit is consistently associated with exacerbated leaf senescence [59,60]. STM196-inoculated plants could survive with a higher proportion of leaf senescence and thus, presented a higher tolerance to leaf photosynthetic damages. Therefore, inoculated plants displayed a delayed and reduced mortality rate during water stress establishment. Leaf senescence is a common way to saving resources [18]. It allows reallocation of nutrients to reproductive organs and reduces water consumption by older and less productive leaves [61]. Leaf senescence is therefore an adaptive trait that may allow plant survival under stressful conditions [61,62]. It has been reported that some microorganisms are able to affect photosynthetic efficiency, especially by an increase of whole-rosette F_v/F_m. For instance, inoculation by the PGPR *Pseudomonas fluorescens* Aur6 strain in *P. halepensis* increased mean F_v/F_m value and lead to the improvement of tree growth under well-watered conditions [33]. The increase in chlorophyll content could participate to the PGPR-triggered improvement of plant photosynthetic performance [31]. Under water stress, a positive correlation between tolerance to water deficit and maintenance of PSII efficiency has been observed in rice inoculated by an arbuscular mycorrhizal fungus [63]. By contrast, it has been recently shown that inoculation by the PGPR *Bukholderia phytofirmans* PsJn strain induces a higher number of senescent leaves in *A. thaliana* at flowering under well watered conditions [64]. Here in accordance with this finding, we found that plants inoculated by PGPR could survive with more critical physiological status.

The improvement of tolerance to leaf damages by STM196-inoculation could be related to a delayed dehydration of tissues and an improved tolerance to low water status. PSII efficiency and leaf relative water content were tightly related, as previously reported by Woo et al., [22]. Traits related to leaf water status are often measured in response to rhizobacteria and drought. In response to PGPR-inoculation, it is widely accepted that rhizobacteria increase leaf water content that leads to increase plant resistance under water deprivation (*e.g.*, [65,66,67]). Here, inoculation by STM196 led to delayed leaf dehydration and then, at the maximum of stress severity, inoculated plants displayed a higher tolerance to low water status. Contrary to common findings, we show that STM196-inoculated plants were more likely to survive at very low water status compared to non-inoculated plants. Moreover, during water-stress establishment, STM196-inoculated plants displayed a lower decline of F_v/F_m for a given leaf water content, and non-inoculated plants began to die at lower soil humidity compared to non inoculated plants. Delayed leaf dehydration induced by STM196-inoculation could explain the delayed mortality. Dehydration delay and dehydration-tolerance are important in survival strategy [52]. These involve traits that increase access to water and decrease water losses and could result from osmolytes accumulation [68,69], changes in stomatal conductance [70] and a large and deep root system [71]. It has been reported that inoculation by *Bacillus spp.* could alleviate negative effects of drought by affecting osmo-regulation through increasing osmoprotectors such as proline, sugars and free amino acids [72]. In the case of STM196 strain, our previous studies under moderate water deficit have shown that inoculation improves *A. thaliana*'s strategy of water saving by a developmental slowdown, a two-fold increase in root biomass and a significant decrease of transpiration rate related to an increase of ABA concentration in the leaf [29]. ABA plays a crucial role in plant responses to water stress and is involved in water loss regulation by control of stomatal closure. Modifications in leaf ABA content by STM196-inoculation could participate to delay and improve tolerance to dehydration and may be a cause of a better survival of plants under severe water stress. Moreover, it has been showed that changes in ABA content could also play a crucial role in the carbon remobilization from senescing leaves of drought-stressed plants [61]. Some other bacteria have also the capacity to modulate ABA metabolism in plants. For instance, recent work showed that inoculation by *Bacillus licheniformis* induces delayed water losses in grapevine that was correlated to an increase of ABA in leaf tissues [73]. Inoculation by STM196 may allow plants to be more efficient to cope with water scarcity in soils.

After rewatering, plant processes such as photosynthesis [23], transpiration [74], plant water status and growth [25] progressively recover their potential. Leaf growth rate followed the variation of soil water availability, and thus its decrease occurred progressively during water stress establishment. Upon rewatering, surviving plants resumed their growth and developed new leaves. We showed that inoculation by STM196 induced a better growth rate after rewatering and led to a large increase in biomass at flowering. Inoculated plants reached a similar biomass at flowering than non-stressed plants. This is due to an increase in both the number and size of leaves. This result was in accordance with our previous findings under moderate water deficit [29], where inoculation by STM196 allowed a 2-fold increase in plant biomass related to an increase in number and size of individual leaves. However, contrary to the findings under moderate water deficit, improvement of plant biomass by STM196 was not related to a delayed flowering time after rewatering.

STM196 may therefore allow a better conservation of leaf water content during stress establishment and help maintaining physiological integrity in a dried state, and then a better growth recovery when soil conditions become suitable for plant growth. The underlying physiological and molecular processes that could be involved in cells viability and growth potential remain to be elucidated.

Conclusion

Overall our findings indicate that inoculation by *Phyllobacterium brassicaceraum* STM196 strain reinforced the survival strategy of *A. thaliana* under conditions of severe water stress. STM196 induced a better tolerance to leaf damages through delayed leaf dehydration during water stress establishment that could allow a better conservation of cell integrity and thus, growth recovery when soil conditions became favorable again. Remarkably, STM196 allowed a production of plant biomass similar to non-stressed plants. Improvement of plant tolerance to water stress is a real challenge for crop breeding, especially under global climate change. The use of plant-bacteria interactions to enhance plant tolerance to abiotic stresses in the field offers valuable and promising prospects in addition or in complement to the classical strategies of genetic selection.

Supporting Information

File S1 Supporting information. Table S1, Soil chemical properties of the compost (Neuhaus N2), soil and two mixtures of both. Mixture 1 was sampled before experimentation and mixture 2 was sampled after experimentations. nd: not determined. Soil analysis was performed by ALFA Agricultural Service and Research Building, Soil Testing Laboratory of Auburn University. Figure S1, Soil water potential during soil drying. Soil water potential was determined using a potentiometer (WP4-T dewpoint meter, Decagon Devices, Pullman, WA 99163, USA) during soil drying (from 0.35 to 0.06 g H_2O g^{-1} dry soil). Figure S2, Growth of *P. brassicacearum* STM196 strain is not affected by soil water deficit. Growth of STM196 strain was represented by cfu/mg of soil under well-watered condition (WW) and water deficit (WD). Data are means (±SE) of 3 replicates. Figure S3, Whole-rosette mean F_v/F_m is not affected by inoculation under well watered condition (WW). Mean F_v/F_m of non-inoculated plants (NI; closed squares) and inoculated plants (I; open squares) during time courses. Data are means (±SE) of 3–32 plants. Grey points represent individuals for each condition (NI; closed symbols and I; open symbols). Figure S4, *P. brassicacearum* STM196 induces a delayed decrease of F_v/F_m in response to WD. Relationship between whole-rosette F_v/F_m and leaf relative water content of non-inoculated (NI; closed symbols) and STM196-inoculated (I; open symbols) plants under well watered (WW) and water deficit (WD; $6\%_p$) during soil drying (35%, 20%, 10% and 6%) and after rewatering (20%r, 35%r and 35%r at flowering). The dashed line represents the 90%-mortality threshold. Surviving plants with mean F_v/F_m values above the threshold, are represented by triangles (n = 3–10 and n = 3–19 for NI and I plants, respectively) and perishing plants, with mean F_v/F_m below the threshold, are represented by upside-down triangles (n = 3–9 and n = 3 for NI and I plants, respectively). Figure S5, Effect of *P. brassicacearum* STM196 strain and water deficit on growth, physiology and development of *A. thaliana* at flowering. A) Dry mass of rosette leaves, B) days to flowering and C) leaf relative water content of non-inoculated (NI) and inoculated (I) plants under well watered (WW) and severe water deficit (WD; $6\%_p$). Data are means (±SE) of 11–27 plants. Different letters indicate significant differences following Kruskal-Wallis test ($P<0.05$).

Acknowledgments

The authors are grateful to Gaëlle Rolland, Alexis Bediee, Crispulo Balsera and Garance Koch for help during the experiments. We thank Antoine LeQuéré and Isabelle Babic for providing us with *gfp*-tagged and rifampin-tetracycline-resistant STM196 strain. We also thank two anonymous reviewers for helpful comments on this paper.

Author Contributions

Conceived and designed the experiments: JB DV F. Vasseur BT. Performed the experiments: JB DV MD ML. Analyzed the data: JB DV F. Vasseur. Contributed reagents/materials/analysis tools: DV F. Varoquaux BT. Contributed to the writing of the manuscript: JB F. Vasseur F. Varoquaux BT DV.

References

1. Grayson M (2013) Agriculture and drought. Nature 501: S1–S1.
2. McDowell N, Pockman WT, Allen CD, Breshears DD, Cobb N, et al. (2008) Mechanisms of plant survival and mortality during drought: why do some plants survive while others succumb to drought? New Phytologist 178: 719–739.
3. Lugtenberg B, Kamilova F (2009) Plant-growth-promoting rhizobacteria. Annu Rev Microbiol 63: 541–556.
4. Yang J, Kloepper JW, Ryu CM (2009) Rhizosphere bacteria help plants tolerate abiotic stress. trends in Plant Science 14: 1–4.
5. Selosse MA, Baudoin E, Vandenkoornhuyse P (2004) Symbiotic microorganisms, a key for ecological success and protection of plants. Comptes rendus biologies 327: 639–648.
6. Lucy M, Reed E, Glick BR (2004) Applications of free living plant growth-promoting rhizobacteria. Antonie Van Leeuwenhoek 86: 1–25.
7. Babalola OO (2010) Beneficial bacteria of agricultural importance. Biotechnol Lett 32: 1559–1570.
8. Lawlor DW (2012) Genetic engineering to improve plant performance under drought: physiological evaluation of achievements, limitations, and possibilities. J Exp Bot 64: 83–108.
9. Thomas AK, Bettina MJE, Amy B, Melvin TT, Bouchra EIO, et al. (2009) Tolerance to low leaf water status of tropical tree seedlings is related to drought performance and distribution. Functional Ecology 23: 93–102.
10. Moore JP, Le NT, Brandt WF, Driouich A, Farrant JM (2009) Towards a systems-based understanding of plant desiccation tolerance. Trends in Plant Science 14: 110–117.
11. Kranner I, Beckett RP, Wornik S, Zorn M, Pfeifhofer HW (2002) Revival of a resurrection plant correlates with its antioxidant status. Plant Journal 31: 13–24.
12. Verslues PE, Agarwal M, Katiyar-Agarwal S, Zhu J, Zhu JK (2006) Methods and concepts in quantifying resistance to drought, salt and freezing, abiotic stresses that affect plant water status. Plant J 45: 523–539.
13. Farooq M, Wahid A, Kobayashi N, Fujita D, Basra SMA (2009) Plant drought stress: Effects, mechanisms and management sustainable agriculture. In: Lichtfouse E, Navarrete M, Debaeke P, Véronique S, Alberola C, editors: Springer Netherlands. pp. 153–188.
14. Gruszka Vendruscolo EC, Schuster I, Pileggi M, Scapim CA, Correa Molinari HB, et al. (2007) Stress-induced synthesis of proline confers tolerance to water deficit in transgenic wheat. Journal of Plant Physiology 164: 1367–1376.
15. Sakamoto A, Murata N (2002) The role of glycine betaine in the protection of plants from stress: clues from transgenic plants. Plant Cell Environ 25: 163–171.
16. Elbein AD, Pan YT, Pastuszak I, Carroll D (2003) New insights on trehalose: a multifunctional molecule. Glycobiology 13: 17R–27R.
17. McDowell NG (2011) Mechanisms linking drought, hydraulics, carbon metabolism, and vegetation mortality. Plant Physiol 155: 1051–1059.
18. Chaves MM (1991) Effects of water deficits on carbon assimilation Journal of Experimental Botany 42: 1–16.
19. Lim PO, Kim HJ, Nam HG (2007) Leaf senescence. Annu Rev Plant Biol 58: 115–136.
20. Murchie EH, Lawson T (2013) Chlorophyll fluorescence analysis: a guide to good practice and understanding some new applications. J Exp Bot: ert208.
21. Jansen M, Gilmer F, Biskup B, Nagel KA, Rascher U, et al. (2009) Simultaneous phenotyping of leaf growth and chlorophyll fluorescence via GROWSCREEN FLUORO allows detection of stress tolerance in *Arabidopsis thaliana* and other rosette plants. Functional Plant Biology 36: 902–914.
22. Woo NS, Badger MR, Pogson BJ (2008) A rapid, non-invasive procedure for quantitative assessment of drought survival using chlorophyll fluorescence. Plant Methods 4: 27.
23. Galmes J, Abadia A, Cifre J, Medrano H, Flexas J (2007) Photoprotection processes under water stress and recovery in Mediterranean plants with different growth forms and leaf habits. Physiologia Plantarum 130: 495–510.
24. Xu Z, Zhou G, Shimizu H (2009) Are plant growth and photosynthesis limited by pre-drought following rewatering in grass? Journal of Experimental Botany 60: 3737–3749.
25. Lechner L, Pereyra-Irujo GA, Granier C, Aguirrezabal LA (2008) Rewatering plants after a long water-deficit treatment reveals that leaf epidermal cells retain their ability to expand after the leaf has apparently reached its final size. Ann Bot 101: 1007–1015.
26. Figueiredo MVB, Burity HA, Martínez CR, Chanway CP (2008) Alleviation of drought stress in the common bean (*Phaseolus vulgaris* L.) by co-inoculation with *Paenibacillus polymyxa* and *Rhizobium tropici*. Applied Soil Ecology 40: 182–188.
27. Belimov AA, Dodd IC, Hontzeas N, Theobald JC, Safronova VI, et al. (2009) Rhizosphere bacteria containing 1-aminocyclopropane-1-carboxylate deaminase increase yield of plants grown in drying soil via both local and systemic hormone signalling. New Phytol 181: 413–423.
28. Liu F, Xing S, Ma H, Du Z, Ma B (2013) Cytokinin-producing, plant growth-promoting rhizobacteria that confer resistance to drought stress in *Platycladus orientalis* container seedlings. Applied microbiology and biotechnology: 1–10.
29. Bresson J, Varoquaux F, Bontpart T, Touraine B, Vile D (2013) The PGPR strain *Phyllobacterium brassicacearum* STM196 induces a reproductive delay and physiological changes that result in improved drought tolerance in Arabidopsis. New Phytologist 200: 558–569.
30. Kohler J, Hernandez JA, Caravaca F, Roldan A (2008) Plant-growth-promoting rhizobacteria and arbuscular mycorrhizal fungi modify alleviation biochemical mechanisms in water-stressed plants. Functional Plant Biology 35: 141–151.
31. Wang CJ, Yang W, Wang C, Gu C, Niu DD, et al. (2012) Induction of drought tolerance in cucumber plants by a consortium of three plant growth-promoting rhizobacterium strains. PLoS One 7.
32. Heidari M, Golpayegani A (2012) Effects of water stress and inoculation with plant growth promoting rhizobacteria (PGPR) on antioxidant status and photosynthetic pigments in basil (*Ocimum basilicum L.*). Journal of the Saudi Society of Agricultural Sciences 11: 57–61.
33. Rincon A, Valladares F, Gimeno TE, Pueyo JJ (2008) Water stress responses of two Mediterranean tree species influenced by native soil microorganisms and inoculation with a plant growth promoting rhizobacterium. Tree Physiol 28: 1693–1701.
34. Timmusk S, Timmusk K, Behers L (2013) Rhizobacterial plant drought stress tolerance enhancement: towards sustainable water resource management and food security. Journal of Food Security 1: 6–10.
35. Dimkpa C, Weinand T, Asch F (2009) Plant-rhizobacteria interactions alleviate abiotic stress conditions. Plant Cell Environ 32: 1682–1694.
36. Mantelin S, Saux MF, Zakhia F, Bena G, Bonneau S, et al. (2006) Emended description of the genus *Phyllobacterium* and description of four novel species associated with plant roots: *Phyllobacterium bourgognense* sp. nov., *Phyllobacterium ifriqiyense* sp. nov., *Phyllobacterium leguminum* sp. nov. and *Phyllobacterium brassicacearum* sp. nov. International journal of systematic and evolutionary microbiology 56: 827–839.
37. Bertrand H, Nalin R, Bally R, Cleyet-Marel JC (2001) Isolation and identification of the most efficient plant growth-promoting bacteria associated with canola (*Brassica napus*). Biology and Fertility of Soils 33: 152–156.
38. Larcher M, Muller B, Mantelin S, Rapior S, Cleyet-Marel JC (2003) Early modifications of *Brassica napus* root system architecture induced by a plant growth-promoting *Phyllobacterium* strain. New Phytologist 160: 119–125.
39. Mantelin S, Desbrosses G, Larcher M, Tranbarger TJ, Cleyet-Marel JC, et al. (2006) Nitrate-dependent control of root architecture and N nutrition are altered by a plant growth-promoting *Phyllobacterium sp.* Planta 223: 591–603.
40. Contesto C, Milesi S, Mantelin S, Zancarini A, Desbrosses G, et al. (2010) The auxin-signaling pathway is required for the lateral root response of Arabidopsis to the rhizobacterium *Phyllobacterium brassicacearum*. Planta 232: 1455–1470.
41. Galland M, Gamet L, Varoquaux F, Touraine B, Desbrosses G (2012) The ethylene pathway contributes to root hair elongation induced by the beneficial bacteria *Phyllobacterium brassicacearum* STM196. Plant Science 190: 74–81.
42. Kechid M, Desbrosses G, Rokhsi W, Varoquaux F, Djekoun A, et al. (2013) The *NRT2.5* and *NRT2.6* genes are involved in growth promotion of Arabidopsis by the plant growth-promoting rhizobacterium (PGPR) strain *Phyllobacterium brassicacearum* STM196. New Phytologist 198: 514–524.
43. Granier C, Aguirrezabal L, Chenu K, Cookson SJ, Dauzat M, et al. (2006) PHENOPSIS, an automated platform for reproducible phenotyping of plant responses to soil water deficit in *Arabidopsis thaliana* permitted the identification of an accession with low sensitivity to soil water deficit. New Phytol 169: 623–635.
44. Hoagland DR, Arnon DI (1950) The water-culture method for growing plants without soil. California Agricultural Experiment Station Circular 347: 1–32.
45. Boyes DC, Zayed AM, Ascenzi R, McCaskill AJ, Hoffman NE, et al. (2001) Growth stage-based phenotypic analysis of Arabidopsis: A model for high throughput functional genomics in plants. Plant Cell 13: 1499–1510.
46. Maxwell K, Johnson GN (2000) Chlorophyll fluorescence - a practical guide. Journal of Experimental Botany 51: 659–668.
47. Fabre J, Dauzat M, Negre V, Wuyts N, Tireau A, et al. (2011) PHENOPSIS DB: an Information System for Arabidopsis thaliana phenotypic data in an environmental context. BMC Plant Biology 11.
48. Cheng HP, Walker GC (1998) Succinoglycan is required for initiation and elongation of infection threads during nodulation of alfalfa by *Rhizobium meliloti*. Journal of bacteriology 180: 5183–5191.
49. Halvorson H, Ziegler N (1933) Application of statistics to problems in bacteriology: I. A means of determining bacterial population by the dilution method. Journal of Bacteriology 25: 101.
50. R Development Core Team (2009) R: a language and environment for statistical computing. Vienna, Austria: R Foundation for Statistical Computing. 409 p.
51. Rivero RM, Kojima M, Gepstein A, Sakakibara H, Mittler R, et al. (2007) Delayed leaf senescence induces extreme drought tolerance in a flowering plant. Proceedings of the National Academy of Sciences of the United States of America 104: 19631–19636.

52. Tyree MT, Engelbrecht BMJ, Vargas G, Kursar TA (2003) Desiccation tolerance of five tropical seedlings in Panama. Relationship to a field assessment of drought performance. Plant Physiology 132: 1439–1447.
53. Skirycz A, Vandenbroucke K, Clauw P, Maleux K, De Meyer B, et al. (2011) Survival and growth of Arabidopsis plants given limited water are not equal. Nature Biotechnology 29: 212–214.
54. Suarez R, Wong A, Ramirez M, Barraza A, Orozco MD, et al. (2008) Improvement of drought tolerance and grain yield in common bean by overexpressing trehalose-6-phosphate synthase in rhizobia. Molecular Plant-Microbe Interactions 21: 958–966.
55. Rodriguez-Salazar J, Suarez R, Caballero-Mellado J, Iturriaga G (2009) Trehalose accumulation in *Azospirillum brasilense* improves drought tolerance and biomass in maize plants. Fems Microbiology Letters 296: 52–59.
56. Ehlert B, Hincha DK (2008) Chlorophyll fluorescence imaging accurately quantifies freezing damage and cold acclimation responses in Arabidopsis leaves. Plant Methods 4: 12.
57. Gray GR, Hope BJ, Qin XQ, Taylor BG, Whitehead CL (2003) The characterization of photoinhibition and recovery during cold acclimation in *Arabidopsis thaliana* using chlorophyll fluorescence imaging. Physiologia Plantarum 119: 365–375.
58. Sperdouli I, Moustakas M (2012) Spatio-temporal heterogeneity in *Arabidopsis thaliana* leaves under drought stress. Plant Biology 14: 118–128.
59. Wingler A, Purdy S, MacLean JA, Pourtau N (2006) The role of sugars in integrating environmental signals during the regulation of leaf senescence. J Exp Bot 57: 391–399.
60. Diaz C, Purdy S, Christ A, Morot-Gaudry JF, Wingler A, et al. (2005) Characterization of markers to determine the extent and variability of leaf senescence in Arabidopsis. A metabolic profiling approach. Plant Physiol 138: 898–908.
61. Munné-Bosch S, Alegre L (2004) Die and let live: leaf senescence contributes to plant survival under drought stress. Functional Plant Biology 31: 203–216.
62. Ono K, Nishi Y, Watanabe A, Terashima I (2001) Possible mechanisms of adaptive leaf senescence. Plant Biology 3: 234–243.
63. Ruiz-Sanchez M, Aroca R, Munoz Y, Polon R, Ruiz-Lozano JM (2010) The arbuscular mycorrhizal symbiosis enhances the photosynthetic efficiency and the antioxidative response of rice plants subjected to drought stress. Journal of Plant Physiology 167: 862–869.
64. Poupin MJ, Timmermann T, Vega A, Zuniga A, Gonzalez B (2013) Effects of the plant growth-promoting bacterium *Burkholderia phytofirmans* PsJN throughout the life cycle of *Arabidopsis thaliana*. PLoS One 8: e69435.
65. Arzanesh MH, Alikhani HA, Khavazi K, Rahimian HA, Miransari M (2011) Wheat (*Triticum aestivum L.*) growth enhancement by *Azospirillum* sp under drought stress. World Journal of Microbiology & Biotechnology 27: 197–205.
66. Creus CM, Sueldo RJ, Barassi CA (2004) Water relations and yield in *Azospirillum*-inoculated wheat exposed to drought in the field. Canadian Journal of Botany 82: 273–281.
67. Marulanda A, Barea JM, Azcon R (2009) Stimulation of plant growth and drought tolerance by native microorganisms (AM fungi and bacteria) from dry environments: mechanisms related to bacterial effectiveness. Journal of Plant Growth Regulation 28: 115–124.
68. Morgan PW (1990) Effects of abiotic stresses on plant hormone systems. In: Wiley-Liss I, editor. in: Stress Responses in plants: adaptation and acclimation mechanisms. pp. pp. 113–146.
69. Taiz LZE (2006) Plant Physiology; Ed. t, editor. Massachusetts: Sinauer Associates Inc. Publishers.
70. Farquhar GD, Sharkey TD (1982) Stomatal Conductance and Photosynthesis. Annual Review of Plant Physiology 33: 317–345.
71. Padilla FM, Pugnaire FI (2007) Rooting depth and soil moisture control Mediterranean woody seedling survival during drought. Functional Ecology 21: 489–495.
72. Vardharajula S, Ali SZ, Grover M, Reddy G, Bandi V (2011) Drought-tolerant plant growth promoting *Bacillus spp.*: effect on growth, osmolytes, and antioxidant status of maize under drought stress. Journal of Plant Interactions 6: 1–14.
73. Salomon MV, Bottini R, de Souza Filho GA, Cohen AC, Moreno D, et al. (2013) Bacteria isolated from roots and rhizosphere of Vitis vinifera retard water losses, induce abscisic acid accumulation and synthesis of defense-related terpenes in in vitro cultured grapevine. Physiologia Plantarum.
74. Bengtson C, Falk SO, Larsson S (1977) The after-effect of water stress on transpiration rate and changes in abscisic acid content of young wheat plants. Physiologia Plantarum 41: 149–154.

17

Systems Responses to Progressive Water Stress in Durum Wheat

Dimah Z. Habash[1*¤], Marcela Baudo[1], Matthew Hindle[2], Stephen J. Powers[2], Michael Defoin-Platel[2], Rowan Mitchell[2], Mansoor Saqi[2], Chris Rawlings[2], Kawther Latiri[†3], Jose L. Araus[4], Ahmad Abdulkader[5], Roberto Tuberosa[6], David W. Lawlor[1], Miloudi M. Nachit[7]

1 Plant Biology and Crop Science, Rothamsted Research, Harpenden, United Kingdom, **2** Computational and Systems Biology, Rothamsted Research, Harpenden, United Kingdom, **3** Laboratoire D'agronomie, National Agricultural Research Institute of Tunisia, Ariana, Tunisia, **4** Dept. of Vegetal Biology, Faculty of Biology, Barcelona, Spain, **5** Biotechnology Department, General Commission for Scientific Agricultural Research, Damascus, Syria, **6** Dept. of Agroenvironmental Science and Technology, University of Bologna, Bologna, Italy, **7** Biodiversity and Integrated Gene Management Program, International Center for Agricultural Research in the Dry Areas, Rabat, Morocco

Abstract

Durum wheat is susceptible to terminal drought which can greatly decrease grain yield. Breeding to improve crop yield is hampered by inadequate knowledge of how the physiological and metabolic changes caused by drought are related to gene expression. To gain better insight into mechanisms defining resistance to water stress we studied the physiological and transcriptome responses of three durum breeding lines varying for yield stability under drought. Parents of a mapping population (Lahn x Cham1) and a recombinant inbred line (RIL2219) showed lowered flag leaf relative water content, water potential and photosynthesis when subjected to controlled water stress time transient experiments over a six-day period. RIL2219 lost less water and showed constitutively higher stomatal conductance, photosynthesis, transpiration, abscisic acid content and enhanced osmotic adjustment at equivalent leaf water compared to parents, thus defining a physiological strategy for high yield stability under water stress. Parallel analysis of the flag leaf transcriptome under stress uncovered global trends of early changes in regulatory pathways, reconfiguration of primary and secondary metabolism and lowered expression of transcripts in photosynthesis in all three lines. Differences in the number of genes, magnitude and profile of their expression response were also established amongst the lines with a high number belonging to regulatory pathways. In addition, we documented a large number of genes showing constitutive differences in leaf transcript expression between the genotypes at control non-stress conditions. Principal Coordinates Analysis uncovered a high level of structure in the transcriptome response to water stress in each wheat line suggesting genome-wide co-ordination of transcription. Utilising a systems-based approach of analysing the integrated wheat's response to water stress, in terms of biological robustness theory, the findings suggest that each durum line transcriptome responded to water stress in a genome-specific manner which contributes to an overall different strategy of resistance to water stress.

Editor: Pilar Hernandez, Institute for Sustainable Agriculture (IAS-CSIC), Spain

Funding: DH, RT, JA, AA, KL, and MN gratefully acknowledge support from the European Union FPVI programme (INCO-CT-509136) which funded material, overheads and salaries. MH was supported by a Lawes Agricultural Trust studentship. DH, CR, SP, MD-P, RM, MB, and MS acknowledge strategic funding support from the Biotechnology and Biological Sciences Research Council (BBSRC) of the UK, and CR was also supported by BBSRC grant BB/F006039/. MN received support from the International Center for Agricultural Research in the Dry Areas. The funders had no role in study design, data collection and analysis, decision to publish, or preparation of the manuscript.

Competing Interests: The authors have declared that no competing interests exist.

* Email: Dimah.Habash@securewheat.com

¤ Current address: Securewheat Research, Securewheat, St Albans, United Kingdom

† Deceased.

Introduction

Durum wheat is a major staple crop in the Mediterranean basin as a source of semolina for the production of pasta, couscous, burghul and other local products. It is cultivated primarily under rain-fed conditions across a wide range of environments varying in the profile and quantity of rainfall with drought as a major abiotic stress [1,2]. Regional to global-scale climate models predict, with high confidence, that the Mediterranean region will face soil drying in the future as temperatures rise and this may increase incidents of agricultural drought [3,4,5]. This will have serious negative implications for crop productivity and food security and thus crop yield and its component traits are targets for wheat breeders in their efforts to design cultivars for drought-prone environments as a means to ameliorate the climatic problems [6]. However, crop yield and drought resistance are complex genetic traits exacerbated by the fact that drought is often encountered with other abiotic and biotic constraints and these factors have contributed to the slow progress, so far, in breeding for relative drought resistance.

To address these challenges, various national and international efforts have been proposed to enhance wheat production by integrating technologies and approaches in plant science to underpin breeding efforts [7]. Thus, improving our understanding of the crop's responses to drought stress has emerged as a major target. Genetic studies on durum wheat have identified quantitative trait loci (QTL) for yield and yield component traits as a first important step in understanding and deconstructing the complex genetic factors defining drought resistance [8,9,6]. In parallel, studies of plants exposed to water stress under controlled environmental conditions enable the mechanistic understanding of the complex responses, adaptations, acclimation and recovery from stresses at the molecular, biochemical and physiological levels [10,11,12]. Genomic tools and 'omic' technologies have recently been employed to screen for the genes, proteins and metabolites responsive to water stress, and have identified key molecular elements [13,14]. Physical stress signals activate a series of molecular cascades and signalling components, both abscisic acid (ABA) and non-ABA-dependent, responsive to abiotic stresses in plants [15]. Common features have been identified in Arabidopsis and grass species [16] which have enabled the initial modelling of gene network responses under water stress [17]. These studies form a mechanistic basis for our knowledge of how physical signals are transduced to biochemical processes which ultimately translate into adjustments of metabolism and physiology to stress. Whilst genetic manipulation of key candidate genes, arising from such studies, has resulted in some form of resistance to water stress in individual transgenic plants, transferring such benefits from model species to improve crop performance under drought has been markedly less successful [18]. There is an urgent need to improve our understanding of drought effects on crops by combining comparisons of cultivars of different field performance with detailed analysis of their resistance mechanisms under well-defined plant water deficits. In addition, it is increasingly argued that systems-based approaches are necessary to integrate knowledge across genetics, genomics, breeding and plant physiology to tackle the complex genetic traits defining drought resistance [19,20,21]. Focusing system studies on crop species *per se* is also advocated considering the importance of genetic backgrounds and species-specific differences in adaptations to stress and for the immediate development of drought resistant crop cultivars, as demonstrated in bread *versus* durum wheat by [22].

Durum wheat grown in the Mediterranean is susceptible to post-anthesis terminal drought, a phase of growth which directly affects yield. Thus wheat breeders have targeted this type of drought by selecting for cultivars with a fast growth rate that rapidly accumulate biomass before flowering as a water stress avoidance strategy [2]. In addition, they have also developed wheat germplasm of promising yield potential by applying a double gradient selection strategy under both scarce and optimal water environments [8]. These efforts have been relatively successful in terms of yield improvement but the physiological and molecular basis of these drought resistant cultivars remains complex and unclear. Our aim in this study was to dissect the responses of drought resistant durum breeding lines to water stress in an attempt to identify molecular and physiological properties defining stress resistance and thus to build knowledge to accelerate the breeding effort. We chose wheat lines from the breeding programme at the International Centre for Agricultural Research in the Dry Areas (ICARDA) with good yield under irrigated and drought prone environments established in multi-year trials in the field (unpublished observations). Our study targets the post-anthesis developmental phase which corresponds to the period when plants are susceptible to terminal drought under field conditions in the Mediterranean basin. We applied rigorous criteria for the development of the water stress experiments, both in design and measurements, following [23]. High-throughput gene expression technology was applied in parallel with physiological measurements to profile durum lines at a range of water deficits (water stress transients) designed to dissect the systems responses to stress. Mathematical and bioinformatics methods were used to deconstruct the complex datasets and results are presented in formats amenable for further mining and exploration. Analysis was undertaken on datasets at equivalent leaf relative water content (RWC) to establish evidence of differential plant genome stress responses amongst the three wheat lines. Results demonstrate a global transcriptome response to water stress in all three lines that has a systems feature of high levels of coordination and uncovered within that key genome-specific differences. These results are discussed using a systems-based framework of biological robustness theory and how it could impact the search for new breeding strategies for drought resistance in durum wheat.

Materials and Methods

Plant material

Two durum wheat [*Triticum turgidum* L. ssp. *durum* (Desf.) Husn.] parents plus one recombinant inbred line (RIL) from the Cham1×Lahn mapping population developed at ICARDA were selected for this study because of their combined drought resistance and yield potential established in multiple year × site field trials in the Mediterranean region. Lahn was characterised by high yield potential under favourable rain environments, Cham1 exhibited relatively higher yield under drought, and RIL2219 was established as being more drought resistant than the best parent in addition to showing good yield stability (unpublished observations). Data collected in multiple field trials (30 over four years) showed RIL2219 closer to Cham1 in its overall phenology so that it reached heading 1.2 and 6 days earlier, on average, than Cham1 and Lahn respectively; it's grain fill duration was 1.69 and 2.69 days later than Cham1 and Lahn respectively; and it was similar to cham1 for days to maturity with both reaching it 5 days earlier than Lahn (unpublished observations).

Plant growth and experimental design

Seeds were sown in peat-free soil-based compost enriched with slow release fertilizer (Osmocote, Scotts UK Professional, Ipswich, Suffolk) and seedlings were vernalized for 3 weeks to synchronise growth and flowering. Three plants per line were transplanted into 3.5 L pots and grown under controlled chamber conditions of 25°C day/20°C night, 14 hours photoperiod, 800 $\mu mol^{-2} \cdot s^{-1}$ photon flux density and 65% relative humidity. Pots were watered uniformly each day until the start of the experiment. Plants were arranged using a randomised block design with 3 biological replicates per combination of line, treatment (control or stressed) and time point of the stress. At anthesis, plants were subjected to four independent time series water deficit experiments (termed water stress transients) by withholding water from pots for a period of up to 6 days with parallel controls being well watered, all measurements were taken at 10 am on a daily basis following the initiation of stress. Flag leaves were sampled from 3 experiments for molecular and biochemical studies and from one representative experiment for photosynthetic studies, one week after flowering (Zadoks scale 70). For each experiment, a flag leaf was sampled for RNA extraction and ABA content, another was taken for photosynthetic and osmotic adjustment measurements and a third for leaf relative water content (%RWC) and water potential.

Measurements of plant water status

The water status of the plants was monitored by flag leaf % RWC (fresh mass – dry mass)/(turgid mass – dry mass) × 100), following [24] and leaf water potential was measured with a pressure chamber, after [25]. Difference between pot mass in the morning and afternoon gave the total plant plus soil water loss.

Physiological and biochemical measurements

Flag leaf transpiration rate, stomatal conductance to water, and CO_2 assimilation rate were measured at 800 μmol m^2 s^{-1} PFD using portable infrared gas analysers (LICOR LI-64 LI-COR Biosciences, Nebraska, USA and SIRAS PP Systems, Norfolk, UK). Plant height, tiller number, development stage and leaf area were measured to monitor growth. Osmotic potential and adjustment was measured using a vapour pressure Osmometer (5100C, Wescor, USA) on samples of sap from frozen, thawed, macerated and centrifuged leaves. ABA was quantified by GC-MS, essentially as described in [26], except that freeze-dried leaf samples were homogenised in 80% methanol-water using a Polytron homogeniser, after which [3-methyl-$^2H_3$]ABA (20–25 ng) was added as internal standard.

RNA isolation and array hybridisation

Individual flag leaves, one leaf per biological sample, were sampled for RNA extraction and transcript expression studies. For all three wheat lines, 3 biological replicates per line were taken at days 1, 2, 3, 4, 5 (all 3 lines) and 6 (RIL2219 only) of the stress plus day 0 and 1 from the well watered controls to give a total of 66 samples. Leaves were frozen in liquid N_2 and total RNA was extracted using TRIZOL Reagent (Invitrogen, UK) following the manufacturer's recommendations. Total RNA was DNase treated with TURBO DNase enzyme (Ambion) and cleaned using RNeasy columns (Qiagen) following the manufacturer's instructions. High quality RNA was submitted to transcriptome profiling at Bristol University (http://www.bristol.ac.uk/biology/research/transcriptomics/). Expression analysis was carried out on Affymetrix GeneChip Wheat genome arrays using Affymetrix GeneChip Eukaryotic One cycle target labelling reagents and protocols (701028 Rev 4). Five μg of total RNA was first reverse transcribed using a T7-Oligo(dT) promotor primer in a first strand cDNA synthesis reaction, followed by RNase H mediated 2nd strand cDNA synthesis. The double stranded cDNA was purified and served as a template in a subsequent in vitro transcription (IVT) reaction carried out in the presence of T7 RNA polymerase and a biotinylated nucleotide analog/ribonucleotide mix to produce complimentary biotinylated RNA (cRNA). The biotinylated cRNA targets were cleaned up, fragmented and hybridised to Wheat Genome arrays for 16 hours in a GeneChip Affymetrix Hybridisation oven 640. Hybridised probe arrays were then washed and stained with a streptavidin-conjugated fluorescent staining solution followed by antibody amplification of the fluorescent signal in a GeneChip fluidics 450 station using the appropriate fluidics scripts for the array format. Following washing and staining, arrays were scanned using the GeneChip scanner 3000. Cell intensity data was automatically computed from the image data using GeneChip Operating Software (GCOS) under default settings. The quality and integrity of both total RNA and labelled and fragmented cRNA was assessed on an Agilent bioanalyser RNA Nano assay.

Data processing and modelling

Microarray data are available in the ArrayExpress database (www.ebi.ac.uk/arrayexpress) under accession number E-MTAB-2410. Microarray data initial analysis was carried out using GenespringGX 8.0 (Agilent Technologies, Inc., Santa Clara, California, USA). The set of array raw data files (.cel files) were pre-processed using the gcRMA algorithm [27] to standardise the mean and variance of expression. Probesets were filtered to remove those with mean signal values less than 10, which were deemed not expressed. All probeset signal data (hereafter referred to as transcript or probe expression) from the filtered set were then exported for ANOVA and multivariate statistical analysis using GenStat (2010, thirteenth edition, VSN International Ltd, Hemel Hempstead, UK). To account for the time differences in wheat line responses to the stress and to present the transcript expression results in terms of equivalent leaf RWC, data were interpolated according to suggestions for handling results of time series experiments [28]. Firstly, the logistic relationship of RWC over time (days) was used to facilitate interpolation of transcript expression on the $\log_2$ scale with respect to leaf RWC at proposed levels of 90, 82, 74, 66, 58 and 50; these levels of RWC were chosen to reflect a range of water deficit, whilst maintaining 6 observed occasions. RWC was then modelled on time (days of stress), fitting a logistic curve to the RWC data collected from each wheat line, using nonlinear regression, $RWC(time) = a+b/[1+\exp((time - c)/d)]$, for estimable parameters a, b, c, and d. For each transcript by replicate (statistical block) by wheat line combination, the change in transcript expression data over the time course was interpolated using Lagrange polynomials [29,30] which enabled the transcript expression at the time points corresponding to the proposed RWC levels to be calculated. The results were further analysed by ANOVA and $\log_2$ transformation which considered the main effect of, and interaction between, the line and RWC factors, and also accounted for the block effect. The transcripts were filtered into groups given significant F-test ($p<0.01$, to ensure well-defined groups) results, after correcting p-values for multiple testing and false-discovery [31] and least significant difference (LSD) values ($p = 0.05$) were used to compare means. Principal Coordinates Analysis (PCo) was applied to the RWC-interpolated data for the 19062 probes derived from 54 samples (three replicates of each of six RWC levels- 0–5 days- by three wheat lines) [32]. The probes proposed to be responsible for the separation observed in a given dimension were found by correlating the coordinates of the points in that dimension to transcript expression values, using an approximate F-test and extracting those transcripts with the greatest (most statistically significant) F-values. Physiological and biochemical measures were modelled with respect to RWC using linear or non-linear regression, testing whether single, parallel or separate lines (or curves) were required (F-tests) for the wheat lines. Linear regression was used for leaf water potential, water loss and osmotic adjustment. Gompertz curves, of the form $y = C - C[\exp(\exp(-k(RWC - m))]$, were used for rate of CO_2 assimilation, transpiration rate and stomatal conductance, with parameters C being the asymptote, m the inflection point and k the exponential rate separate, or common, for each wheat line. ABA content showed an asymmetric rise to a peak, therefore a rational functions model was used, $\log(ABA) = a+[(b+cRWC)/(1+ dRWC+fRWC^2)]$, with estimable parameters a, b, c, d and f, taking ABA on the log (to base e) scale to account for greater variability in this measure.

To estimate the contribution of allelic differences in gene sequences to probe hybridisation, for each perfect match probe the significance of the ratio changes across lines was calculated in proportion to the corresponding mean probe-set abundance. Robust Multiarray Averaging (RMA) background correction and constant normalisation was applied to the probe intensities [33,27]. The probe intensities were transformed using the $\log_2$ of the ratio of the probe to the mean of the corresponding probe-set intensity. For probe-set intensities the same probe normalisation and background corrections were applied, together with perfect match correction and summarization using median polish [33,27]. The significance of the presence of allelic difference for lines was determined from two-way ANOVA of line and time factors, using the p-value for the significance of the difference across lines.

Annotation and MapMan pathways analysis

The wheat Affymetrix chip was designed with 11 probe sets per gene and the majority of the analysis in this study focused on the average expression value of the 11 probes in a probe set. In our analysis, we did not filter out any probes or identify unigenes, thus we present results for each probe set signal intensity that represent changes that could be due to unigenes, splice variants, homeoalleles or several members of a gene family. Probes were assigned to orthologous genes and metabolic bins, or pathways, using the MapMan programme following the standard protocol [34,35], bin # 35 was unassigned annotation. When referring to probe intensity or transcript expression of genes from a particular metabolic pathway, from now onwards in the text, this refers to an annotated orthologous gene. qPCR was done on eight genes and results showed similar changes to Affymetrix probe expression as a function of the stress transient in all lines (unpublished observations).

Results

Physiological responses of wheat lines to water stress

Cessation of watering and loss of water from the soil for 6 days progressively decreased flag leaf RWC and water potential for all wheat lines (Figure 1 A): the initial RWC was identical for the three cultivars. It was not possible to accurately measure RWC for Lahn on day 6 as it was severely wilted. RWC decreased more slowly in RIL2219 than in either parent, as demonstrated by its larger RWC after day 3, related to the slower rate of total water loss from its pots both as a function of time and of RWC (Figure 1A, 1F). The relation between RWC and water potential of the flag leaves was linear, and identical, in the three cultivars (Figure 1B) and therefore RWC was used as the reference for tissue water status. As the rates of water loss and decrease in RWC differed significantly between the cultivars, all subsequent analyses were made with reference to RWC. RIL2219 had higher photosynthetic rate (Figure 1 C), instantaneous transpiration rate (Figure1 D) and a larger stomatal conductance (Figure 1 E) in the early phases of water depletion (from ca. 100 to 80% RWC). However, these decreased substantially in all cultivars (and marginally more so in RIL2219) as RWC fell from ca. 80 to 60% and where we also observed that RIL2219 and Cham1 maintained a higher osmotic adjustment for equivalent RWC when compared to Lahn (Figure 1 G). Furthermore, the ABA content of turgid leaves was larger in RIL2219 than Cham1 or Lahn notably at the start of the experiments in leaves at equivalent RWC, but increased similarly as RWC decreased, rising greatly and reaching a maximum at about 65% (Figure 1 H) after which there was a general decline. There were no statistically significant ($p<0.05$, F-test) difference in flowering time, leaf area or plant height between the lines in any of the experiments (unpublished observations).

Dissection of transcriptome responses to progressive water stress

Flag leaf RNA samples from three wheat lines subjected to the transient of water stress were submitted for the study of global transcript expression using the wheat Affymetrix gene expression chip. Initial filtering showed that 19062 probes (32% of total probe sets) were modulated in value in response to water stress, genotype, or in combination, in all lines during the stress transient. In the first three days of the stress transient after watering was stopped, probes showed moderate changes in expression and this increased sharply as the stress progressed and leaf RWC declined with some delay shown by RIL 2219 (Figure 2). To analyse probe expression responses as a function of equivalent stress amongst the three wheat lines, this delay was taken into consideration by interpolation of expression values as a function of leaf RWC. ANOVA was then applied to 19062 probes across the three lines and RWC transient, and the results are available for each probe as mean expression values and LSDs for comparisons (File S1). This analysis identified four independent groups showing statistically significant changes with respect to *Line* only (932 probes in File S2), *Stress* only (2579 probes in File S3), *Line+Stress* independently (7644 probes in File S4), *Line* × *Stress* interaction (7699 probes in File S5), and a fifth group of 208 probes showing *no statistically significant effect* of either factor (File S6). A detailed analysis of probes annotated to metabolic pathways significantly ($p<0.05$, LSD) altered from 90-50% leaf RWC for each of the above ANOVA groups was done using the MapMan programme. Some probes annotated to the same gene model and subsequently allocated to more than one of the latter ANOVA groups may be a result of hybridisation to closely related sequences such as gene families, homeoalleles or paralogs.

Equivalent transcript responses in all lines

ANOVA identified probes in *Stress* (File S3), *Line+Stress* (File S4) and *Line* × *Stress* (File S5) groups that were statistically altered in expression in response to the water stress transient when compared to controls. Systematic analysis of each of these datasets using MapMan showed similar general trends of metabolic pathway responses to water stress and these are summarized in the following sections with focus on the *Line+Stress* (File S4) group as a representative. The 7644 probes in this latter group showed initial differences in expression between the wheat lines at the onset of the stress but this did not influence the profile of response further into the stress transient as RWC decreased; the lines showed similar and parallel stress responses. Probes annotated to specific metabolic pathways (bins) significantly altered from 90-50% leaf RWC are documented in individual spreadsheets in File S4 and Table 1 to enable further mining. Analysis revealed that all represented metabolic pathway bins showed some probes with significant ($p<0.05$, LSD) over-expression or under-expression of varying orders of magnitude at each leaf RWC (Figure 3). The profiles of the changes in expression of metabolic pathways are complex and varied as leaf RWC decreases, and we focus in this paper on general trends and particularly on changes in regulatory pathways, photosynthetic, primary and secondary metabolism (see following sections).

Early changes. The experimental design uncovered pathways specifically altered early in the stress response with an average of 26% of annotated probes showing changed expression when leaf RWC decreased from 90 to 82% (File S4 spreadsheet summary of early changes). Of those, 247 probes annotated across most of the metabolic bins showed a reduction in expression (File S4 early changes underexpressed spreadsheet) and an even larger number of probes (624) increased in expression with bias towards RNA and protein processing, signalling and transport (File S4 early changes overexpressed spreadsheet). Results also showed that 73% of all annotated probes showed changes in expression as the

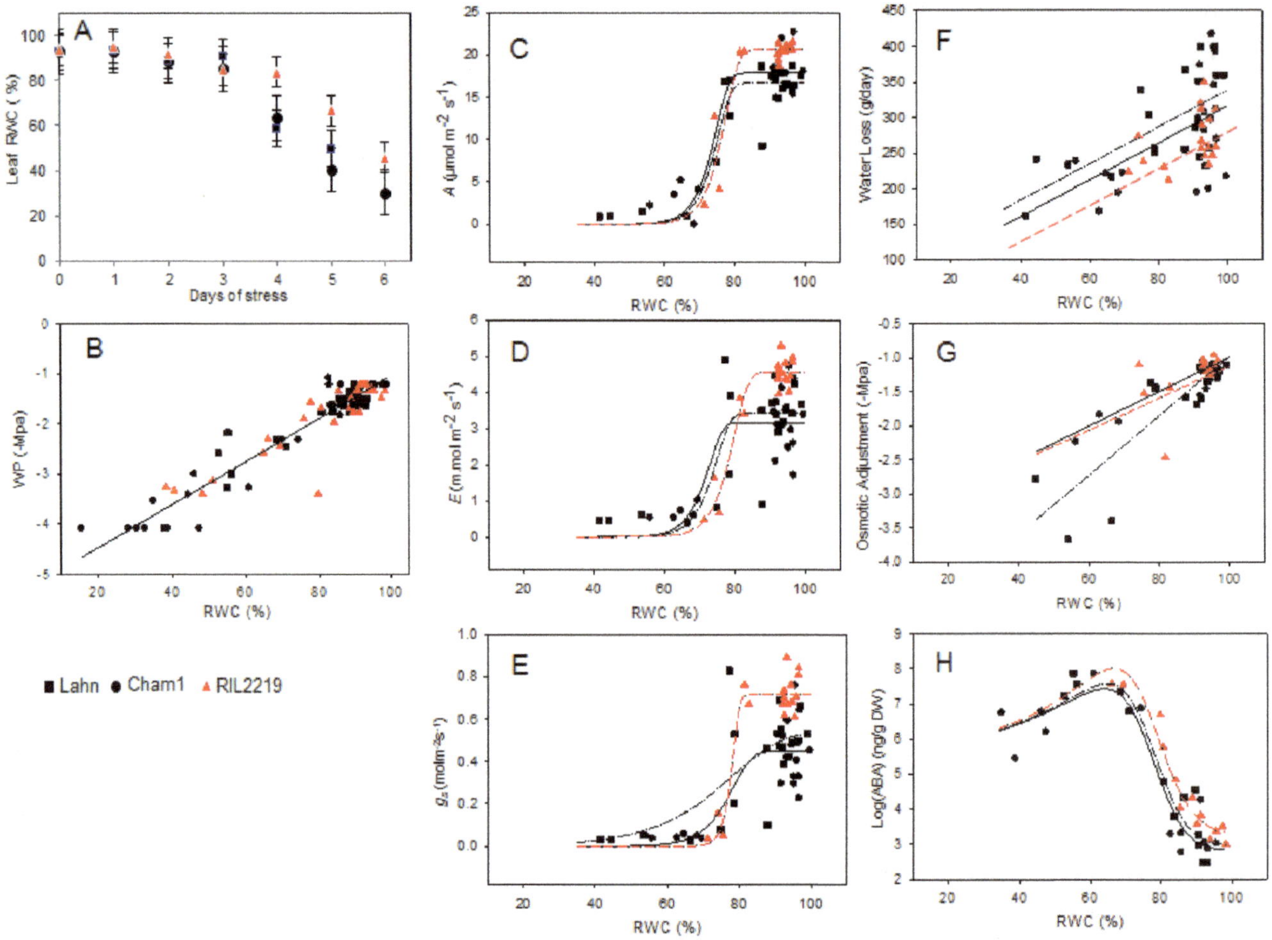

Figure 1. Physiological and biochemical parameters of leaves during water deficit. Plant water status was measured by flag leaf %RWC as a function of days of stress (A), leaf water potential (B) and total plant water loss (F) during a stress transient of 0–6 days for three wheat lines Lahn (■), Cham1 (●) and RIL2219 (▲). Leaf photosynthetic and biochemical parameters were CO_2 assimilation (C), transpiration (D), stomatal conductance (E), osmotic adjustment (G) and ABA content (H). Linear regression analysis gave a single line relationship for leaf water potential, parallel lines ($p < 0.05$, F-test) for water loss, and separate lines ($p < 0.05$, F-test) for osmotic adjustment. Using the Gompertz curve, separate C and k parameters ($p < 0.05$, F-tests) were required for CO_2 assimilation, whereas for transpiration and stomatal conductance only separate C parameters ($p < 0.05$, F-test) were required; (C–H) red line, black dashed line and black solid line are for RIL2219, Lahn and Cham1 respectively. For ABA, separate c parameters were significant ($p < 0.05$, F-test) in the rational functions model.

leaf RWC declined to 74% therefore defining a transition to the next stage of stress response.

Regulatory pathways. The largest number of probes affected by the stress transient were annotated to regulatory pathways with several members showing changes at the very early stage of the stress (82%leaf RWC) (Figure 4). Across the time course of the stress transient, probes annotated to DNA repair and synthesis (chromatin structure: histone, retrotransposon/transposase.hat-like transposase) were identified as significantly altered compared to un-stressed controls with the majority showing under-expression (File S4 bin 28). RNA processing (ribonucleases, RNA helicase, siRNA methyltransferase, splicing), regulation of transcription (transcription factor families or TFs) and RNA binding were particularly affected (File S4 bin 27). Probes annotated to several TFs showed modulation during the whole stress transient (Figure 5), and members belonging to AP2-EREBP, ARF, bHLH, bZIP, C2C2, MADS, Histone ATase, NAC, SET families and MYB showed changes at the earliest stage of stress at 82% leaf RWC with a MYB4 transcript (ta.26049.1.s1_a_at) giving one of the highest measured increases in expression (Figure 5). At 74% leaf RWC, 78% of TF transcripts showed changes in expression profiles and by the final stress point of 50% leaf RWC, 13 showed small changes specific only to that stress point and belonged to ARF, ALFIN, HB MADS, WRKY and histone HDA5 families (Figure 5, File S4 bin 27).

A large number of transcripts in protein modification specifically protein assembly and cofactor ligation, protein degradation (cysteine protease, aspartate protease, autophagy, serine protease, metalloprotease, AAA type, ubiquitin, ubiquitin E1/E2/E3/ubiquitin ubiquitin/ubiquitin ubiquitin protease/ubiquitin proteasome), protein folding, protein glycosylation, post-translational modification and protein targeting were also altered during the stress transient (Figure 4, and File S4 bin 29) with amino acid activation and protein synthesis transcripts showing predominant down-regulation (Figure 3). Transcripts for signalling components, especially a large number of calcium- and G-protein-related and various receptor kinases (leucine rich repeats classes II\ III\VI\-VIII-1\X\XI\VIX, DUF 26, legume-lectin, *Catharanthus roseus*-like RLK1, wheat LRK-10 like, proline extension like, wall associated kinase, S-locus glycoprotein like) and others included

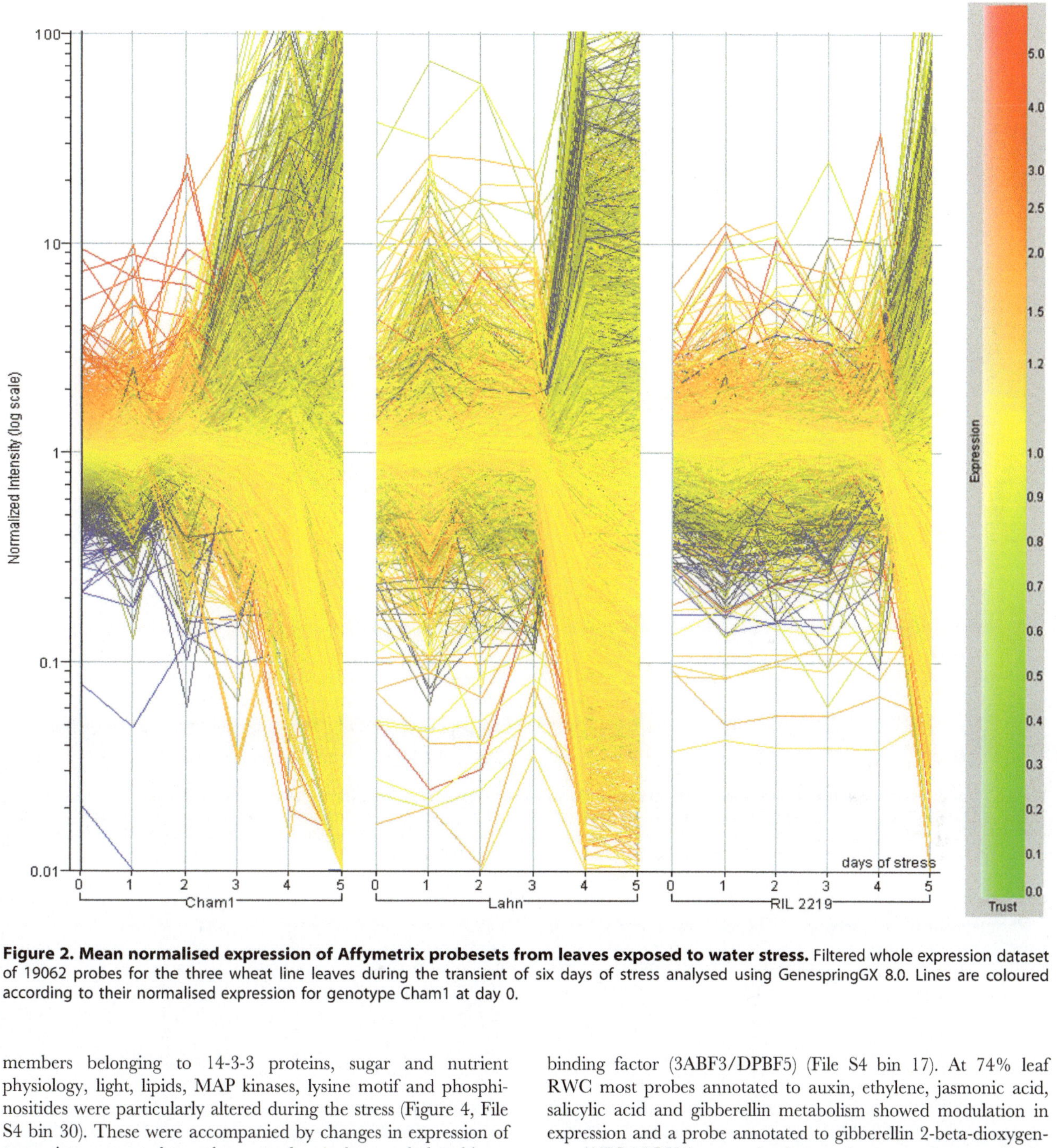

Figure 2. Mean normalised expression of Affymetrix probesets from leaves exposed to water stress. Filtered whole expression dataset of 19062 probes for the three wheat line leaves during the transient of six days of stress analysed using GenespringGX 8.0. Lines are coloured according to their normalised expression for genotype Cham1 at day 0.

members belonging to 14-3-3 proteins, sugar and nutrient physiology, light, lipids, MAP kinases, lysine motif and phosphinositides were particularly altered during the stress (Figure 4, File S4 bin 30). These were accompanied by changes in expression of transcripts annotated to redox control; ascorbate and glutathione, dismutases and catalases, glutaredoxins, peroxiredoxin and thioredoxin with most changes measured at 66% leaf RWC (Figure 4, File S4 bin 21).

Hormones and stress-specific responses. Probes belonging to each of the major categories of hormone metabolism were modulated in expression as a result of water stress with the majority showing over-expression (Figure 4, File S4 bin 17). They showed close to average change (25%) at the early stress point of 82% RWC except transcripts of abscisic acid (ABA) where 60% of probes showed early phase changes. Nine out of ten probes annotated to ABA metabolism showed over-expression across the transient with the largest change annotated to signal transduction DNA binding transcription activity ABA-responsive element-binding factor (3ABF3/DPBF5) (File S4 bin 17). At 74% leaf RWC most probes annotated to auxin, ethylene, jasmonic acid, salicylic acid and gibberellin metabolism showed modulation in expression and a probe annotated to gibberellin 2-beta-dioxygenase (ATGA2OX1) gave the largest increase amongst all the hormone classes measured at the end of the stress transient (Figure 4, File S4 bin 17; Table 1). Water stress also modulated the expression of a group of probes designated to biotic and abiotic stress-specific responses with 29% of all probes changed at 82% leaf RWC (File S4 bin 20 stress). This included universal stress proteins, genes that respond to cold, heat, touch, salt/drought stress (dehydration-responsive protein-related, drought-responsive family protein, early-responsive to dehydration protein-related/ ERD protein-related, DREPP plasma membrane polypeptide family protein, EARLY-RESPONSIVE TO DEHYDRATION 3,4 (ERD3,4)) and a large number annotated to heat shock proteins (HSP: 17, 18, 22, 21, 23, 70, 81, 83, 88, 91, 98, 101, DNAJ, GFA2 and MTHSC70-1).

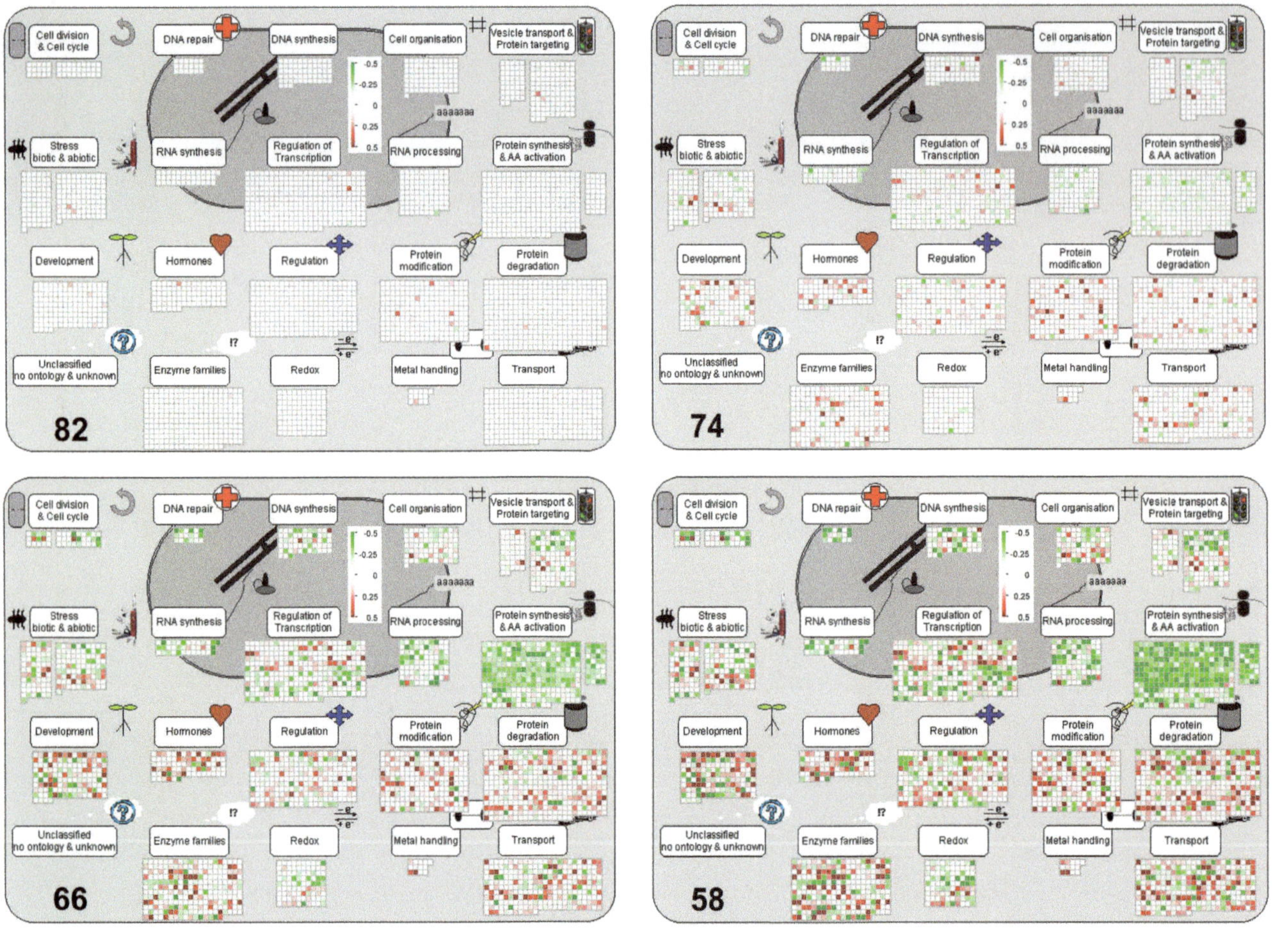

Figure 3. Overview of cell function transcript expression during the water stress transient. Transcript profile changes, at decreasing leaf %RWC (82,74,66,58), were similar across the stress transient for the three wheat lines in probes taken from MapMan bins 15,17,20,21,26,27,28,29,30,31,33,34 in the *Line+Stress* ANOVA group dataset (File S4). Results were visualised in MapMan and the colour scale for transcripts that were increased or decreased in abundance were denoted as red and green, respectively.

Photosynthesis and energy. The majority of probes annotated to photosynthesis were down-regulated during the stress transient with transcripts annotated to LHCII, PS1 and ATP synthase changing early at 82% leaf RWC (Figure 6, File S4 bin 1). Most other transcripts showed changes at 74% leaf RWC, they were annotated to light reactions (LHCII, PS2 and PS1 polypeptide subunits), Rubisco small and large subunit, plastoquinone dehydrogenase complex, ATP synthase beta chain, ferredoxin 3 (ATFD3), serine hydroxymethyltransferase (SHM4), cyclic electron flow (NDHH), phosphoglycerate kinase 1 (PGK1), triosephosphate isomerase (TIM) and phosphoribulokinase (PRK). A few were up-regulated and belonged to phosphoribulokinase (PGK), ATP synthase (VHA-A), subunit of ATP synthase (ATPA), cyclic electron flow (PIFI, NAD2.2) and PS2 polypeptide subunits. This predominant down-regulation of expression of photosynthetic transcripts was accompanied by a reduction in the expression of several probes annotated to tetrapyrrole biosynthesis (Figure 6, File S4 bin 19). Water stress also modulated the expression of transcripts of several enzymes annotated to glycolysis such as pyruvate kinase, phosphofructokinases, phosphopyruvate hydratase (LOS2), and gave increases specifically in phosphoenolpyruvate carboxylase (ATPPC3), fructose-2,6-bisphosphate 2-phosphatase (F2KP) and fructose-bisphosphate aldolase; with phosphofructokinase (PFK2) transcript showing the highest increase measured through the whole transient. Meanwhile phosphoenolpyruvate carboxylase 1 (ATPPC1), pyrophosphate-fructose-6-P phosphotransferase, triose-phosphate isomerase (ATCTIMC) and glyceraldehyde-3-phosphate dehydrogenase (GAPC1, 2) declined (File S4 bin 4). Changes in transcripts of the tricarboxylic acid cycle occurred mainly at 74% leaf RWC, with large changes in carbonic anhydrase (File S4 bin 8) and specific increases in probes annotated to gluconeogenesis at 82%leaf RWC: malate synthase, citrate synthase, malate dehydrogenase PMDH1 and pyruvate dikinase (PPDK) (File S4 bin 6). Various transcripts of mitochondrial electron transport also showed changes, such as ATP synthesis complex I and IV, cyt C and F1-ATPase, with specific increases in alternative oxidase (File S4 bin 9).

Sugar, amino acid, lipid and secondary metabolism. Water stress increased the expression of probes annotated to sucrose degradation (sucrose synthase SUS3, 4; fructokinase pfkB-type; hexokinase HXK1), some sucrose invertases (cell wall, neutral and vacuolar) and starch degradation (beta-amylase BMY8/BAM3 and alpha amylase ATAMY1). This trend was accompanied by large decreases in the expression of probes annotated to alpha-glucan phosphorylase 2, cell wall invertases such as beta-fructofuranosidase (ATFRUCT3,5), starch synthesis (synthase, branching, transporter and D-fructose-1,6-bisphosphate

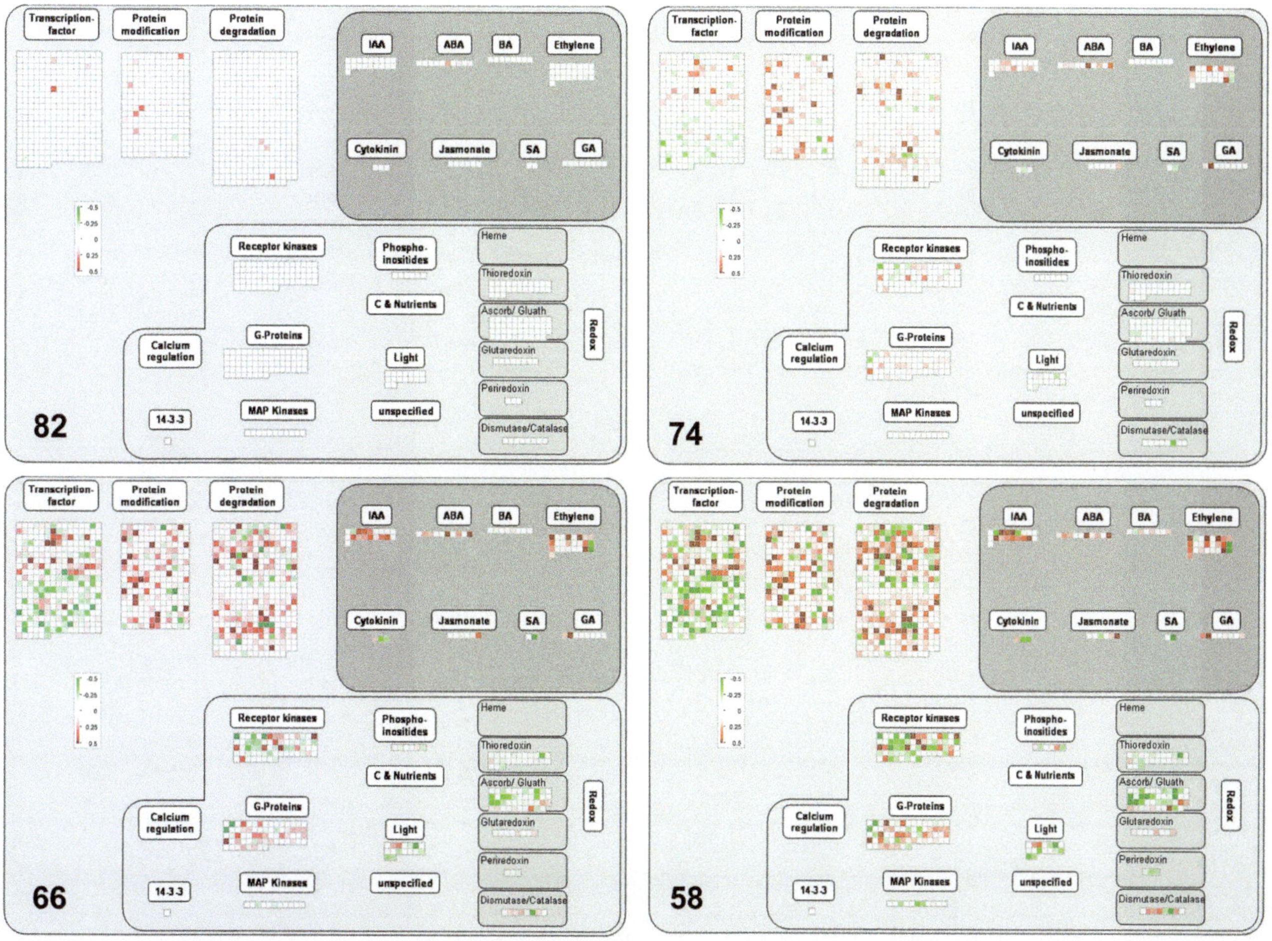

Figure 4. Expression of genes involved in regulatory pathways during water stress. Response of transcripts annotated to MapMan bins 17, 21, 27, 29 and 30 at decreasing leaf %RWC during the water stress transient. ANOVA Dataset and visualisation as for Figure 3.

1-phosphohydrolase) (Figure 6, Figure S1, File S4 bin2) and by changes in minor sugar metabolism (File S4 bin3). Various probes annotated to cell wall metabolism, including degradation and synthesis, showed large levels of change especially early in the stress at small water deficit of 82% RWC (Figure 6, File S4 bin10). Water stress produced major modulations in expression of transcripts of amino acid metabolism in both synthesis and degradation: there were decreases in the synthesis of arginine, histidine, threonine, lysine, leucine, cysteine, methionine, aspartate, asparagine, branched chain amino acids, serine, tryptophan, precursors for aromatic amino acids. There were however increases in transcripts for the synthesis of cysteine, methionine, glutamate decarboxylase 1, 4-aminobutyrate transaminase (GABA-T), aspartate transaminase, proline synthase (P5CS2), some branched-chain aminotransferase, acetolactate synthase small subunit and various aromatic amino acids. These changes were accompanied by decreases in transcripts for the degradation of histidine, GCN5-related N-acetyltransferase, threonine, methionine, lysine, asparagine, tyrosine and tryptophan and increases in those involved in the degradation of arginine, valine, branched-chain group, methionine, lysine, tyrosine and tryptophan (Figure 6, File S4 bin 13).

Water stress caused down-regulation of transcripts of nitrate reductase (NIA2) and increases in those for glutamine synthetase (cytosolic Gln1;4), glutamate dehydrogenase (GDH2) and glutamate synthase (GLT1) (Figure 6, File S4 bin12). Changes in the expression of probes annotated to lipid metabolism were also measured for biosynthetic enzymes of fatty acid desaturation, fatty acid synthesis and elongation, steroids, squalene and sphingolipid metabolism, lipid degradation (beta oxidation, lipases, lysophospholipases) lipid transfer proteins and phospholipid synthesis (Figure 6 and File S4 bin11). Probes were also altered in nucleotide metabolism with overall down-regulation in purine and ATP synthesis, whilst some coding for pyrimidine and CTP synthase were up-regulated (Figure 6, File S4 bin 23). Probes annotated to several secondary metabolic pathways were affected by the stress with predominant increases in expression of phenylpropanoids, lignin biosynthesis, flavanoids, betaine, and wax metabolism (Figure 7, File S4 bin16).

Cell organisation, transport and development. Probes annotated to cell cycle, cell division, cell organisation, development and vesicle transport were modulated by water stress (Figure 3, File S4 bin 31, 33) with probes annotated to late embryogenesis abundance showing over-expression at the earlier phases of the stress (File S4 bin 33). A large number of transcripts annotated to transport showed changes in expression such as those belonging to ABC transporters and multidrug resistance proteins; transporters for : amino acids, ammonium, calcium, sugars, sucrose, metals, cyclic nucleotide or calcium regulated channels, peptides and oligopeptides, phosphate, potassium, sulphate and

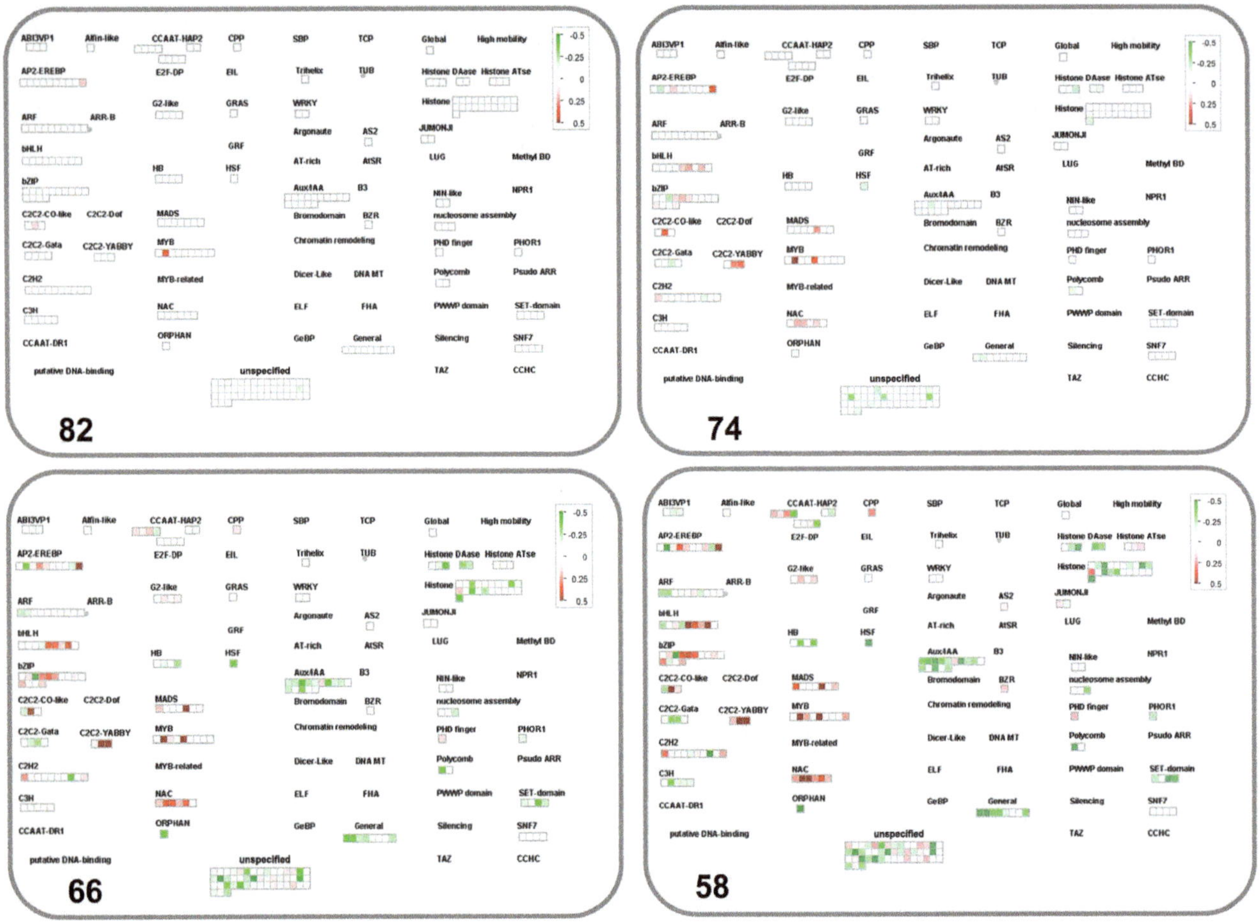

Figure 5. Changes in RNA regulation with focus on transcription during water stress. Expression of probes annotated to transcription from MapMan bin 27 and 28 at decreasing leaf %RWC during the water stress transient. ANOVA Dataset and visualisation as for Figure 3.

unspecified ions; H+ transporting pyrophosphatase; Major Intrinsic Proteins (PIP, TIP); metabolite transporters at the envelope and mitochondrial membrane; NDP-sugars at the ER; p- and v-ATPases, p- and v-ATPases.H+-transporting two-sector ATPase (File S4 bin 34).

Another group of probes, assigned to miscellaneous enzyme families involved in various metabolic processes, were modulated in response to the stress and included several members of cytochrome P450 which showed relatively large changes in expression (File S4 bin 26). Annotated probes from all metabolic bins sorted by the largest increase in expression measured at the last time point of the stress are summarised in Table 1. Probes contributing to most differences across the whole stress response were identified using PCo, see below.

Genotype-dependent responses

The ANOVA group *Line* × *Stress* comprised 7699 probes documenting an interaction between the two independent variables wheat line (two parents and RIL 2219) and stress (transient); probes that showed statistically significant responses to the stress and also measured differences between the lines in the way they responded to the water deficit. The global nature of the stress response, in all three lines, is demonstrated by the many pathways affected and by the large number of affected probes annotated to regulatory pathways presented in File S5 and Figure 8. These were essentially similar to those documented by the other ANOVA groups *Stress* (File S3) and *Line+Stress* (File S4) in showing down-regulation of photosynthesis, reconfiguration of primary and secondary metabolism and a predominant transcript modulation of regulatory pathways for RNA and protein modification. Nevertheless, each of the 7699 probes showed a statistically significant difference in expression between the lines in terms of absolute value, starting value and/or pattern of change across the transient. These are documented for each probe and partitioned into MapMan metabolic bins highlighting probes showing changes early in the stress transient for each line (File S5 early changes sorted for each line spreadsheets, Table 2). Inspection of the early phase of the stress at 82% leaf RWC showed the greatest changes in gene expression had occurred in Lahn, with 10% less changing in Cham1 and even 60% less in RIL 2219 with the latter being the most conserved. Graphical plots of the two parent's transcripts showed distinctly different profile responses to water stress, at the same leaf %RWC, across several metabolic pathways, whilst RIL 2219 combined features derived from either parent and also showed some distinct responses either in the trend or in the magnitude of the transcript expression (Figure 8). Moreover, this plot uncovers a more responsive transcriptome in Lahn, with at least one probe per bin showing a higher expression value than either Cham1 or

Table 1. Probes with the largest over-expression or under-expression values achieved at the final stress point. Data represents the ratio of change in expression at the final stress level of 50% relative to that at 90 leaf %RWC taken from the ANOVA group dataset stress+line (File S4). Transcript profile changes, as a function of leaf %RWC, were similar across the stress transient for the three wheat lines and probes were annotated to MapMan pathways and bins.

MapMan Bin#	MapMan Pathway	Affymetrix Probe	MapMan Annotation	Expression Change 50:90
1	Photosynthesis	ta.3252.1.s1_at	PS2 polypeptide subunits	1.34
2	Sugar metabolism	ta.11114.1.a1_at	ATBETAFRUCT4/VAC-INV (VACUOLAR INVERTASE)	0.93
3	Minor sugars	ta.20649.2.s1_a_at	ATTPS6 (Arabidopsis thaliana trehalose phosphatase/synthase 6)	0.83
4	Glycolysis	taaffx.80151.1.s1_at	PFK2 phosphofructokinase family protein	1.08
5	Fermentation	taaffx.734.1.s1_x_at	ALDH7B4 (ALDEHYDE DEHYDROGENASE 7B4)	0.93
6	Gluconeogenesis	ta.23970.1.a1_at	MASY malate synthase	1.59
7	OPP	ta.9372.1.s1_at	transaldolase	0.57
8	Tricarboxylic acid cycle	ta.2022.2.s1_at	aconitate hydratase, cytoplasmic, putative	0.35
9	Mitochondrial electron transport	taaffx.121853.1.s1_at	AOX1C (alternative oxidase 1C)	0.70
10	Cell wall	taaffx.7177.1.s1_s_at	glycosyl hydrolase family 3 protein	1.91
11	Lipid metabolism	ta.9528.1.a1_at	oxidoreductase, acting on the CH-CH group of donors	0.91
12	N metabolism	ta.1870.1.s1_a_at	GDH2 (GLUTAMATE DEHYDROGENASE 2)	0.78
13	Amino acid metabolism	taaffx.95414.1.s1_at	ATBCAT-3/BCAT3 (BRANCHED-CHAIN AMINOTRANSFERASE 3)	1.50
14	S assimilation	ta.11132.1.s1_at	adenylylsulfate kinase	0.14
15	Metal handling	ta.17991.1.s1_x_at	binding, chelation and storage	0.49
16	Secondary metabolism	ta.11147.1.a1_at	FAH1 (FERULATE-5-HYDROXYLASE 1)	1.11
17	Hormones	taaffx.12175.1.s1_at	ATGA2OX1 (GIBBERELLIN 2-OXIDASE 1)	1.60
18	Cofactors and vitamins	ta.13745.1.s1_a_at	LIP1 (LIPOIC ACID SYNTHASE 1	0.10
19	Tetrapyrrole biosynthesis	ta.11708.1.s1_at	HEMD; uroporphyrinogen-III synthase	0.08
20	Abiotic stress	ta.1316.2.s1_x_at	universal stress proteins (USP)family protein	1.26
20	Stress (biotic and abiotic)	ta.13785.1.s1_at	(CHIB1) acidic endochitinase	1.66
21	Redox	ta.5610.1.s1_at	AHB1 (ARABIDOPSIS HEMOGLOBIN 1)	1.04
22	Polyamine synthesis	ta.3005.1.s1_a_at	ATAIH/EMB1873 (AGMATINE IMINOHYDROLASE)	0.13
23	Nucleotide metabolism	taaffx.97704.1.a1_at	EMB2742 (EMBRYO DEFECTIVE 2742)	0.73
24	Degradation of xenobiotics	ta.8571.1.s1_a_at	lactoylglutathione lyase family protein/glyoxalase I family protein	1.83
25	C1 metabolism	ta.20549.1.s1_x_at	FDH (FORMATE DEHYDROGENASE	0.32
26	Misc enzyme families	ta.11025.1.a1_at	FAD-binding domain-containing protein	1.41
27	RNA processing	ta.26049.1.s1_a_at	MYB4 (myb domain protein 4); transcription factor	1.39
28	DNA synthesis	taaffx.128682.1.s1_at	ENDO4 (ENDONUCLEASE 4)	1.45
29	Protein modification	ta.3507.1.s1_at	SCPL16 (serine carboxypeptidase-like 16)	1.64
30	Signalling	ta.4334.1.s1_at	CPK6 (CALCIUM-DEPENDENT PROTEIN KINASE 6)	1.55
31	Cell cycle, division, vesicle transport and organisation	ta.214.1.s1_at	TUB6 (BETA-6 TUBULIN)	1.37
33	Development	ta.13396.1.s1_at	late embryogenesis abundant	1.87
34	Transport	ta.9295.1.a1_at	PGP21 (P-GLYCOPROTEIN 21); ATPase, coupled to transmembrane	1.42
1	Photosynthesis	ta.22984.2.s1_x_at	LHB1B2 (Photosystem II light harvesting complex gene 1.5); chlorophyll binding	−1.28
2	Sugar metabolism	ta.1197.1.s1_x_at	ATPHS2/PHS2 (ALPHA-GLUCAN PHOSPHORYLASE 2)	−1.04
3	Minor sugars	ta.12952.2.s1_at	pfkB-type carbohydrate kinase family protein	−0.65
4	Glycolysis	ta.9096.2.s1_x_at	ATPPC1 (PHOSPHOENOLPYRUVATE CARBOXYLASE 1)	−0.25
5	Fermentation	ta.28355.2.s1_x_at	ADH2 (ALCOHOL DEHYDROGENASE 2)	−0.52
6	Gluconeogenesis	ta.37.1.s1_at	PMDH2 (PEROXISOMAL NAD-MALATE DEHYDROGENASE 2)	−0.36
7	OPP	ta.29329.1.s1_at	phosphogluconate dehydrogenase	−0.56
8	Tricarboxylic acid cycle	taaffx.123182.1.s1_at	carbonic anhydrase family protein	−1.35

Table 1. Cont.

MapMan	MapMan	Affymetrix	MapMan	Expression Change
Bin#	**Pathway**	**Probe**	**Annotation**	**50:90**
9	Mitochondrial electron transport	ta.1147.3.s1_at	NDH-M (SUBUNIT NDH-M OF NAD(P)H:PLASTOQUINONE DEHYDROGENASE COMPLEX)	−0.56
10	Cell wall	ta.25823.1.a1_at	MUM4 | MUM4 (MUCILAGE-MODIFIED 4)	−1.12
11	Lipid metabolism	ta.4873.1.s1_at	FAD8 (FATTY ACID DESATURASE 8)	−1.32
12	N metabolism	taaffx.12494.1.s1_at	nitrogen regulation family protein	−0.58
13	Amino acid metabolism	ta.23170.1.a1_s_at	DHDPS2 (DIHYDRODIPICOLINATE SYNTHASE)	−1.06
14	S assimilation	ta.19092.1.s1_x_at	SIR (sulfite reductase)	−0.11
15	Metal handling	taaffx.86907.1.s1_at	heavy-metal-associated domain-containing protein / copper chaperone (CCH)-related	−0.13
16	Secondary metabolism	taaffx.12557.1.a1_at	terpenoid metabolism	−1.08
17	Hormones	ta.3526.1.s1_at	NFD5 (NUCLEAR FUSION DEFECTIVE 5)	−0.99
18	Cofactors and vitamins	ta.8003.1.s1_at	UbiE/COQ5 methyltransferase family protein	−0.70
19	Tetrapyrrole biosynthesis	ta.9574.1.s1_at	GUN5 (GENOMES UNCOUPLED 5)	−1.02
20	Abiotic stress	ta.3155.1.s1_at	GFA2 (GAMETOPHYTIC FACTOR 2); heat shock protein binding	−0.80
20	Stress (biotic and abiotic)	ta.27335.1.s1_at	ATMLO1/MLO1 (MILDEW RESISTANCE LOCUS O 1)	−0.84
21	Redox	taaffx.12477.1.s1_at	thioredoxin family protein.	−1.20
22	Polyamine synthesis	ta.24743.2.s1_a_at	SPDS1 (SPERMIDINE SYNTHASE 1)	−0.15
23	Nucleotide metabolism	ta.3179.1.s1_x_at	thymidylate kinase family protein	−1.20
24	Degradation of xenobiotics	ta.7031.3.s1_a_at	GLX2-2 (GLYOXALASE 2-2)	−0.10
25	C1 metabolism	ta.4843.1.a1_at	dihydropterin pyrophosphokinase	−0.73
26	Misc enzyme families	ta.385.2.s1_at	ATPRMT11/PRMT11 (ARABIDOPSIS ARGININE METHYLTRANSFERASE 11)	−1.20
27	RNA processing	ta.26907.1.s1_at	ribonuclease T2 family protein	−1.20
28	DNA synthesis	ta.13372.1.s1_at	hydrolase, alpha/beta fold family protein	−1.06
29	Protein modification	ta.19412.1.s1_at	ribosomal protein L20 family protein	−1.13
30	Signalling	ta.7378.4.s1_at	ATARCA (Arabidopsis thaliana Homolog of the Tobacco ArcA); nucleotide binding	−1.19
31	Cell cycle, division, vesicle transport and organisation	ta.14519.1.s1_x_at	regulators of chromosome condensation (RCC1family).	−0.89
33	Development	ta.7378.4.s1_at	ATARCA (Arabidopsis thaliana Homolog of the Tobacco ArcA); nucleotide binding	−1.19
34	Transport	taaffx.8804.2.s1_at	PIP2;5/PIP2D (plasma membrane intrinsic protein 2;5); water channel	−1.21

RIL2219; the only exceptions being for bins 10 (cell wall), 15 (metal handling), 16 (secondary metabolism) and 17 (Hormones).

Identifying probes responsible for most differences between lines. To explore patterns and sources of variation responsible for differences between wheat lines from those due to the stress transient, the whole RWC-interpolated dataset of 19062 probes was submitted to PCo analysis. The analysis identified three major PCos and revealed clear zoning of each line across the stress transient and also between the lines with little evidence of randomness, demonstrating the structure, global nature and tight regulation of the response to water stress (Figure 9). PCo1 accounted for 47.84% of the total variation in the sample-to-sample similarities and separated the samples with regards to increasing stress for all three lines (Figure 9 A, B). The plot suggests that each line transcriptome transitions smoothly across the stress in the direction of the arrows and the circles highlight possible system functional states Figure 9 A. PCo2, explaining 15.01% of the variance, separated the data as regards line-to-line differences, in particular separating Lahn from Cham1 and RIL 2219, these latter two lines showing more similar coordinates on this axis (Figure 9 A, C). PCo3, accounting for 9.95% of the variance, also pulled out line-to-line differences, but this time extracting differences for Cham1 from the other two lines (Figure 9 B, C). PCo2 and PCo3 therefore uncovered the global ways in which each of the two parents Cham1 (PCo2) and Lahn (PCo3) are similar to, or different from, the offspring, RIL 2219 which in turn combines transcriptome patterns derived from both parents in terms of its response to decreased RWC. The plot of PCo1 *vs.* PCo2 shows that RIL 2219 and Cham1 are closer, or more similar, throughout the transient, and that all lines become more similar as the stress progresses to the severe point at 50 leaf % RWC (Figure 9 A). Furthermore, the plot of PCo1 *vs.* PCo3 (Figure 9 B) showed that the response of Cham1 differed from the other two lines as RWC fell from 82 to 58%. Plotting PCo2 *vs.* PCo3 (Figure 9C) revealed that PCo3 can be used to consider relative within-line stability as the stress increases, three distinct

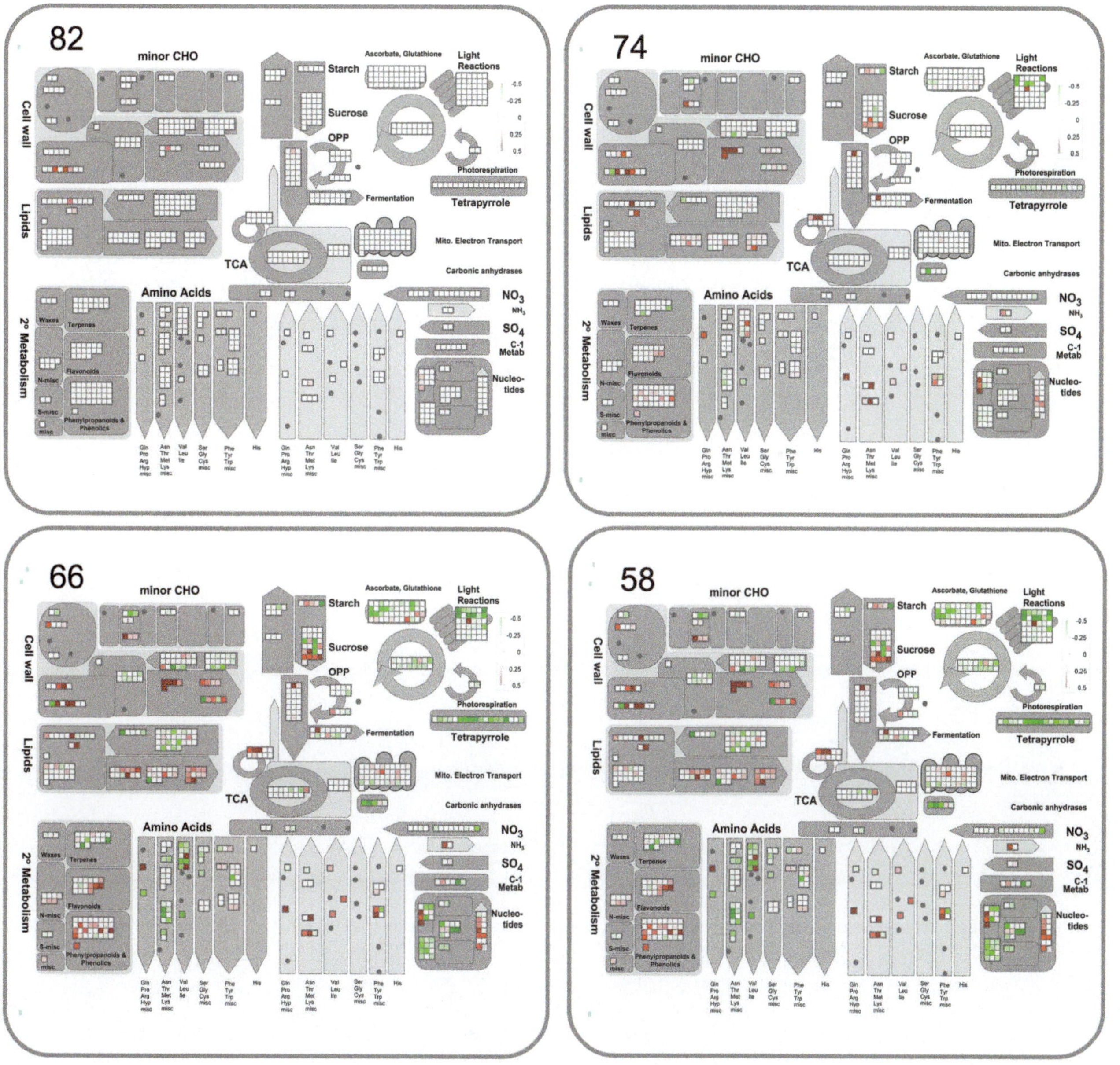

Figure 6. Overview of transcription of genes annotated to primary metabolism during water stress. Expression of probes from MapMan bins 1, 2, 3, 4, 5, 6, 7, 8, 9, 10, 11, 12, 13, 14, 16, 19, 21, 23 and 25, 30 at decreasing leaf %RWC during the water stress transient. ANOVA Dataset and visualisation as for Figure 3.

line-specific circle clouds of points are shown, the most disparate set being for Lahn (more susceptible parent) and the most clustered set being for Cham1 (more resistant parent) apart from its point at 50% RWC, which is a considerable distance away from the others. The set of points for the RIL 2219 is more similar to that of Cham1 than to that of Lahn, viewed from the PCo2 axis. Therefore, it is clear that PCo2 and PCo3 pull out differences between lines at all stress points including the relatively stress-free condition of 90% RWC. Importantly, the spread of coordinates of samples at 90% leaf RWC for all three lines demonstrates the intrinsic differences that exist between the transcriptomes of each line before the onset of water deficit.

The analysis also enabled the identification of the major probes responsible for differences between lines and those due to the stress transient. The 5% of probes most significantly correlated (F-tests) with the PCo coordinates PCo1, PCo2 and PCo3 (File S7) were imported into the MapMan programme for assignment to bins and results showed differences in the distribution of these probes across the metabolic pathways (Figure 10). Thus, PCo1 identified 1–3 probes in annotated bins 1 (photosynthesis), 10 (cell wall), 13 (amino acid metabolism), 23 (nucleotide synthesis), 26 (misc. enzyme families), 30 (signalling), 31 (cell organisation), 33 (development), 34 (transport), and several belonging to regulatory bins 27 (RNA processing), 29 (protein modification) and bin 34 (transport). The probe with the greatest F-value (Ta.3659.1.-

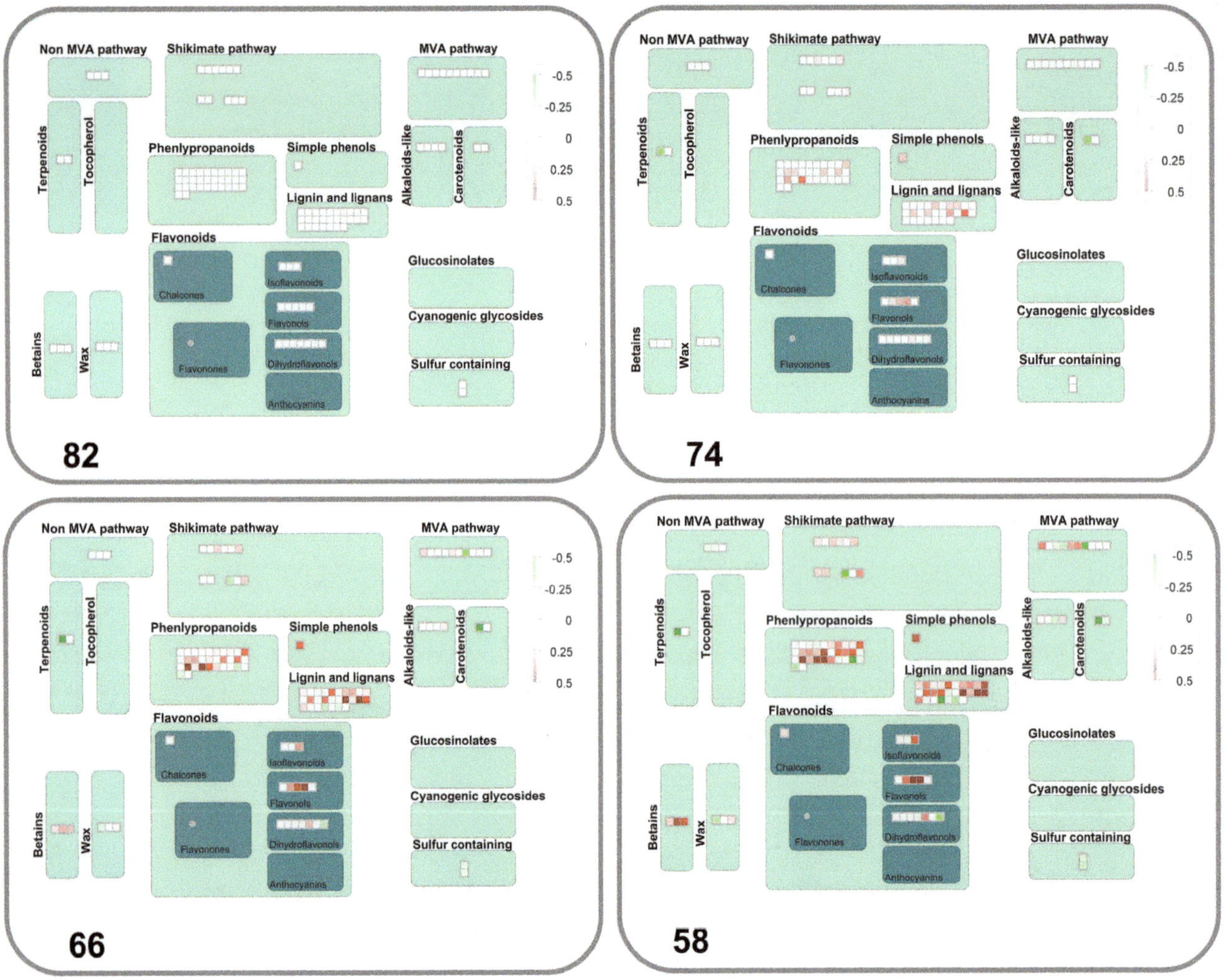

Figure 7. Changes in transcription of genes annotated to secondary metabolism during water stress. Expression of probes annotated to MapMan bin 16 at decreasing leaf %RWC during the water stress transient. ANOVA Dataset and visualisation as for Figure 3.

S1_a_at) gave no annotation using MapMan, so we explored this further using PLEXdb to obtain a putative annotation to heat shock factor-binding protein 1, using rice alignment at Gramene (LOC_Os09g20830.2). PCo2 identified probes in annotated bins 11 (lipid metabolism), 28 (DNA), 29 (protein modification) and 34 (transport); the probe with the highest F-value (taaffx.3401.1.s1_at) was annotated to vacuolar sorting protein 39, having small GTPase regulator activity and association with the TGF beta receptor. Finally, PCo3 identified probes in bins 2 (major sugars), 10 (cell wall), 11 (lipid metabolism), 16 (secondary metabolism), 17 (hormones), 21(redox), 27 (RNA regulation), 29 (protein modification), 30(signalling) and 34 (transport) with the most influential probe ta.16038.1.s1_at annotated to hormone metabolism- ABA inducible, encoding one of five HVA22 homologs in *Arabidopsis*; an ABA- and stress-inducible gene.

Another source of variation that can influence the expression values for probes between the three lines is allelic variation at the gene sequence level. This may affect the binding of a particular probe from a given probe set and the extent of its occurrence in our study was estimated by calculating the ratio of a probe signal to that averaged over the eleven probes in the corresponding probe set. A preliminary analysis was done for probe sets annotated to transcription factors, bin27, for the three ANOVA groups, *line*, *line+stress* and *line x stress* where such differential hybridisation may be an important consideration when comparing expression differences between the lines. The results showed that for the *line only* ANOVA group, 15 probes out of 396 probes (36 probe sets each containing 11 individual probes) showed some evidence of differential hybridisation between the lines but with no stress response (File S8). The analysis also showed that for the *line+stress* ANOVA group, 140 individual probes out of a total of 4862 showed statistically different hybridisation when compared to the average for probe sets; however, this component did not affect the response to water stress since all lines behaved similarly. The group for which such allelic hybridisation may contribute to the differential stress response amongst the three lines is the *line x stress* ANOVA group and results showed that 108 individual probes out of a total of 4807 (File S8) may have such an interaction; further verification in new studies would be required to confirm this.

Probes showing no response to stress

The ANOVA group, *Line*, identified 932 probes that were significantly different in transcript expression between the three

Table 2. Top two probes showing largest changes in expression during the early phase in the stress for each wheat line. Data represents the ratio of change in expression (over and under-expression values) at 82% relative to that at 90 leaf %RWC taken from the ANOVA group dataset *line x stress* (File S5).

MapMan Bin#	MapMan Annotation	Affymetrix Probe	82:90
			Lahn
26	misc.cytochrome P450	ta.3813.1.a1_at	**0.61**
27	RNA.regulation of transcription.Aux/IAA family	ta.10395.1.s1_a_at	**0.47**
34	transport.peptides and oligopeptides	taaffx.65026.1.a1_at	**−0.31**
29	protein.synthesis.ribosomal protein.prokaryotic.mitochondrion.50S subunit.L2	ta.28514.2.s1_at	**−0.46**
			Cham1
16	secondary metabolism.phenylpropanoids.lignin biosynthesis.CAD	ta.2659.1.s1_at	**0.38**
16	secondary metabolism.simple phenols	ta.4455.1.a1_at	**0.35**
34	transport.metal	ta.9290.1.s1_at	**−0.11**
20	stress.biotic	taaffx.108556.1.s1_at	**−0.11**
			RIL2219
17	hormone metabolism.ABA.synthesis-degradation.9-cis-epoxycarotenoid dioxygenase	taaffx.76007.1.s1_at	**0.46**
29	protein.degradation.ubiquitin.E3.SCF.FBOX	taaffx.12903.1.a1_at	**0.26**
8	TCA/org. transformation.carbonic anhydrases	taaffx.37789.1.a1_at	**−0.11**
20	stress.abiotic.heat. Heat shock protein HSP21	ta.202.1.s1_at	**−0.36**

lines but did not show changes in response to the water stress transient (File S2). We also identified 208 probes (File S6) that showed no statistical difference either between lines or in response to the stress but which were selected during the initial filtering step using Genespring GX. These two ANOVA groups could be used to provide potential control genes for future water stress transcript expression studies in durum wheat.

Discussion

Durum wheat gene expression responds to water stress in a global and regulated manner

The stress transient enabled the capture of the complexity of the plant's response to water stress as a first step in dissecting inherent features of the biological system under perturbation. The

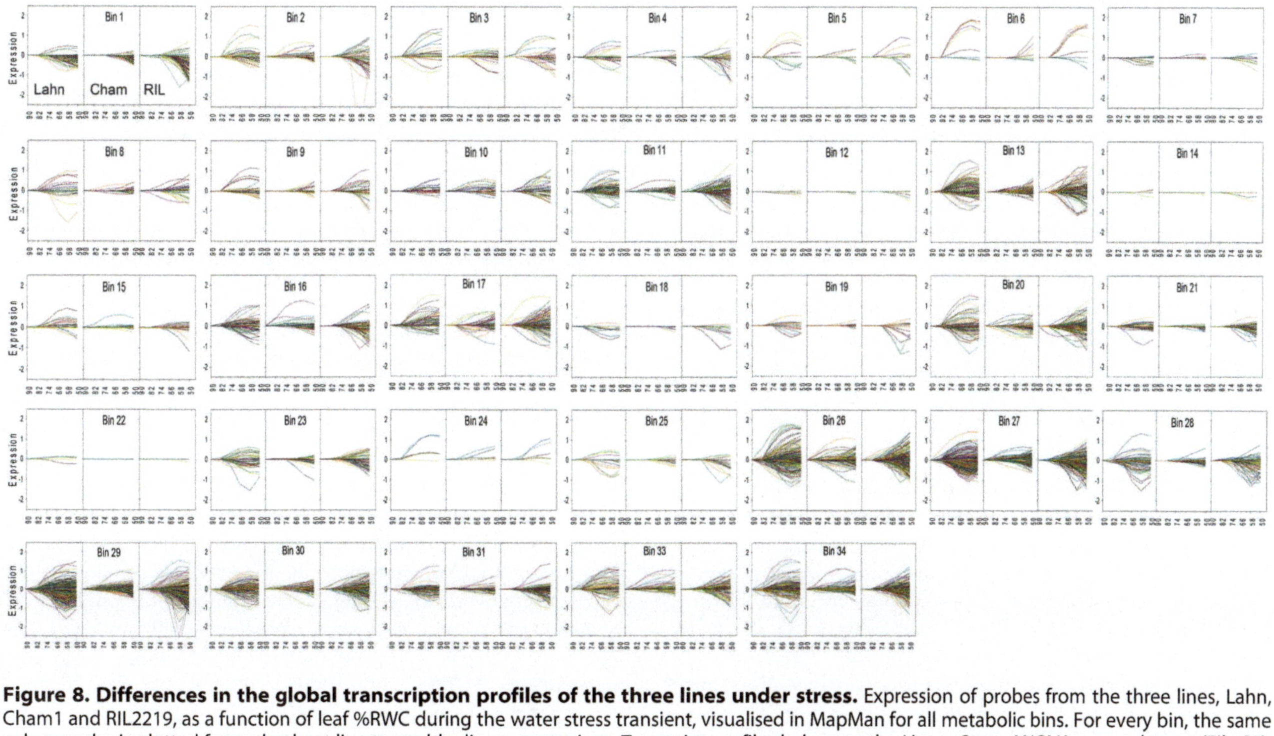

Figure 8. Differences in the global transcription profiles of the three lines under stress. Expression of probes from the three lines, Lahn, Cham1 and RIL2219, as a function of leaf %RWC during the water stress transient, visualised in MapMan for all metabolic bins. For every bin, the same colour probe is plotted for each wheat line to enable direct comparison. Transcript profiles belong to the *Line* × *Stress* ANOVA group dataset (File S5).

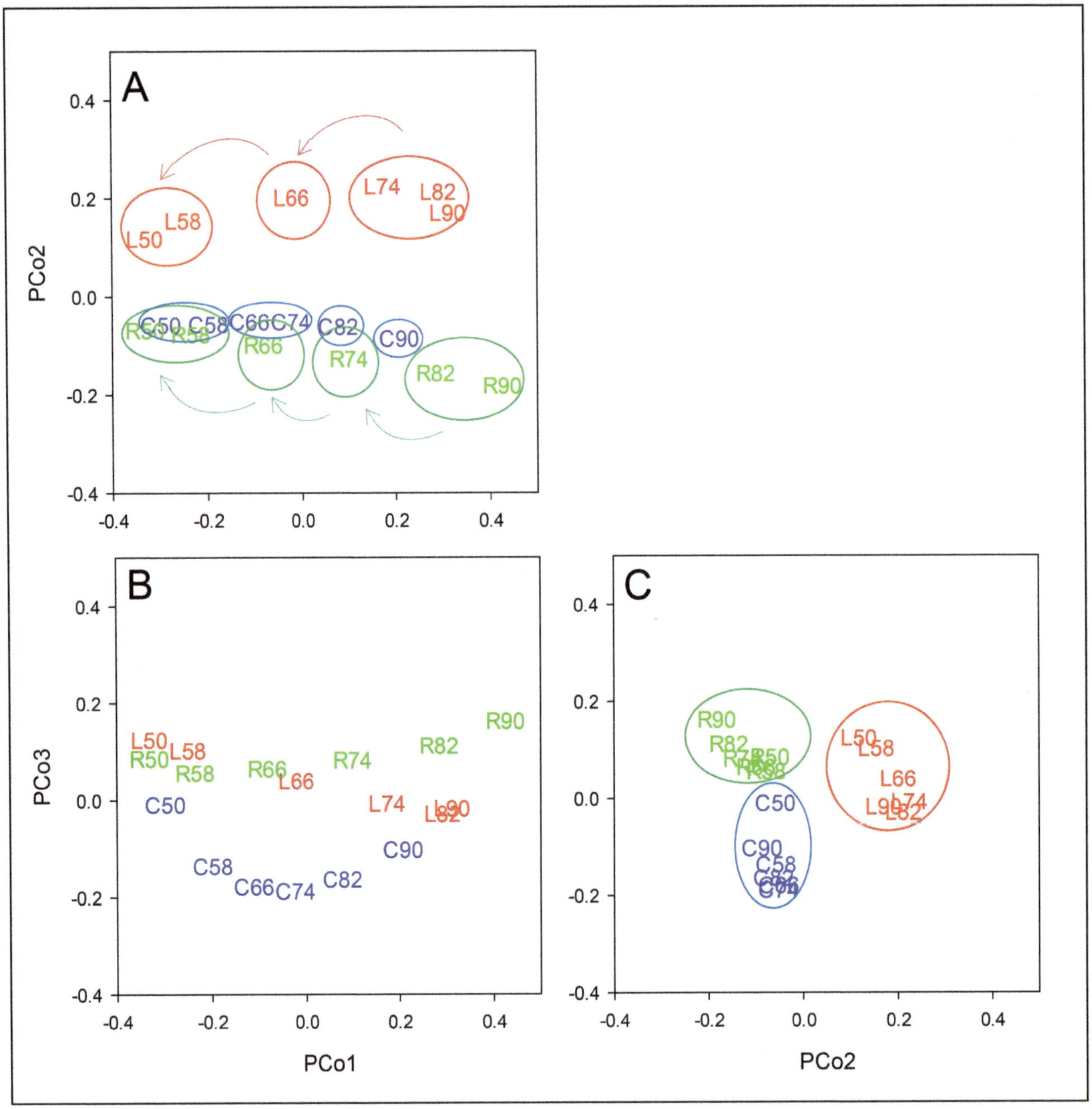

Figure 9. PCo plots for the expression of all 19062 probe dataset. The first three PCos account for 72.8% of the variation represented in the (reduced) similarity matrix for the 18 wheat line by RWC combinations. Visualisation of the combinations in the three dimensions was achieved by plotting PCo1 *vs.* PCo2 (A), PCo1 *vs.* PCo3 (B) and PCo2 *vs.* PCo3 (C). Letters L, C and R represent the lines Lahn, Cham1 and RIL 2219 respectively at 90-50 leaf %RWC values. Arrows on plot A show the direction of the stress transient of decreasing leaf RWC and the free-drawn circles around the points highlight potential functional states for each line, inferred from distance between points.

mathematical models identified all statistically significant responses, without filtering, which uncovered all changes in transcript expression, not just focussing on a particular arbitrary magnitude of change, or exploring only specific pathways; unlike most published studies. The results showed that the durum wheat transcriptome responded in a global manner to progressive water stress as demonstrated by the modulation of a large number of probes annotated to various metabolic pathways in all three lines; this has been observed in previous datasets on various species [36,37,38,39,22,40,41]. Our study also showed evidence of a highly regulated response in the proportionally large number of probes coding for regulatory genes (DNA, RNA and protein modification, transcription factors, hormones and signalling) specifically altering expression at the early phases of the stress. This supports and demonstrates the fundamental components of regulatory networks of sensing, transducing and responding that operate at the genome level when exposed to abiotic stress [16,42]. These are essential features of system robustness that enable the maintenance of function under stress through immediate changes in some metabolic pathways, documented in File S3, File S4, File S5, and also via the launch of downstream pathways such as protective secondary metabolism. Particular attention has been

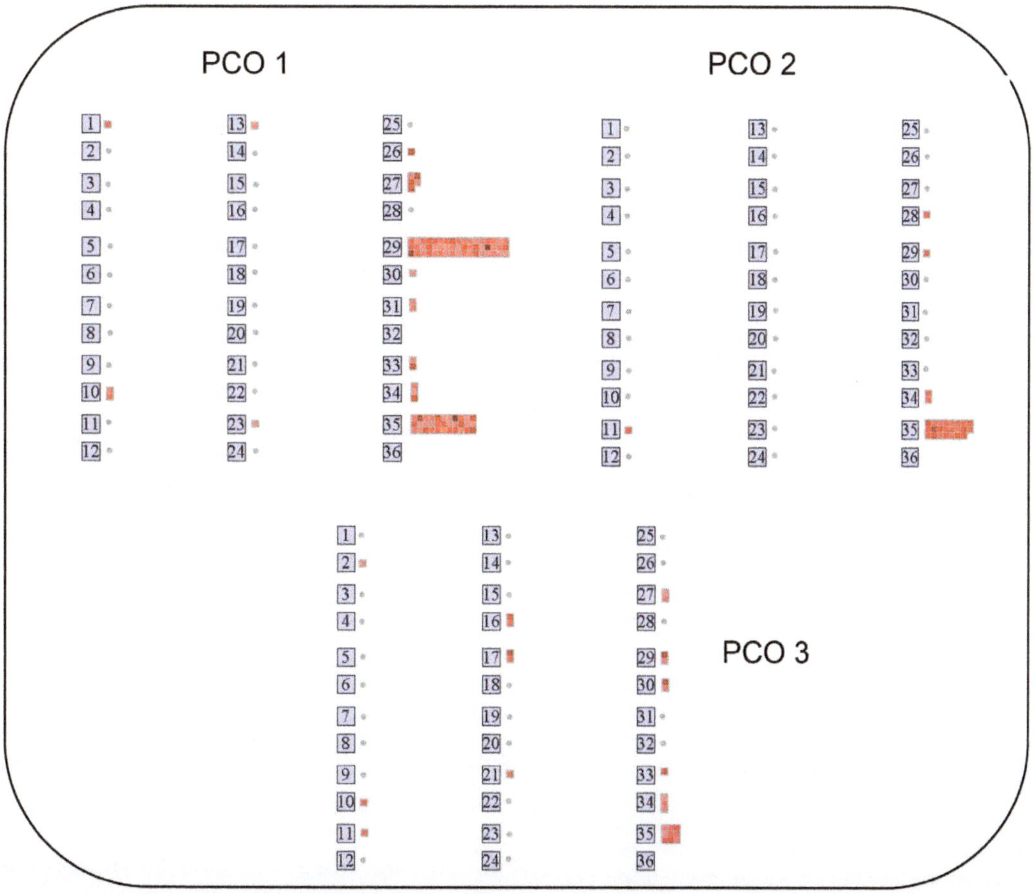

Figure 10. MapMan visualisation of the top PCo probes responsible for differences. 5% of probes with greatest F-values following PCO (data from File S7) showed differences in distribution across all metabolic bins for PCo1, PCo2, and PCo3. For presentation in Mapman, FPCO maximum values were set at 2000, 700 and 450 for PCo1, PCo2, and PCo3 respectively.

given to transcription factors as major control points in the plant system's response to water stress [43] and here we present the first overview of the water stress response profile of several durum wheat family members changing expression over the range of RWC. We identified particular groups of probes annotated to AP2-EREBP, ARF, bHLH, bZIP, C2C2, MADS, Histone ATase, SET and MYB4 families that were altered in leaf transcription early on in the stress transient when no other physiological changes were measurable. This reinforces the importance of transcription factors in the regulation and coordination of transcriptional and metabolic responses which form essential components of regulatory networks. Another interesting result is that of the modulations in transcripts for heat shock proteins throughout the stress transient. Molecular chaperones play a crucial role in the reestablishment of cellular homeostasis under abiotic stress [44] and alterations to heat shock proteins, caused by shared metabolic responses, are common to abiotic stresses such as water deficit and heat and thus share fundamental mechanisms as discussed by [17].

The changes in expression of a large number of probes annotated to protein modification and signalling in addition to hormones further supports this integrated global view of the wheat genome stress response as the system aims to balance key processes such as protein synthesis, cellular maintenance, metabolism, growth and development. Thus, these multifaceted changes in gene expression of the regulatory pathways are best interpreted as essential features of robustness as the system adjusts to perturbations and moves from one functional state to another, as envisaged in Figure 9A. Therefore, the tendency to consider particularly large increases in gene expression as all candidates for 'improving' plant growth under drought is flawed and such individual changes should be understood in the context of robustness with all its elements of degeneracy, feedback, homeostasis and environmental tracking operating within the biological system [45,46]. To establish how this highly regulated response is co-ordinated at the gene expression level across the stress transient, mathematical and bioinformatics models can be used to uncover new gene networks for water stress response in cereals and other species [47,42]. Recently, co-expression network algorithms have been successfully used to explore gene networks, network architecture and control hubs, and to sketch out new hypotheses for further testing [48,49]. Future studies would also enable a comparison of our datasets with similar ones from barley, bread wheat, maize and rice, in order to determine which transcriptome responses are conserved as a common framework amongst cereals and which are species- and environment-specific.

Water stress downregulates photosynthesis and alters transcripts for sugar metabolism

Durum wheat leaves exhibited a decline in photosynthetic parameters, osmotic adjustment and increases in ABA content as the stress progressed through the transient of declining leaf RWC, as expected [50]. This was accompanied by a lowering of photosynthetic, photorespiratory and tetrapyrrole biosynthetic transcripts which reflects a down-regulation of light capture and carbon fixation. Such patterns have been seen in previous studies in dicots, rice, barley and wheat and support a common framework for the plant's photosynthetic responses to water stress [48]. We also found that the lowered expression of photosynthetic transcripts was coupled by increases in expression for sucrose and starch breakdown, and a decrease for starch synthesis. This implies a requirement for soluble sugar availability for respiration, solute homeostasis and/or signalling throughout the plant [51]. The perturbations observed in key regulators of carbon metabolism and glycolysis such as phosphofructokinase, phosphoenolpyrovate carboxylase and pyruvate kinase with a concomitant increase in expression of enzymes of gluconeogenesis and sucrose invertases may also contribute to an increase in sucrose hydrolysis [52] in agreement with analysis of metabolites under water deficit [51]. These results provide some evidence for leaf osmotic adjustment as a strategy to maintain cell turgor pressure [51,53], it might also reflect the need for respiratory substrates as respiration is maintained when photosynthesis slows and ceases.

Water stress reconfigures transcription of primary and secondary metabolic pathways

The complex changes measured in transcripts coding for amino acid and lipid metabolism, GABA-T, and various aminotransferases, demonstrate the reconfiguration of primary metabolism under stress as the system moves between different metabolic 'states of function'. The overall downregulation of transcripts for synthesis of most amino acid groups and an increase in the expression of transcripts involved in amino acid breakdown occurred downstream in the stress response as the plant adjusts metabolism from synthesis towards remobilisation and protective metabolic pathways, in addition to providing substrate for respiration as photosynthesis declines. This is demonstrated by specific changes in transcripts for key enzymes involved in amino acid and inorganic nitrogen assimilation. For example, the down-regulation of nitrate reductase implies a decreased assimilation of inorganic nitrogen and this was accompanied by over-expression of transcript annotated to cytosolic glutamine synthetase, glutamate dehydrogenase and glutamate synthase which in turn would release glutamate and 2 oxoglutarate for recycling, thus accelerating remobilisation and senescence [54], and possibly increasing the production of proline via glutamate. This enables increased remobilisation of nitrogen assimilates as metabolism is adjusting the decline in growth with primary synthesis under stress conditions.

We also measured changes in transcripts coding for secondary metabolism such as phenylpropanoids, lignin biosynthesis, flavonoids, betaine, and wax metabolism, thus providing evidence for a protective adjustment role in the stress response. The increases we observed in the expression of an orthologue for proline biosynthesis, P5CS2 might be considered as providing proline for osmotic adjustment and/or reflecting an imbalance in amino acid regulation. Other adaptive responses such as changes in expression of transcripts for cell wall biosynthesis, measured early on in the stress at small changes in RWC, demonstrate the dynamism of mature cell walls in response to decreased turgor and supports recent studies [55,56,57,11]. We also observed that a range of responsive transcripts to both biotic and abiotic stress were also modulated and this supports the existence of common global gene networks for environmental stress responses; interestingly [35] also noted similar changes in biotic pathogen-related gene expression under water stress in wheat.

Wheat lines show constitutive differences in gene expression

The time series approach allowed the dissection of the complex transcriptional response into the wheat specific component and those specific to changes in RWC, as well as identifying probes showing an interaction between the two. A large number of probes showed inherent or constitutive differences in transcript expression between the wheat lines before the stress had started; all probes in File S2, File S4 and some probes in File S5. This is clearly seen in the PCO plots in Figure 9 where the values at leaf 90%RWC are dispersed across all axes for all three lines. Large systemic differences in gene expression between germplasm within a species, including between siblings and parent, have been shown to be strikingly common in rice [58], emmer wheat [41] and Arabidopsis [59]. It is unclear however, at this stage of study, how these differences contribute towards the ability of a line to resist stress and these results should be coupled to expression QTL studies dissecting GxE to enable further evaluation. Studying the transcriptome of wheat lines under both controlled and field conditions would then enable us to understand the importance of these global differences between lines in terms of the variation in the final integrated responses to the stress in the complex field environment and we are currently evaluating such datasets from field trials. Integrating genomic high-throughput datasets with fine mapped QTLs in future studies would consequently bridge the link between changes at the molecular level with downstream yield-related traits underpinning final grain yield [21].

Differential gene expression responses amongst wheat lines under stress

The three wheat lines responded to water stress in a global framework of common changes of gene expression in regulatory, primary, secondary, and protective metabolic pathways. In addition, we documented thousands of probes that showed significant differences in the transcript response for each wheat line during the stress transient (File S5). These differences were analysed at equivalent leaf RWC reflecting real differences in gene responses between the susceptible and resistant lines which may prove to be the most interesting in terms of targets for genotype-specific responses under water stress. The probe responsible for inducing most variation across the whole dataset in terms of line-to-line stability, derived from PCo3, was annotated to an ABA inducible HVA22E or a late embryogenesis abundant protein which is implicated in a protective stress response by regulating vesicular traffic in stressed cells [60]. This, and other probes identified by PCo, can now be tested in new mechanistic studies to establish whether they are new regulatory control points for drought resistance in durum wheat. An interesting result from our study was that Lahn, the drought sensitive parental line, had a higher number of modulated probes in response to the stress, especially at the initial phase of the transient. This was true at equivalent leaf %RWC, across most pathways, and in comparison to the other two lines, especially RIL2219 which had the lowest number of modulated probes at that stage. An examination of the literature highlighted few comparative analyses in rice [39], a maize drought resistant RIL line with fewer modulated genes under stress [61] whilst drought in maize landraces showed a mixed picture [62]. Therefore, it may also be important to consider the metrics of gene expression changes in addition to the nature of the genes and the networks involved.

In addition to differential gene expression documented between the three lines, we have estimated the contribution of allelic differential hybridisation to the transcripome response. We have identified a small proportion of such individual probes in bins annotated to transcription factor probe sets, most of which did not influence the response of probes to water stress (ANOVA group *line only* and *Line+stress*). Nevertheless, a handful were identified as promising in the ANOVA group *Line x stress* and one in particular, probe # 602715 belonging to the probe set ta.28513.1.s1_s_at and annotated as an Auxin Response Factor family, deserves close verification since it was also identified by PCo analysis as a major candidate for within-line stability under stress (File S7 and S8). These represent possible candidates for marker development and should be further examined in future studies to verify their potential for durum wheat marker assisted breeding.

One of the most striking findings is the co-ordinated and highly structured transcriptome response for each wheat line across the water stress transient as demonstrated by the lack of scatter and high level of structure in Fig. 9C. Each line started at a different point along the axes for the three PCos demonstrating inherent differences in transcript expression across most metabolic pathways and this distinctness is maintained during the transient with some indication of reaching closer values at the final stress point. This provides strong evidence for a distinct cultivar-specific global transcriptome response in each wheat line so that the whole genome is responding in a highly coordinated manner. Very few studies have undertaken such a global analysis and for those that have utilised PCo, a similar interpretation can also be derived [22]. These, together with constitutive differences are important findings and should be factored into current simplistic efforts to identify a singular mechanism for drought resistance and genetic modification of a particular selected metabolic pathway. We argue that a shift in thinking is required to integrate high throughput datasets to advance our understanding of plant responses under stress and that robustness theory may offer such a framework.

Physiological traits differ before and during the stress

The most drought resistant RIL 2219 line had higher photosynthetic and transpiration rates, greater stomatal conductance and ABA content prior to the initiation of stress and was also able to maintain greater RWC and leaf water potential for a longer period than either parent. It is interesting to note that parent Cham1 has relatively larger stomatal conductance when compared to other breeding varieties [63]. The greater rate of leaf transpiration may seem counterintuitive for a drought resistant line, but if coupled with lower leaf area and/or bigger capacity to capture water (for example from a more efficient root system) then it can afford to transpire slightly more and maintain its stomata open for longer to allow entry of CO_2 for fixation. We argue that this could provide an important advantage in the sustained accumulation of assimilate, especially during the fast vegetative growth period where water may be available, and hence increased grain yield if integrated over the whole growth season. In addition, the higher leaf transpiration rate for RIL2219 may also reduce the canopy temperature, this being another added benefit for durum wheat during terminal stress when drought is often compounded with heat stress [64,65,66]. We propose that these combined factors are responsible for RIL2219's inherent greater capacity for carbon assimilation and thus the ability to maintain higher yield stability than either parent under drought in the field, as supported by datasets from a large number of field trials under various drought gradients (unpublished observations). Evidence exists that wheat breeders have inadvertently selected for higher stomatal conductance in their quest for greater yield potential under optimal conditions, and this has also proved to be beneficial under water-limiting Mediterranean environments [67,64,68,69,70,65,71,72,73,74]. There was also evidence for greater osmotic adjustment ability in the two drought resistant lines Cham1 and RIL 2219 than in Lahn. Taken together, improving osmotic adjustment, and increasing carbon fixation and stomatal conductance may define one physiological strategy for success in breeding high yield stability wheat for Mediterranean drought prone environments. Further systems studies are now necessary to link the transcriptome, proteome, metabolome and phenome of wheat to establish how the different levels of function interact to produce a final yield under stress.

Conclusions

We found evidence of a common global framework of physiological, biochemical and gene expression responses in all three wheat lines under water stress. The stress transient identified the predominant modulation of the transcriptome of regulatory pathways early in the stress response, downregulation of photosynthesis, reconfiguration of primary metabolism and the launch of protective downstream pathways at various leaf RWC of relevance to terminal drought experienced by crops under field conditions. We also uncovered genotype-specific transcriptome and physiological differences before and during the stress amongst the three wheat lines which suggests that each wheat line functions within a genome-specific structure that form the basis of different stress-resistance strategies. We interpret the results by ascribing systems–based concepts of robustness as a fundamental framework of the plant's response to perturbation under stress. By targeting our studies to durum wheat lines from the breeding programme at ICARDA, our results should be useful to breeders searching for novel variation and new concepts to test and exploit in wheat cultivar improvement.

Supporting Information

Figure S1 Expression of genes in sugar metabolism during stress. Expression of probes annotated to MapMan bin 2 as a function of decreasing leaf %RWC. ANOVA Dataset and visualisation as for Figure 3.

File S1 ANOVA results of probe expression values of three wheat lines in response to water stress. Whole dataset of 19062 probes interpolated values, pre-ANOVA, for individual replicates at each leaf %RWC level and for the three lines Lahn, Cham1 and RIL2219 in spreadsheet 'Interpolated data at RWC levels'. ANOVA, log (base2), split the whole dataset into 5 independent statistical groups represented by individual spreadsheets: probes showing differences in expression between the lines but not during the stress in spreadsheet 'line only 932 probes'; probes showing differences in expression as a result of the stress transient but not between lines in spreadsheet 'stress only 2579 probes'; probes with differences in expression between the lines at the onset of the stress with the profile changing in the same way during the stress transient in spreadsheet 'line+stress 7644 probes'; probes with differences in expression due to line and stress combined, showing an interaction in spreadsheet 'line x stress 7699 probes'; and probes with no differences due to stress or line after initial filtering in spreadsheet 'no effect 208 probes'. Statistical results are given as least significant difference (LSD) at 5% and 1% levels of significance for each of the ANOVA groups.

File S2 MapMan analysis of ANOVA group *line only* 932 probes. Probes showing statistically different expression values between the three lines Lahn, Cham1 and RIL2219 but with no effect of the water stress. Data were calculated as expression value beyond the relevant LSD (5%) given in File S1 when comparing means for each probe between the three lines Lahn, Cham1 and RIL2219, or otherwise zero when there was no significant difference. Probes are presented by Affymetrix id and MapMan annotation to bin name, pathway and description of gene model.

File S3 MapMan analysis of ANOVA group *stress only* 2579 probes. Probes showing no differences in expression between the three lines Lahn, Cham1 and RIL2219 but changing in response to the stress. Data were calculated as expression value beyond the relevant LSD (5%) given in File S1 when comparing means for stressed (82, 74, 66, 58 and 50% leaf RWC) to control (90% leaf RWC) conditions for each probe, or otherwise zero when there was no significant difference. Probes are presented by Affymetrix id and MapMan annotation to bin name, pathway and description of gene model. Analysis of results are presented in individual spreadsheets for all MapMan bins and sorted for different leaf %RWC. In addition, probes with the greatest levels of over-expression or under-expression measured at the earliest stress point of 82% are highlighted in spreadsheets 'early changes over-expression' and 'early changes under-expression', respectively.

File S4 MapMan analysis of ANOVA group *line+stress* 7644 probes. Probes showing consistent differences in expression between the three lines Lahn, Cham1 and RIL2219 in response to the stress transient –*i.e.* they show a parallel profile of change in transcript response across the stress transient. Data were calculated as expression value beyond the relevant LSD (5%) of data in File S1 when comparing means for stressed (82, 74, 66, 58 and 50% leaf RWC) to control (90% leaf RWC) conditions for each probe, or otherwise zero when there was no significant difference. Data were also calculated as the differences in expression between the three lines. Probes are presented by Affymetrix id and MapMan annotation to bin name, pathway and description of gene model. Analysis of results are presented in individual spreadsheets for all MapMan bins and sorted for different leaf %RWC. In addition, probes with the greatest levels of over-expression or under-expression measured at the earliest stress point of 82% are highlighted in spreadsheets 'early changes over-expression' and 'early changes under-expression', respectively. Percentage changes in probes at 82:90 and 74:90% leaf RWC stress points are also included in spreadsheet 'summary of early changes' and calculated as a percentage of probes changed at 82:90 or 74:90 as a proportion of all probes showing changes in each specific bin across the whole transient.

File S5 MapMan analysis of ANOVA group *line x stress* 7699 probes. Probes showing differences in expression between the three lines Lahn, Cham1 and RIL2219, *and* in response to the stress; they show an interaction. Data were calculated as expression value beyond the relevant LSD (5%) given in File S1 when comparing means for stressed (82, 74, 66, 58 and 50% leaf RWC) to control (90% leaf RWC) conditions for each probe from each of the three lines Lahn, Cham1 and RIL2219, or otherwise zero when there was no significant difference. Probes are presented by Affymetrix id and MapMan annotation to bin name, pathway and description of gene model. Analysis of results are presented in individual spreadsheets for all MapMan bins and sorted for different leaf %RWC for the three lines. In addition, probes from all the bins which showed expression changes at the earliest stress point of 82% are highlighted for the three lines in spreadsheet 'early changes' and sorted for the three lines in subsequent spreadsheets. A summary of two probes showing the largest changes measured at the early stage of 82%RWC for each line is presented in spreadsheet 'summary top changes early'.

File S6 MapMan analysis of ANOVA '*no effect* 208 probes' group. Probes, post initial filtering, showing no difference in expression between the three lines Lahn, Cham1 and RIL2219, or in response to the stress from data in File S1. Probes are presented by Affymetrix id and MapMan annotation to bin name, pathway and description of gene model.

File S7 MapMan annotation of PCo results. Mapman analysis of probes responsible for maximum separation observed, as defined by F-tests for the correlation between probes and principal coordinates (PCos), following PCo applied to the RWC-interpolated data for the 19062 probes derived from 54 samples (three replicates of each of six RWC levels by three wheat lines). Individual spreadsheets show annotation for the probes relevant for the first three principal coordinates retained from the analysis (PCo1, 2, 3). Probes are presented by Affymetrix id and MapMan annotation to bin name, pathway and description of gene model.

File S8 The contribution of allelic differences in gene sequences to probe hybridisation for probes annotated to transcription factors. ANOVA for each perfect match probe and the significance of the ratio changes across lines was calculated in proportion to the corresponding mean probe-set abundance. The name of the probe and P value are given for the corresponding probe set taken from *line only*, *line+stress* and *Line x stress* ANOVA groups datasets from File S3,4,5 respectively focusing on MapMan bin 27.

Acknowledgments

We acknowledge Peter Hedden for help with ABA measurements and ICARDA Durum Wheat Breeding Program for providing the lines from the Cham1xLahn mapping population. This work is dedicated to Kawther Latiri (deceased).

Author Contributions

Conceived and designed the experiments: DH SP KL AA RT DL JA MN. Performed the experiments: MB DL DH. Analyzed the data: MH SP MD-P RM CR MS DH. Contributed reagents/materials/analysis tools: DH KL AA RT JA MN. Wrote the manuscript: DH. Developed the genetic material: MN.

References

1. Nachit MM, Nachit G, Ketata H, Gauch HG, Zobel RW (1992) Use of AMMI and linear regression models to analyze genotype-byenvironment interaction in durum wheat. Theor. Appl. Genet 83: 597–601.
2. Araus JL, Slafer GA, Reynolds MP, Royo C (2002) Plant breeding and drought in C3 cereals: what should we breed for? Ann. Bot. 89, Special No, 925–940.
3. Stocker TF, Qin D, Plattner GK, Alexander LV, Allen SK, et al. (2013) Technical Summary. In: Stocker TF, Qin D, Plattner GK, Tignor M, Allen SK, Boschung J, Nauels A, Xia Y, Bex V and Midgley PM, editors. Climate Change 2013: The Physical Science Basis. Contribution of Working Group I to the Fifth Assessment Report of the Intergovernmental Panel on Climate Change Cambridge University Press, Cambridge, United Kingdom and New York, NY, USA.

4. Bates BC, Kundzewicz ZW, Wu S, PalutikofJP, Eds. (2008) Climate change and water. Technical Paper of the Intergovernmental Panel on Climate Change. IPCC Secretariat, Geneva.
5. Hoerling M, Eischeid J, Perlwitz J, Quan X, Zhang T, et al. (2012) On the Increased Frequency of Mediterranean Drought. J Clim 25: 2146–2161.
6. Habash DZ, Kehel K, Nachit MM (2009) Genomic approaches for designing durum wheat ready for climate change with a focus on drought. J. Exp. Bot. 60: 2805–2815.
7. WHEAT (2011) Global alliance for improving food security and the livelihoods of the resource-poor in the developing world. CIMMYT publication, Mexico.
8. Nachit MM, Elouafi I (2004) Durum adaptation in the Mediterranean dryland: breeding, stress physiology, and molecular markers. In: Rao SC, Ryan J, editors. Challenges and strategies for dryland agriculture. CSSA Special Publication 32. Madison, Wisconsin, USA: Crop Science Society of America Inc., American Society of Agronomy Inc, 203–218.
9. Maccaferri M, Sanguineti MC, Corneti S, Araus JL, Salem MB, et al. (2008) Quantitative Trait Loci for grain yield and adaptation of durum wheat (Triticum durum Desf.) across a wide range of water availability. Genetics 178: 489–511.
10. Moinuddin, Fischer RA, Sayre KD, Reynolds MP (2005). Osmotic Adjustment in Wheat in Relation to Grain Yield under Water Deficit Environments. Agron. J. 97: 1062–1071.
11. Alvarez S, Marsh EL, Schroeder SG, Schachtman DP (2008) Metabolomic and proteomic changes in the xylem sap of maize under drought. Plant Cell Environ. 31: 325–340.
12. Harb A, Krishnan A, Ambavaram MMR, Pereira A (2010) Molecular and Physiological Analysis of Drought Stress in Arabidopsis Reveals Early Responses Leading to Acclimation in Plant Growth. Plant Physiol. 154: 1254–1271.
13. Ahuja I, de Vos R, Bones AM, Hall RD (2010) Plant molecular stress responses face climate change. Trends Plant Sci. 15: 664–674.
14. Mochida K, Shinozaki K (2010) Genomics and Bioinformatics Resources for Crop Improvement. Plant Cell Physiol. 51: 497–523.
15. Yamaguchi-Shinozaki K, Shinozaki K (1994) A novel cis-acting element in an Arabidopsis gene is involved in responsiveness to drought, low temperature, or high-salt stress. Plant Cell 6: 251–264.
16. Nakashima K, Ito Y, Yamaguchi-Shinozaki K (2009) Transcriptional Regulatory Networks in Response to Abiotic Stresses in Arabidopsis and Grasses1. Plant Physiol. 149: 88–95.
17. Shinozaki K, Yamaguchi-Shinozaki K (2007) Gene networks involved in drought stress response and tolerance. J. Exp. Bot. 58: 221–227.
18. Reguera M, Peleg Z, Blumwald E (2011) Targeting metabolic pathways for genetic engineering abiotic stress-tolerance in crops. Biochim. Biophys. Acta. 2012: 186–94.
19. Tuberosa R, Salvi S (2006) Genomics-based approaches to improve drought tolerance of crops. Trends Plant Sci. 11: 405–12.
20. Cattivelli L, Rizza F, Badeck F, Mazzucotelli E, Mastrangelo A, et al. (2008) Drought tolerance improvement in crop plants: An integrated view from breeding to genomics. Field Crops Res. 105: 1–14.
21. Fleury D, Jefferies S, Kuchel H, Langridge P (2010) Genetic and genomic tools to improve drought tolerance in wheat. J. Exp. Bot. 61: 3211–3222.
22. Aprile A, Mastrangelo A, De Leonardis A, Galiba G, Roncaglia E, et al. (2009) Transcriptional profiling in response to terminal drought stress reveals differential responses along the wheat genome. BMC Genomics, 10: 279.
23. Lawlor DW (2013) Genetic engineering to improve plant performance under drought: physiological evaluation of achievements, limitations, and possibilities. J Exp Bot. 64: 83–108.
24. Barr HD, Weatherley PE (1962) A re-examination of the relative turgidity technique for estimating water deficit in leaves. Aust. J. Biol. Sci. 15: 413–428.
25. Scholander PF, Hammel HT, Hemmingsen EA, Bradstreet ED (1964) Hydrostatic pressure and osmotic potential in leaves of mangroves and some other plants. Proc. Natl. Acad. Sci. U.S.A. 52: 119–25.
26. Fernandez-Garcia N, Hernandez M, Casado-Vela J, Bru R, Elortza F, et al. (2011) Changes to the proteome and targeted metabolites of xylem sap in *Brassica oleracea* in response to salt stress. Plant Cell Environ. 34: 821–836.
27. Wu ZJ, Irizarry RA, Gentleman R, Martinez-Murillo F, Spencer F (2004) A model-based background adjustment for oligonucleotide expression arrays. J Am Stat Assoc 99: 909–917.
28. Bar-Joseph Z (2004) Analyzing time series gene expression data. Bioinformatics. 20: 2493–2503.
29. Mastroianni G, Milovanovic G (2008) Interpolation Processes: Basic Theory and Applications. Springer Monographs in Mathematics, Springer-Verlag, Berlin, Germany.
30. Gasca M, Sauer T (2000) On the history of multivariate polynomial interpolation, J Comput Appl Math 122: 23–25.
31. Benjamini Y, Hochberg Y (1995) Controlling the false discovery rate: a practical and powerful approach to multiple testing. J R Stat Soc Series B 57: 289–300.
32. Krzanowski WJ (2000) Principles of multivariate analysis: a user's perspective. Oxford University Press, Oxford, UK.
33. Gautier L, Cope L, Bolstad BM, Irizarry RA (2004) affy–analysis of Affymetrix GeneChip data at the probe level. Bioinformatics 20: 307–315.
34. Thimm O, Bläsing O, Gibon Y, Nagel A, et al. (2004) MAPMAN: a user-driven tool to display genomics data sets onto diagrams of metabolic pathways and other biological processes. Plant J. 37: 914–39.
35. Usadel B, Nagel N, Thimm O, Redestig H, Blaesing OE, et al. (2005) Extension of the Visualization Tool MapMan to Allow Statistical Analysis of Arrays, Display of Coresponding Genes, and Comparison with Known Responses. Plant Physiol 138: 1195–1204.
36. Xue GP, McIntyre CL, Chapman S, Bower NI, Way H, et al. (2006) Differential gene expression of wheat progeny with contrasting levels of transpiration efficiency. Plant Mol. Biol. 61: 863–81.
37. Kilian J, Whitehead D, Horak J, Wanke D, Weinl S, et al. (2007) The AtGenExpress global stress expression data set: protocols, evaluation and model data analysis of UV-B light, drought and cold stress responses. Plant J. 50: 347–63.
38. Talamè V, Ozturk NZ, Bohnert HJ, Tuberosa R (2007) Barley transcript profiles under dehydration shock and drought stress treatments: a comparative analysis. J. Exp. Bot. 58: 229–240.
39. Degenkolbe T, Do PT, Zuther E, Repsilber D, Walther D, et al. (2009) Expression profiling of rice cultivars differing in their tolerance to long-term drought stress. Plant Mol. Biol. 69: 133–53.
40. Ergen NZ, Thimmapuram J, Bohnert HJ, Budak H (2009). Transcriptome pathways unique to dehydration tolerant relatives of modern wheat. Funct. Integr. Genomics 9: 377–396.
41. Krugman T, Chagué V, Peleg Z, Balzergue S, Just J, et al. (2010) Multilevel regulation and signalling processes associated with adaptation to terminal drought in wild emmer wheat. Funct Integr Genomics 10: 167–186.
42. Cramer G R, Urano K, Delrot S, Pezzotti M, Shinozaki K (2011) Effects of abiotic stress on plants: a systems biology perspective. BMC Plant Biol. 11: 163.
43. Hirayama T, Shinozaki K (2010) Research on plant abiotic stress responses in the post- genome era: past, present and future. Plant J. 61: 1041–1052.
44. Wang W, Vinocur B, Shoseyov O, Altman A (2004) Role of plant heat-shock proteins and molecular chaperones in the abiotic stress response. Trends Plant Sci. 9: 244–252.
45. Kitano H (2007) Towards a theory of biological robustness. Mol. Syst. Biol. 3: 137.
46. Whitacre JM (2012) Biological robustness: paradigms, mechanisms, and systems principles. Front Genet. 11; 3: 67.
47. Claverie JM (1999) Computational methods for the identification of differential and coordinated gene expression. Hum. Mol. Genet. 8(10): 1821–32.
48. Pinheiro C, Chaves MM (2011). Photosynthesis and drought: can we make metabolic connections from available data? J. Exp. Bot. 62: 869–882.
49. Swarbreck SM, Defoin-Platel M, Hindle M, Saqi M, Habash DZ (2010) New perspectives on glutamine synthetase in grasses. J. Exp. Bot. 62: 1511–1522.
50. Lawlor DW, Tezara W (2009) Causes of decreased photosynthetic rate and metabolic capacity in water-deficient leaf cells: a critical evaluation of mechanisms and integration of processes. Ann. Bot. 103: 561–579.
51. Xue GP, McIntyre CL, Glassop D, Shorter R (2008) Use of expression analysis to dissect alterations in carbohydrate metabolism in wheat leaves during drought stress. Plant Mol. Biol. 67: 197–214.
52. van Dongen JT, Gupta KJ, Ramírez-Aguilar SJ, Araújo WL, Nunes-Nesi A, et al. (2011) Regulation of respiration in plants: a role for alternative metabolic pathways. J. Plant Physiol. 168: 1434–43.
53. Rosa M, Prado C, Podazza G, Interdonato R, González JA, et al. (2009) Soluble sugars–Metabolism, sensing and abiotic stress. Plant Signal Behav. 4: 388–393.
54. Miflin BJ, Habash DZ (2002) The role of glutamine synthetase and glutamate dehydrogenase in nitrogen assimilation and possibilities for improvement in the nitrogen utilization of crops. J Exp Bot. 53: 979–87.
55. Konno H, Yamasaki Y, Sugimoto M, Takeda K (2008) Differential changes in cell wall matrix polysaccharides and glycoside-hydrolyzing enzymes in developing wheat seedlings differing in drought tolerance. Plant Physiol. 165: 745–54.
56. Moura JC, Bonine CA, de Oliveira Fernandes Viana J, Dornelas MC, Mazzafera P (2010) Abiotic and biotic stresses and changes in the lignin content and composition in plants. J Integr Plant Biol. 52: 360–76.
57. Sasidharan R, Voesenek LACJ, Pierik R (2011) Cell wall modifying proteins mediate plant acclimatization to biotic and abiotic Stresses, CRC Crit Rev Plant Sci 30: 6, 548–562.
58. Moumeni A, Satoh K, Kondoh H, Asano T, Hosaka A, et al. (2011) Comparative analysis of root transcriptome profiles of two pairs of drought-tolerant and susceptible rice near-isogenic lines under different drought stress. BMC Plant Biol. 11: 174.
59. Gan X, Stegle O, Behr J, Steffen JG, Drewe P, et al. (2011) Multiple reference genomes and transcriptomes for Arabidopsis thaliana. Nature 477: 419–423.
60. Guo WJ, Ho TH (2008) An abscisic acid-induced protein, HVA22, inhibits gibberellin-mediated programmed cell death in cereal aleurone cells. Plant Physiol. 147: 1710–22.
61. Zheng J, Fu J, Gou M, Huai J, Liu Y, et al. (2010) Genome-wide transcriptome analysis of two maize inbred lines under drought stress. Plant Mol. Biol. 72: 407–21.
62. Hayano-Kanashiro C, Calderón-Vázquez C, Ibarra-Laclette E, Herrera-Estrella L, Simpson J (2009) Analysis of Gene Expression and Physiological Responses in Three Mexican Maize Landraces under Drought Stress and Recovery Irrigation. PLoS ONE 4: e7531.
63. Monneveux P, Rekika D, Acevedo E, Merah O (2006) Effect of drought on leaf gas exchange, carbon isotope discrimination, transpiration efficiency and productivity in field grown durum wheat genotypes. Plant Sci. 170: 867–872.
64. Fischer RA, Rees D, Sayre KD, Lu ZM, Condon AG, et al. (1998) Wheat Yield Progress Associated with Higher Stomatal Conductance and Photosynthetic Rate, and Cooler Canopies. Crop Sci. 38: 1467–1475.

65. Araus JL, Casadesus J, Bort J (2001) Recent Tools for the Screening of Physiological Traits Determining Yield. In Reynolds, MP, Ortiz-Monasterio JI, McNab A (eds.). Application of Physiology in Wheat Breeding. Mexico, D.F.: CIMMYT.
66. Reynolds MP, Nagarajan S, Razzaque MA, Ageeb OAA (2001) Heat tolerance. In Reynolds, MP, Ortiz-Monasterio JI, McNab A (eds.). Application of Physiology in Wheat Breeding. Mexico, D.F.: CIMMYT. 88–100.
67. Blum A (1996) Yield potential and drought resistance: are they mutually exclusive? In:. Reynolds MP, Rajaram S, McNab A (eds.). Increasing Yield Potential in Wheat: Breaking the Barriers. Mexico, D.F.: CIMMYT. 76–89.
68. Delgado MI, Reynolds MP, Larque-saavedra A, Nava T (1994) Mexico, Genetic diversity for photosynthesis in wheat under heat-stressed environments and its relationship to productivity. Wheat Special Report No. 30, CIMMYT, Mexico.
69. Brennan JP, Condon AG, Van Ginkel M, Reynolds MP (2007) An economic assessment of the use of physiological selection for stomatal aperture-related traits in the CIMMYT wheat breeding programme. J Agric Sci. 145: 187–194.
70. Condon AG, Richards RA, Rebetzke GJ, Farquhar GD (2004) Breeding for high water-use efficiency. J. Exp. Bot. 55: 2447–2460.
71. Giunta F, Motzo R, Pruneddu G (2008) Has long-term selection for yield in durum wheat also induced changes in leaf and canopy traits? Field Crops Res. 106: 68–76.
72. Bahar B, Yildirim M, Barutcular C (2009) Relationships between stomatal conductance and yield components in spring durum wheat under Mediterranean conditions. Not Bot Horti Agrobo 37: 45–48.
73. Richards RA, Rebetzke GJ, Watt M, Condon AG, Spielmeyer W, et al. (2010) Breeding for improved water productivity in temperate cereals: phenotyping, quantitative trait loci, markers and the selection environment. Functional Plant Biology, 2010, 37, 85–97.
74. Khamssi NN, Najaphy A (2012) Comparison of photosynthetic components of wheat genotypes under rain-fed and irrigated conditions. Photochem. Photobiol. 88: 76–80.

18

AtRD22 and AtUSPL1, Members of the Plant-Specific BURP Domain Family Involved in *Arabidopsis thaliana* Drought Tolerance

Vokkaliga Thammegowda Harshavardhan[1], Le Van Son[2,3], Christiane Seiler[1], Astrid Junker[4], Kathleen Weigelt-Fischer[4], Christian Klukas[5], Thomas Altmann[4], Nese Sreenivasulu[1,6], Helmut Bäumlein[2], Markus Kuhlmann[1]*

1 Research Group Abiotic Stress Genomics, Interdisciplinary Center for Crop Plant Research (IZN), Halle (Saale), Germany, and Leibniz Institute of Plant Genetics and Crop Plant Research (IPK) Gatersleben, Stadt Seeland, OT Gatersleben, Germany, **2** Research Group Gene Regulation, Leibniz Institute of Plant Genetics and Crop Plant Research (IPK) Gatersleben, Stadt Seeland, OT Gatersleben, Germany, **3** National Key Laboratory of Gene Technology, Institute of Biotechnology Vietnam, Academy of Science and Technology, Hanoi, Vietnam, **4** Research Group Heterosis, Leibniz Institute of Plant Genetics and Crop Plant Research (IPK) Gatersleben, Stadt Seeland, OT Gatersleben, Germany, **5** Research Group Image Analysis, Leibniz Institute of Plant Genetics and Crop Plant Research (IPK) Gatersleben, Stadt Seeland, OT Gatersleben, Germany, **6** Grain Quality and Nutrition Center, International Rice Research Institute (IRRI), Metro Manila, Philippines

Abstract

Crop plants are regularly challenged by a range of environmental stresses which typically retard their growth and ultimately compromise economic yield. The stress response involves the reprogramming of approximately 4% of the transcriptome. Here, the behavior of *AtRD22* and *AtUSPL1*, both members of the *Arabidopsis thaliana* BURP (BNM2, USP, RD22 and polygalacturonase isozyme) domain-containing gene family, has been characterized. Both genes are up-regulated as part of the abscisic acid (ABA) mediated moisture stress response. While *AtRD22* transcript was largely restricted to the leaf, that of *AtUSPL1* was more prevalent in the root. As the loss of function of either gene increased the plant's moisture stress tolerance, the implication was that their products act to suppress the drought stress response. In addition to the known involvement of AtUSPL1 in seed development, a further role in stress tolerance was demonstrated. Based on transcriptomic data and phenotype we concluded that the enhanced moisture stress tolerance of the two loss-of-function mutants is a consequence of an enhanced basal defense response.

Editor: Jin-Song Zhang, Institute of Genetics and Developmental Biology, Chinese Academy of Sciences, China

Funding: The work was funded by Vietnamese government. The junior Research Group Abiotic Stress Genomics (MK and NS) is funded by IZN (Interdisciplinary Center for Crop Plant Research, Halle (Saale), Germany. VTH and CS are funded by BMBF (GABI-GRAIN: Integrative genomics approach for exploring seed quality and yield under terminal drought; FKZ 0315041A and FKZ 0315041C) and IB-BMBF (Ind09/526). The funders had no role in study design, data collection and analysis, decision to publish, or preparation of the manuscript.

Competing Interests: The authors have declared that no competing interests exist.

* Email: kuhlmann@ipk-gatersleben.de

Introduction

Abiotic stress factors such as moisture stress, salinity, extreme temperature and variable light intensity can disturb plant metabolism and growth. It has been estimated that crop yield losses due to such stresses lie in the order of 50% [1], so increasing the resilience of crop plants will be an important contributor to yield stability. Abiotic stress affects both photosynthesis and photorespiration, as well as having an impact on the energy and redox status of the plant cell. One of the most damaging products of stress is the group of compounds referred to as reactive oxygen species (ROS). The plant response also includes the induced synthesis of certain enzymes and low molecular weight compounds associated with antioxidant activity, redox regulators, chaperones such as heat shock proteins and late embryogenesis abundant proteins, water and ion transporters, the production of compatible osmolytes to maintain cellular water content and the fine tuning of proteolysis involved in programmed cell death. Numerous attempts, with varying levels of success, have been made to genetically engineer the production of some of these components with a view to enhancing abiotic stress tolerance [2].

The phytohormone abscisic acid (ABA) is involved in the regulation of expression of many stress-responsive genes, although other stress responsive genes are known to be regulated in an ABA-independent manner [3–5]. The regulation of several stress-inducible genes is mediated by the interaction of bZIP transcription factors (ABFs) with ABA-response elements (ABRE) in target gene promoters [6].

The *Arabidopsis thaliana* gene *RESPONSIVE TO DEHYDRATION22* (*AtRD22*), originally identified as a gene which responded to dehydration [7,8], encodes a member of the BURP protein family, members of which share a highly conserved BURP domain at their C terminus [9]), sometimes also referred to as a U domain [10,11]. BURP proteins appear to be plant-specific; some examples are the *Vicia faba* unknown seed protein (USP) [10], the

Brassica napus microsporogenesis-specific protein BNM2 [12–14], the *Panicum maximum* apomixis-specific protein [15] and the wheat pollen protein RAFFTIN [16]. The soybean genome harbors 23 BURP protein encoding genes, of which 17 are responsive to stress [17–20]. The 18 *Populus trichocarpa* BURP family members fall into five recognizable sub-families [21]. There are 15 related genes in maize, one of which is specifically expressed in the root cortex parenchyma [22] while the sorghum genome harbors 11 homologs [23]. The ectopic expression of the *A. thaliana* gene *AtUSPL1* has been shown to distort seed development and to alter the morphology of seed lipid vesicles [24].

AtRD22 is up-regulated by moisture stress, salinity stress and exogenously supplied ABA [8] and its induction has been used as a marker for abiotic stress [25–29]. The heterologous expression in both *A. thaliana* and rice of the soybean gene *GmRD22* enhances salinity stress tolerance [30]. Members of the BURP family are up-regulated by stress in rice [31], soybean [30] and maize [23]. Members of the BURP family have been described in relation to stress conditions. In cotton, an RD22-like protein interacts with an α-expansin and the over expression of both proteins simultaneously promotes growth and fruit weight [32].Remarkably, the expression programmes active during stress response partially overlap with gene expression during early embryogenesis and seed desiccation [24]. Most likely ancient stress response genes have been recruited to protect seed tissue from dehydration stress in drying seeds [33,34]. Here, a combination of genetic, molecular and physiological approaches has been applied to isolate and characterize T-DNA insertion mutants of *AtRD22* and *AtUSPL1*.

Materials and Methods

Plant material and growth conditions

A. thaliana seeds (ecotype Col-0 and the three T-DNA insertion lines SALK_146066 (*rd22-1*), WiscDsLox481 484P12 (*rd22-2*) and SALK_022325 (*uspl1*)) obtained from the European Arabidopsis Stock Center (NASC) were stratified before imbibing them for three days at 4°C in the dark. The resulting germinated seedlings were grown in soil for four weeks under 60% relative humidity, a constant temperature of 22°C and under a 16 h photoperiod (light intensity of 120 μmol m^{-2} s^{-1}). Drought stress was applied by a. active dehydration under low humidity for 1–5 days and b. withholding water for 1–5 days; the relative soil water content was monitored using an HH2 moisture meter (delta-T devices, Cambridge, UK). Moisture stress was also mimicked in two week old seedlings by transferring them for three days on a MS basal medium containing one of 150 mM or 300 mM NaCl, 100 μM ABA, 4% w/v trehalose, 4% w/v sorbitol, 4% w/v glucose, 4% w/v fructose, 4% w/v sucrose, 300 mM mannitol, 15% w/v PEG 6000 or 4% w/v PEG 20000.

Detailed information is provided in Figure 1/Figure S4. Primers used for genotyping are listed in Table S1. Drought stress was applied by withdrawal of water and relative soil water content [%] was controlled with HH2 moisture meter (delta-T devices Ltd, Cambridge, UK).

Plasmid construction and plant transformation

Standard protocols [35] were used to prepare plasmids for Gateway Cloning Technology (Invitrogen GmbH, Karlsruhe, Germany). To achieve its ectopic expression, the *AtUSPL1* coding sequence was cloned into the pBENDER GATEWAY vector (a gift from B. Weisshaar, MPI, Cologne).

A 962 bp fragment harboring the *AtUSPL1* promoter was amplified, re-sequenced and inserted into the *Sph*I/*Hind*III

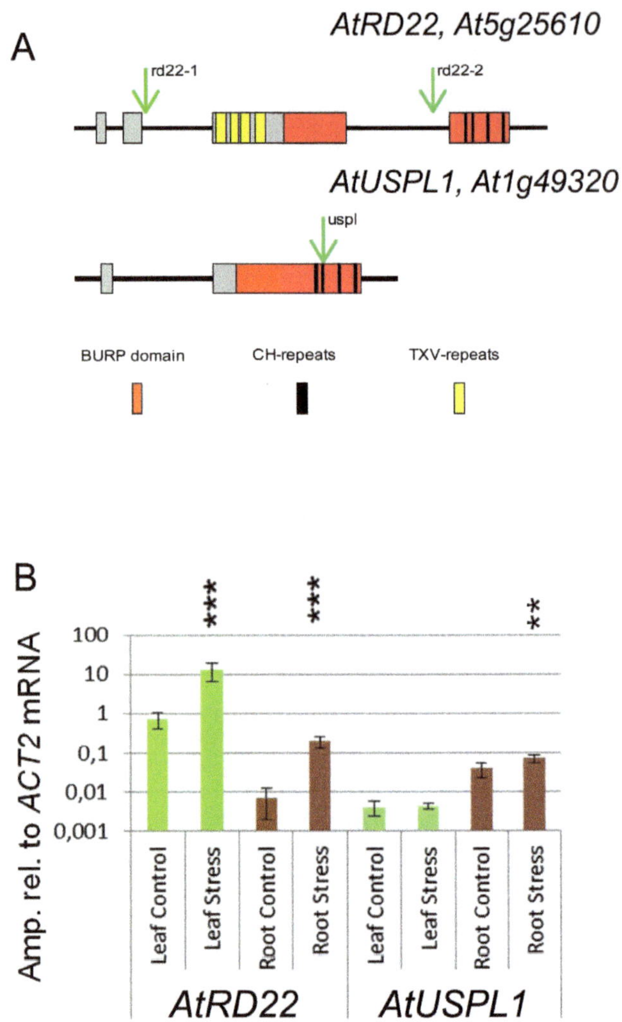

Figure 1. *AtRD22* and *AtUSPL1*, members of the BURP gene family. A. Scheme of BURP-domain containing proteins AtRD22 and AtUSPL1. BURP (named after BNM2, USP, RD22 and Polygalacturonase isozyme) proteins are identified by their C-terminal BURP-domain (red). The BURP domains contain 4 CH-repeats (black). In comparison to AtUSPL1, AtRD22 contains an additional motif, the TXV repeats (yellow), in AtUSPL1 no repetitive domain structure can be found. AGI ID is given in brackets. Position of T-DNA insertions for the used loss-of function mutants is indicated by a green arrow. B. Quantification of *AtRD22* and *AtUSPL1* mRNA in leaves and roots. The bar diagram indicates the amplification of *AtRD22* (left) and *AtUSPL1* mRNA relative to *ACT2* mRNA (amp. rel. to *ACT2* mRNA). Leaf (green) and root (brown) tissue of well watered and drought stressed (2% RWC) plants (N = 5) was analyzed. Asterisks indicate significant differences (T-test: *** $p<0.01$, ** = $p<0.05$).

cloning site of pBIN101 (Clontech) in front of the *uidA GUS* reporter gene. Stable transformation of *A. thaliana* was performed following [36] and subsequent GUS staining to detect transgene expression following [37].

Automated plant phenotyping and image analysis

After 14 days of pre-cultivation pots (diameter 10 cm) were transferred to an automated phenotyping system composed of a conveyor belt transport system and three imaging chambers for plant imaging in the visible (390–750 nm) and NIR (1400–1510 nm with detection peak at 1452 nm) wavelength ranges as well as capturing fluorescence emission (520–750 nm) as well as a watering and weighing station (LemnaTec 3D Scanalyzer; LemnaTec, GmbH, Aachen, Germany) installed in a climate controlled plant cultivation chamber (environmental conditions: 8 h/16 h night/day, 20/22°C (night/day), relative air humidity of 60% and 240 µmol m^{-2} s^{-1} light intensity). Ten blocks (corresponding to ten biological replicates) with each 15 plants (2×5 genotypes under stress conditions, 1×5 under control conditions) were used. Half of the blocks were placed on opposite sides of the cultivation chamber and pots within one block were randomized in order to account for lane or block effects. For reasons of acclimation, plants were grown under control conditions for another week (till day 21 after sowing) before stress application was started. Automated watering was performed every 2nd day to reach defined target values of the pot weight (70% field capacity for control, no watering for soil drought stress). Relative soil water content [%] was controlled in parallel to pot weight with HH2 moisture meter (delta-T devices, Cambridge, UK). The growth rate was calculated as Relative Growth Rate (RGR) = (log(LA[tn+1]) - log(LA[tn]))/((tn+1) - tn) following [38]; Leaf area (LA), timepoint n (tn). Statistical significant growth differences were estimated by Welch t-test from values for plant area determined from top view (visual [39]). Detailed information on calculation of NIR intensities and Cameras used are provided in Table S3.

Estimation of chlorophyll content

Chlorophyll and pheophytin (a chlorophyll degradation product were extracted from the aerial tissue of 6–7 seedlings using a 1:1 acetone:DMSO mixture, and the absorbance of the extract was recorded at 663, 645 and 553 nm [40] using the UVIKON XS 60/99-90286 spectrophotometer. The experiments comprised three independent replicates. Since the chlorophyll content in the mutant plants grown under non-stressed conditions differed from that present in wild type plants, a relative value was calculated to derive the effect of the stress treatment (Figure S6C). Chlorophyll and pheophytin contents were estimated according to [41], and subjected to a one way ANOVA.

RNA isolation qRT-PCR and Northern analysis

RNA was isolated using an RNeasy kit (Qiagen, Hilden, Germany). For the purposes of Northern blotting, 10 µg RNA was loaded into each lane of a 1.2% w/v agarose, 15% v/v formaldehyde gel, electrophoresed, then transferred passively onto a Hybond N+ membrane (Amersham, GE Healthcare, Waukesha, USA) using 10× SSC as the transfer buffer. The RNA was cross-linked to the membrane by UV irradiation. An *AtUSPL1* probe was amplified from *A.thaliana* genomic DNA using primer pair USPa/b (sequences given in Table S1) and labeled with α-^{32}P dCTP via random priming (Ready Prime Labeling, Pharmacia, GE Healthcare, Waukesha, USA). Membrane/probe hybridizations were carried out in Church hybridization solution [42] at 65°C with pre-hybridization for 6 h and hybridization for 16 h. The membrane was washed for 15 min twice each in 2xSSC, 0.1% SDS, 0.5xSSC, 0.1% SDS and 0.1xSSC, 0.1% SDS at 65°C. Signals were detected and quantified with a Bio-Imaging analyser BAS2000 or X-ray film (Fuji Photo Film Co. Ltd., Tokyo, J). Quantitative real time PCRs (qRT-PCR) were run following [43], in order to assess transcript abundances in leaf and root tissue of six week old plants (five replicate RNA extractions per biological sample, and three technical replications per RNA extract). All primer sequences are given in Table S1.

RNA Isolation, target synthesis and microarray hybridization

Total RNA was isolated from the aerial tissue of two week old seedlings using the TRIzol reagent (Invitrogen) and RNAeasy columns (Qiagen) (see Figure S7 for detailed information). The integrity of the RNA was monitored by agarose gel electrophoresis, and its concentration estimated using a NanoDrop device (Peqlab, Erlangen, Germany), following the manufacturer's protocol. The integrity of the RNA was further confirmed using an Agilent 2100 Bioanalyzer in conjunction with the RNA 6000 Nano assay (Agilent Technologies, Böblingen, Germany). RNA was processed for use on an Affymetrix Arabidopsis ATH1 Genome Array (ATLAS Niolabs GmbH, Berlin, Germany).

Microarray data analysis

The ATH1 chip, which assays 22,392 unique genes, was used to contrast the transcriptomes of wild type Col-0, *rd22* and *uspl1* plants grown under both control and moisture stressed conditions. Probe set to target gene mappings were taken from the TAIR Web site: ftp://ftp.arabidopsis.org/home/tair/Microarrays/Affymetrix/affy_ATH1 _array_elements2010-12-20.txt. The microarray experiment and basic data interpretation was performed by ATLAS biolabs GmbH (Berlin, Germany) by GeneChip Operating System (GCOS) 1.4. To ensure reliability of the analyses, each GeneChip experiment was performed with two biological replicates. After logarithmic transformation of the data, the average expression for all experimental samples for this probe set was subtracted from each individual expression value, thus leading to a positive value in case of above-average expression levels and a negative value in case of below-average expression levels. GeneSpring GX software (Agilent) was used to gene-wise normalize the expression data. To identify potentially differentially expressed genes, the fold changes >2 and <2 of expression values were identified. This was done separately for the Col-0 and mutant series at different conditions. For the Col-0 and respective mutant sample, a simple moderated t-test was performed and P values were corrected using the Benjamini and Hochberg [44] false discovery rate control, applying standard limma procedures. Differentially expressed genes, between Col-0 and mutant samples were identified for both conditions using the limma nestedF procedure, applying a significance threshold of 0.5 in combination with Benjamini-Hochberg false-discovery rate control and a minimal log2-fold change value of 2. A functional categorization of the differentially transcribed genes was derived using Mapman software [45].

Results

The *BURP* gene family

The *A. thaliana* genome contains five *BURP* family genes; these include both *AtUSPL1 (At1g49320)* and *AtRD22 (At5g25610)* (Figure 1A), but also three genes encoding proteins sharing similarity with the tomato non-catalytic β-subunit of polygalacturonase. The latter have been proposed to be designated as *AtPG1* (*At1g60390*), *AtPG2* (*At1g70370*) and *AtPG3* (*At1g23760*). Based on an alignment of related sequences extracted from various species, the family can be subdivided into eight sub-families [23], according to which *AtRD22* belongs to the ATRD22-like subgroup, *AtUSPL1* to the *BNM2*-like sub-family, and *AtPG1-3* to the PG1β-like subfamily. The *A. thaliana* genome has no representative of either the V-VIII or the VfUSP like sub-groups. The N-terminal regions of *AtPG1-3* each include sequences encoding 21 FXXY–N_{9-11} repeats (of unknown function) [24]. AtRD22 contains four TXV-repeats, while AtUSPL1 has no repetitive features. Based on the domain structure and phenotype this study is restricted to the functional analysis of *AtRD22* and *AtUSPL1*.

Transcription of *AtRD22* and *AtUSPL1* during development and in response to stress

Some *Arabidopsis BURP* genes were described to be preferentially expressed in early embryogenesis [24], their involvement in stress response was less obvious. An exception of this is *AtRD22*, which was shown previously to be induced under drought treatments [8].

The transcription profiles of *AtRD22* and *AtUSPL1* were recovered from the Genevestigator database ([46] Figure S1A) and verified using qRT-PCR and Northern blotting. Archival microarray data suggested that while *AtRD22* transcript is abundant throughout plant development in the aerial part of the plant, that of *AtUSPL1* is low and is restricted to the root. In non-stressed plants, *AtRD22* transcription is highest in the leaf, and particularly so in the guard cells [46]. *AtUSPL1* transcript is most abundant in the root (Figure 1B, Figure S1B), but is also detectable in the aerial part of the plant early in development. In the root, *AtUSPL1* transcription is stimulated by exposure to either mannitol or NaCl, as is that of *AtRD22* in the leaf [47]. Here, moisture stress up-regulated *AtRD22* transcription was detected, particularly in the leaf. *AtUSPL1* transcript was detectable in the root, and its level was enhanced by the imposition of moisture stress (Figure 1B). *AtUSPL1* expression was also assayed by tracking GUS expression produced by the *pAtUSPL1::GUS* transgene. GUS activity was detected in the young leaf, in the hypocotyl and in the stem (Figure 2A). In the silique, its expression was only detectable in the mature seed funiculum (Figure 2B), while in the developing flower and stem, a low level of expression was observed (Figure 2C). GUS activity was especially strong in the root tip (Figure 2D). Compared to the strong *AtRD22* promoter activity in aerial parts of the plant [26], *ProAtUSPL1* shows strong transcriptional activity in the root tissue.

According to *in silico* data, treatment with ABA strongly induces *AtRD22* in the leaf, and *AtUSPL1* responds similarly in the root. *AtRD22* is also inducible by exposure to mannitol, glucose, nitrate, high levels of illumination, high temperature and salinity (Figure S2). The gene is down-regulated when the plant is treated with paclobutrazol, an inhibitor of gibberellin synthesis, as well as with cycloheximide, a general inhibitor of protein synthesis, and with syringolin, an inhibitor of cell proliferation.

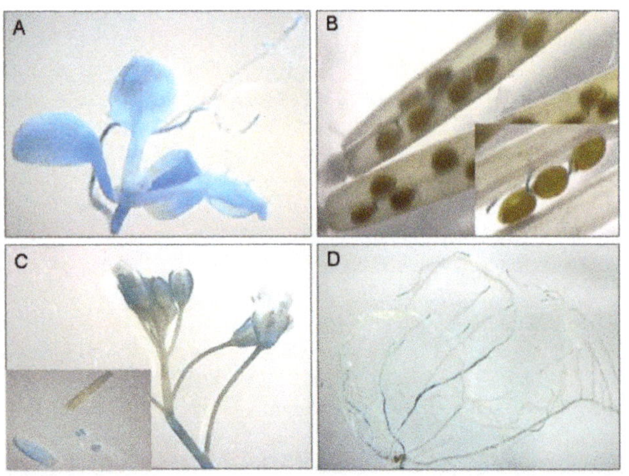

Figure 2. Histochemical *ProAtUSPL1::GUS* activity in transgenic Arabidopsis plants. The *AtUSPL1* promoter activity was determined by histochemical localisation of GUS activity derived from the transgenic *ProATUSPL1::GUS* reporter gene. Activity indicated by blue colour can be seen in A) seedling; B) in funiculus of mature seeds; C) in flowers and stems; and D) in roots.

The behavior of *AtRD22* and *AtUSPL1* T-DNA insertion mutants

In wild type plants, the expression of the two selected *BURP*-gene family members was confirmed by quantitative RT real-time PCR (Figure 1B) and microarray analysis of the aerial part and among them *AtRD22* transcript is abundant under stress treatments (Figure S3). The loss of function mutants achieved by T-DNA insertions were analysed for characterization of *AtRD22* and *AtUSPL1* genes in the functional context of drought stress tolerance (Figure S4A). For *AtRD22* two independent T-DNA insertion alleles were used: *rd22-1* and *rd22-2*. For *AtUSPL1* the mutant line *uspl1* was used. To analyse the functional loss of both members of the BURP-gene family the *rd22-1/uspl1* double mutant was analysed. In the two analysed T-DNA insertion lines *rd22-1* and *uspl1* no mRNA of the respective gene was detectable by semi-quantitative PCR and Northern analysis (Figure S4B), which is corroborated by microarray analysis (Figure S4C). Therefore we assume that the used T-DNA mutants represent a loss of function mutation of the respective gene.

Loss of AtRD22 and AtUSPL1 lead to enhanced drought tolerance

The response of the loss-of-function *AtRD22 and AtUSPL1* T-DNA insertion mutants to moisture stress is summarized in Figure 3 and Figure S4/S5. Two independent *AtRD22* (*rd22-1* and *rd22-2*) and one *AtUSPL1* (*uspl1*) mutants were analysed, along with the *rd22-1/uspl1* double mutant. In both *rd22-1* and *uspl1,* transcript of the mutated gene was not detected based on either semi-quantitative PCR or Northern blotting (Figure S4B), corroborating the prediction of microarray analysis (Figure S4C). When subjected to moisture stress for 2–3 days, wild type plants became discolored as a result of an accumulation of anthocyanin and their growth ceased, whereas both the single and double mutant plants remained green and showed no evidence of any growth retardation (Figure 3A). Exposure to a longer period of moisture stress discriminated between wild type and the mutants in a similar manner (Figure S5A).

Detection of plant growth via automated phenotyping

To quantify the drought response of the mutant plants an automated phenotyping platform (LemnaTec) was used. The growth of 30 plants of control (Col-0), *rd22-1*, *rd22-2*, *uspl1* and *rd22-1/uspl1* each under defined control and drought stress condition was analysed. Drought stress was applied from 21 DAS by complete withdrawal of water. To monitor the growth of the plants, leaf area was estimated from top view images ([39,48], Figure S5B, Table S2) and used for calculation of relative growth rates [38]. An elevated growth rate of the *rd22-1* single mutant plant as well as for the double mutant plants compared to wild type could be found in the early phase of the experiment (22 DAS, Figure 3B). The first significant drought related difference in growth rate was detectable after 7 days without watering (28 DAS, Figure S5B). A significant difference in the growth rate of the *rd22/uspl1* double mutant plants compared to the wild type plants was detectable almost throughout the entire experiment initially from day 24 till day 32 after sowing under stress conditions (Figure 3B, Figure S5B).

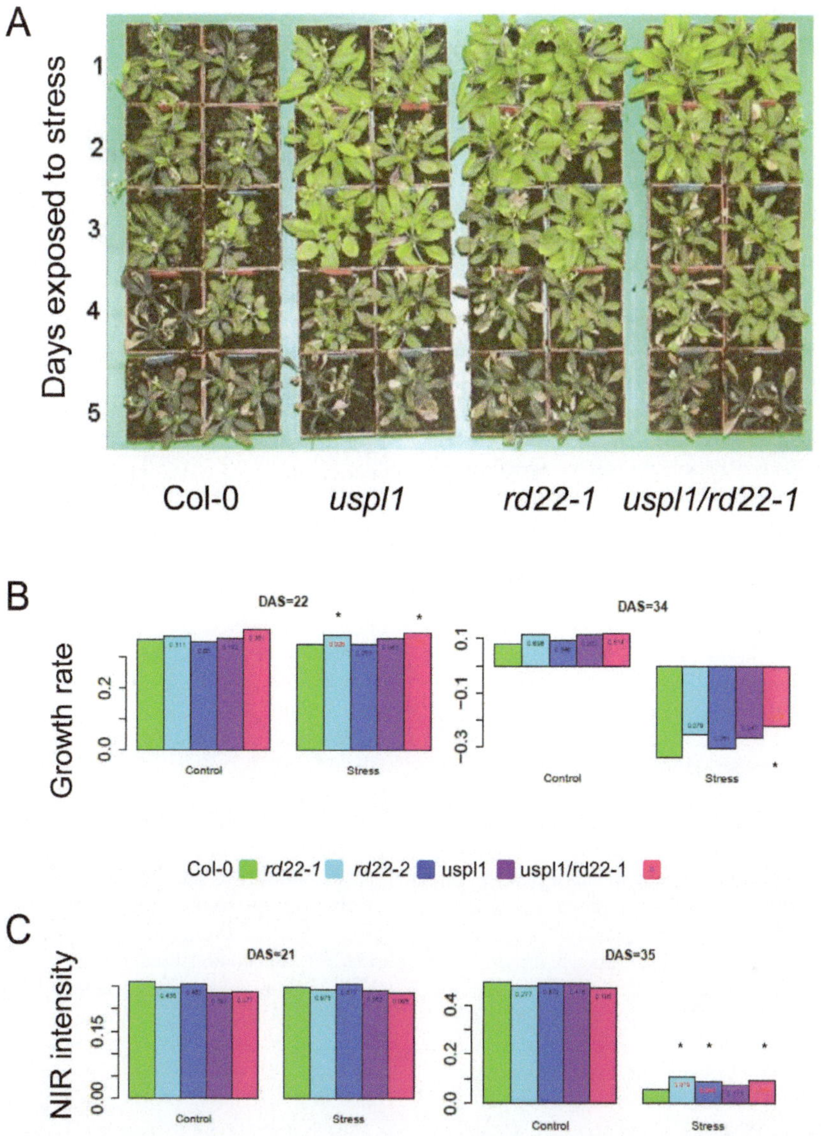

Figure 3. Influence of drought stress treatment on single and double loss of function mutants. A) Four weeks old wild type (Col-0), single and double mutant plants (*rd22-1* and *uspl1, rd22-1/uspl1*) were drought stressed for 1–5 days before they were returned to climate chamber conditions. Time of stress treatment in days is indicated left. B) Growth rates of plants under control and drought stress conditions. Bars indicate the growth rates at 22 days after sowing (DAS) for early phase of drought stress and 34 for the late phase of drought stress. For application of drought stress stop of watering started at 21 DAS. Wild type (Col-0): green bar; *rd22-1*: bright blue bar; *rd22-2*: dark blue bar; *uspl1*: purple bar; *rd22-1/uspl1* pink bar. Original data: Figure S5 B, Table S2, Asterisks indicate significant differences ($p<0.01$). C) NIR reflection as a water content-related parameter. Bars indicate the NIR intensity at 21 days after sowing (DAS) for start of experiment and at 35 DAS for the end of experiment. Wild type (Col-0): green bar; *rd22-1*: bright blue bar; *rd22-2*: dark blue bar; *uspl1*: purple bar; *rd22-1/uspl1* pink bar. $N_{control}=5$, $N_{stress}=10$ plants. Original data in Table S3.

Plant senescence as consequence of the drought stress was monitored by quantifying the ratio of yellow to green pixels. Leaves of the mutant plants *rd22-1*, *uspl1* and the double mutant showed a lower accumulation of yellow stained material, indicative for reduced senescence (33 DAS, Figure S5C). In order to analyse the relative water status of the plants the reflected near-infrared (NIR) radiation (1450 nm) from leaves was detected [49]. NIR intensity is calculated as 1 – NIR reflectance and illustrates the relation to the water content of the leaves. The drought stress related decrease of the NIR intensity was detectable at 33 DAS (Table S3). All genotypes analysed showed a similar level of detectable NIR reflectance in the beginning of the experiment and continuously under control conditions. At the end of the drought stress exposure *rd22* and *rd22/uspl1* double mutant plants showed a higher NIR intensity, indicating higher water content in the leaves (Figure 3C).

Plant growth response to salinity and osmotic stress and exposure to ABA

To further investigate the role of *AtRD22* and *AtUSPL1* during salinity stress and correlating responses on the transcriptional level, wild type and mutant plants were grown on plates with the respective treatments. Two-weeks-old seedlings were transferred for three days to MS basal medium containing one of the following stress inducing compounds: 150 mM and 300 mM NaCl, 100 μM ABA, 4% trehalose, 4% sorbitol, 4% glucose, 4% fructose, 4%

sucrose, 300 mM manitol, 15% PEG 6000 and 4% PEG 20000. Although these conditions are artificially mimicking drought, they were chosen to achieve a uniform plant response to the stimulus. When exposed to either 150 mM NaCl or 100 μM ABA, only minor signs of stress symptoms were apparent, but in the former case, the growth of both the wild type and single mutant plants was retarded (Figure S6A). Wild type and *rd22-1* plants exhibited the least extent of leaf bleaching, while the double mutant and particularly the *uspl1* single mutant, were more visibly affected. Exposure to the various sugars had only a mild effect on plant growth, but a general tendency was for the mutant plants to be more vigorous than the wild type ones. Therefore NaCl 150 mM conditions were chosen for further transcriptome and chlorophyll analysis.

Chlorophyll and pheophytin content

After a four day exposure to moisture stress, the quantity of chlorophyll a and b in the *rd22-1* mutant and the double mutant leaves was less than that in non-stressed plants (Figure S6C). Since the contents of chlorophyll a and chlorophyll b were strongly correlated with one another, subsequent measurements considered the total chlorophyll content. Pheophytin is a degradation product of chlorophyll, that accumulates during senescence, dark [50] and salt stress [51]. In plants grown in the presence of 300 mM NaCl, the content of chlorophyll (Figure 4A) fell sharply, while that of pheophytin rose (Figure 4B). However when challenged by a lesser level of stress (100 μM ABA or 150 mM NaCl), while the chlorophyll content was reduced, there was no measurable increase in pheophytin content (Figure 4C). Chlorophyll content in the *rd22-1* and *uspl1* mutants was more severely reduced in the presence of either 150 mM NaCl or 100 μM ABA. The *uspl1* mutant was the most compromised genotype with respect to chlorophyll content when the medium was supplemented with fructose, while the *rd22-1* and the double mutant plants out-performed the wild type and *uspl1* mutant plants when the stress was imparted by PEG. To clearly indicate this different chlorophyll degradation in the *rd22-1* single and double mutant plants the amount [%] of total clorophyll is indicated relative to the amount of clorophyll at control conditions (Figure 4C). Compared to the amount of chlorophyll under control conditions the *rd22-1* and *uspl1* mutant plants show the strongest reduction upon 150 mM NaCl and 100 μM ABA. In addition, the reduction of chlorophyll in the *uspl1* mutant with fructose show a major difference compared to the other genotypes. In the *rd22-1* and *rd22-1/uspl1* mutant plants the amount of chlorophyll is higher on fructose, 150 mM NaCl and PEG supplemented media compared to the wild type and *uspl1* mutant plants. Taken into account that the chlorophyll content in the mutant in the unstressed condition varies from the wild type a relative value is indicated to show the change due to the treatment (Figure S6B).

The conclusion was that AtRD22 acts to suppress chlorophyll degradation under moisture stress.

Transcriptome analysis

The transcriptional responses to exposure to 150 mM NaCl and to 4% w/v trehalose overlapped by about 10% (Figure S6A). The 150 mM NaCl treatment resulted in a changed transcript abundance for 913 of the 22,392 genes represented on the ATH1 chip. Salinity stress and both the trehalose and sorbitol treatments marginally up-regulated *AtRD22*, while the other *BURP* family genes were hardly affected (Figure S6D). The four genes showing the greatest change in transcript abundance between wild type and mutant plants grown under non-stressed conditions (Table 1) were *At4g33720* (encoding CAP, a PR protein under the control of the DREB2A tanscription factor, [52]: up-regulated by 12 fold in both *rd22-1* and *uspl1* compared to wild type); *At5g49700* (encoding a DNA-binding protein associated with moisture stress [53]: up-regulated by 56 fold in *rd22-1* and by 40 fold in *uspl1*); *At2g27550* (encoding ATC, a systemic inhibitor of floral initiation [54]: up-regulated by 23 fold in both *rd22-1* and *uspl1*); and *At2g44380* (encoding a DC1-domain containing protein, involved in the abiotic stress response, [55]: up-regulated by 22 fold in both *rd22-1* and *uspl1*). The full set of differentially transcribed genes detected in response to the various treatments is given in Table S4 (see also Figure S7). A total of 77 genes displayed differential transcription between wild type and *rd22-1* plants grown under non-stressed conditions; 31 out of these are associated with either biotic or abiotic regulatory pathways (Figure S7A) and included eight encoding a peroxidase putatively involved in H_2O_2 degradation. The absence of AtUSPL1 only re-programmed 18 genes, of which five (one encoding a peroxidase) are associated with either biotic or abiotic regulatory pathways (Figure S7B). Exposure to 150 mM NaCl resulted in the up-regulation of a number of genes associated with either moisture stress or pathogen defense (Table 2). In *rd22-1*, out of the 764 differentially transcribed genes, 231 fell into one of these two categories (Figure S7C), while in the *uspl1* mutant, only 18 genes were up-regulated, of which five were associated with either the biotic or the abiotic stress response (Figure S7D). In the latter mutant, *AtRD22* was strongly up-regulated in the presence of salinity stress. When the stress was applied by the addition of trehalose to the medium, 55 of the 171 differentially transcribed genes identified in the *rd22-1* mutant were associated with either the abiotic or the biotic stress response (Figure S7E). Over 50% (111 out of 212) of the genes responding to trehalose treatment reacted similarly to salinity stress (Figure S3A, Table S4). The set of common up- regulated genes included the transcription factor genes *AtMYB15* (*At3g23250*) and *WRKY33* (*At2g38470*), *LPT3* and *LPT4* (encoding lipid transporters), *At1g35910* (a putative trehalose-6-phosphate phosphatase), *At1g61120* (terpene synthase 04), *At1g16850* (unknown function) and *At1g78410* (VQ motif-containing protein). Although fewer genes were induced by the trehalose (212) than by the 150 mM NaCl (913) treatment, the conclusion was that the two stress agents must affect a similar class of gene. The response to both stress agents also included the down-regulation of several photosynthesis-related genes.

Discussion

RD22 and USPL1 have suppressor function during drought stress

AtRD22 and *AtUSPL1* were transcribed predominantly in, respectively, the leaf and the root. The induction of both genes by moisture stress generated a slight growth reduction. While the induction by moisture stress of *AtRD22* has been noted previously, this was not the case for *AtUSPL1*, most likely due to a concentration on the short-term stress response. Microarray-based transcriptomic analyses have shown that exposure to either NaCl or trehalose induces not only the up-regulation of salinity, trehalose and moisture stress responsive genes, but also the down-regulation of photosynthesis-related genes. The latter is implied by both the induction of leaf chlorosis and senescence and the measured fall in the chlorophyll content of the leaves of stressed plants. The present data confirmed that both salinity and trehalose activate the ABA-mediated moisture stress response. [56–58]. The sole *BURP* gene family to be up- regulated in the aerial part of the plant was *AtRD22,* consistent with observations based on the expression of the transgene *pAtRD22::GUS* [26].

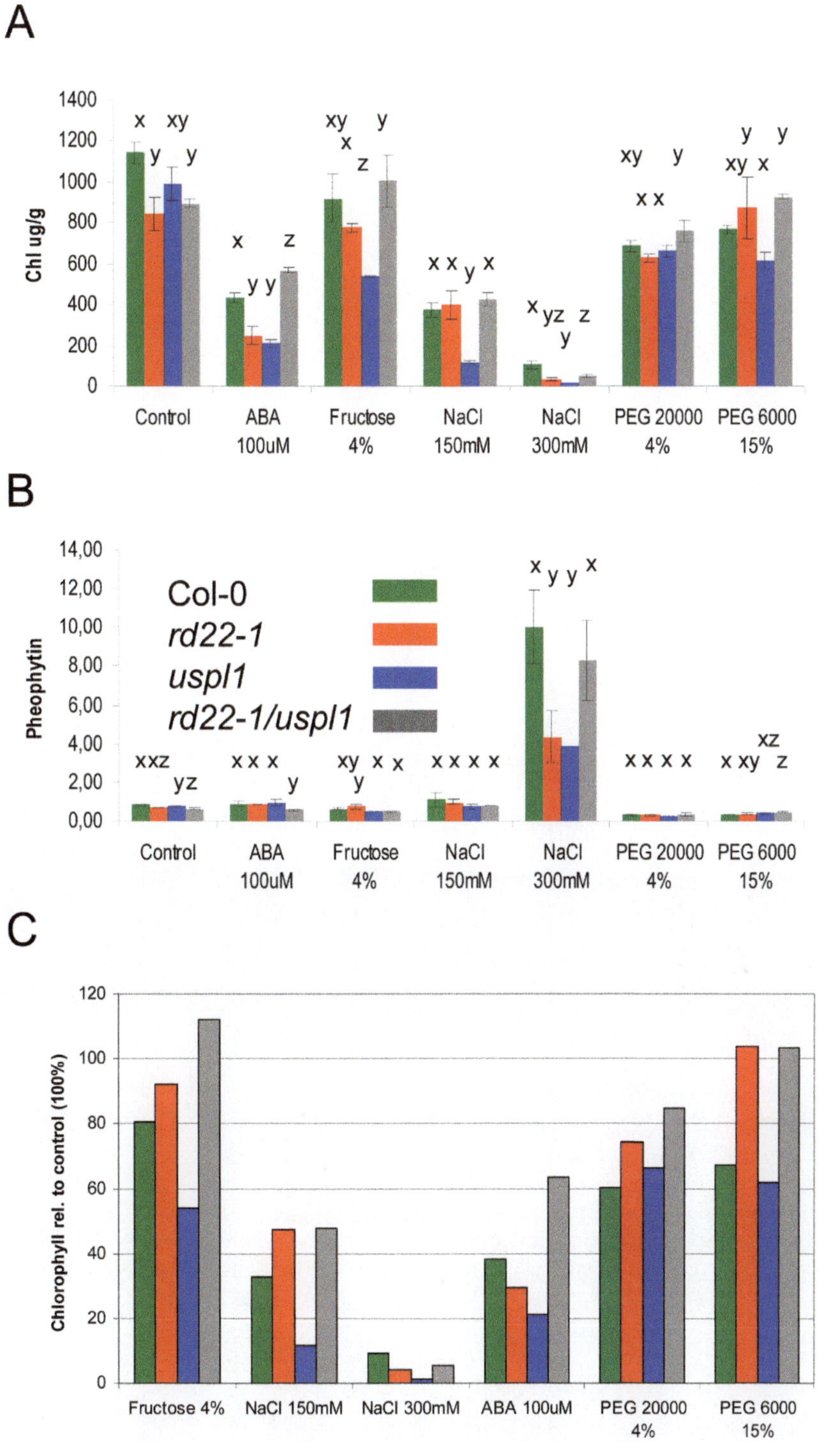

Figure 4. Chlorophyll and pheophytin content in single and double *BURP* mutants. The bars represent the total A) chlorophyll and B) pheophytin content in leaves from wild type (Col-0, green bar), single and double mutant plants (*rd22-1*, red bar, *uspl1*, blue bar and *rd22-1/uspl1*, grey bar). Chlorophyll a and b was determined separately (Figure S6A) and subsumed as total chlorophyll content. Error bar represents standard error. N = 5–6 plants in duplicate. Statistical analysis was performed by oneway ANOVA at alpa = 0.05 Tukey post hoc test: same letters indicate no difference, different letters indicate significant difference. C) The bars show the total chlorophyll content [%] relative to unstressed control plants.

The root specificity of *AtUSPL1* expression was confirmed both by the behavior of the *pAtUSPL1::GUS* transgene and by Northern hybridization experiments.

The T-DNA insertion mutants of both *AtRD22* and *AtUSPL1* produced no detectable relevant transcript, so each was taken as a genuine loss-of- function allele. In both cases, a reduced photosynthesis phenotype was exhibited, as reflected by a fall in

Table 1. Top differential expressed genes in A) *rd22-1* and B) *uspl1* at standard growth conditions in the aerial part of 2 week old seedlings.

A) ***rd22***

Probe Set ID	fold change	Gene	AGI ID	put. Function	Probe Set ID	fold change	Gene	AGI ID	put. Function
257162_s_at	59		At3G24290	amm. transporter	265364_at	−45		At2G13330	Transposon
248564_at	56		At5G49700	DNA binding	246908_at	−40	RD22	At5G25610	Disease Defense
249052_at	47	PDF1.2	At5G44420	Disease Defense	251677_at	−38	ORG3	At3G56980	DNA bind
245276_at	41	ATHB-2	At4G16780		257453_at	−30		At1G65130	hydrolase
265102_at	39		At1G30870	Peroxidase	261684_at	−29		At1G47400	
259276_at	38		At3G01190	PER27 Peroxidase	265665_at	−28		At2G27420	
263096_at	38	AHB1	At2G16060		249645_at	−26	THI2.2	At5G36910	toxin receptor
264567_s_at	37		At1G05250	Peroxidase	266472_at	−26			
254044_at	35	XTR9	At4G25820	hydrolase	254170_at	−22		At4G24430	lyase
254200_at	34		At4G24110		247252_at	−22		At5G64770	

B) ***uspl1***

Probe Set ID	fold change	Gene	AGI ID	put. Function	Probe Set ID	fold change	Gene	AGI ID	put. Function
247718_at	42	LTP4	At5G59310	Lipid transfer	252795_at	−38			
248564_at	40		At5G49700	DNA bind	246948_at	−23	CYP71B12	At5G25130	Cytochr. P450
261226_at	31	ATEXPA11	At1G20190	Expansin	250207_at	−22	CHS	At5G13930	
253667_at	30		At4G30170	peroxidase	256940_at	−19		At3G30720	
252374_at	24	ARR5	At3G48100	transc. Reg.	267168_at	−2		At2G37770	aldo/keto red.
254979_at	23		At4G10550	subtilase					
254832_at	23		At4G12490	protinase inhibitor					
266209_at	23	ATC	At2G27550	A.th centroradialis					
267385_at	22		At2G44380	DC domain					
247717_at	22	LTP3	At5G59320	Lipid transfer					

Complete list of differential expressed genes in Table S4.

Table 2. Differential expressed genes in A) *rd22-1* and B) *uspl1* on medium containing 150mM NaCl in the aerial part of 2 week old seedlings.

A) *rd22* on NaCl		
up regulated genes		
Probe Set ID	**Fold change**	**Gene Symbol**
247215_at	40	**PROPEP3**
266017_at	38	
252984_at	34	**ELI3-2**
252487_at	33	
256627_at	33	
246340_s_at	32	**FAMT**
264005_at	31	AGP2
257206_at	29	
267035_at	26	**AGT3**
259975_at	25	
253872_at	24	RD26
254101_at	24	**AMY1**
266267_at	23	**ATGSTU4**
256245_at	22	**HSP70**
255502_at	21	
247308_at	20	
256436_at	20	
266142_at	11	
256603_at	11	
258791_at	10	**PR4**
264415_at	6	**RAP2.6**
257517_at	4	
258277_at	4	**PAD3**
260919_at	3	
249481_at	2	
down regulated		
Probe Set ID	**Fold change**	**Gene Symbol**
246366_at	−27	
256772_at	−22	**BGAL1**
250366_at	−21	
261118_at	−20	
261684_at	−20	
267264_at	−17	**SCPL11**
258497_at	−11	**COL2**
261413_at	−11	**PLL5**
246908_at	−7	RD22
247450_at	−2	
261351_at	−2	
266363_at	−1	
B) *uspl1* on NaCl		
up regulated genes		
Probe Set ID	Fold change	Gene Symbol
261684_at	42	
258497_at	21	**COL2**
266363_at	20	
247308_at	18	

Table 2. Cont.

B) ***uspl1*** **on NaCl**		
247215_at	15	**PROPEP3**
257206_at	14	
249481_at	13	
261118_at	12	
246908_at	11	RD22
256245_at	11	**HSP70**
261351_at	11	
255502_at	11	
250366_at	11	
264415_at	10	**RAP2.6**
266267_at	10	**ATGSTU4**
256225_at	10	
256603_at	4	
258277_at	2	**PAD3**
266017_at	2	
256627_at	1	
267035_at	1	**AGT3**
down regulated		
Probe Set ID	**Fold change**	**Gene Symbol**
246366_at	−27	
266142_at	−24	
258791_at	−23	**PR4**
267264_at	−20	**SCPL11**
254101_at	−17	**AMY1**
259975_at	−15	
246340_s_at	−15	**FAMT**
256436_at	−15	
256772_at	−14	**BGAL1**
257517_at	−14	
260919_at	−13	
252487_at	−13	
252984_at	−11	**ELI3-2**
264005_at	−11	AGP2
247450_at	−11	
253872_at	−11	RD26
261413_at	−7	**PLL5**

both chlorophyll content under non-stressed growing conditions and photosynthesis-associated gene transcript abundance in stressed plants. Both under prolonged moisture stress mimicking conditions and actual moisture stress, stress tolerance was enhanced in all of the mutants. The mutants' performance (and particularly that of the double mutant), as measured by either the development of leaf area or by the rate of leaf tissue senescence, showed that they were more tolerant of moisture stress than were wild type plants. NIR intensity appeared to reliably reflect relative water content, which was higher in each of the mutants than in wild type.

Given that photosynthesis is clearly compromised by moisture stress, it was not surprising that the transcription of several photosynthesis-associated genes was altered when the plants were exposed to moisture stress. Although the mutant plants' transcriptomes were indistinguishable from that of the wild type with respect to photosynthesis-related genes, nevertheless their leaves contained less chlorophyll than the leaves of wild type plants raised under non-stressed conditions. The reduction in chlorophyll content in the mutants induced by exposure to either NaCl or PEG was less severe than that in wild type plants. Notably, under moisture stress conditions, the leaves of double mutant plants retained more chlorophyll than those of wild type ones. The retention of chlorophyll can be expected to support a higher rate of photosynthesis, so that less transpiration is required to generate a given quantity of assimilate. Since the plant's capacity to retain its water is improved, its water content under moisture stress conditions was greater (Figure 3C).

The observed up-regulation of peroxidase encoding genes in the mutants suggested a secondary effect of the loss-of-function mutations. The transcriptomic data implied that the oxidative state within the mutant plants differed from that within the wild type, with a knock-on effect on gene expression in both ABA-dependent and ABA-independent pathways [59]. Both the phenotype of the mutants as well as the transcriptional response of both genes suggested that under moisture stress conditions, their products exerted a suppressor function. Similar conclusions have been drawn regarding the effect of mutations of genes encoding SnRK2 [60] the effect of which is an almost complete abolition of the ABA response. In particular, *AtRD22* is not transcribed in these mutants when the growing medium is supplemented with ABA reflecting moisture stress condition. Under normal growing conditions, the *SnRK2* mutants exhibit reduced growth, which (along with plant survival) is improved by the addition of ABA to the growing medium. The inference is that the BURP domain containing proteins AtRD22 and AtUSPL1 act as suppressors of the ABA-mediated moisture stress response.

RD22 act as suppressor predominantly in the leaf, while USPL1 act in the root

Although both *AtRD22* and *AtUSPL1* exhibited organ-specific transcription, the loss of function of each gene induced a comparable improvement in the plant's tolerance to moisture stress, at least at the level of the vigour of the leaves. The sensitivity of the microarray platform rules out any trans-silencing effects of either T-DNA insertion on other members of the BURP gene family. Since this sort of silencing has been associated with several T-DNA insertion mutations [61], the assumption is that both *AtRD22* and *AtUSP1* are involved in a holistic moisture stress response. The microarray analysis demonstrated that the two loss-of-function mutations resulted in the induction of a partially overlapping set of genes in the aerial part of the plant, even though *AtUSPL1* was transcribed specifically in the root. Not only was the tolerance of moisture stress enhanced in both mutants, but also their chlorophyll content was reduced. As the lack of *AtRD22* and *AtUSP1* led to an enhanced tolerance to moisture stress, the proposed suppressor function of the BURP domain-containing proteins during an episode of moisture stress includes an organ-specific component. Both proteins are part of an ABA mediated moisture stress response pathway, with AtRD22 functioning mainly in the aerial part of the plant and AtUSPL1 in the root.

Increased drought resistance might be correlated to an increased defence gene response

The transcriptomic analysis of the two mutants revealed an increased transcript abundance compared to the wild type levels with respect to various genes associated with the response to biotic stress (for example, the gene *PDF1.2* was up-regulated by nearly 50 fold). This class of genes was differentially transcribed both under non-stressed and moisture stressed conditions. Several peroxidase encoding genes were also up-regulated. Peroxidases are known to represent an integral component of the plant's hypersensitive response [62]. H_2O_2 is also known as signal molecule during the stress response [63,64]. A loss of control over the production of H_2O_2 might result in the described reduction of chlorophyll content [65,66] and increased water content under drought stress conditions. Such regulation would subsequently lead to better performance of the mutant plants under drought stress.

The up-regulation of peroxidase encoding genes implies a level of linkage between the ABA-mediated moisture stress response and defense against pathogen invasion. Such a connection has been proposed in a suggested model for the function of OCP3, a homeodomain transcription factor. The loss of *OCP3* function results in an enhanced tolerance to moisture stress and at the same time an increased sensitivity to ABA. The abundance of *AtRD22* transcript in the *ocp3* mutant is not different to that in the wild type, and the plant's susceptibility to *Botrytis cinerea* infection is reduced [67]. The *ocp3* mutant was initially identified and named after the phenotype of constitutive over expression of a cationic peroxidase [68]. Here, *AtOCP3* was marginally down regulated when the plants were challenged by NaCl, but there was no transcriptional difference between wild type and either of the two mutants. The implication is that the BURP-containing proteins act to enhance the plant's moisture stress response via the up-regulation of peroxidase encoding genes. In soybean, a direct interaction between RD22 and cell wall-localized peroxidases has been described [69]. The up- regulation of H_2O_2 detoxifying enzymes enhances moisture stress tolerance [70] as well as explaining the link with the pathogen defense response.

A reduced chlorophyll content, in conjunction with an elevated level of transcription of defense response genes under non-stressed growing conditions, suggests that the mutant plants are primed to mount a stronger and/or more rapid set of measures to prevent the accumulation of ROS. A change in the accumulation of H_2O_2 would not only have an impact on the defense response [71], but also on the response to moisture stress [63,72]. Due to the overall elevated expression of H_2O_2-detoxifying enzymes such responses could be limited, leading to an enhanced drought resistance. Our study provides the first functional approach to investigate *BURP*-domain encoding gene function in addition to the previously published structural comparisons.

Supporting Information

Figure S1 Expression profile of the *Arabidopsis thaliana BURP* gene family. A) Expression profile of the *Arabidopsis thaliana BURP* gene family. Data obtained from Genevestigator database (Zimmerman *et al.*, 2004). Relative expression of *AtRD22* (red) and *AtUSPL1* (blue) is given for the different developmental stages of Arabidopsis life cycle (left to right: germinating seed, seedling, Young rosette, developed rosette, bolting, young flower, developed flower, flowers and siliques, mature siliques, senescence). B) Expression of *AtUSPL1* confirmed by Northern Blot analysis. Expression of *AtUSPL1* was determined from root, leaf, shoot, young silique and total flower tissue of *Arabidopsis thaliana* (*Col-0*) plants. For equal loading of the RNA samples probing of the membrane with specific probe against housekeeping mRNA of *AtROC1* (rotamase cyclophilin, renamed in *AtCYP1*) was performed. C) Expression of *AtRD22* obtained from Arabidopsis eFP Browser (Winter *et al.*, 2007). The expression of selected stimuli (Cold: 4°C, Osmotic: 300 mM Mannitol, Salt: 150 mM NaCl and Drought: air steam 15 min) is displayed for the aerial as well as the hypogeic part of the plant. *AtRD22* expression is induced in the aerial part of the plant after applying osmotic, salt stress and slightly increased after loss of water due to airstream treatment. D) Expression of *AtUSPL1* obtained from Arabidopsis eFP Browser (Winter *et al.*, 2007). The expression of selected stimuli (Cold: 4°C, Osmotic: 300 mM mannitol, Salt: 150 mM NaCl and Drought: air steam 15 min) is displayed for the aerial as well as the hypogeic part of the plant. *AtUSPL1* expression is induced in the hypogeic part of the plant after applying osmotic, salt stress and slightly increased after loss of water due to airstream treatment.

Figure S2 Expression of the BURP domain containing gene family in *Arabidopsis thaliana*. Expression analysis of *AtRD22* (*At5G25610*) and *AtUSPL1* (*At1G49320*) obtained from Genevestigator database (Zimmerman et al., 2004) displaying induced and reduced expression after different conditions and stresses. Displayed are only changes in expression upon stress/ treatment above threefold with a statistic significance $p<0.001$. Red indicates up-regulation; Green indicates down-regulation.

Figure S3 Comparison of gene expression in Arabidopsis wild type plants grown on 150 mM NaCl and 4% trehalose supplemented MS medium. BURP gene family mRNA in *Col-0* under selected stress conditions. Bars indicate the expression pattern obtained by microarray analysis using ATH1 chip: *AtRD22* (red): 246908_at; *AtUSPL1* (blue): 262388_at; *AtPG1* (dark green): 265131_at; *AtPG2* (green): 264277_at; *AtPG3* (bright green): 264315_at. Displayed is the rel. Abundance of mRNA.

Figure S4 Characterization of BURP mutant plants. A) Scheme of the *AtRD22* and *AtUSPL1* gene model. In blue the exon-intron structure within the coding region of the respective gene is given. The protein structure refers to Figure 1. In the encoded protein parts are given in aminoacids [aa]. And the size of the fulllenght protein is given below the gene description. Position of T-DNA insertions of used mutant plants are indicated by black lines. Mutant alleles for *rd22*: *rd22-1* (SALK_146066) and *rd22-2* (WiscDsLox481-484P12). Mutant alleles for *uspl1*: *uspl1*: (SALK_022325). The T-DNA insertion line SALK_146066 (*rd22-1*) is based on pROK2 conferring kanamycin resistance and the WiscDsLox481-484P12 (*rd22-2*) is based on pWiscDs-Lox conferring phosphinotricin (BASTA) resistance. *uspl1* T-DNA insertion lines SALK_022325 (referred to as *uspl1*, based on pROK2 conferring kanamycin resistance from Nottingham Arabidopsis Stock Centre) was analyzed. The position of the T-DNA insertion in *At1G49320 (AtUSPL1)* was determined by PCR and subsequent sequencing. The position of the T-DNA insertion are depicted and confirmed by PCR. Double mutant *rd22-1/uspl1* line was generated by crossing SALK_146066 and SALK_022325 and identified in the F_3 generation by PCR. Kanamycin and phosphinotricin resistance of the plants was tested on germination medium (1 MS salts; 10 g/l sucrose) plates with 40 mg/l kanamycin or 20 mg/l glufosinate-ammonium under long day conditions. B) Analysis of used T-DNA insertion mutants. The absence of *AtRD22* and *AtUSPL1* mRNA in homozygous *rd22* and *uspl1* mutant plants was determined by Northern Blot analysis (left) and semi quantitative RT-PCR. C) Analysis of BURP-gene family mRNA in *rd22-1* and *uspl1* mutants by microarray analysis on MS medium. Bars indicate the rel. expression signal obtained by microarray analysis using ATH1 chip from each single experiment: *AtRD22* (red): 246908_at; *AtUSPL1* (blue): 262388_at; *AtPG1* (dark green): 265131_at; *AtPG2* (green): 264277_at; *AtPG3* (bright green): 264315_at.

Figure S5 Increased drought stress resistance of the *rd22* and *uspl1* mutant plants. A) The plants were drought stressed by withdrawal of water. Top: day 0 (80% RWC in the soil); Bottom: appearance of plants after 8 days without watering. B) Top: Projected area of wild type and mutant plants under control conditions and drought stress (dotted line) obtained by lemnatec phenotyping; Drought stress was started 21 days after sawing (DAS). Middle: Growth rates calculated based on Poorter and Lewis 1986 for individual days. Wild type (Col-0): green line; *rd22-1*: bright blue line; *rd22-2*: dark blue line; *uspl1*: purple line; *rd22-1/uspl1*pink line (+/- s.e.m.). Bottom: Statistical analysis or growth rates at 28 DAS. Asterisks indicate significant differences ($p<0.05$) between control and stress. C) Estimation of senescence after drought stress. The graph indicates the ration of yellow to green pixels in the plant area of the analysed top view images from day 33. Wild type (Col-0): green bar; *rd22-1*: bright blue bar; *rd22-2*: dark blue bar; *uspl1*: purple bar; *rd22-1/uspl1*: pink bar. $N_{control} = 5$, $N_{stress} = 10$ plants. Asterisks indicate significant differences ($p<0.05$).

Figure S6 A) Differentially expressed genes (Col-0) between 150 mM NaCl containing medium and standard growth conditions sorted by relation to pathway (Mapman). Top: Differentially expressed genes (Col-0) between 150 mM NaCl containing medium and standard growth conditions sorted by relation to pathway (Mapman). The different numbers indicate different categories/pathways and are described in the table. Bottom: Differentially expressed genes (Col-0) between standard growth conditions and 4% trehalose containing medium. The bar diagram indicates the percentage of common (red) and inverse (yellow) regulated genes upon salt (grey) and sugar (blue) treatment. Approximately half of the genes induced by 4% trehalose treatment are also reacting on 150 mM NaCl treatment. B) Influence of ABA and NaCl on single and double loss of function mutants. Growth phenotypes of wild type (Col-0), single and double mutant plants (*rd22-1* and *uspl1, rd22-1/uspl1*) on standard MS-medium and 150 mM NaCl, 300 mM NaCl and 100 µM ABA supplemented MS-Medium. Two week old seedlings were transferred for 3 days to the MS basal and supplemented medium. C) Chlorophyll and pheophytin content of wild type and *rd22-1*, *uspl1* and *rd22-1/uspl1* mutant plants on different supplemented media. Two week old seedlings were transferred to the MS basal + one of the following stress treatments for 4 days: 100 µM ABA, 4% fructose, 150 mM NaCl, 300 mM NaCl, 4% PEG 20000 and 15% PEG 6000. From Top to bottom: Chlorophyll a and b content. The error bar represents standard error. Content was estimated from 5-6 plants in duplicate. Statistical analysis was performed by oneway ANOVA at alpa = 0.05 Tukey post hoc test: same letters indicate no difference, different letters indicate significant difference. Chlorophyll a and b and pheophytin content [%] relative to unstressed control plants.

Figure S7 Differentially expressed genes in the *rd22-1* and *uspl1* plants grown on control, 150 mM NaCl and 4% trehalose containing medium. List of used categories is given in Figure 3 A. Top: Display of top regulated category (Bin 20, biotic and abiotic stress pathways) of differential regulated genes. Bottom: Overview of all differentially regulated genes mapped to categories (Bins) by MAPMAN. For RNA extraction tissue was grinded in liquid nitrogen and the homogenized powder was added to 1 ml TRIZOL and incubated at RT for 5 min. Samples were centrifuged at 10.000 rpm for 10 min and the supernatant was transferred to a new tube. 200 µl of chloroform were added and incubated at room temperature for 2–3 min. Samples were again centrifuged as described above and the aqueous supernatant was transferred to the QIAshredder column and centrifuged for 30 s at 10000 rpm. 350 µl of RLT buffer (plus β-mercaptoethanol) and 250 µl of absolute ethanol were added to the flow-through and passed through an RNAeasy spin column. All the following steps were performed as described in the manufacturer's protocol followed by in-column DNAse digestion. A) Schematic display of differentially expressed genes in *rd22-1* by

MAPMAN. 31 out of 77 differential regulated genes are mapping to biotic and abiotic stress pathways. B) Schematic display of differentially expressed genes in *uspl1* by MAPMAN. 5 out of 18 differential regulated genes are mapping to biotic and abiotic stress pathways. C) Schematic display of differentially expressed genes in *rd22-1* on 150 mM NaCl by MAPMAN. 231 out of 764 differentially regulated genes are mapping to biotic and abiotic stress pathways. D) Schematic display of differentially expressed genes in *uspl1* on 150 mM NaCl by MAPMAN. 7 out of 12 differentially regulated genes are mapping to biotic and abiotic stress pathways. E) Schematic display of differentially expressed genes in *rd22* on 150 mM NaCl by MAPMAN. 55 out of 171 differentially regulated genes are mapping to biotic and abiotic stress pathways.

Table S1 Primer information.

Table S2 Average projected plant area (mm2) from top images.

Table S3 Average near-infrared intensity as observed from top images, obtained using a Nir 300 camera from VDS Vosskühler (now Allied Vision Technologies). High values indicate relative high water content.

Table S4 List of differentially expressed genes.

Acknowledgments

We thank Elke Liemann and Jana Lorenz for excellent technical assistance, Dijun Chen for support with statistical analysis of lemnatec data, Ingo Mücke for his support using the automated phenotyping system.

Author Contributions

Conceived and designed the experiments: NS HB MK TA. Performed the experiments: VTH CS LVS KWF. Analyzed the data: AJ CK. Wrote the paper: MK AJ TA.

References

1. Bray EA, Bailey-Serres J, Weretilnyk E (2000) Responses to abiotic stresses. Biochemistry and Molecular Biology of Plants, Rockville, MD, USA: American Society of Plant Physiologists. pp.1158–1203.
2. Sreenivasulu N, Sopory SK, Kishor PBK (2007) Deciphering the regulatory mechanisms of abiotic stress tolerance in plants by genomic approaches. Gene 388: 1–13.
3. Agarwal PK, Agarwal P, Reddy MK, Sopory SK (2006) Role of DREB transcription factors in abiotic and biotic stress tolerance in plants. Plant Cell Rep 25: 1263–1274.
4. Hasegawa PM, Bressan RA, Zhu JK, Bohnert HJ (2000) Plant Cellular and Molecular Responses to High Salinity. Annu Rev Plant Physiol Plant Mol Biol 51: 463–499.
5. Yamaguchi-Shinozaki K, Shinozaki K (2006) Transcriptional regulatory networks in cellular responses and tolerance to dehydration and cold stresses. Annu Rev Plant Biol 57: 781–803.
6. Kasuga M, Liu Q, Miura S, Yamaguchi-Shinozaki K, Shinozaki K (1999) Improving plant drought, salt, and freezing tolerance by gene transfer of a single stress-inducible transcription factor. Nature Biotechnology 17: 287–291.
7. Yamaguchi-Shinozaki K, Shinozaki K (1992) A novel Arabidopsis DNA binding protein contains the conserved motif of HMG-box proteins. Nucleic Acids Res 20: 6737.
8. Yamaguchi-Shinozaki K, Shinozaki K (1993) The plant hormone abscisic acid mediates the drought-induced expression but not the seed-specific expression of rd22, a gene responsive to dehydration stress in Arabidopsis thaliana. Mol Gen Genet 238: 17–25.
9. Hattori J, Boutilier KA, van Lookeren Campagne MM, Miki BL (1998) A conserved BURP domain defines a novel group of plant proteins with unusual primary structures. Mol Gen Genet 259: 424–428.
10. Baumlein H, Boerjan W, Nagy I, Bassuner R, Van Montagu M, et al. (1991) A novel seed protein gene from Vicia faba is developmentally regulated in transgenic tobacco and Arabidopsis plants. Mol Gen Genet 225: 459–467.
11. Wohlfarth T, Braun H, Kirik V, Kolle K, Czihal A, et al. (1998) Regulation and evolution of seed globulin genes. Journal of Plant Physiology 152: 600–606.
12. Boutilier KA, Gines MJ, DeMoor JM, Huang B, Baszczynski CL, et al. (1994) Expression of the BnmNAP subfamily of napin genes coincides with the induction of Brassica microspore embryogenesis. Plant Mol Biol 26: 1711–1723.
13. Teerawanichpan P, Xia Q, Caldwell SJ, Datla R, Selvaraj G (2009) Protein storage vacuoles of Brassica napus zygotic embryos accumulate a BURP domain protein and perturbation of its production distorts the PSV. Plant Molecular Biology 71: 331–343.
14. Treacy BK, Hattori J, Prudhomme I, Barbour E, Boutilier K, et al. (1997) Bnm1, a Brassica pollen-specific gene. Plant Molecular Biology 34: 603–611.
15. Chen LZ, Guan LM, Seo M, Hoffmann F, Adachi T (2005) Developmental expression of ASG-1 during gametogenesis in apomictic guinea grass (Panicum maximum). Journal of Plant Physiology 162: 1141–1148.
16. Wang AM, Xia Q, Xie WS, Datla R, Selvaraj G (2003) The classical Ubisch bodies carry a sporophytically produced structural protein (RAFTIN) that is essential for pollen development. Proceedings of the National Academy of Sciences of the United States of America 100: 14487–14492.
17. Batchelor AK, Boutilier K, Miller SS, Hattori J, Bowman LA, et al. (2002) SCB1, a BURP-domain protein gene, from developing soybean seed coats. Planta 215: 523–532.
18. Datta N, Lafayette PR, Kroner PA, Nagao RT, Key JL (1993) Isolation and Characterization of 3 Families of Auxin down-Regulated Cdna Clones. Plant Molecular Biology 21: 859–869.
19. Granger C, Coryell V, Khanna A, Keim P, Vodkin L, et al. (2002) Identification, structure, and differential expression of members of a BURP domain containing protein family in soybean. Genome 45: 693–701.
20. Xu HL, Li YX, Yan YM, Wang K, Gao Y, et al. (2010) Genome-scale identification of Soybean BURP domain-containing genes and their expression under stress treatments. Bmc Plant Biology 10.
21. Shao YH, Wei G, Wang L, Dong Q, Zhao Y, et al. (2011) Genome-Wide Analysis of BURP Domain-Containing Genes in Populus trichocarpa. Journal of Integrative Plant Biology 53: 743–755.
22. Held BM, John I, Wang H, Moragoda L, Tirimanne TS, et al. (1997) Zrp2: a novel maize gene whose mRNA accumulates in the root cortex and mature stems. Plant Mol Biol 35: 367–375.
23. Gan D, Jiang H, Zhang J, Zhao Y, Zhu S, et al. (2011) Genome-wide analysis of BURP domain-containing genes in maize and sorghum. Mol Biol Rep 38: 4553–4563.
24. Van Son L, Tiedemann J, Rutten T, Hillmer S, Hinz G, et al. (2009) The BURP domain protein AtUSPL1 of Arabidopsis thaliana is destined to the protein storage vacuoles and overexpression of the cognate gene distorts seed development. Plant Molecular Biology 71: 319–329.
25. Abe H, Yamaguchi-Shinozaki K, Urao T, Iwasaki T, Hosokawa D, et al. (1997) Role of arabidopsis MYC and MYB homologs in drought- and abscisic acid-regulated gene expression. Plant Cell 9: 1859–1868.
26. Iwasaki T, Yamaguchishinozaki K, Shinozaki K (1995) Identification of a Cis-Regulatory Region of a Gene in Arabidopsis-Thaliana Whose Induction by Dehydration Is Mediated by Abscisic-Acid and Requires Protein-Synthesis. Molecular and General Genetics 247: 391–398.
27. Park MY, Chung MS, Koh HS, Lee DJ, Ahn SJ, et al. (2009) Isolation and functional characterization of the Arabidopsis salt-tolerance 32 (AtSAT32) gene associated with salt tolerance and ABA signaling. Physiol Plant 135: 426–435.
28. Sanchez JP, Chua NH (2001) Arabidopsis PLC1 is required for secondary responses to abscisic acid signals. Plant Cell 13: 1143–1154.
29. Song HM, Zhao RM, Fan PX, Wang XC, Chen XY, et al. (2009) Overexpression of AtHsp90.2, AtHsp90.5 and AtHsp90.7 in Arabidopsis thaliana enhances plant sensitivity to salt and drought stresses. Planta 229: 955–964.
30. Wang HM, Zhou L, Fu YP, Cheung MY, Wong FL, et al. (2012) Expression of an apoplast-localized BURP-domain protein from soybean (GmRD22) enhances tolerance towards abiotic stress. Plant Cell and Environment 35: 1932–1947.
31. Ding Z, Li S, An X, Liu X, Qin H, et al. (2009) Transgenic expression of MYB15 confers enhanced sensitivity to abscisic acid and improved drought tolerance in Arabidopsis thaliana. J Genet Genomics 36: 17–29.
32. Xu B, Gou JY, Li FG, Shangguan XX, Zhao B, et al. (2013) A Cotton BURP Domain Protein Interacts With -Expansin and Their Co-Expression Promotes Plant Growth and Fruit Production. Molecular Plant 6: 945–958.
33. Schallau A, Kakhovskaya I, Tewes A, Czihal A, Tiedemann J, et al. (2008) Phylogenetic footprints in fern spore- and seed-specific gene promoters. Plant Journal 53: 414–424.
34. Sreenivasulu N, Radchuk V, Strickert M, Miersch O, Weschke W, et al. (2006) Gene expression patterns reveal tissue-specific signalling networks controlling programmed cell death and ABA-regulated maturation in developing barley seeds (vol 47, pg 310, 2006). Plant Journal 47: 987–987.

35. Orkin S (1990) Molecular-Cloning - a Laboratory Manual, 2nd Edition - Sambrook, J, Fritsch, Ef, Maniatis, T. Nature 343: 604–605.
36. Bechthold N, Elli J, Pelletier G (1993) *In planta Agrobacterium*-mediated transfer by infiltration of adult *Arabidopsis thaliana* plants. CR Acad Sci Paris 316: 871–883.
37. Jefferson RA, Kavanagh TA, Bevan MW (1987) Gus Fusions - Beta-Glucuronidase as a Sensitive and Versatile Gene Fusion Marker in Higher-Plants. Embo Journal 6: 3901–3907.
38. Poorter H, Lewis C (1986) Testing Differences in Relative Growth-Rate - a Method Avoiding Curve Fitting and Pairing. Physiologia Plantarum 67: 223–226.
39. Klukas C, Chen D, Pape JM (2014) IAP: an open-source information system for high-throughput plant phenotyping. Plant Physiol.
40. Govind G, Harshavardhan VT, Patricia JK, Dhanalakshmi R, Senthil Kumar M, et al. (2009) Identification and functional validation of a unique set of drought induced genes preferentially expressed in response to gradual water stress in peanut. Mol Genet Genomics 281: 591–605.
41. Hiscox JD, Israelstam GF (1979) Method for the Extraction of Chlorophyll from Leaf Tissue without Maceration. Canadian Journal of Botany-Revue Canadienne De Botanique 57: 1332–1334.
42. Church GM, Gilbert W (1984) Genomic sequencing. Proceedings of the National Academy of Sciences of the United States of America 81: 1991–1995.
43. Seiler C, Harshavardhan VT, Reddy PS, Hensel G, Kumlehn J, et al. (2014) Abscisic acid flux alterations result in differential abscisic acid signaling responses and impact assimilation efficiency in barley under terminal drought stress. Plant Physiol 164: 1677–1696.
44. Benjamini Y, Hochberg Y (1995) Controlling the False Discovery Rate - a Practical and Powerful Approach to Multiple Testing. Journal of the Royal Statistical Society Series B-Methodological 57: 289–300.
45. Thimm O, Blasing O, Gibon Y, Nagel A, Meyer S, et al. (2004) MAPMAN: a user-driven tool to display genomics data sets onto diagrams of metabolic pathways and other biological processes. Plant Journal 37: 914–939.
46. Zimmermann P, Hirsch-Hoffmann M, Hennig L, Gruissem W (2004) GENEVESTIGATOR. Arabidopsis microarray database and analysis toolbox. Plant Physiology 136: 2621–2632.
47. Winter D, Vinegar B, Nahal H, Ammar R, Wilson GV, et al. (2007) An "Electronic Fluorescent Pictograph" Browser for Exploring and Analyzing Large-Scale Biological Data Sets. Plos One 2.
48. Klukas C, Pape JM, Entzian A (2012) Analysis of high-throughput plant image data with the information system IAP. J Integr Bioinform 9: 191.
49. Munns R, James RA, Sirault XR, Furbank RT, Jones HG (2010) New phenotyping methods for screening wheat and barley for beneficial responses to water deficit. J Exp Bot 61: 3499–3507.
50. Schelbert S, Aubry S, Burla B, Agne B, Kessler F, et al. (2009) Pheophytin Pheophorbide Hydrolase (Pheophytinase) Is Involved in Chlorophyll Breakdown during Leaf Senescence in Arabidopsis. Plant Cell 21: 767–785.
51. Santos CLV, Campos A, Azevedo H, Caldeira G (2001) In situ and in vitro senescence induced by KCl stress: nutritional imbalance, lipid peroxidation and antioxidant metabolism. Journal of Experimental Botany 52: 351–360.
52. Qin F, Sakuma Y, Tran LSP, Maruyama K, Kidokoro S, et al. (2008) Arabidopsis DREB2A-interacting proteins function as RING E3 ligases and negatively regulate plant drought stress-responsive gene expression. Plant Cell 20: 1693–1707.
53. Catala R, Ouyang J, Abreu IA, Hu Y, Seo H, et al. (2007) The Arabidopsis E3 SUMO ligase SIZ1 regulates plant growth and drought responses. Plant Cell 19: 2952–2966.
54. Huang NC, Jane WN, Chen J, Yu TS (2012) Arabidopsis thaliana CENTRORADIALIS homologue (ATC) acts systemically to inhibit floral initiation in Arabidopsis. Plant Journal 72: 175–184.
55. Li C, Lv J, Zhao X, Ai X, Zhu X, et al. (2010) TaCHP: a wheat zinc finger protein gene down-regulated by abscisic acid and salinity stress plays a positive role in stress tolerance. Plant Physiol 154: 211–221.
56. Barrero JM, Rodriguez PL, Quesada V, Piqueras P, Ponce MR, et al. (2006) Both abscisic acid (ABA)-dependent and ABA-independent pathways govern the induction of NCED3, AAO3 and ABA1 in response to salt stress. Plant Cell Environ 29: 2000–2008.
57. Hwang IW, Goodman HM (1995) An Arabidopsis-Thaliana Root-Specific Kinase Homolog Is Induced by Dehydration, Aba, and Nacl. Plant Journal 8: 37–43.
58. Rosado A, Schapire AL, Bressan RA, Harfouche AL, Hasegawa PM, et al. (2006) The Arabidopsis tetratricopeptide repeat-containing protein TTL1 is required for osmotic stress responses and abscisic acid sensitivity. Plant Physiology 142: 1113–1126.
59. Shinozaki K, Yamaguchi-Shinozaki K (1992) [Plant genes induced by drought stress and ABA]. Tanpakushitsu Kakusan Koso 37: 1190–1199.
60. Fujii H, Verslues PE, Zhu JK (2011) Arabidopsis decuple mutant reveals the importance of SnRK2 kinases in osmotic stress responses in vivo. Proc Natl Acad Sci U S A 108: 1717–1722.
61. Daxinger L, Hunter B, Sheik M, Jauvion V, Gasciolli V, et al. (2008) Unexpected silencing effects from T-DNA tags in Arabidopsis. Trends in Plant Science 13: 4–6.
62. Almagro L, Gomez Ros LV, Belchi-Navarro S, Bru R, Ros Barcelo A, et al. (2009) Class III peroxidases in plant defence reactions. J Exp Bot 60: 377–390.
63. Desikan R, Cheung MK, Bright J, Henson D, Hancock JT, et al. (2004) ABA, hydrogen peroxide and nitric oxide signalling in stomatal guard cells. Journal of Experimental Botany 55: 205–212.
64. Hancock JT, Desikan R, Neill SJ (2001) Role of reactive oxygen species in cell signalling pathways. Biochem Soc Trans 29: 345–350.
65. Kar RK, Choudhuri MA (1987) Possible Mechanisms of Light-Induced Chlorophyll Degradation in Senescing Leaves of Hydrilla-Verticillata. Physiologia Plantarum 70: 729–734.
66. Hynninen PH, Kaartinen V, Kolehmainen E (2010) Horseradish peroxidase-catalyzed oxidation of chlorophyll a with hydrogen peroxide Characterization of the products and mechanism of the reaction. Biochimica Et Biophysica Acta-Bioenergetics 1797: 531–542.
67. Ramirez V, Coego A, Lopez A, Agorio A, Flors V, et al. (2009) Drought tolerance in Arabidopsis is controlled by the OCP3 disease resistance regulator. Plant Journal 58: 578–591.
68. Coego A, Ramirez V, Gil MJ, Flors V, Mauch-Mani B, et al. (2005) An Arabidopsis homeodomain transcription factor, OVEREXPRESSOR OFCATIONIC PEROXIDASE 3, mediates resistance to infection by necrotrophic pathogens. Plant Cell 17: 2123–2137.
69. Wang YN, Liu C, Li KX, Sun FF, Hu HZ, et al. (2007) Arabidopsis EIN2 modulates stress response through abscisic acid response pathway. Plant Molecular Biology 64: 633–644.
70. Miao Y, Lv D, Wang P, Wang XC, Chen J, et al. (2006) An Arabidopsis glutathione peroxidase functions as both a redox transducer and a scavenger in abscisic acid and drought stress responses. Plant Cell 18: 2749–2766.
71. Mullineaux P, Ball L, Escobar C, Karpinska B, Creissen G, et al. (2000) Are diverse signalling pathways integrated in the regulation of arabidopsis antioxidant defence gene expression in response to excess excitation energy? Philos Trans R Soc Lond B Biol Sci 355: 1531–1540.
72. Neill S, Barros R, Bright J, Desikan R, Hancock J, et al. (2008) Nitric oxide, stomatal closure, and abiotic stress. J Exp Bot 59: 165–176.

19

Experimental Comparison of the Reproductive Outcomes and Early Development of the Offspring of Rats Given Five Common Types of Drinking Water

Hui Zeng[1], Wei-qun Shu[1]*, Ji-an Chen[1], Lin Liu[2], Da-hua Wang[1], Wen-juan Fu[1], Ling-qiao Wang[1], Jiao-hua Luo[1], Liang Zhang[1], Yao Tan[1], Zhi-qun Qiu[1], Yu-jing Huang[1]

1 Department of Environmental Hygiene, College of Preventive Medicine, Third Military Medical University, Chongqing, P. R. China, **2** The Lundberg-Kienlen Lung Biology and Toxicology Laboratory, Department of Physiological Sciences, Oklahoma State University, Stillwater, Oklahoma, United States of America

Abstract

Tap water (unfiltered), filtered tap water and processed bottled water (purified water, artificial mineralized water, or natural water) are now the five most widely consumed types of drinking water in China. However, the constituents (organic chemicals and inorganic ingredients) of the five waters differ, which may cause them to have different long-term health effects on those who drink them, especially sensitive children. In order to determine which type of water among the five waters is the most beneficial regarding reproductive outcomes and the developmental behaviors of offspring, two generations of Sprague–Dawley rats were given these five waters separately, and their reproductive outcomes and the developmental behaviors of their offspring were observed and compared. The results showed that the unfiltered tap water group had the lowest values for the maternal gestation index (MGI) and offspring's learning and memory abilities (OLMA); the lowest offspring survival rate was found in the purified water group; and the highest OLMA were found in the filtered tap water group. Thus, the best reproductive and offspring early developmental outcomes were found in the group that drank filtered tap water, which had the lowest levels of pollutants and the richest minerals. Therefore, thoroughly removing toxic contaminants and retaining the beneficial minerals in drinking water may be important for both pregnant women and children, and the best way to treat water may be with granular activated carbon and ion exchange by copper zinc alloy.

Editor: Nick Ashton, The University of Manchester, United Kingdom

Funding: This project was supported by National Natural Science Foundation of China (No. 81001234) and the Project of Chongqing Municipal Health Bureau (Grant No. 2012-2-447). The funders had no role in study design, data collection and analysis, decision to publish, or preparation of the manuscript.

Competing Interests: The authors have declared that no competing interests exist.

* Email: xm0630@sina.com

Introduction

Global environmental and economic changes have led to the diversification of human drinking water. Traditional tap water is the most popular drinking water in the world. The addition of chlorine to tap water is one of the most common treatments to ensure its bacteriological quality. However, tap water remains susceptible to biological or chemical contamination [1]: if the water contains organic matter, this may produce disinfection by-products (DBPs), especially trichloromethane (THMs), in the water [2–4]. In addition, heavy metals such as lead and copper can be leached from pipes into the potable water stream [5–8]. Therefore, unpleasant tastes such as a chlorine flavor, DBPs and lead exposure in tap water may be the most common reasons driving people to choose alternative drinking water options such as bottled water or filtered tap water.

Bottled water's consumption has been steadily growing for the past 30 years. In 2011, the consumption was approximately 40,000 million liters in China (ranked number 1), 32,500 million liters in the United State of America (ranked number 2) and 262 billion liters in total around the world (90 countries) [9]. Three major types of bottled water are sold in Chinese groceries and supermarkets: bottled purified water, bottled mineralized water, and bottled natural water [10]. Bottled purified water, including distilled water, demineralized water, deionized water and reverse osmosis water, is usually tap water that has been treated by a series of filtration processes to remove nearly all minerals and electrolytes, disinfected by ozone or chlorine and finally packaged in a bottle (table 1) [11]. Thus, the purified water in theory is only H_2O. However, the purified water tastes bad and may not quench thirst [12]. In order to improve the taste, small quantities of mineral salts such as potassium chloride and magnesium sulfate are added to the purified water, resulting in mineralized (or low-mineral drinking) water (table 1) [11]. Bottled natural water comes from high-quality underground or surface water sources. This water is also treated by serial filtration, usually disinfected by ozone and then packaged in bottles *in situ* (table 1) [11]. As such, bottled natural water generally contains certain amounts of minerals. Thus, it is clear that different bottled waters contain different minerals, and the mineral levels in these bottled waters are lower than those in the tap water.

In a previous study that reported that drinking water is an important source of essential elements such as Ca and Mg [13], Sabatier suggested that magnesium and calcium in water is more bioavailable to a higher content (from 40% to 60%) than the magnesium and calcium obtained through diet because calcium

Table 1. The drinking water treatment process for the five drinking waters in China.

	water treatment process
tap water	surface water → preliminary sedimentation → coagulation (aluminium polychlorid) → sedimentation → filtration → disinfected by chlorine → clean water tank → water pipes → user
bottled purified water	municipal tap water → quarts sand filtration → activated carbon → reverse osismis or nano filtration (0.0001 μm) → disinfected by ozone → packaged by plastic bottle(disinfect by chlorine) → user
bottled mineralized water	municipal tap water → quarts sand filtration → activated carbon → reverse osismis or nano filtration (0.0001 μm) → minerals added → disinfected by ozone → packaged by plastic bottle(disinfected by chlorine) → user
bottled natural water	surface/underground water → quarts sand filtration → activated carbon → ultrafiltration (0.001–0.1 μm)→ disinfect by ozone → packaged by plastic bottle(disinfected by chlorine) → user
filtered tap water	municipal tap water → activated carbon → KDF filtration → user

and magnesium are mainly present as the simple ions Ca^{2+} and Mg^{2+} in water [14]. Furthermore, Gillies reported that tap water supplies 10% of the average individual's zinc intake [15]. Additionally, consumers want to have a drinking water option that has sufficient quantities of beneficial minerals but no pollutants, and filtered tap water may meet these requirements. Water filtration via a terminal water processor can not only remove chlorine and other impurities [16] but also significantly improve the taste and odor of public tap water. Therefore, it is suitable for home or anywhere where the water quality is poor. Currently, Chinese water pollution is widespread, and water filters are used in more and more residential buildings and private kitchens to improve public tap water quality. At present, more than 15% of the families in Beijing, Guangzhou and Shanghai have a household water purifier. Many materials can be used for water filtration: food-grade cocoanut active charcoal (CAC) and kinetic degradation fluxion (KDF) are the most popular in China. CAC can remove residual chlorine; KDF is a high-purity copper-zinc formulation that uses redox (oxidation/reduction) to remove chlorine, lead, mercury, iron, and hydrogen sulfide from water. The process also has mild anti-bacterial, algicidal, and fungicidal effects [17–18].

Therefore, the constituents in all five of these popular types of water are different and may have different biological effects in humans. Several studies have reported that many factors in drinking water have negative effects on human reproduction and development. For example, epidemiologic studies have reported that low calcium and magnesium intake from drinking water significantly increase the risk of delivering a very low birth weight baby [19–20]. Rats given water containing high levels of zinc have deficits in spatial and working memory [21]. Drinking purified water cannot obviously affect the rats' reproductive outcomes, but it can induce the occurrence of development retardation in the offspring [22]. Eliminating harmful effects on humans over their lifetime and in future generations is the ultimate goal of improving drinking water. However, up to now, there have been no comparative studies regarding the effects of drinking water on the reproductive system and offspring development. In this study, the reproductive and developmental outcomes of rats given the five types of water consumed most widely in China were compared to address which drinking water is the best for pregnant women and infants.

Materials and Methods

Ethics Statement

The Sprague–Dawley rats (150 females, 70~90 g; 75 males, 110~130 g) used in our experiments were obtained from the Laboratory Animal Center, Third Military Medical University (Chongqing, China) and were treated humanely according to the criteria outlined in the "Guide for the Care and Use of Laboratory Animals" prepared by the National Academy of Sciences. The

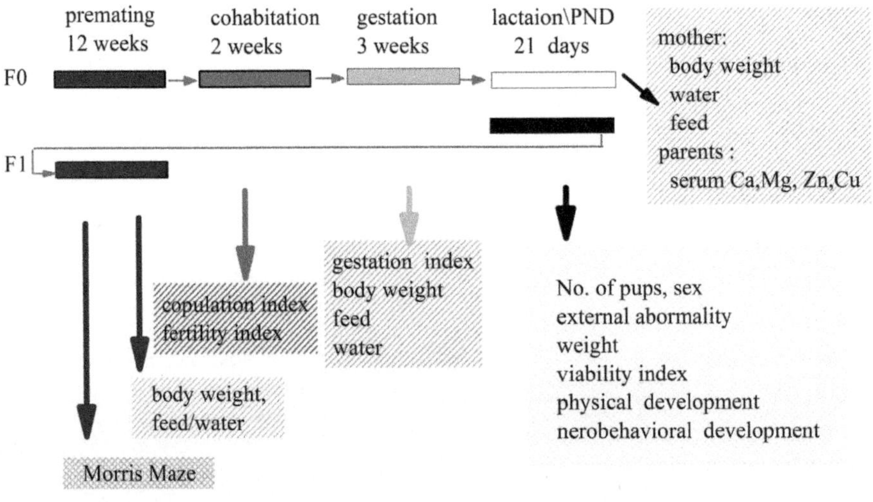

Figure 1. Schematic illustration of the study design (PND, postnatal day).

Table 2. The water quality indices of the five drinking waters.

	bPW	bMW	bNW	FTW	TW	WHO Guideline (2011)	GB5749-2006 in China	LDL	Unit
pH	6.8	6.8	7.55	7.72	7.57	6.5–8.5	6.5–8.5	0	
TDS	1.2	10.9	87.2	291	229	<1000	<1000	0.1	mg/L
TH_{CaCO3}	0.8	2.3	69.6	202.4	200.3	——	<450	0.05	mg/L
COD_{Mn}	0.5	0.6	0.6	0.6	1.0	——	<3	0.05	mg/L
Potassium	<0.5	3.4	<0.5	2.1	2.5	——	——	0.5	mg/L
Sodium	0.1	0.1	0.1	17.0	12.4	——	200	0.1	mg/L
Calcium	0.04	0.02	0.04	40.4	52.9	——	——	0.01	mg/L
Magnesium	0.02	0.4	0.02	11.0	12.7	——	——	0.01	mg/L
Zinc	0.01	0.01	0.01	0.03	0.07	——	<1.0	0.01	mg/L
Copper	0.05	0.05	0.06	0.05	0.05	<1.0	<1.0	0.01	mg/L
Iron	<0.01	<0.01	<0.01	<0.01	0.13	<0.3	<0.3	0.01	mg/L
Mercury	<0.0001	0.0002	<0.0001	0.0001	0.0003	<0.006	<0.001	0.0001	mg/L
Arsenic	<0.01	<0.01	<0.01	<0.01	0.01	<0.01	<0.01	0.01	mg/L
Lead	<0.005	<0.005	<0.005	<0.005	<0.005	<0.01	<0.01	0.005	mg/L
Nitrite (nitrogen)	0.002	<0.001	0.001	0.026	<0.001	<0.9	<0.005	0.001	mg/L
Nitrate (nitrogen)	<0.5	<0.5	0.5	0.8	1.2	<11	<10	0.5	mg/L
Fluoride	<0.1	<0.1	<0.1	0.2	0.2	<1.5	<1.0	0.1	mg/L
Perchlormethane	0.0013	0.0009	0.0011	<0.0001	<0.0001	——	<0.002	0.0001	mg/L
Trichlormethane	0.021	0.033	0.015	<0.001	0.029	<0.3	<0.06	0.001	mg/L

Abbreviation: bMW, bottled mineralized water; bNW, bottled natural water; bPW, bottled purified water; COD, Chemical Oxygen Demand; FTW, filtered tap water; TDS, Total Dissolved Solids; TH, Total hardness; TW, tap water; WHO, world health organization; LDL, lowest detectable limit; —— not establishing guideline values.

Table 3. The reproductive outcomes of the rat parents given the five drinking waters.

	bPW	bMW	bNW	FTW	TW	P
No. of rats (male/female)	30/15	30/15	30/15	30/15	30/15	
Copulation index (%)a, male	100 (15/15)	100 (15/15)	100 (15/15)	100 (15/15)	100 (15/15)	1.00
Copulation index (%)a, female	100 (30/30)	100 (30/30)	97 (29/30)	100 (30/30)	100 (30/30)	1.00
Maternal Fertility index (%)b	67 (20/30)*	100 (30/30)	83 (24/29)	73 (22/30)	100 (30/30)	0.00
Maternal Gestation index (%)c	100 (20/20)*	100 (30/30)*	100 24/24)*	100 (22/22)*	87 (26/30)	0.01
Gestation length (days)d	22 (0)	22 (0)	22 (0)	22 (0)	22 (1)	0.13

aCopulation index (%) = (no. of animals with successful copulation/no. of animals paired) × 100.
bFertility index (%) = (no. of animals with pregnant/no. of animals with successful copulation) ×100.
cGestation index (%) = (no. of females that delivered live pups/no. of pregnant females) ×100.
dValues are given as the median (interquartile range).
*$p<0.05$, statistically significant difference from TW group.
Abbreviation: bMW, bottled mineralized water; bNW, bottled natural water; bPW, bottled purified water; FTW, filtered tap water; TW, tap water.

protocol was approved by the Committee on the Ethics of Animal Experiments of Chongqing Experimental Animal Management Center. All serum sample collections were performed under sodium pentobarbital anesthesia, and all efforts were made to minimize suffering.

This study is part of a non-profit project supported by the National Natural Science Foundation of China. All necessary permits were obtained for the described field studies and approved by Chongqing Municipal Health Bureau. The study location is not privately owned, and the field studies did not involve endangered or protected species.

Drinking water and diet

All the animals were acclimated to the laboratory environment for 1 week before the beginning of the study and had free access to food and water.

The tap water used was the municipal water of Chongqing city, which originated from the Yangzi River and was disinfected by liquid chlorine in the water treatment plant. The filtered tap water was municipal tap water filtered by a KDF-CAC-purifier. The tap water and filtered tap water were collected every day for the animals to drink. Three types of bottled water were purchased from supermarkets at a single time point, and each of the same type of bottled water was produced at the same time and was of a single origin. Furthermore, the quality guarantee period for the bottled water is 12 months, which ensured that the samples did not expire before the completion of the animal experiment. One box of bottled water was randomly selected from each type of bottled water to analyze the water quality parameters. The tap water and filtered tap water were collected once at the same time to analyze the water quality parameters. The water quality parameters were determined according to Chinese GB/5750-2006 [23].

The rat feed was prepared bi-monthly in the Laboratory Animal Center of the Third Military Medical University (license number: SCX 2007-018) and strictly followed the GB 14924-2001 in China for experimental animal feed nutrition (calcium 1.11%, magnesium 0.22%).

Parent's reproductive procedure

The animals were randomly divided into 5 groups, and each group was given one type of water from 28 days of age of the parents to 8 weeks of age of the pups (Fig. 1). The reproductive procedures were practiced according to the procedure modified from OECD 415 and GB15193.15-2003 in China [24–25]. At 119 days of age, two females were paired with one male (2:1) from the same water group for 14 days until mating was confirmed by a copulatory plug or sperm in a vaginal rinse. The mating day was recorded as gestation day (GD) 0. The mated females were weighed on GD 0, 6, 12 and 18 and on lactation day (LD) 0, 4, 7, 14 and 21. Water and diet consumption were also recorded on the same days. The day of offspring birth was identified as postnatal day (PND) 0.

Parents' serum calcium, magnesium, phosphorus, copper and zinc levels

Following lactation, parental rats were anesthetized by i.p. pentobarbital, and blood was collected by heart puncture into heparinized syringes. The total serum concentrations of calcium, magnesium phosphorus, and copper were determined in an Olympus AU 600 auto-analyzer(Japan) using the commercial kits of the same brand. Serum zinc was determined in an air–acetylene flame atomic absorption spectrometer (Puxi TAS-986, Peking).

Parents' serum estradiol and testosterone levels

Serum estradiol and testosterone levels were determined in a radioimmunoassay counter (GC-911γ, Zhongjia photoelectric company) using commercial kits (Northern Beijing Institute of Biotechnology).

Offspring studies

All the pregnant rats were allowed to give birth and nurture their offspring normally. On PND 0, the pups were examined for gross malformations, and the numbers of live and stillborn pups were recorded. Each litter was examined twice daily for survival. On PND 4, 8 surviving pups (four males and four females, if possible) were retained randomly, and the remaining pups were culled from each nest. Each pup was weighed on PND 0, 4, 7, 14 and 21.

All behavioral development parameters of the pups were assessed between 9:00 and 10:00 a.m. Pups were separated from the mothers for the time of observation and then immediately

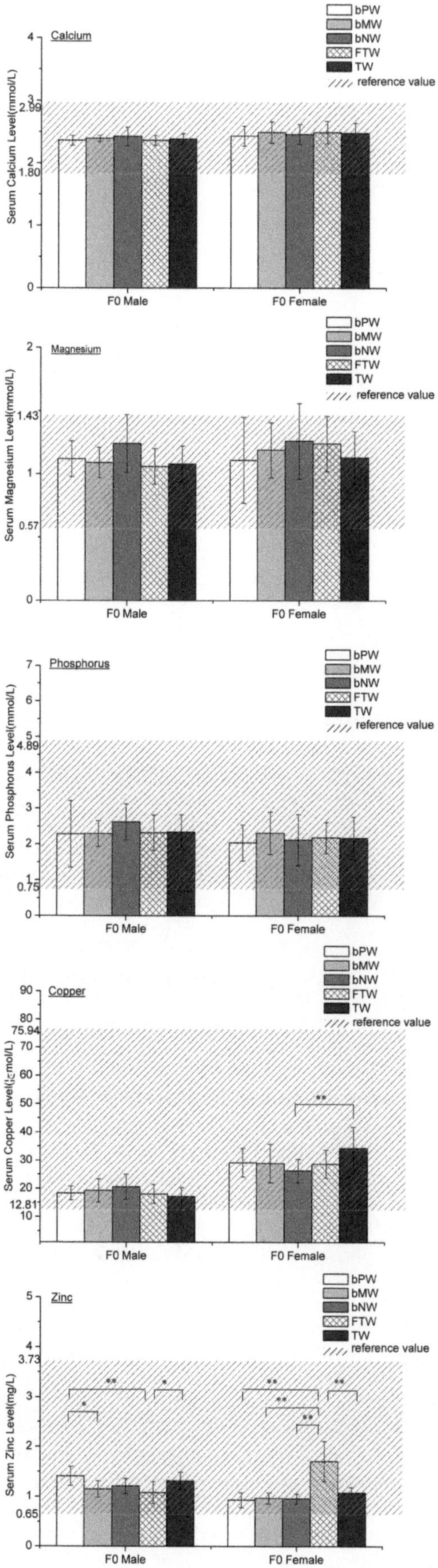

Figure 2. The serum mineral levels of the maternal rats after consuming the five waters ($\bar{x}\pm s$). Statistically significant differences between groups TW, bPW, bMW, bNW, and FTW are marked with asterisks: *p<0.05 and **p<0.01.

returned to their home cages. Pinna detachment, incisor eruption, eye opening, cliff avoidance and surface righting [26–27] were recorded according to the method previously described by our group [22].

The acquisition of spatial learning and memory was assessed via three components in the Morris water maze (MWM) (including hidden platform acquisition, probe trial and subsequent visible platform test) according to a modified version of the procedure of Morris [28]. Ten pups were chosen randomly from ten different litters per group for the MWM. On PND 28, all of them were tested in the MWM [22]. In the hidden platform acquisition test, the rat was given two trials a day for 5 days with an inter-trial period of 15 min, and the time to reach the escape platform was measured. Thus, the learning test refers to the time to reach a hidden platform on successive trials over 5 days. The probe trial was conducted on the sixth day. The platform was removed from the pool, and each rat was allowed to swim for 90 s in the water maze. The number of platform area crossings was recorded, and the memory test refers to the number of goal crossings (traversing the actual location of the escape platform). In the visible platform test, each rat was placed in the pool to find the visible platform in 90 s, and the time to find the platform and the swimming velocity were observed and measured.

Statistical analysis

Following an assessment for homogeneity of variance, the data for quantitative and continuous variables (e.g., body weights) collected from all the rats were analyzed by one-way ANOVA. Data were transformed to achieve approximate normality if the data were not normally distributed. Data are presented as medians (interquartile range). For the hidden platform test, a repeated-measure-ANOVA was used to determine the significance of the difference among the groups. The frequency data were analyzed by a nonparametric χ^2-test. The Pearson bivariate correlation analysis test was used to analyze the relationship between the levels of minerals (in water and serum) and reproductive or neurobehavioral parameters. Unless otherwise noted, the presented data are the mean values ± standard errors of the means (SEM). Statistical analyses were performed using SPSS 20.0 Statistical Software. In all the experiments, p<0.05 was considered to be statistically significant.

Results

The constituents of the five types of drinking waters

Table 2 shows that the levels of the constituents in the 5 waters were all within the levels established by the GB5749-2006 in China[29] and the WHO guidelines (2011). The three bottled waters all contained relatively low levels of total dissolved solids (TDS) and total hardness (TH), indicating that the levels of inorganic components were lower in the bottled waters. Macro elements such as calcium, magnesium, and sodium were higher in the tap water and the filtered tap water than in the three bottled waters. Furthermore, the calcium and magnesium ratio was 1:20 in mineralized water but 2:1 in natural water and 4:1 in tap water or filtered tap water. Zinc, another important component, was highest in tap water. Fluoride and nitrate concentrations were also higher in tap water and filtered tap water than in the three bottled waters. Arsenic is a toxic element in water, and its level in tap

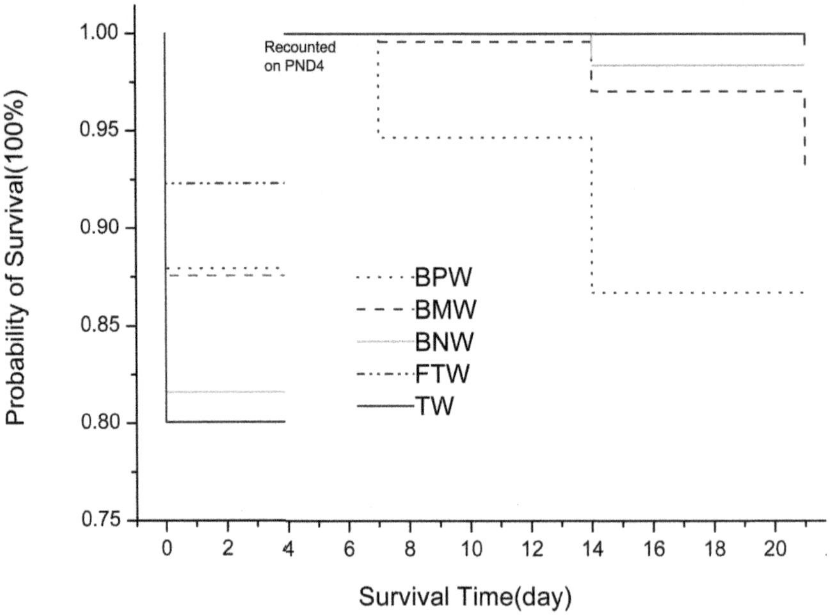

Figure 3. The survival plot of the F1 pups on lactation day 0, 4, 7, 14 and 21. Statistically significant differences between groups TW, bPW, bMW, bNW, and FTW on LD 0, LD 4 and LD 21 ($p<0.05$). The survival rats were recounted because 8 surviving pups (4 males and 4 females, if possible) were retained randomly, and the remaining pups were culled from each nest on PND 4.

water reached the limit of GB5749-2006 in China. Chemical oxygen demand (COD), an indicator of the level of organic components, was the highest in tap water among five waters (Table 2). Trichloromethane, a conventional water quality index of the DBPs, was highest in bottled mineralized water and lowest in the filtered tap water. Perchloromethane, another conventional water quality index in Chinese GB5749-2006 that may originate from raw water, was found at high levels in three types of water.

Parents' body weight, feed consumption and water consumption

No significant differences in body weight, water consumption or feed consumption were observed in maternal rats in the premating, mating, gestation and lactation periods (data are included in Data S1).

Parents' reproductive parameters and serum levels of estradiol and testosterone

Table 3 shows that the maternal fertility index in the bPW group was significantly lower than that in the TW group, but the maternal gestation index in the TW group was statistically lower than in the other groups. There were no significant differences in the copulation index (male and female), fertility index (females) or the gestation length (females). Furthermore, there were also no significant differences in the levels of testosterone (male) and estradiol (females) among all groups (data are included in Data S1).

Parental serum concentrations of calcium, magnesium, phosphorus, copper and zinc

There were no significant differences in parental serum levels of calcium, magnesium or phosphorus among all five groups (Fig. 2). The maternal serum copper levels in the bNW group were statistically lower than in the TW group (Fig. 2). The paternal serum zinc levels in the FTW group were lower than in the TW and bPW groups, but the levels in the bPW group were statistically significantly higher than in the bMW group. Furthermore, the maternal serum zinc levels in the FTW group were the highest among the five groups (Fig. 2).

Offspring's developmental parameters

On LD 0, there were no significant differences in external malformation, litter size, pup weight, sex ratio (data are included in Data S1) and survival rate (Fig. 3). On LD 4, survival rates in the TW and bNW groups were significantly lower than that in the FTW group (Fig. 3). On LD 21, the survival rate in the bPW group was the lowest among the 5 groups (Fig. 3). From LD 0 to LD 21, there were also no significant differences in pups' body weights, physiological development (pinna detachment, incisor eruption, and eye opening), and reflex development (surface righting and cliff avoidance reflex) (data are included in Data S1).

Offspring's learning and memory ability after lactation

During the learning period (days 1–5), the time to reach the platform in the place navigation test was shorter day by day in all groups (Fig. 4A), indicating that all the pups could generate space allocation memory regarding the platform. Based on the repeated measure ANOVA, the TW group showed a longer time to reach the platform than the bPW, bMW, bNW, and FTW groups (Fig. 4A). The pups' memory can be tested by the spatial probe test, and the number of goal crossings was recorded to denote the memory ability. Based on one-way ANOVA, the FTW group exhibited statistically significantly higher memory ability than the TW group (Fig. 4B). There were no significant differences in the visible platform trial (negative data not shown).

The relationship of water constituents, serum mineral levels and reproductive and developmental parameters

Table 4 shows that the maternal gestation index was negatively associated with water COD ($p = 0.005$), zinc ($p = 0.016$), and arsenic levels ($p = 0.000$). It also showed that the pups' memory

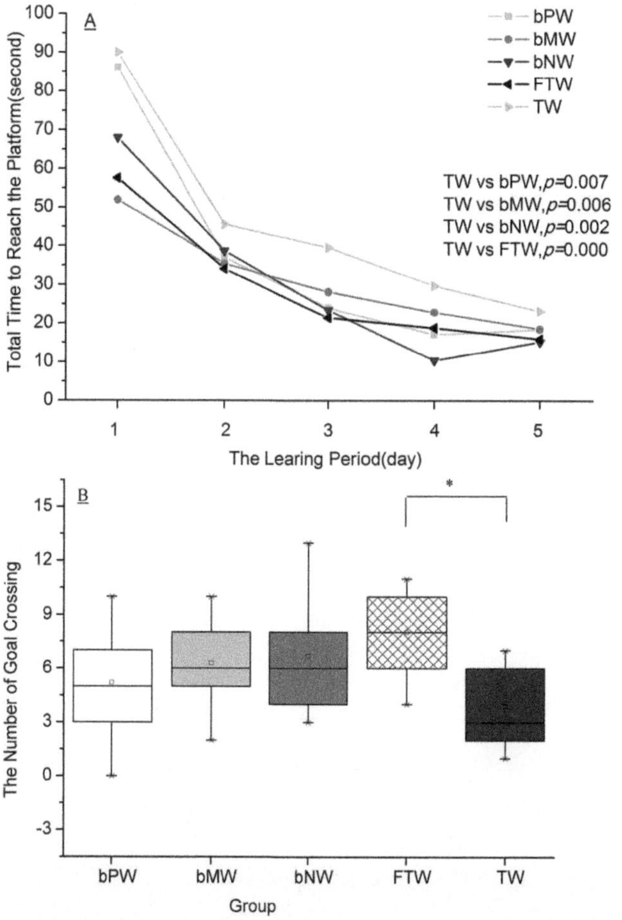

Figure 4. The pups' learning (A) and memory ability (B) in the place navigation test after the mothers drank each of the five waters (median). The data showed that the time to reach the platform of rat pups in the TW group was statistically longer than that of rat pups in the bPW, bMW, bNW and FTW groups, and the total number of rats to reach the platform in the FTW group was statistically higher than in the TW group. Statistically significant differences in the TW and FTW groups are marked with asterisks: *$p<0.05$.

ability was also negatively associated with water COD ($p = 0.012$), arsenic ($p = 0.009$), and perchloromethane levels ($p = 0.012$) but was positively associated with maternal serum zinc levels ($p = 0.027$). Furthermore, maternal estradiol was negatively associated with perchloromethane ($p = 0.042$).

Maternal serum copper and zinc levels were significantly associated with water constituents, including calcium and magnesium. However, maternal serum calcium and magnesium levels had no correlation with water constituents. No correlation was observed between other water constituents and other maternal reproductive or pup developmental parameters.

Discussion

Water constituents

Different water treatment procedures may result in different water compositions. In this study, the data showed the different TDS levels in five types of water. There were no minerals or very low mineral levels in the bottled purified water and bottled mineralized water, which suggested that the "mineralized water" on the market appears not to actually be mineralized water. The calcium and magnesium ratio also indicated that the "mineralized" water did not meet the natural rule. The TDS and TH were also lower in bottled natural water, which indicated that the filtration treatment removed the majority of the minerals. In contrast, house water filters removed only minimal amounts of beneficial minerals but effectively removed toxic metals by ion exchange. It should be noted that the organic constituent was detected in all five waters. Trichloromethane and perchloromethane were the most common DBPs detected in all three bottled waters, and the levels were very similar, indicating that the main source of organic pollutants may be the plastic bottle and cap, which were always disinfected by chlorine. To our knowledge, COD is commonly used to indirectly measure the amount of organic compounds in water [29] because it is difficult or impossible to identify each contaminant in water, especially the trace organic contaminants. The present study showed that the COD of the tap water was the highest, although it did not exceed the GB19298-2003 and GB5749-2006 in China [30–31]. The high COD in tap water indicated that trichloromethane was not the only organic pollutant. Previous studies in our laboratory reported that 30 types of non-volatile organic pollutants were detected in the tap water that came from the same water plant, while 46 types of non-volatile organic pollutants were detected in the source waters, which were the Yangzi River, in 2000–2001 [32]; 50 types of non-volatile organic pollutants were detected in the same source water in the year 2005 [33]. The water treatment technology in that water plant has remained the same to the present day. However, because the pollution of the water source has not stopped, the number and variety of pollutants may increase. Fortunately, it is obvious that trichloromethane and other pollutants were effectively removed from the tap water through the use of a house water filter. Thus, the three bottled waters were just soft water with low TDS, the tap water was full of not only minerals but also pollutants, and the filtered tap water was full of minerals but with fewer pollutants. Furthermore, the bottled mineralized water did not meet the natural rule with regard to the ratio of calcium to magnesium.

Water constituents exert effect on reproduction

Previous studies have shown that macro elements such as calcium and magnesium and trace elements such as zinc and copper have great impact on reproduction [34–36]. In the present study, the data showed varied levels of such mineral elements in the five drinking waters (Table 2) and the lowest female gestation index but a high fertility index in the TW group (Table 3). These results suggest that embryo implantation in early pregnancy may be affected. Macro elements in the body such as calcium, magnesium and phosphorus are fully supplied in the diet, but when the micro elements are marginally deficient in the diet, water becomes the main source [37]. In this study, the calcium level was 1.11% and the magnesium level was 0.22% in the rat diet, and those levels can fulfill the rats' requirements for growth and development. Therefore, the serum levels of calcium and magnesium were not significantly different between the 5 groups, and the correlation analysis also showed no significant relationship between the levels of water macro elements (calcium, magnesium and phosphorus) and reproductive or developmental parameters. Micro elements such as zinc may be derived from the galvanized plumbing materials. In the present study, zinc levels were highest in the tap water. Unfortunately, we did not determine the zinc content in the feed, as nutritional requirements may be fully met by the rats' feed. However, the data also showed that the serum zinc level was statically different in all 5 groups, and there were no correlations between the serum zinc level and the water zinc level.

Table 4. Correlation analysis of the relationship among water constituents, serum mineral levels, maternal reproductive and pups' developmental parameters.

	Maternal Serum Calcium	Maternal Serum Magnesium	Maternal Serum Phosphorus	Maternal Serum Copper	Maternal Serum Zinc	Maternal Serum estradiol	Paternal serum Testosterone	Maternal Fertility Index	Maternal Gestation Index	Pup's Memory Ability
Water TDS	0.075	0.053	−0.002	0.163	0.675**	0.196	−0.047	0.020	−0.450	0.09
Water COD_{Mn}	0.061	−0.070	0.020	0.356**	−0.009	−0.049	−0.172	0.669	−0.975**	−0.305**
Water Calcium	0.072	−0.035	0.003	0.308**	0.500**	0.141	−0.092	0.220	−0.738	−0.096
Water Magnesium	0.077	−0.025	0.007	0.291**	0.551**	0.157	−0.085	0.189	−0.683	−0.061
Water Zinc	0.055	−0.082	−0.001	0.372**	0.201	0.038	−0.133	0.437	−0.943**	−0.261
Water Copper	−0.042	0.148	−0.042	−0.260**	−0.243	−0.099	0.028	−0.059	0.250	0.118
Water Arsenic	0.034	−0.112	−0.004	0.381**	−0.079	−0.054	−0.150	0.567	−1.00**	−0.368**
Water Trichlormethane	−0.105	−0.011	−0.041	−0.256*	−0.597*	−0.157	0.104	0.715	−0.406	−0.010
Water Perchlormethane	0.009	−0.127	0.060	0.181	−0.652**	−0.237*	−0.130	−0.295	−0.001	−0.353*
Maternal Serum Calcium	1.00	0.718**	0.671**	0.229*	−0.047	0.038	—	0.144	0.049	−0.166
Maternal Serum Magnesium	0.718**	1.00	0.734**	0.088	−0.001	0.067	—	−0.101	0.442	−0.130
Maternal Serum Phosphorus	0.671**	0.734**	1.00	0.005	−0.086	0.075	—	−0001	0.378	−0.132
Maternal Serum Copper	0.229*	0.088	0.005	1.00	0.165	−0.154	—	0.879**	−0.826	−0.079
Maternal Serum Zinc	−0.047	−0.001	−0.086	0.165	1.00	−0.227	—	0.176	−0.225	0.316**

*, $p<0.05$,
**, $p<0.01$ indicate significant difference level.
No statistically correlation between other water constituents and pups' development parameters, between maternal serum elements and pups' development parameters (data not shown).

The correlation analysis also showed that the zinc level in water was significantly and negatively correlated to the maternal gestation index, although maternal serum zinc levels had no significant relation to the maternal gestation index. Serum zinc values vary diurnally, decrease after meals, and appear to be related to gender and age [37]. More than 90% of zinc is stored in the muscle, and only 10% is present in the serum. Because the zinc content of the muscle was not determined in the current study, we cannot rule out the possibility that the differences in the serum zinc levels are partly attributed to zinc mobilization from muscle tissue. It appears that there was a complicated link between zinc levels in the water and serum. The relationship between water zinc levels and reproductive index parameters requires further study. Thus, the micro elements such zinc, but the not macro elements in the water, may affect the rats' reproduction.

Clearly, toxic chemical elements such as arsenic exert negative reproductive effects [38]. In the present study, we found that only the arsenic levels in tap water reached the limit of the GB19298-2003 and GB5749-2006 in China. A correlation analysis showed that arsenic level in the water was negatively associated with maternal gestation index. However, the arsenic values were very close to each other, suggesting that arsenic levels may have not been sufficient to decrease maternal reproductive capacity in the TW group but may have increased the effects of other factors; these findings require further study.

Organic contaminants in water also have clear reproductive toxicity [39–40]. There was a negative correlation between water COD and maternal gestation index parameters. It may be possible that the lowest MGI in the TW group may have been the result of organic pollutants in the tap water other than trichloromethane, which was present in similar levels in each of the 4 waters. The organic pollutants including DBPs in the tap water may also exert an effect on reproductive function, as 40 types of trace organic pollutants have been detected in tap water, and the organic extract can increase endometrial thickness in rats [41–42]. How the mixture exerts these effects require further study.

Water constituents exert effects on development

Previous studies have suggested that zinc can enhance learning and memory ability [43] and that organic pollutants [44] and arsenic [45–46] in drinking water can decrease learning and memory abilities. The results from the present study are consistent with these findings: organic pollutants and arsenic were the highest in the TW group, and the learning ability of the pups in the TW group was statistically lower than in the other groups; organic pollutants were the lowest and the maternal serum zinc level was the highest in the FTW, and the memory ability of the pups in the FTW group was the highest among the 5 groups. A correlation analysis also showed that the pups' memory ability was positively correlated to maternal serum zinc levels and negatively correlated to COD_{mn}, arsenic and perchloromethane levels in water. Thus, the better memory ability in the FTW group may be due to the near absence of organic compounds or toxic metals and the rich mineral content in the filtered tap water. However, the reason for the lower learning ability in the TW group may be complicated because COD_{mn}, arsenic and perchloromethane levels were all below the Chinese water standards GB5749-2006 and WHO Guidelines (2001) for tap water; in particular, the arsenic levels were very close in all five waters, which suggests that the effect on learning ability is most likely due to other pollutants we did not detect or a complex combination of all pollutants. These results require further study.

Survival rates on PND 4 and PND 21 can be used as indicators of the maternal nutritional status and maternal instinct. On PND 0, survival rates were not significantly different between the 5 groups. However, on PND 4, survival rates in all 5 groups decreased, and the survival rates in the TW and bNW groups were statistically lower than in the FTW group, which indicated that the maternal nutrition may have been deficient in the TW and bNW groups because the mothers nursed all pups. After PND 4, the pups were culled to 6–8 pups per litter in order to assure each mother had enough nutrition to nurse all pups. Therefore, the survival rate increased on PND 21. However, the survival rate in the bPW group was the lowest among the 5 groups, suggesting that maternal nutrition in the bPW group may have been insufficient.

Overall, among the five drinking waters, filtered tap water had the lowest levels of pollutants, had the highest hardness, was the richest in minerals, and showed the best benefit for maternal reproductive parameters and pups' development parameters. Removing toxic contaminants and maintaining minerals are both important for drinking water, especially when consumed by pregnant women and children. Granular activated carbon and ion exchange by copper zinc alloy may be the best way to treat water, but how water produced with these methods exerts beneficial effects needs further study. This result may have important implications for the selection of healthy drinking water and for water plants to optimize their treatment processes, as organic pollutants and toxic metals (arsenic) in drinking water may decrease maternal reproductive parameters.

Author Contributions

Conceived and designed the experiments: WS LL. Performed the experiments: HZ JC DW WF LW JL LZ YT ZQ YH. Analyzed the data: HZ LZ. Contributed reagents/materials/analysis tools: HZ JC DW WF LW JL LZ YT ZQ YH. Wrote the paper: HZ WS.

References

1. Rosa C, Franco F, Paolo T. (2011) Drinking water quality: Comparing inorganic components in bottled water and Italian tap water. J Food Compost Anal 24: 184–193.
2. Richardson SD. (2003) Disinfection by-products and other emerging contaminants in drinking water. Trends Analyt Chem 22: 666–684.
3. Wolf A, Bergmann A, Wilken RD, Gao X, Bi Y, et al. (2013) Occurrence and distribution of organic trace substances in waters from the Three Gorges Reservoir, China. Environ Sci Pollut Res Int 20: 7124–39.
4. Richardson SD, Plewa MJ, Wagner ED, Schoeny R, DeMarini DM. (2007) Occurrence, genotoxicity, and carcinogenicity of regulated and emerging disinfection byproducts in drinking water: a review and roadmap for research. Mutat Res 636: 178–242.
5. Clement M, Seux R, Rabarot S. (2000) A practical model for estimating total lead intake from drinking water. Water Res 34: 1533–1542.
6. Cartier C, Nour S, Richer B, Deshommes E, Prévost M. (2012) Impact of water treatment on the contribution of faucets to dissolved and particulate lead release at the tap. Water Res 46: 5205–5216.
7. Kim EJ, Herrera JE, Huggins D, Braam J, Koshowski S. (2011) Effect of pH on the concentrations of lead and trace contaminants in drinking water: A combined batch, pipe loop and sentinel home study. Water Res 45: 2763–2774.
8. Fertmann R, Hentschel S, Dengler D, Janßen U, Lommel A. (2004) Lead exposure by drinking water: an epidemiological study in Hamburg, Germany. Int J Hyg Environ Health 207: 235–244.

9. Richard H. (2012) global bottled water congress and market trends. Available at: http://zenithinternational.com/pdf/events/00101slides.pdf. Accessed 8 April 2013.
10. Deng XQ, Liang JH, Cheng CY. (2001) A superficial view on the development of bottled water market at home and abroad. Food and Fermentat ion Industries 27(4): 70–74.
11. No author. (2008) bottled water: the question of the source. Popular Standardization 9: 6–9. Available at: http://www.cqvip.com/QK/96243X/200809/28415476.html. Accessed 8 April 2013.
12. Frantisek K. Health risks from drinking demineralised water. Available at: http://www.who.int/water_sanitation_health/dwq/nutrientschap12.pdf. Accessed 8 April 2013.
13. Widdowson EM. (1944) Health significance of drinking water calcium and magnesium Minerals in human nutrition. British Medical Bulletin 10: 221–222.
14. Sabatier M, Arnaud MJ, Kastenmayer P, Rytz A, Barclay DV. (2002) Meal effect on magnesium bioavailability from mineral water in healthy women. American Journal of Clinical Nutrition 75: 65–71.
15. Gillies ME, Paulin HV. (1982) Estimations of daily mineral intakes from drinking water. Human nutrition:applied nutrition 36: 287–292.
16. Xu W, Wang DX, Zhang JW, Shen JY. (2011) Function Comparison of Four Types of Household Water Purifier. Occup and HeMth 27: 2004–2005.
17. Teakle JT. (1997) Pretreatment of reverse osmosis systems with KDF55 process medium. Michigan:KDF Fluid Treatment, Inc pp: 1–5.
18. Xiong R, Liu W, Xi X, Xiao S. (2004) Application and amelioration prospect of copper-zinc alloy in water treatment. Industrial Safety and Dust Control 30: 5–8.
19. Yang CY, Chiu HF, Tsai SS, Chang CC, Sung FC. (2002) Magnesium in drinking water and the risk of delivering a child of very low birth Weight. Magnes Res 15: 207–213.
20. Yang CY, Chiu HF, Chang CC, Wu TN, Sung FC. (2002) Association of very low birth weight with calcium levels in drinking water. Environ Res 89: 189–194.
21. Linkous DH, Smith LN, Conko KM, Jones BF, Flinn JM. (2004) Differential effects of zinc association in drinking waters on spatial memory. Met Ions Biol Med 8: 390–395.
22. Zeng H, Shu WQ, Zhao Q, Chen Q. (2008) Reproductive and neurobehavioral outcome of drinking purified water under magnesium deficiency in the rat's diet. Food Chem Toxicol 46: 1495–1502.
23. Ministry of Health of China, Standardization Administration of China. National Standard of the People's Republic of China—standard examination methods for drinking water (GB/T5750-2006). Issued on December 29, 2006. Implemented on July 1, 2007.
24. OECD Publishing, 1983. Test No. 415: One-Generation Reproduction Toxicity Study. In: OECD Guidelines for the Testing of Chemicals/Section 4: Health Effects. Available: http://www.oecdbookshop.org/oecd/display.asp?lang=EN&sf1=identifiers&st1=5lmqcr2k7nnq. Accessed 24 August 2013.
25. Ministry of Health of China, Standardization Administration of China. National Standard of the People's Republic of China—reproductive study (GB15193.15-2003). Issued on September 24, 2003. Implemented on May 1, 2004.
26. Zbinden G. (1981) Experimental methods in behavioral teratology. Arch Toxicol 48: 69–88.
27. Pantaleoni GC, Fanini D, Sponta AM, Palumbo G, Giorgi R, et al. (1988) Effects of maternal exposure to polychlorobiphenyls(PCBs) on F1 generation behavior in the rat. Fundam Appl Toxicol 11: 440–449.
28. Morris R. (1984) Developments of water-maze procedure for studying spatial learning in the rat. J Neurosci Methods 11: 47–60.
29. Devi R, Dahiya RP. (2006) Chemical oxygen demand (COD) reduction in domestic wastewater by fly ash and brick kiln ash. Water Air Soil Pollut 174: 33–46.
30. Ministry of Health of China, Standardization Administration of China. National Standard of the People's Republic of China—hygienic standard of bottled water for drinking (GB19298-2003). Issued on September 24, 2003. Implemented on May 1, 2004.
31. Ministry of Health of China, Standardization Administration of China. National Standard of the People's Republic of China—standards for drinking water quality (GB 5749–2006). Issued on December 29, 2006. Implemented on July 1, 2007(replacing GB 5749–1985).
32. Tian HJ, Shu WQ, Zhang XK, Wang YM, Cao J. (2002) Organic pollutants in source water in Jialing and Yangze River (Chongqing Section). Resources and Environment in the Yangtze basin 24(4): 226–239.
33. Guo ZS, Luo CH, Zhang WD, Lu Y, Sun J, et al. (2006) The analysis of the persistent organic pollution in the Three Gorges Region in Chongqing. Environmental Monitoring in China 22(4): 45–48.
34. Mao YX. (2003) Analysis magnesium ions content in normal pregnancy and abnormal pregnancy. Chinese Journal of Birth Health and Heredity 11: 73–75.
35. Cheng F, Wang Y. (2012) The correlation analysis on pregnant women's and neonate's serum levels of calcium,ferrum,zinc,copper and weight. Journal of Chongqing Medical University 37: 89–91.
36. Dutt B, Mills CF. (1960) Reproductive Failure in Rats Due to Copper Deficiency. J Comp Pathol 70: 120–125.
37. Maret W, Sandstead HH. (2006) Zinc requirements and the risks and benefits of zinc supplementation. J Trace Elem Med Biol 20: 3–18.
38. Bloom MS, Fitzgerald EF, Kim K, Neamtiu I, Gurzau ES. (2010) Spontaneous pregnancy loss in humans and exposure to arsenic in drinking water. Int J Hyg Environ Health 213: 401–413.
39. Yang CY, Cheng BH, Tsai SS, Wu TN, Lin MC, et al. (2000) Association between chlorination of drinking water and adverse pregnancy outcome in Taiwan. Environ Health Perspect 108: 765–768.
40. Wright JM, Schwartz J, Dockery DW. (2003) Effect of trihalomethane exposure on fetal development. Occup Environ Med 60: 173–180.
41. Cao B, Ren Q, Cui ZH, Li P, Cao J. (2009) Evaluation of reproductive toxicity in rats caused by organic extracts of Jialing River water of Chongqing, China. Environ Toxicol Pharmacol 27: 357–365.
42. Ren Q, Cao B, Qiu ZQ, Zhao Q, Shu WQ. (2005) A study of toxicity of organic extract of tap water on male reproductive system in mice. Acta Academiae Medicinae Militaris Tertiae 27: 1012–1015.
43. Nakashima AS, Dyck RH. (2009) Zinc and cortical plasticity. Brain Res Rev 59: 347–373.
44. Janulewicz PA, White RF, Winter MR, Weinberg JM, Gallagher LE, et al. (2008) Risk of learning and behavioral disorders following prenatal and early postnatal exposure to tetrachloroethylene (PCE)-contaminated drinking water. Neurotoxicol Teratol 30: 175–185.
45. Jing J, Zheng G, Liu M, Shen X, Zhao F, et al. (2012) Changes in the synaptic structure of hippocampal neurons and impairment of spatial memory in a rat model caused by chronic arsenite exposure. Neurotoxicology 33: 1230–1238.
46. O'Bryant SE, Edwards M, Menon CV, Gong G, Barber R. (2011) Long-term low-level arsenic exposure is associated with poorer neuropsychological functioning: a Project Frontier study. Int J Environ Res Public Health 8: 861–874.

Permissions

All chapters in this book were first published in PLOS ONE, by The Public Library of Science; hereby published with permission under the Creative Commons Attribution License or equivalent. Every chapter published in this book has been scrutinized by our experts. Their significance has been extensively debated. The topics covered herein carry significant findings which will fuel the growth of the discipline. They may even be implemented as practical applications or may be referred to as a beginning point for another development.

The contributors of this book come from diverse backgrounds, making this book a truly international effort. This book will bring forth new frontiers with its revolutionizing research information and detailed analysis of the nascent developments around the world.

We would like to thank all the contributing authors for lending their expertise to make the book truly unique. They have played a crucial role in the development of this book. Without their invaluable contributions this book wouldn't have been possible. They have made vital efforts to compile up to date information on the varied aspects of this subject to make this book a valuable addition to the collection of many professionals and students.

This book was conceptualized with the vision of imparting up-to-date information and advanced data in this field. To ensure the same, a matchless editorial board was set up. Every individual on the board went through rigorous rounds of assessment to prove their worth. After which they invested a large part of their time researching and compiling the most relevant data for our readers.

The editorial board has been involved in producing this book since its inception. They have spent rigorous hours researching and exploring the diverse topics which have resulted in the successful publishing of this book. They have passed on their knowledge of decades through this book. To expedite this challenging task, the publisher supported the team at every step. A small team of assistant editors was also appointed to further simplify the editing procedure and attain best results for the readers.

Apart from the editorial board, the designing team has also invested a significant amount of their time in understanding the subject and creating the most relevant covers. They scrutinized every image to scout for the most suitable representation of the subject and create an appropriate cover for the book.

The publishing team has been an ardent support to the editorial, designing and production team. Their endless efforts to recruit the best for this project, has resulted in the accomplishment of this book. They are a veteran in the field of academics and their pool of knowledge is as vast as their experience in printing. Their expertise and guidance has proved useful at every step. Their uncompromising quality standards have made this book an exceptional effort. Their encouragement from time to time has been an inspiration for everyone.

The publisher and the editorial board hope that this book will prove to be a valuable piece of knowledge for researchers, students, practitioners and scholars across the globe.

List of Contributors

Joshua S. Gruber, Ayse Ercumen and John M. Colford, Jr.
Division of Epidemiology, University of California, Berkeley, California, United States of America

Gabriel J. Bellante, Scott L. Powell and Rick L. Lawrence
Department of Land Resources and Environmental Sciences, Montana State University, Bozeman, Montana, United States of America

Kevin S. Repasky
Department of Electrical and Computer Engineering, Montana State University, Bozeman, Montana, United States of America

Tracy Dougher
Department of Plant Sciences and Plant Pathology, Montana State University, Bozeman, Montana, United States of America

Sami Khanal
School of Environment and Natural Resources, Ohio State University, Wooster, OH, United States of America

Robert P. Anex
Dept. of Biological Systems Engineering, University of Wisconsin-Madison, Madison, Wisconsin, United States of America

Christopher J. Anderson and Daryl E. Herzmann
Dept. of Agronomy, Iowa State University, Ames, Iowa, United States of America

Ciara E. O'Reilly, Robert M. Hoekstra, Robert Quick and Tracy Ayers
Division of Foodborne, Waterborne, and Environmental Diseases, Centers for Disease Control and Prevention, Atlanta, Georgia, United States of America

Ethel V. Taylor
Division of Foodborne, Waterborne, and Environmental Diseases, Centers for Disease Control and Prevention, Atlanta, Georgia, United States of America
Epidemic Intelligence Service, Centers for Disease Control and Prevention, Atlanta, Georgia, United States of America

Ribka Fantu and Sisay Alemayehu Abayneh
Centers for Disease Control and Prevention, Addis Ababa, Ethiopia

Barbara Marston
Division of Global HIV/AIDS, Centers for Disease Control and Prevention, Atlanta, Georgia, United States of America

Yordanos B. Molla, Tegene Sewnet and Fitsum Abebe
Michael Dejene Public Health Consultancy Services, Addis Ababa, Ethiopia

Joan Laur and Uwe G. Hacke
Department of Renewable Resources, University of Alberta, 442 Earth Sciences Building, Edmonton, Alberta, Canada

Ali Alsamawi, Joy Murray and Manfred Lenzen
Integrated Sustainability Analysis, School of Physics, University of Sydney, Sydney, New South Wales, Australia

Daniel Moran
Programme for Industrial Ecology, Faculty of Engineering, Norwegian University of Science and Technology, Trondheim, Trondheim, Norway

Keiichiro Kanemoto
Institute of Decision Science for a Sustainable Society, Kyushu University, Fukuoka, Fukuoka, Japan

Bing Wang, Fenxiang Wen, Jiangtao Wu and Xiaojun Wang
College of Environmental Science and Resources, Shanxi University, Taiyuan, China

Yani Hu
Library, Hebei University of Science and Technology, Shijiazhuang, China

Kevin E. McCluney and John L. Sabo
School of Life Sciences, Arizona State University, Tempe, AZ, United States of America

Yang Du, Yanhui Gao, Fangang Meng, Shoujun Liu, Zhipeng Fan, Junhua Wu and Dianjun Sun
Center for Endemic Disease Control, Chinese Center for Disease Control and Prevention, Harbin Medical University; Key Lab of Etiology and Epidemiology, Education Bureau of Heilongjiang Province & Ministry of Health (23618504), Harbin, China

Marcos D. Robles, Robert M. Marshall and Jeanmarie A. Haney
The Nature Conservancy Center for Science and Public Policy, Tucson, Arizona, United States of America

Frances O'Donnell
Northern Arizona University, Flagstaff, Arizona, United States of America

Edward B. Smith
The Nature Conservancy in California, Sacramento, California, United States of America

David F. Gori
The Nature Conservancy in New Mexico, Santa Fe, New Mexico, United States of America

Marco Lauteri
Institute of Agro-Environmental and Forest Biology, National Research Council, Porano, Italy

Matthew Haworth and Mauro Centritto
Trees and Timber Institute (IVALSA - Istituto per la valorizzazione del legno e delle specie arboree), National Research Council, Florence, Italy

Rachid Serraj
Consultative Group on International Agricultural Research, Independent Science and Partnership Council Secretariat, Food and Agriculture Organisation of the United Nations, Rome, Italy
International Rice Research Institute, Los Banos, Philippines

Maria Cristina Monteverdi
Forestry Research Centre, National Research Council, Arezzo, Italy

Enkai Xu, Guoqiang Fan, Suyan Niu, Zhenli Zhao, Minjie Deng and Yanpeng Dong
Institute of Paulownia, Henan Agricultural University, Jinshui Area, Zhengzhou, Henan, P.R. China
College of Forestry, Henan Agricultural University, Jinshui Area, Zhengzhou, Henan, P.R. China

Michio Murakami and Taikan Oki
Institute of Industrial Science, The University of Tokyo, Meguro, Tokyo, Japan
Japan Science and Technology Agency, Core Research for Evolutionary Science and Technology (CREST), Chiyoda, Tokyo, Japan

Babatunde Bello, Xueyan Zhang, Chuanliang Liu, Zhaoen Yang, Zuoren Yang, Qianhua Wang, Ge Zhao and Fuguang Li
State Key Laboratory of Cotton Biology, Cotton Research Institute, Chinese Academy of Agricultural Sciences, Beijing, China

Anne-Cyrielle Genard-Zielinski
Institut Méditerranéen de Biodiversité et d'Ecologie marine et continentale (IMBE) Aix Marseille Université, CNRS, IRD, Avignon Université, Technopô le Arbois-Méditerranée, France
Laboratoire des Sciences du Climat et de l'Environnement (LSCE-IPSL), Unité Mixte CEA-CNRS-UVSQ (Commissariat á l'Energie Atomique, Centre National de la Recherche Scientifique, Université de Versailles Saint-Quentin-en-Yvelines), Gif-sur-Yvette, France

Elena Ormeño and Catherine Fernandez
Institut Méditerranéen de Biodiversité et d'Ecologie marine et continentale (IMBE) Aix Marseille Université, CNRS, IRD, Avignon Université, Technopô le Arbois-Méditerranée, France

Christophe Boissard
Laboratoire des Sciences du Climat et de l'Environnement (LSCE-IPSL), Unité Mixte CEA-CNRS-UVSQ (Commissariat á l'Energie Atomique, Centre National de la Recherche Scientifique, Université de Versailles Saint-Quentin-en-Yvelines), Gif-sur-Yvette, France

Justine Bresson and Marc Labadie
Laboratoire d9Ecophysiologie des Plantes sous Stress Environnementaux (LEPSE), UMR759, Institut National de la Recherche Agronomique-SupAgro, Montpellier, France
Laboratoire des Symbioses Tropicales et Méditerranéennes (LSTM), UMR113, Université Montpellier 2-IRD-CIRAD-INRA-SupAgro, Montpellier, France

François Vasseur
Laboratoire d9Ecophysiologie des Plantes sous Stress Environnementaux (LEPSE), UMR759, Institut National de la Recherche Agronomique-SupAgro, Montpellier, France
Max PlanckInstitute for Developmental Biology, Tübingen, Germany

Myriam Dauzat and Denis Vile
Laboratoire d9Ecophysiologie des Plantes sous Stress Environnementaux (LEPSE), UMR759, Institut National de la Recherche Agronomique-SupAgro, Montpellier, France

Fabrice Varoquaux and Bruno Touraine
Laboratoire des Symbioses Tropicales et Méditerranéennes (LSTM), UMR113, Université Montpellier 2-IRD-CIRAD-INRA-SupAgro, Montpellier, France

Dimah Z. Habash, Marcela Baudo and David W. Lawlor
Plant Biology and Crop Science, Rothamsted Research, Harpenden, United Kingdom

Matthew Hindle, Stephen J. Powers, Michael Defoin-Platel, Rowan Mitchell, Mansoor Saqi and Chris Rawlings
Computational and Systems Biology, Rothamsted Research, Harpenden, United Kingdom

Kawther Latiri
Laboratoire D'agronomie, National Agricultural Research Institute of Tunisia, Ariana, Tunisia

Jose L. Araus
Dept. of Vegetal Biology, Faculty of Biology, Barcelona, Spain

Ahmad Abdulkader
Biotechnology Department, General Commission for Scientific Agricultural Research, Damascus, Syria

Roberto Tuberosa
Dept. of Agroenvironmental Science and Technology, University of Bologna, Bologna, Italy

Miloudi M. Nachit
Biodiversity and Integrated Gene Management Program, International Center for Agricultural Research in the Dry Areas, Rabat, Morocco

Vokkaliga Thammegowda Harshavardhan, Christiane Seiler and Markus Kuhlmann
Research Group Abiotic Stress Genomics, Interdisciplinary Center for Crop Plant Research (IZN), Halle (Saale), Germany, and Leibniz Institute of Plant Genetics and Crop Plant Research (IPK) Gatersleben, Stadt Seeland, OT Gatersleben, Germany

Le Van Son
Research Group Gene Regulation, Leibniz Institute of Plant Genetics and Crop Plant Research (IPK) Gatersleben, Stadt Seeland, OT Gatersleben, Germany
National Key Laboratory of Gene Technology, Institute of Biotechnology Vietnam, Academy of Science and Technology, Hanoi, Vietnam

Astrid Junker, Kathleen Weigelt-Fischer and Thomas Altmann
Research Group Heterosis, Leibniz Institute of Plant Genetics and Crop Plant Research (IPK) Gatersleben, Stadt Seeland, OT Gatersleben, Germany

Christian Klukas
Research Group Image Analysis, Leibniz Institute of Plant Genetics and Crop Plant Research (IPK) Gatersleben, Stadt Seeland, OT Gatersleben, Germany

Nese Sreenivasulu
Research Group Abiotic Stress Genomics, Interdisciplinary Center for Crop Plant Research (IZN), Halle (Saale), Germany, and Leibniz Institute of Plant Genetics and Crop Plant Research (IPK) Gatersleben, Stadt Seeland, OT Gatersleben, Germany
Grain Quality and Nutrition Center, International Rice Research Institute (IRRI), Metro Manila, Philippines

Helmut Bäumlein
Research Group Gene Regulation, Leibniz Institute of Plant Genetics and Crop Plant Research (IPK) Gatersleben, Stadt Seeland, OT Gatersleben, Germany

Hui Zeng, Wei-qun Shu, Ji-an Chen, Da-hua Wang, Wen-juan Fu, Ling-qiao Wang, Jiao-hua Luo, Liang Zhang, Yao Tan, Zhi-qun Qiu and Yu-jing Huang
Department of Environmental Hygiene, College of Preventive Medicine, Third Military Medical University, Chongqing, P. R. China

Lin Liu
The Lundberg-Kienlen Lung Biology and Toxicology Laboratory, Department of Physiological Sciences, Oklahoma State University, Stillwater, Oklahoma, United States of America

Index

A

Abiotic Stress Factors, 225
Antiretroviral Therapy (ART), 40
Arabidopsis Tolerance, 192, 235
Atomic Radiation, 150-151, 160-161

B

Biofuel Crop Production, 30, 35, 37

C

Cell Oxidation, 180, 190
Climate Simulations, 30-31, 39
Climate Variability, 107-108, 115-116, 119
Coliform Bacteria, 1, 11
Comparison Clients, 40, 42, 44
Complex Signaling, 145, 162

D

Diarrhea Pathogens, 1
Diarrheal Risk, 1
Dietary Intake, 150, 160
Diffusional Constraints, 122, 131
Downy Oak, 180-181
Drought Stressed Rice, 122
Drought Tolerance, 57, 122-123, 131-135, 137, 140, 145-149, 162-164, 167, 173, 177-179, 192, 202-203, 223, 225
Durum Wheat, 148, 204-205, 217-224

E

Ecosystem Services, 107
Environmental Factors, 71, 76, 79-82
Expression Analysis, 53, 55, 57, 134, 148, 169, 174, 206, 223, 236

F

Fecal Contamination, 1-3, 11-12
Feedstock Production, 30, 34, 38
Forest Thinning, 107-108, 111, 113, 115-116, 119
Fukushima City, 150-161

G

Gas Exchange, 49, 57-58, 123-125, 132-133, 177, 179, 181, 184-185, 187-188, 223
Geologic Carbon Sequestration (GCS), 15
Gossypium Hirsutum, 162, 165, 173, 177-178
Green Fluorescent Protein (GFP), 162
Ground Water, 37, 59, 62-63, 68
Growth Recovery, 192, 197-198, 201

H

Hilly-gully Loess Plateau, 71
Hiv-infected Persons, 40, 47
Household Drinking Water, 1-3, 5, 9, 11-14, 44
Human Activities, 83
Hyperspectral Detection, 15, 27
Hyperspectral Imaging System, 15
Hypothyroidism, 96-98, 100, 102-103, 106

I

Inexpensive Interventions, 40, 47
Intervention Clients, 40, 42, 44-45
Iodine Deficiency, 96-98, 103-106
Iodine Deficiency Disorders (IDD), 96
Isoprene Emissions, 180-181, 183-184, 187-191

L

Landscape Change, 30
Leaf Hydraulic Conductance, 49-50, 52-53, 55, 57-58
Low Temperature Tolerance, 162-163, 174

M

Mechanical Thinning, 107-116, 118
Mesophyll Layer, 122, 130-131
Microscopic Observations, 134-135
Modern Genetics, 162
Morphological Responses, 122, 133, 173
Mutualistic Bacteria, 192

N

Nuclear Accident, 150, 159-161

O

Opportunistic Infections, 40, 47

P

Paulownia, 134-135, 142, 146-147
Phenolic-like Compounds, 180
Photosynthetic Capacity, 122, 132-133, 193, 199

Pine Forests, 107-118, 120
Plant Growth-promoting
Rhizobacteria (PGPR), 192
Ploidy-dependent Differences, 134
Populus Trichocarpa, 49-50, 57, 138, 226, 237
Potential Evapotranspiration (PET), 30
Precipitation (P), 30-31
Preventive Health Interventions, 40

Q
Quantitative Analysis, 59-60, 65, 81

R
Radionuclides, 150-157, 159-161
Relative Risks (RR), 1
Remote Sensing, 15-16, 18, 22, 26-29, 120
Reproductive Outcomes, 239-240, 242
Riparian Arthropod Communities, 83, 95

S
Small Catchment, 71-72, 77, 79, 81-82
Soil Conservation Service (SCS), 59
Soil Water Content, 71, 77-78, 80-82, 118, 135, 137, 193, 226-227
Streamflow Impacts, 30
Sub-stomatal Pore, 122
Subsurface Co2 Leak, 15
Supplementary Defense
System, 180
Surface Runoff Model, 59
Systems Responses, 204

T
Tap Water, 239
Terrestrial Ecosystems, 83
Terrestrial Water Cycle, 71
Thyroid Cancers, 150, 157, 159, 161
Thyroid Dysfunction, 96, 105
Tokyo Electric Power, 150
Transcriptome-wide Profiling, 134
Transgenic Arabidopsis, 145, 162, 167, 169, 173, 178, 228

V
Vertical Distribution, 71
Vertical Profiles, 71-76, 78
Visual Modflow, 59, 62, 67, 69-70

W
Water Channel Proteins, 49, 58
Water Conservation, 59-60, 66-70, 82
Water Deficit, 57, 122-123, 131-132, 137, 143, 147, 167, 173, 179, 192-194, 196-197, 199-202, 205-206, 208, 211-212, 215, 219-220, 223
Water Limitation, 180-181, 190
Water Resources, 14, 37-39, 59, 69-71, 83-84, 89-90, 92, 94-95, 120, 153
Water Stress, 15-17, 19, 21-22, 24, 26-28, 30, 34, 37, 49-50, 53, 55, 57-58, 71, 107, 119, 132-133, 143, 147, 164, 169, 177, 179-205, 207, 209-212, 214-217, 219-221
Water Stressed Vegetation, 15, 22
Water-dependent Habitats, 107
Watershed-scale Runoff, 107
Weather Research Forecast (WRF), 30-31
Whole-plant Hydraulic Conductance (KPLANT), 49